THE PHILOSOPHY AND PHYSICS OF NOETHER'S THEOREMS

In 1918, Emmy Noether, in her paper 'Invariante Variationsprobleme', proved two theorems (and their converses) on variational problems that went on to revolutionise theoretical physics. One hundred years later, the mathematics of Noether's theorems continues to be generalised, and the physical applications of her results continue to diversify. This centenary volume brings together world-leading historians, philosophers, physicists, and mathematicians in order to clarify the historical context of this work, its foundational and philosophical consequences, and its myriad physical applications. Suitable for advanced undergraduate and graduate students and professional researchers, this is a go-to resource for those wishing to understand Noether's work on variational problems and the profound applications which it finds in contemporary physics.

JAMES READ is Associate Professor in the Faculty of Philosophy, University of Oxford, and a Tutorial Fellow in Philosophy at Pembroke College, Oxford. He is also currently Treasurer of the British Society for Philosophy of Science. He has won several international prizes in the philosophy of physics, including the Hanneke Janssen Prize and Clifton Memorial Prize.

NICHOLAS J. TEH is Associate Professor of Philosophy at the University of Notre Dame. His primary research area is the intersection of philosophy, physics, and mathematics. He has been awarded several prestigious grants for his research, including two NSF fellowships.

THE PHILOSOPHY AND PHYSICS OF NOETHER'S THEOREMS

A Centenary Volume

Edited by

JAMES READ
Pembroke College, University of Oxford

NICHOLAS J. TEH
University of Notre Dame

CAMBRIDGE
UNIVERSITY PRESS

CAMBRIDGE
UNIVERSITY PRESS

Shaftesbury Road, Cambridge CB2 8EA, United Kingdom

One Liberty Plaza, 20th Floor, New York, NY 10006, USA

477 Williamstown Road, Port Melbourne, VIC 3207, Australia

314–321, 3rd Floor, Plot 3, Splendor Forum, Jasola District Centre, New Delhi – 110025, India

103 Penang Road, #05–06/07, Visioncrest Commercial, Singapore 238467

Cambridge University Press is part of Cambridge University Press & Assessment, a department of the University of Cambridge.

We share the University's mission to contribute to society through the pursuit of education, learning and research at the highest international levels of excellence.

www.cambridge.org
Information on this title: www.cambridge.org/9781108486231

DOI: 10.1017/9781108665445

First published 2022

A catalogue record for this publication is available from the British Library.

Library of Congress Cataloging-in-Publication Data
Names: Teh, Nicholas J., editor. | Read, James, 1991– editor.
Title: The philosophy and physics of Noether's theorems : a centenary volume / Nicholas Teh, James Alexander Mabyn Read.
Description: New York, NY : Cambridge University Press, 2022. | Includes bibliographical references and index.
Identifiers: LCCN 2021057283 (print) | LCCN 2021057284 (ebook) | ISBN 9781108486231 (hardback) | ISBN 9781108665445 (epub)
Subjects: LCSH: Noether's theorem. | Symmetry (Physics) | Conservation laws (Physics)
Classification: LCC QC793.3.S9 P45 2022 (print) | LCC QC793.3.S9 (ebook) | DDC 539.7/25–dc23/eng/20220120
LC record available at https://lccn.loc.gov/2021057283
LC ebook record available at https://lccn.loc.gov/2021057284

ISBN 978-1-108-48623-1 Hardback

Contents

List of Contributors *page* vii

Introduction 1
JAMES READ AND NICHOLAS J. TEH

1 The Noether Theorems in Context 4
YVETTE KOSMANN-SCHWARZBACH

2 Felix Klein and Emmy Noether on Invariant Theory and Variational Principles 25
DAVID E. ROWE

3 Moscow, Oxford, or Princeton: Emmy Noether's Move from Göttingen (1933) 52
TOMOKO L. KITAGAWA

4 Getting to the Bottom of Noether's Theorem 66
JOHN C. BAEZ

5 BV Quantisation in Perturbative Algebraic QFT: Fundamental Concepts and Perspectives 100
KASIA REJZNER

6 Divergence Invariant Variational Problems 134
PETER J. OLVER

7 Do Symmetries 'Explain' Conservation Laws? The Modern Converse Noether Theorem *vs* Pragmatism 144
HARVEY R. BROWN

8 Noether's First Theorem and the Energy-Momentum Tensor Ambiguity Problem 169
MARK ROBERT BAKER, NIELS LINNEMANN, AND CHRIS SMEENK

9 Noether's Theorems and Energy in General Relativity 197
SEBASTIAN DE HARO

10 Geometric Objects and Perspectivalism 257
JAMES READ

11 Substantive General Covariance and the Einstein–Klein Dispute: A Noetherian Approach 274
LAURENT FREIDEL AND NICHOLAS J. TEH

12 Noether Charges, Gauge-Invariance, and Non-Separability 296
HENRIQUE GOMES

13 Observability, Redundancy, and Modality for Dynamical Symmetry Transformations 322
DAVID WALLACE

14 The Gauge Argument: A Noether Reason 354
HENRIQUE GOMES, BRYAN W. ROBERTS, AND JEREMY BUTTERFIELD

Index 377

Contributors

John C. Baez,
University of California, Riverside

Mark Robert Baker,
University of Western Ontario and St. Francis Xavier University

Harvey R. Brown,
University of Oxford

Jeremy Butterfield,
University of Cambridge

Sebastian De Haro,
University of Amsterdam

Laurent Freidel,
Perimeter Institute for Theoretical Physics

Henrique Gomes,
University of Cambridge

Tomoko L. Kitagawa,
University of Oxford

Yvette Kosmann-Schwarzbach,
Paris

Niels Linnemann,
University of Western Ontario and Universität Bremen

Peter J. Olver,
University of Minnesota

James Read,
University of Oxford

Kasia Rejzner,
University of York

Bryan W. Roberts,
London School of Economics and Political Science

David E. Rowe,
Johannes Gutenberg-Universität Mainz

Chris Smeenk,
University of Western Ontario

Nicholas J. Teh,
University of Notre Dame

David Wallace,
University of Pittsburgh

Introduction

JAMES READ AND NICHOLAS J. TEH

Few articles can reasonably be described as epoch-making. Einstein's 'Zur Elektrodynamik bewegter Körper' (1905) is undoubtedly one such; Turing's 'On Computable Numbers, with an Application to the Entscheidungsproblem' (1936) is undoubtedly another. But standing equally tall among these ranks should surely be the article to which this volume – and so much else besides – owes its existence: Emmy Noether's 'Invariante Variationsprobleme' (1918). In that one article, Noether proved two theorems (and their converses), forging links between symmetries and conserved quantities which were to go on – whether by her intentions or not – to constitute the bedrock of modern theoretical physics.

But – perhaps surprisingly, perhaps not – the significance of an epoch-maker is not always recognised in the moment. Whether this is so in the case of Einstein is debatable; it is certainly true for Turing – and arguably even more so for Noether. Strikingly, the significance of what Noether proved in her 1918 article was not well appreciated until as late as the 1970s: only at that point were all of the theorems of the 1918 piece widely understood; and only at that point did they begin to be generalised and applied in substantially novel ways. Since then, progress has not stopped, and this volume – born out of an international 2018 centenary conference held at the London campus of the University of Notre Dame – represents the next episode in the same continuation. Bringing together historians, physicists, mathematicians, and philosophers, the volume constitutes the cutting edge of our understanding of (the application of) Noether's seminal work on variational problems, one hundred years on from her original article.

Why do we add 'the application of' in parentheses above? It is now well-known that Noether remarked little on the physical applications of her mathematical results; see, for example, the contributions of Kosmann-Schwarzbach and Rowe in this volume. In light of this, one should distinguish the creativity of Noether's methods – the creativity of her *technique* – from the creativity of their application – the creativity of the physical *representations* effected on the basis of her methods. In the post-1970s literature, the former are at least relatively well understood (albeit still not completely; see, for example, the contributions of Baez and Olver to this volume, which continue to add to such understanding); not so for the latter, in the case of which we are only beginning to explore a rich orchard of fruits.

The contributions to this volume pursue a number of distinct threads on both Noether's techniques and their applications to physics; these we will summarise here as succinctly as possible. We begin with Noether's history: both regarding specifically her work on variational problems, and more generally. In Chapter 1, **Yvette Kosmann-Schwarzbach** reviews this historical context of 'Invariante Variationsprobleme', from its prehistory, to its doldrums in the mid-twentieth century, to (as alluded to above) its revival in the post-1970s literature. Following on from this, in Chapter 2 **David Rowe** focuses on the interactions between Noether and Felix Klein in the years surrounding the appearance of her 'Invariante Variationsprobleme', and specifically on the role of differential invariants in Noether's two theorems. In Chapter 3, **Tomoko Kitagawa** focuses on another specific episode highlighted by Kosmann-Schwarzbach: namely, Noether's deliberations preceding her move to Bryn Mawr College.

Having presented this updated Noether history, we turn to the mathematics of her theorems, both generalisations and applications. In Chapter 4, **John Baez** illuminates the content of Noether's (first) theorem in the Hamiltonian context by pursuing a (Jordan–Lie) algebraic – rather than the traditional geometric – approach. In Chapter 5, **Kasia Rejzner** continues this study of Noetherian themes from an algebraic point of view (this time via homological algebras), by exploring the 'BV formalism' – an extension of the BRST prescription, in which auxiliary fields enjoying rigid symmetries are introduced, and in which the Noether charges associated with those symmetries are then quantised – from the point of view of perturbative algebraic quantum field theory.

Next, we turn to more philosophical questions regarding the explanatory arrow running from symmetries to conservation laws which is often (misleadingly, our authors would have it!) taken to be an important moral drawn from Noether's theorems. In Chapter 6, **Peter Olver** considers the significance of the fact that one can define infinitely many inequivalent Lagrangians invariant under a stipulated set of variational symmetries: should these Lagrangians be understood as encoding 'equivalent' physics – and if so, why? In Chapter 7, **Harvey R. Brown** questions the reasons for which, in light of the converse of Noether's first theorem, symmetries are often indeed considered to have this explanatory priority over conservation laws. In Chapter 8, **Mark Baker**, **Niels Linnemann**, and **Chris Smeenk** deploy the under-appreciated work of Bessel-Hagen in order to demonstrate how the converse of Noether's first theorem can be used to resolve ambiguities over what should be regarded as the 'physical' energy-momentum tensor in field theories.

The next three chapters address the significance of Noether's theorems in the context in which they were originally developed: the foundations of general relativity. In Chapter 9, **Sebastian de Haro** provides both a crystal-clear survey of the role of Noether's theorems in considerations of gravitational energy in general relativity, as well as a substantial novel contribution to recent philosophical discussions regarding the status of gravitational energy in that theory, including quasi-local notions. In Chapter 10, **James Read** continues these discussions, arguing that the pseudotensorial quantities obtained on application of Noether's theorems to general relativity are best interpreted physically from within the framework of 'perspectival realism'. Finally, in Chapter 11, **Laurent Freidel** and **Nicholas J. Teh** apply Noether's theorems in order to shed new light on three infamously vexed notions in the foundations of spacetime theories: (i) general covariance, (ii) the Principle of Relativity, and (iii) the status of conserved charges (including, again, gravitational energy).

Leading on from Teh and Freidel's contribution, the following pair of chapters address the question of the empirical significance of symmetry transformations in physics. In Chapter 12, **Henrique Gomes** brings to bear the symmetry (in)variance of Noether charges upon the *à la mode* question of whether and when gauge transformations (such as the familiar $U(1)$ transformations in electromagnetism) have 'direct empirical significance'. In Chapter 13, **David Wallace** develops a systematic account of when symmetry transformations in physics can relate empirically distinct states of affairs, again by considering subsystem-environment decompositions, and relates this to the content of Noether's theorems.

The final topic addressed by our authors is the gauge argument – i.e., the well-known (but infamously problematic!) principle by which new physical fields are introduced into one's physics in order to restore invariance under a certain set of transformations. In Chapter 14, **Henrique Gomes**, **Bryan Roberts**, and **Jeremy Butterfield** use Noether's second theorem to shed light on the gauge argument.

The above should suffice to convince the reader that the field of matters Noetherian is rich and blossoming. In a recent interview, Edward Witten defined beautiful mathematics as that which 'gives you more out than you put in'.[1] Given that, by their own admission, the contributions to this volume still only begin to scratch the surface of the potential stock of applications for Noether's theorems, the results of 'Invariante Variationsprobleme' surely constitute beautiful mathematics (and, via their applications, mathematical physics) in the highest degree. Insofar as physicists, mathematicians, and philosophers are united in their attraction to exquisite things, there is every reason to think that the epoch-making status of Noether's results will endure, and that 'Invariante Variationsprobleme' has secured its rightful place amongst the Immortals.

Acknowledgements

First and foremost, we are profoundly grateful to Bryan Roberts, both for his substantial work in organising the centenary conference out of which this volume developed, and for his editorial role in the early stages of the preparation of this volume – neither the conference nor this piece would have been possible without him. Second, we thank William Wolf for his very efficient work preparing the index for this volume. Finally, we are, of course, extremely grateful to all authors and reviewers of this volume: the result is a celebration of Noether's two theorems of which we can be proud. We also acknowledge the generous support of NSF Grant 1947155 and JTF Grant 61521 while working on this volume.

[1] https://youtu.be/O3isFuQ2q2A. Accessed 8 July 2021.

1

The Noether Theorems in Context

YVETTE KOSMANN-SCHWARZBACH

> Methodi in hoc libro traditæ, non solum maximum
> esse usum in ipsa analysi, sed etiam eam ad resolutionem
> problematum physicorum amplissimum subsidium afferre.
>
> *Leonhard Euler (1744)*

'The methods described in this book are not only of great use in analysis, but are also most helpful for the solution of problems in physics.' Replacing 'in this book' by 'in this article', the sentence that Euler wrote in the introduction to the first supplement of his treatise on the calculus of variations in 1744 applies equally well to Emmy Noether's 'Invariante Variationsprobleme', published in the *Nachrichten von der Königlichen Gesellschaft der Wissenschaften zu Göttingen, Mathematisch-physikalische Klasse* in 1918.

1.1 Introduction

In this talk,[1] I propose to sketch the contents of Noether's 1918 article, 'Invariante Variationsprobleme', as it may be seen against the background of the work of her predecessors and in the context of the debate on the conservation of energy that had arisen in the general theory of relativity.[2]

Situating Noether's theorems on the invariant variational problems in their context requires a brief outline of the work of her predecessors, and a description of her career, first in Erlangen, then in Göttingen. Her 1918 article will be briefly summarised. I have endeavoured to convey its contents in Noether's own vocabulary and notation with minimal recourse to more recent terminology. Then I shall address these questions: how original was Noether's 'Invariante Variationsprobleme'? How modern were her use of Lie groups and her introduction of generalised vector fields? And how influential was her article? To this end, I shall sketch its reception from 1918 to 1970.

[1] This text is a revised version of the lecture I delivered at the international conference, 'The Philosophy and Physics of Noether's Theorems', a centenary conference on the 1918 work of Emmy Noether, London, 5 October 2018.

[2] An English translation of Noether's article together with an account of her work and the history of its reception, from Einstein to Deligne, may be found in my book, *The Noether Theorems, Invariance and Conservation Laws in the Twentieth Century*, translated by Bertram E. Schwarzbach (Kosmann-Schwarzbach 2010). This book contains an extensive bibliography, only a small part of which is reproduced in the list of references below. It is an expanded version of my earlier book in French, *Les Théorèmes de Noether, Invariance et lois de conservation au XXe siècle* (2004).

For many years, there was practically no recognition of either of these theorems. Then multiple references to 'the Noether theorem' or 'Noether's theorem' – in the singular – began to appear, referring either to her first theorem, in the publications of those mathematicians and mathematical physicists who were writing on mechanics – who ignored her second theorem – or to her second theorem by those writing on general relativity and, later, on gauge theory. I shall outline the curious transmission of her results, the history of the mathematical developments of her theory, and the ultimate recognition of the wide applicability of 'the Noether theorems'. To conclude, in the hope of dispelling various misconceptions, I shall underline what Noether was *not*, and I shall reflect on the fortune of her theorems.

1.2 A Family of Mathematicians

Emmy Noether was born to a Jewish family in Erlangen (Bavaria, Germany) in 1882. Her life was described in Hermann Weyl's obituary (Weyl 1935). In a manuscript *curriculum vitae*, written for official purposes *circa* 1917, she described herself as 'of Bavarian nationality and Israelite confession'.[3] She died in Bryn Mawr (Pennsylvania) in the United States in 1935, after undergoing an operation. Why she had to leave Germany in 1933 to take up residence in America is clear from the chronology of the rise of the Nazi regime in Germany and its access to power and has, of course, been told in the many accounts of her life that have been published,[4] while numerous and sometimes fanciful comments have appeared in print and in the electronic media in recent years.

She was the daughter of the renowned mathematician, Max Noether (1844–1921), professor at the University of Erlangen. He had been a privatdozent, then an 'extraordinary professor' in Heidelberg before moving to Erlangen in 1875, and was eventually named an 'ordinary professor' in 1888. Her brother, Fritz, was born in 1884 and studied mathematics and physics in Erlangen and Munich. He became professor of theoretical mechanics in Karlsruhe in 1902 and submitted his *Habilitation* thesis in 1912. Later, he became professor in Breslau, from where he, too, was forced to leave in 1933. He emigrated to the Soviet Union and was appointed professor at the University of Tomsk. Accused of being a German spy, he was jailed and shot in 1941.

1.3 The Young Emmy Noether

Emmy Noether first studied languages in order to become a teacher of French and English, a suitable profession for a young woman. But from 1900 on, she studied mathematics, first in Erlangen with her father, then audited lectures at the university. For the winter semester in 1903–4, she travelled to Göttingen to audit courses at the university. At that time, new regulations were introduced which enabled women to matriculate and take examinations. She then chose to enroll at the University of Erlangen, where she listed

[3] Declaring one's religion was compulsory in Germany at the time.

[4] The now classical biographies of Noether can be found in the book written by Auguste Dick (1970), translated into English in 1981, and in the volumes of essays edited by James W. Brewer and Martha K. Smith (1981), and by Bhama Srinivasan and Judith D. Sally (1983).

mathematics as her only course of study,[5] and in 1907, she completed her doctorate under the direction of Paul Gordan (1837–1912), a colleague of her father. Here I open a parenthesis: One should not confuse the mathematician Paul Gordan, her 'Doktorvater', with the physicist Walter Gordon (1893–1939). The 'Clebsch–Gordan coefficients' in quantum mechanics bear the name of Noether's thesis adviser together with that of the physicist and mathematician Alfred Clebsch (1833–72). However, the 'Klein–Gordon equation' is named after Walter Gordon and the physicist Oskar Klein (1894–1977) who, in turn, should not be confused with the mathematician Felix Klein, about whom more will be said shortly.

1.4 Noether's 1907 Thesis on Invariant Theory

Noether's thesis at Erlangen University, entitled 'Über die Bildung des Formensystems der ternären biquadratischen Form' ("On the Construction of the System of Forms of a Ternary Biquadratic Form"), dealt with the search for the invariants of those forms (i.e., homogeneous polynomials) which are ternary (i.e., in 3 variables) and biquadratic (i.e., of degree 4). An extract of her thesis appeared in the *Sitzungsberichte der Physikalisch-medizinischen Societät zu Erlangen* in 1907, and the complete text was published the following year in the *Journal für die reine und angewandte Mathematik* (Crelle's Journal). She later distanced herself from her early work as employing a needlessly computational approach to the problem.

After 1911, her work in algebra was influenced by Ernst Fischer (1875–1954), who was appointed professor in Erlangen upon Gordan's retirement in 1910. Noether's expertise in invariant theory revealed itself in the publications in 1910, 1913, and 1915 that followed her thesis, and was later confirmed in the four articles on the invariants of finite groups that she published in 1916 in the *Mathematische Annalen*. She studied in particular the determination of bases of invariants that furnish an expansion with integral or rational coefficients of each invariant of the group, expressed as a linear combination of the invariants in the basis.

At Erlangen University from 1913 on, Noether occasionally substituted for her ageing father, thus beginning to teach at the university level, but not under her own name.

1.5 Noether's Achievements

Her achievement of 1918, whose centenary was duly celebrated in conferences in London and Paris, eventually became a central result in both mechanics and field theory and, more generally, in mathematical physics, though her role was rarely acknowledged before 1950 and, even then, only a truncated part of her article was cited. On the other hand, her articles on the theory of ideals and the representation theory of algebras published in the 1920s made her world famous. Her role in the development of modern algebra was duly recognised by the mathematicians of the twentieth century, while they either considered her work on invariance principles to be an outlying and negligible part of her work or, more often, ignored

[5] On this, as well as on other oft-repeated facts of Noether's biography, see Dick (1970), English translation, p. 14.

it altogether. In fact, the few early biographies of Noether barely mention her work on invariant variational problems, while both past and recent publications treat her fundamental contributions to modern algebra. I shall not deal with them here. They are, and will no doubt continue to be, celebrated by all mathematicians.

1.6 In Göttingen: Klein, Hilbert, Noether, and Einstein

In 1915, the great mathematicians Felix Klein (1849–1925) and David Hilbert (1862–1943) invited Noether to Göttingen in the hope that her expertise in invariant theory would help them understand some of the implications of Einstein's newly formulated general theory of relativity. In Göttingen, Noether took an active part in Klein's seminar. It was in her 1918 article that she solved a problem arising in the general theory of relativity and proved 'the Noether theorems'. In particular, she proved and vastly generalised a conjecture made by Hilbert concerning the nature of the law of conservation of energy. Shortly afterwards, she returned to pure algebra.

At the invitation of Hilbert, Einstein had come to Göttingen in early July 1915 to deliver a series of lectures on the general theory of relativity, which is to say, on the version that preceded his famous paper, 'Die Feldgleichungen der Gravitation' ("The Field Equations of Gravitation"), of November of that year. Noether must have attended these lectures. It is clear from Hilbert's letter to Einstein of 27 May 1916 that she had by then already written some notes on the subject of the problems arising in the general theory of relativity:

> My law [of conservation] of energy is probably linked to yours; I have already given Miss Noether this question to study.

Hilbert adds that, to avoid a long explanation, he has appended to his letter 'the enclosed note of Miss Noether'. On 30 May 1916, Einstein answered him in a brief letter in which he derived a consequence of the equation that Hilbert had proposed 'which deprives the theorem of its sense', and then asks, 'How can this be clarified?' He continues,

> Of course it would be sufficient if you asked Miss Noether to clarify this for me.[6]

Thus, her expertise was conceded by both Hilbert and Einstein as early as her first year in Göttingen, and was later acknowledged more explicitly by Klein when he re-published his articles of 1918 in his collected works (Klein 1921), a few years before his death.

1.7 Noether's Article of 1918

In early 1918, Noether published an article on the problem of the invariants of differential equations in the *Göttinger Nachrichten*, 'Invarianten beliebiger Differentialausdrücke' ("Invariants of Arbitrary Differential Expressions"), which was presented by Klein at the meeting of the *Königliche Gesellschaft der Wissenschaften zu Göttingen* (Royal Göttingen

[6] Einstein, *Collected Papers*, 8A, nos. 222 and 223.

Scientific Society) of 25 January. It was then, in the winter and spring of 1918, that Noether discovered the profound reason for the difficulties that had arisen in the interpretation of the conservation laws in the general theory of relativity. Because she had left Göttingen for a visit to Erlangen to see her widowed and ailing father, her correspondence remains and yields an account of her progress in this search. In her postcard to Klein of 15 February, she already sketched her second theorem, but only in a particular case. It is in her letter to Klein of 12 March that Noether gave a preliminary formulation of an essential consequence of what would be her second theorem, dealing with the invariance of a variational problem under the action of a group which is a subgroup of an infinite-dimensional group. On 23 July, she presented her results to the *Mathematische Gesellschaft zu Göttingen* (Göttingen Mathematical Society). The article which contains her two theorems is 'Invariante Variationsprobleme' ("Invariant Variational Problems"). On 26 July, Klein presented it at the meeting of the more important – because it was not restricted to an audience of pure mathematicians – Göttingen Scientific Society, and it was published in the *Nachrichten* (Proceedings) of the Society of 1918, on pages 235–47. A footnote on the first page of her article indicates that 'The definitive version of the manuscript was prepared only at the end of September.'

1.8 What Variational Problems Was Noether Considering?

We consider variational problems which are invariant under a continuous group (in the sense of Lie). ... What follows thus depends upon a combination of the methods of the formal calculus of variations and of Lie's theory of groups.[7]

Noether considers a general n-dimensional variational problem of order κ for an $\mathbb{R}^\mu$-valued function, where n, κ, and μ are arbitrary integers, defined by an integral,

$$I = \int \cdots \int f\left(x, u, \frac{\partial u}{\partial x}, \frac{\partial^2 u}{\partial x^2}, \ldots, \frac{\partial^\kappa u}{\partial x^\kappa}\right) dx,$$

where $x = (x_1, \ldots, x_n) = (x_\lambda)$ denote the independent variables, and where $u = (u_1, \ldots, u_\mu) = (u_i)$ are the dependent variables. In footnotes, she states her conventions and explains her abbreviated notations: 'I omit the indices here, and in the summations as well whenever it is possible, and I write $\frac{\partial^2 u}{\partial x^2}$ for $\frac{\partial^2 u_\alpha}{\partial x_\beta \partial x_\gamma}$, etc', and 'I write dx for $dx_1 \ldots dx_n$ for short'.

Noether then states her two theorems:

In what follows we shall examine the following two theorems:

I. If the integral I is invariant under a [group] $\mathfrak{G}_\rho$, then there are ρ linearly independent combinations among the Lagrangian expressions which become divergences – and conversely, that implies the invariance of I under a [group] $\mathfrak{G}_\rho$. The theorem remains valid in the limiting case of an infinite number of parameters.

[7] I cite the English translation of Noether's article that appeared in *The Noether Theorems* (2010).

II. If the integral I is invariant under a [group] $\mathfrak{G}_{\infty\rho}$ depending upon arbitrary functions and their derivatives up to order σ, then there are ρ identities among the Lagrangian expressions and their derivatives up to order σ. Here as well the converse is valid.[8]

Noether proves the direct part of both theorems in section 2, then the converse of theorem I in section 3 and that of theorem II in section 4. In section 2, she assumes that the action integral $I = \int f dx$ is invariant. Actually, she assumes a more restrictive hypothesis, the invariance of the integrand, $f dx$, which is to say, $\delta(f dx) = 0$. This hypothesis is expressed by the relation

$$\bar{\delta} f + \mathrm{Div}(f \cdot \Delta x) = 0.$$

Here Div is the divergence of vector fields and $\bar{\delta} f$ is the variation of f induced by the variation

$$\bar{\delta} u_i = \Delta u_i - \sum \frac{\partial u_i}{\partial x_\lambda} \Delta x_\lambda.$$

Thus, Noether introduced the evolutionary representative, $\bar{\delta}$, of the vector field δ, and $\bar{\delta} f$ is the Lie derivative of f in the direction of the vector field $\bar{\delta}$. What she introduced, with the notation $\bar{\delta}$, is a generalised vector field, which is not a vector field in the usual sense, on the trivial vector bundle $\mathbb{R}^n \times \mathbb{R}^\mu \to \mathbb{R}^n$. In fact, if

$$\delta = \sum_{\lambda=1}^{n} X^\lambda(x) \frac{\partial}{\partial x^\lambda} + \sum_{i=1}^{\mu} Y^i(x,u) \frac{\partial}{\partial u^i},$$

then $\bar{\delta}$ is the vertical generalised vector field

$$\bar{\delta} = \sum_{i=1}^{\mu} \left(Y^i(x,u) - X^\lambda(x) u^i_\lambda \right) \frac{\partial}{\partial u^i},$$

where $u^i_\lambda = \frac{\partial u^i}{\partial x^\lambda}$. It is said to be 'generalised' because its components depend on the derivatives of the $u^i(x)$. It is said to be 'vertical' because it contains no terms in $\frac{\partial}{\partial x^\lambda}$.[9]

By integrating by parts, Noether obtains the identity

$$\sum \psi_i \, \bar{\delta} u_i = \bar{\delta} f + \mathrm{Div}\, A,$$

where the ψ_i's are the 'Lagrangian expressions', i.e., the components of the Euler–Lagrange derivative of f, and A is linear in $\bar{\delta} u$ and its derivatives. In view of the invariance hypothesis which is expressed by $\bar{\delta} f + \mathrm{Div}(f \cdot \Delta x) = 0$, this identity can be written as

$$\sum \psi_i \, \bar{\delta} u_i = \mathrm{Div}\, B, \quad \text{with} \quad B = A - f \cdot \Delta X.$$

8 In a footnote, Noether announces that she will comment on 'some trivial exceptions' in the next section of her article.

9 The evolutionary representative of an ordinary vector field has also been called the vertical representative. Both terms are modern. Noether does not give $\bar{\delta}$ a name. An arbitrary vertical generalised vector field is written locally, $Z = \sum_{i=1}^{\mu} Z^i \left(x, u, \frac{\partial u}{\partial x}, \frac{\partial^2 u}{\partial x^2}, \cdots \right) \frac{\partial}{\partial u^i}$.

Therefore, B is a conserved current for the Euler–Lagrange equations of f, and the proof of the direct part of Theorem I is complete: the equations $\text{Div } B = 0$ are the conservation laws that are satisfied when the Euler–Lagrange equations $\psi_i = 0$ are satisfied.

Noether then proves the converse of Theorem I. The existence of ρ 'linearly independent divergence relations' implies the invariance under a (Lie) group of symmetries of dimension ρ, by passing from the infinitesimal symmetries to invariance under their flows, provided that the vector fields Δu and Δx are ordinary vector fields. In the general case, the existence of ρ linearly independent conservation laws yields the infinitesimal invariance of f under a Lie algebra of infinitesimal symmetries of dimension ρ. Equivalence relations have to be introduced to make these statements precise.

Theorem II deals with a symmetry group depending on arbitrary functions, such as the group of diffeomorphisms of the spacetime manifold and, more generally, the groups of all gauge theories that would be developed, beginning with the article of Chen Ning Yang and Robert L. Mills, 'Conservation of Isotopic Spin and Isotopic Gauge Invariance' (1954). Noether showed that to such symmetries there correspond identities satisfied by the variational derivatives, and conversely. The assumption is that 'the integral I is invariant under a [group] $\mathfrak{G}_{\infty\rho}$ depending upon arbitrary functions and their derivatives up to order σ'. In other words, Noether assumes the existence of ρ infinitesimal symmetries of the variational integral, each of which depends linearly on an arbitrary function p (depending on $\lambda = 1, 2, \ldots, \rho$) of the variables $x_1, x_2, \ldots, x_n$, and its derivatives up to order σ. Such a symmetry is defined by a vector-valued linear differential operator, $\mathcal{D}$, of order σ, with components $\mathcal{D}_i$, $i = 1, 2, \ldots, \mu$. Noether then introduces, without giving it a name or a particular notation, the adjoint operator, $(\mathcal{D}_i)^*$, of each of the $\mathcal{D}_i$'s. By construction, $(\mathcal{D}_i)^*$ satisfies

$$\psi_i \, \mathcal{D}_i(p) = (\mathcal{D}_i)^*(\psi_i) \, p + \text{Div } \Gamma_i,$$

where Γ_i is linear in p and its derivatives. The symmetry assumption and, again, an integration by parts imply

$$\sum_{i=1}^{\mu} \psi_i \, \mathcal{D}_i(p) = \text{Div } B.$$

This relation implies

$$\sum_{i=1}^{\mu} (\mathcal{D}_i)^*(\psi_i) \, p = \text{Div} \left(B - \sum_{i=1}^{\mu} \Gamma_i \right).$$

Since p is arbitrary, by Stokes's theorem and the Du Bois–Reymond lemma,

$$\sum_{i=1}^{\mu} (\mathcal{D}_i)^*(\psi_i) = 0.$$

Thus, for each $\lambda = 1, 2, \ldots, \rho$, there is a differential relation among the components ψ_i of the Euler–Lagrange derivative of the Lagrangian f that is identically satisfied.

Noether explains the precautions that must be taken – the introduction of an equivalence relation on the symmetries – for the converse to be valid. She then observes that each identity may be written as $\sum_{i=1}^{\mu} a_i \psi_i = \text{Div}\, \chi$, where χ is defined by a linear differential operator acting on the ψ_i's. She shows that each divergence, Div B, introduced above, is equal to the divergence of a quantity C, where C vanishes once the Euler–Lagrange equations, $\psi_i = 0$, are satisfied. Furthermore, from the equality of the divergences of B and C, it follows that

$$B = C + D$$

for some D whose divergence vanishes identically, which is to say, independently of the satisfaction of the Euler–Lagrange equations or 'off-shell' in modern terminology. These are the conservation laws that Noether called *improper divergence relations*. In the modern terminology, there are two types of trivial conservation laws. If the quantity C itself, and not only its divergence, vanishes on $\psi_i = 0$, then C is a *trivial conservation law of the first kind*. If the divergence of D vanishes identically, i.e., whether or not $\psi_i = 0$, then D is a *trivial conservation law of the second kind* or Div D is a *null divergence*.

1.9 Hilbert's Conjecture, Groups, and Relativity

The last section of Noether's article deals with Hilbert's conjecture. He had asserted, without proof, in early 1918 that, in the case of general relativity, 'the energy equations do not exist at all', that is, there are no proper conservation laws:

> Indeed I claim that for *general* relativity, that is, in the case of the *general* invariance of the Hamiltonian function, energy equations which, in your sense, correspond to the energy equations of the orthogonally invariant theories, do not exist at all; I can even call this fact a characteristic feature of the general theory of relativity.[10]

Noether shows that the situation is better understood 'in the more general setting of group theory'. She explains the apparent paradox that arises from the consideration of the finite-dimensional subgroups of groups that depend upon arbitrary functions. She emphasises the conclusion of her argument by setting it as follows, with italics in the original:

> *Given I invariant under the group of translations, then the energy relations are improper if and only if I is invariant under an infinite group which contains the group of translations as a subgroup.*

Noether concludes by quoting in her final footnote Klein's striking formula from page 287 of his 1910 paper, 'Über die geometrischen Grundlagen der Lorentzgruppe':

> The term 'relativity' as it is used in physics should be replaced by 'invariance with respect to a group'.

1.10 How Original Were Noether's Two Theorems?

Noether's article did not appear in a vacuum. Analysing the contributions of her predecessors requires a detailed development.[11] Here, I shall give only a very brief account of some

[10] Klein (1918, p. 477), citing Hilbert.
[11] See *The Noether Theorems* (2010, pp. 29–39).

of the most important points of this history. Lagrange, in his *Méchanique analitique* (1788), claimed that his method for deriving 'a general formula for the motion of bodies' yields 'the general equations that contain the principles, or theorems known under the names of the *conservation of kinetic energy*, of the *conservation of the motion of the center of mass*, of the *conservation of the momentum of rotational motion*, of the *principle of areas*, and of the *principle of least action*'.[12] In the second edition of his *Mécanique analytique*, in 1811, as a preliminary to his treament of dynamics, he presented a detailed history of the diverse 'principes ou théorèmes' (principles or theorems) formulated before his *Mécanique*, thus recognising the contributions of his predecessors in the discovery of these principles – Galileo, Huyghens, Newton, Daniel Bernoulli, Maupertuis, Euler, the Chevalier Patrick d'Arcy and d'Alembert – and in this second edition, he explicitly observed a correlation between these principles of conservation and invariance properties. After Lagrange, the correlation between invariances and conserved quantities was surveyed by Jacobi in several chapters of his *Vorlesungen über Dynamik*, lectures delivered in 1842–3 but published only posthumously in 1866. The great advances of Sophus Lie (1842–99) – his theory of continuous groups of transformations that was published in articles and books that appeared between 1874 and 1896 – became the basis of all later developments, such as the work of Georg Hamel (1877–1954) on the calculus of variations and mechanics in 1904, and the publication of Gustav Herglotz (1881–1953) on the 10-parameter invariance group of the [special] theory of relativity in 1911. In her 1918 article, Noether cited Lie very prominently, as his name appears three times in the eight lines of the introductory paragraph, but with no precise reference to his published work. Both Hamel and Herglotz were cited by her. In her introduction, she also referred to publications, all of them very recent, by '[Hendrik] Lorentz and his students (for example, [Adriaan Daniel] Fokker), [Hermann] Weyl, and Klein for certain infinite groups' and, in a footnote, she wrote, 'In a paper by [Adolf] Kneser that has just appeared (Math. Zeitschrift, vol. 2), the determination of invariants is dealt with by a similar method.' In fact, while Noether was completing the definitive version of her manuscript, in August 1918, Kneser had submitted an article, 'Least Action and Galilean Relativity', in which he used Lie's infinitesimal transformations and, as Noether would do, emphasised the relevance of Klein's Erlangen program, but he did not treat questions of invariance. Noether stressed the relation of her work to 'Klein's second note, *Göttinger Nachrichten*, 19 July 1918', stating that her work and Klein's were 'mutually influential' and referring to it for a more complete bibliography. In section 5 of her paper, she cited an article, 'On the Ten General Invariants of Classical Mechanics' by Friedrich Engel (1861–1941), that had appeared two years earlier. Indeed, scattered results in classical and relativistic mechanics, tying together properties of invariance and conserved quantities, had already appeared in the publications of Noether's predecessors which she acknowledged. However, none of them had discovered the general principle contained in her Theorem I and its converse. Her Theorem II and its converse were completely new. In the expert opinion of

[12] Lagrange (1788, p. 182), italics in the original.

the theoretical physicist Thibaut Damour,[13] the second theorem should be considered the most important part of her article. It is certainly the most original.

1.11 How Modern Were Noether's Two Theorems?

What Noether simply called 'infinitesimal transformations' are, in fact, vast generalisations of the ordinary vector fields, and they are now called generalised vector fields. They would eventually be re-discovered, independently, in 1964 by Harold H. Johnson, then at the University of Washington, who called them 'a new type of vector fields', and in 1965 by Robert Hermann (1931–2020). They appeared again in 1972 as Robert L. Anderson, Sukeyuki Kumei, and Carl Wulfman published their 'Generalization of the Concept of Invariance of Differential Equations. Results of Applications to Some Schrödinger Equations' in *Physical Review Letters*. In 1979, R. L. Anderson, working at the University of Georgia in the United States, and Nail Ibragimov (1938–2018), then a member of the Institute of Hydrodynamics at the Siberian branch of the USSR Academy of Sciences in Novosibirsk – such east-west collaboration was rare at the time – in their monograph, *Lie-Bäcklund Transformations in Applications*, duly citing Klein and Noether while claiming to generalise 'Noether's classical theorem', called them 'Lie-Bäcklund transformations', a misleading term because Albert V. Bäcklund (1845–1922) did not introduce this vast generalisation of the concept of vector fields, only infinitesimal contact transformations. The concept of a generalised vector field is essential in the theory of integrable systems which became the subject of intense research after 1970. On this topic, Noether's work is modern, half a century in advance of these re-discoveries. Peter Olver's book, *Applications of Lie Groups to Differential Equations* (1986a), is both a comprehensive handbook of the theory of generalised symmetries of differential and partial differential equations, and the reference for their history, while his article of the same year on 'Noether's theorems and systems of Cauchy–Kovalevskaya type' is an in-depth study of the mathematics of Noether's second theorem. His article (Olver 2018), written for the centenary of Noether's article, stresses the importance of her invention of the generalised vector fields.

In Göttingen, Noether had only one immediate follower, Erich Bessel-Hagen (1898–1946), who was Klein's student. In 1921, he published an article in the *Mathematische Annalen*, entitled 'Über die Erhaltungssätze der Elektrodynamik' ("On the Conservation Laws of Electrodynamics"), in which he determined in particular those conservation laws that are the result of the conformal invariance of Maxwell's equations. There, Bessel-Hagen recalls that it was Klein who had posed the problem of 'the application to Maxwell's equations of the theorems stated by Miss Emmy Noether about two years ago regarding the invariant variational problems' and he writes that, in the present paper, he formulates the two Noether theorems 'slightly more generally' than they had been formulated in her article. How did he achieve this more general result? By introducing the concept of 'divergence symmetries' which are infinitesimal transformations which leave the Lagrangian invariant

13 Damour is a professor at the Institut des Hautes Études Scientifiques and a member of the Académie des Sciences de l'Institut de France.

up to a divergence term, or 'symmetries up to divergence'. They correspond not to the invariance of the Lagrangian $f dx$, but to the invariance of the action integral $\int f dx$, i.e., instead of satisfying the condition $\delta(f dx) = 0$, they satisfy the weaker condition $\delta(f dx) = \operatorname{Div} C$, where C is a vectorial expression. Noether's fundamental relation remains valid under this weaker assumption, provided that $B = A - f \cdot \Delta x$ is replaced by $B = A + C - f \cdot \Delta x$. Immediately after he stated that he had proved the theorems in a slightly more general form than Noether had, Bessel-Hagen added: 'I owe these [generalised theorems] to an oral communication by Miss Emmy Noether herself'. We infer that, in fact, this more general type of symmetry was also Noether's invention. Bessel-Hagen's acknowledgment is evidence that, to the question, 'Who invented divergence symmetries?', the answer is: Noether.

1.12 How Influential Were Noether's Two Theorems?

The history of the reception of Noether's article in the years 1918–70 is surprising. She submitted the 'Invariante Variationsprobleme' for her *Habilitation*, finally obtained in 1919, but she never referred to her article in any of her subsequent publications. I know of only one mention of her work of 1918 in her own writings, in a letter she sent eight years later to Einstein, who was then an editor of the journal *Mathematische Annalen*. In this letter, which is an informal referee report, she rejects a submission 'which unfortunately is by no means suitable' for the journal, on the grounds that 'it is first of all a restatement that is not at all clear of the principal theorems of my "Invariante Variationsprobleme" (Gött[inger] Nachr[ichten], 1918 or 1919), with a slight generalization – the invariance of the integral up to a divergence term – which can actually already be found in Bessel-Hagen (Math[ematische] Ann[alen], around 1922)'.[14]

I found very few early occurrences of Noether's title in books and articles. While Hermann Weyl, in *Raum, Zeit, Materie*, first published in 1918, performed computations very similar to hers, he referred to Noether only once, in a footnote in the third (1919) and subsequent editions. It is clear that Richard Courant must have been aware of her work because a brief summary of a limited form of both theorems appears in all German, and later English editions of 'Courant–Hilbert', the widely read treatise on methods of mathematical physics first published in 1924. It is remarkable that we found so few explicit mentions of Noether's results in searching the literature of the 1930s. In 1936, the little-known physicist Moisei A. Markow (1908–94), who was a member of the Physics Institute of the USSR Academy of Sciences in Moscow, published an article in the *Physikalische Zeitschrift der Sowjetunion* in which he refers to 'the well-known theorems of Noether'. Markow was a former student of Georg B. Rumer (1901–85), who had been an assistant of Max Born in Göttingen from 1929 to 1932. Rumer, in 1931, had proved the Lorentz invariance of the Dirac operator but did not allude to any associated conservation laws, while in his articles on the general theory of relativity published in the *Göttinger Nachrichten* in 1929 and 1931, he

[14] For a facsimile, a transcription, and a translation of Noether's letter, see Kosmann-Schwarzbach (2010, pp. 161–5), and see comments on this letter, Kosmann-Schwarzbach (2010), pp. 51–2.

cited Weyl but never Noether. Similarly, it seems that V. A. Fock (1898–1974) never referred to Noether's work in any of his papers to which it was clearly relevant, such as his celebrated 'Zur Theorie des Wasserstoffatoms' (On the Theory of the Hydrogen Atom) of 1935. Was it because, at the time, papers carried few or no citations? Or because Noether's results were considered to be 'classical'? The answers to both questions are probably positive, this paucity of citations being due to several factors.

An early, explicit reference to Noether's publication is found in the article of Ryoyu Utiyama (Utiyama 1916–90), then in the department of physics of Osaka Imperial University, 'On the Interaction of Mesons with the Gravitational Field. I', which appeared in *Progress of Theoretical Physics* (Utiyama 1947), four years before he was awarded the PhD. His paragraph I begins with the 'Theory of invariant variation' for which he cites both Noether's 1918 article and page 617 of Pauli's 'Relativitätstheorie' (1921). Following Noether closely, he proves the first theorem, introducing 'the substantial variation of any field quantity', which he denotes by δ^* – i.e., what Noether had denoted by $\bar{\delta}$ – and also treats the case where the dependent variables 'are not completely determined by [the] field equations but contain r undetermined functions'. This text dates, in fact, to 1941, as the author reveals in a footnote on the first page: 'This paper was published at the meeting[s] of [the] Physico-mathematical Society of Japan in April 1941 and October 1942, but because of the war the printing was delayed'. Such a long delay in the publication of this scientific paper is one example – among many – of the influence of world affairs on science. It appears that this publication is a link in the chain leading from Noether's theorems to the development, by the physicists, of the gauge theories, where the variations of the field variables depend on arbitrary functions. Episodes in this history, told by Utiyama himself, were published in Lochlainn O'Raifeartaich's book (1997), from which we learn that, although Utiyama published his important paper 'Invariant Theoretical Interpretation of Interaction' in the *Physical Review* only in 1956, two years after the famous article of Yang and Mills, he had worked independently and had treated more general cases, showing that gauge potentials are in fact affine connections. In this paper, Utiyama gave only six references: one is (necessarily) to the publication of Yang and Mills, another is to his own 1947 paper, clearly establishing the link from his previous work to the present one, and another reference is to page 621 of Pauli (1921). This time, however, a reference to Pauli serves as a reference to Noether, so that her name does not appear.

In later developments, in the Soviet Union in 1959, Lev S. Polak published a translation of Noether's 1918 article into Russian and, in 1972, Vladimir Vizgin published a historical monograph whose title, in English translation, is *The Development of the Interconnection between Invariance Principles and Conservation Laws in Classical Physics*, in which he analysed both of Noether's theorems. At that time, new formulations of Noether's first theorem had started to appear with the textbook of Israel M. Gel'fand and Sergeĭ V. Fomin on the calculus of variations, published in Moscow in 1961, which contains a modern presentation of Noether's first theorem – although not yet using the formalism of jets as would soon be the case – followed by a few lines about her second theorem. This book appeared in an English translation two years later. In the 1970s, Gel'fand published several articles with Mikhael Shubin, Leonid Dikiĭ (Dickey), Irene Dorfman, and Yuri Manin on the 'formal calculus of variations', not mentioning Noether because they dealt mainly with the Hamiltonian formulation of the problems, while Manin's 'Algebraic Theory of Nonlinear Differential Equations' (1978) as well as Boris Kupershmidt's 'Geometry of Jet Bundles and the Structure of Lagrangian and Hamiltonian Formalisms' (1980) both contain a 'formal Noether theorem', which is a modern, generalised version of her first theorem. A few years earlier already, in the article 'Lagrangian Formalism in the Calculus of Variations' (1976), Kupershmidt had presented an invariant approach to the calculus of variations in differentiable fibre bundles, and Noether's first theorem was formulated for the Lagrangians of arbitrary finite order.

Further research in geometry in Russia yielded new genuine generalisations of the concepts introduced by Noether and of her results. Alexandre Vinogradov (1938–2019), who had been a member of Gel'fand's seminar in Moscow, left the Soviet Union for Italy in 1990 and the second part of his career was at the University of Salerno. Beginning in 1975, Vinogradov, together with Joseph Krasil'shchik – who worked in Moscow, then for several years in the Netherlands, and again in Moscow at the Independent University – published extensively on symmetries, at a very general and abstract level, greatly generalising Noether's formalism and results, and on their applications, a theory fully expounded in their book (Krasil'shchik and Vinogradov 1997).

Searching for other lines of transmission of Noether's results, one finds that in the early 1960s Enzo Tonti (later professor at the University of Trieste) translated Noether's article into Italian, but his translation has remained in manuscript. It was transmitted to Franco Magri in Milan who, in 1978, wrote an article in Italian where he clearly set out the relation between symmetries and conservation laws for non-variational equations, a significant development, but he did not treat the case of operators defined on manifolds. In France, Jean-Marie Souriau (1922–2012) was well aware of 'les méthodes d'Emmy Noether' which he cited as early as 1964, on page 328 of his book, *Géométrie et relativité*. In 1970, independently of Bertram Kostant (1928–2017), Souriau introduced the concept of a momentum map. The conservation of the momentum of a Hamiltonian action is the Hamiltonian version of Noether's first theorem. Souriau called that result 'le théorème de Noether symplectique', although there is nothing Hamiltonian or symplectic in Noether's article! Souriau's fundamental work on symplectic geometry and mechanics was based on Lagrange, as he himself claimed, but it was also a continuation of Noether's theory.

1.13 From General Relativity to Cohomological Physics

The history of the second theorem – the improper conservation laws – belongs to the history of general relativity. In the literature on the general theory, the improper conservation laws which are 'trivial of the second kind' are called 'strong laws', while the conservation laws obtained from the first theorem are called 'weak laws'. The strong laws play an important role in basic papers of Peter G. Bergmann in 1958, of Andrzej Trautman in 1962, and of Joshua N. Goldberg in 1980. While the second theorem, which explained in which cases such improper conservation laws would exist, had been known among relativists since the early 1950s, it became an essential tool in the non-Abelian gauge theories that were developed by the mathematical physicists, following the publication of Utiyama's 1956 paper that generalised the Abelian theory of Yang and Mills (1954) to not necessarily commutative gauge groups.

The identities that were proved by Noether in her second theorem are at the basis of Jim Stasheff's 'cohomological physics'. They appeared already in his lecture at Ascona (Stasheff 1997). Later, in Stasheff's article with Ronald Fulp and Thomas Lada (2003), 'Noether's Variational Theorem II and the BV Formalism', Noether's identities associated with the infinitesimal gauge symmetries of a Lagrangian theory appear as the antighosts in the Batalin–Vilkovisky construction for the quantisation of Lagrangians with symmetries. The validity of Noether's second theorem is extended to ever more general kinds of symmetries, interpreting physicists' constructions in gauge theories of increasing complexity.

1.14 Have the Noether Theorems Been Generalised?

Whether the Noether theorems have been generalised has a straightforward answer: except for Bessel-Hagen (and we have seen that his generalisation was certainly suggested and probably entirely worked out by Noether herself), they were not, until the 1970s. Until then, the so-called generalisations were all due to physicists and mathematicians who had no direct knowledge of her article but still thought that they were generalising it, while in fact they were generalising the truncated and restricted version of her first theorem that had appeared in Edward L. Hill's article, 'Hamilton's Principle and the Conservation Theorems of Mathematical Physics', in 1951.

In the late 1970s and early 1980s, using different languages, both linguistically and mathematically, Olver in Minneapolis and Vinogradov in Moscow made great advances in the Noether theory. Equivalence classes were defined for symmetries on the one hand and for conservation laws on the other, bringing precision to the formulation of Noether's results. In order to set up a one-to-one map between symmetries and conservation laws, it is appropriate to first consider the enlarged class of the divergence symmetries (which are the infinitesimal transformations that leave the Lagrangian invariant up to a divergence term, i.e., the concept of symmetry to be found in Bessel-Hagen's article of 1921). Define a divergence symmetry of a differential equation to be trivial if its evolutionary representative vanishes on the solutions of the equation or if its divergence vanishes identically, independently of the field equations, and consider the equivalence classes of divergence symmetries modulo the trivial symmetries. Recall the definition of the trivial conservations laws of the first and of the second kind and consider the equivalence classes of conservation laws modulo the trivial ones. Restrict the consideration of Lagrangians to those whose system of Euler–Lagrange equations is 'normal', meaning roughly that the highest-order partial derivatives of the unknown functions are expressed in terms of all the other derivatives. Then one can formulate what can be called 'the Noether–Olver–Vinogradov theorem' which took the following form both rigorous and simple, first in the article of Luis Martínez-Alonso (1979), then in versions of increasing generality, *circa* 1985: For Lagrangians such that the Euler–Lagrange equations are a normal system, Noether's correspondence induces a one-to-one map between equivalence classes of divergence symmetries and equivalence classes of conservation laws.

Concerning the extension to non-variational equations of Noether's correspondence between symmetries and conservation laws, we find the early work of Magri (1978), who showed that, if $\mathcal{D}$ is a differential operator and $V\mathcal{D}$ is its linearisation, searching for the restriction of the kernel of the adjoint $(V\mathcal{D})^*$ of $V\mathcal{D}$ to the solutions of $\mathcal{D}(u) = 0$ is an algorithmic method for the determination of the conservation laws for a possibly non-variational equation, $\mathcal{D}(u) = 0$. For an Euler–Lagrange operator, the linearised operator is self-adjoint. Therefore, this result generalises the first Noether theorem. This idea is to be found later and much developed in the work of several mathematicians, most notably Vinogradov, Toru Tsujishita, Ian Anderson, and Olver.

Meanwhile, Noether's theory was being set in the modern language of differential geometry and generalised. Trautman, in his 'Noether Equations and Conservation Laws' (1967), followed by 'Invariance in Lagrangian Systems' (1972), was the first to present even a part of Noether's article in the language of manifolds, fibre bundles, and, in particular, the jet bundles that had been defined and studied around 1940 by Charles Ehresmann (1905–79) and his student Jacques Feldbau (1914–45), and by Norman Steenrod (1910–71). In 1970, Stephen Smale published the first part of his article 'Topology and Mechanics' in which he proposed a geometric framework for mechanics on the tangent bundle of a manifold. Hubert Goldschmidt and Shlomo Sternberg wrote a landmark paper in 1973 in which they formulated the Noether theory, for first-order Lagrangians, in an intrinsic, geometric fashion. Jerrold Marsden published extensively on the theory and important applications of Noether's correspondence between invariance and conservation, from 1974 until his death at the early age of 68 in 2010. In the 1970s, several other authors contributed to the 'geometrisation' of Noether's first theorem, notably Jedrzej Śniatycki, Demeter Krupka, and Pedro García. The ideas that permitted the recasting of Noether's theorems in geometric form and their genuine generalisation were first of all that of smooth differentiable manifold (i.e., manifolds of class C^∞), and then the concept of a jet of order k of a mapping ($k \geq 0$) defined as the collection of the values of the components in a local system of coordinates of a vector-valued function and of their partial derivatives up to the order k, the concept of manifolds of jets of sections of a fibre bundle, and finally of jets of infinite order. The manifold of jets of infinite order of sections of a fibre bundle is not defined directly but as the inverse limit of the manifold of jets of order k, when k tends to infinity. It was Vinogradov (1977) who showed that generalised vector fields are nothing other than ordinary vector fields on the bundle of jets of infinite order of sections of a bundle. Both Lagrangians and conservation laws then appear as special types of differential forms. The divergence operator may be interpreted as a horizontal differential, one that acts on the independent variables only, and thus one obtains a cohomological formulation of Noether's first theorem. The study of the exact sequence of the calculus of variations, and of the variational bicomplex, which constitutes a vast generalisation of Noether's theory, was developed in 1975 and later by Włodzimierz Tulczyjew in Warsaw, Paul Dedecker in Belgium, Vinogradov in Moscow, Tsujishita in Japan, and, in the United States, by Olver (1986b) and by Ian Anderson, as well as by Kupershmidt (1976, 1980).

In the discrete versions of the Noether theorems, the differentiation operation is replaced by a shift operator. The independent variables are now integers, and the integral is replaced by a sum, $\mathcal{L}[u] = \sum_n L(n, [u])$, where $[u]$ denotes $u(n)$ and finitely many of its shifts. The variational derivative is expressed in terms of the inverse shift. A pioneer was John David Logan, who published 'First Integrals in the Discrete Variational Calculus' in 1973. Much more recent advances on the discrete analogues of the Noether theorems, an active and important field of research, may be found in a series of papers by Peter Hydon and Elizabeth Mansfield (2001), published since 2001, including a discrete version of the second theorem.

1.15 Were the Noether Theorems Ever Famous?

Whereas both theorems were analyzed by Vizgin in his 1972 monograph on invariance principles and conservation laws in classical physics, it appears that the existence of the first and second theorems in one and the same publication was not expressed in written form in any language other than Russian before the first edition of Olver's book in 1986 and his contemporaneous article where 'Noether's theorems' appear in the title. At roughly the same time, one can find 'theorems', in the plural, in a few other publications: in Hans A. Kastrup's contribution to *Symmetries in Physics (1600–1980)*, the text of a 1983 communication finally printed in 1987 in this extremely rich collection of essays, and in my mathematical paper, 'Sur les théorèmes de Noether', presented in Marseille-Luminy in 1985 at the 'Journées relativistes' organised by André Lichnerowicz, which also appeared in 1987. Then came David Rowe's survey (Rowe 1999).

Fame came eventually. I quote from Gregg Zuckerman's 'Action Principles and Global Geometry' (1987):

> E. Noether's famous 1918 paper, 'Invariant variational problems' crystallized essential mathematical relationships among symmetries, conservation laws, and identities for the variational or 'action' principles of physics. ... Thus, Noether's abstract analysis continues to be relevant to contemporary physics, as well as to applied mathematics.[15]

Therefore, approximately 70 years after her article had appeared in the *Göttingen Nachrichten*, fame came to Noether for this (very small) part of her mathematical *œuvre*. In the 20 page contribution of Pierre Deligne and Daniel Freed to the monumental treatise, *Quantum Fields and Strings: A Course for Mathematicians* (1999), she was credited not only with 'the Noether theorems' but also with 'Noether charges' and 'Noether currents'. For as long as gauge theories had been developing, these terms had, in fact, been in the vocabulary of the physicists, such as Utiyama, Yuval Ne'eman (1999), or Stanley Deser whose discussion of 'the conflicting roles of Noether's two great theorems' and 'the physical impact of Noether's theorems' continues to this day in articles (Deser 2019) and preprints. At the end of the twentieth century, the importance of the concepts she had introduced was

[15] Here Zuckerman cites Olver's book.

finally recognised, and her name was attached to them by mathematicians and mathematical physicists alike.

1.16 In Lieu of Conclusion

One can read in a text published as late as 2003 by a well-known philosopher of science that 'Noether's theorems can be generalised to handle transformations that depend on the $u^{(n)}$ as well.' Any author who had only glanced at Noether's paper, or read parts of Olver's book, would have been aware that Noether had already proved her theorems under that generalised assumption. This, in fact, is one of the striking and important features of Noether's 1918 article. Therefore, *caveat lector*! It is better to read the original than to rely on second-hand accounts. For my part, I shall not attempt to draw any philosophical conclusions from what I have sketched here of Noether's 'Invariante Variationsprobleme', its genesis, its consequences, and its influence, because I want to avoid the mistakes of a non-philosopher, of the kind that amateurs make in all fields.[16]

It is clear that Noether was not a proto-feminist. She was not a practicing Jew. Together with her father, she converted to Protestantism in 1920, which did not protect her from eventual dismissal from the University of Göttingen by the Nazis. She was not an admirer of American democracy, and her sympathies were with the Soviet Union. Even though her 1918 work was clearly inspired by a problem in physics, she was never herself a physicist and did not return to physics in any of her subsequent publications. She never explored the philosophical underpinnings or outcomes of her work – in a word, she was not a philosopher. She was a generous woman admired by her colleagues and students, and a great mathematician.

While the Noether theorems derive from the algebraic theory of invariants developed in the nineteenth century – a chapter in the history of pure mathematics – it is clear from the testimony of Noether herself that the immediate motivation for her research was a question that arose in physics, at the time when the new general theory of relativity was emerging – a fact that she stated explicitly in her 1918 article. The results of this article have indeed become – in increasingly diverse ways which deserve to be much more fully investigated than time and space permitted – a fundamental instrument for mathematical physicists. On the one hand, these results are essential parts of the theories of mechanics and field theory and many other domains of physical science, and on the other, in a series of mainly separate developments, her results have been generalised by pure mathematicians to highly abstract levels, but that was not accomplished in her lifetime. Had she lived longer, she would have witnessed this evolution and the separate, then re-unified, paths of mathematics and physics, and we are free to imagine that she would have taken part in the mathematical discoveries that issued from her 23 page article.

[16] For a philosophical outlook, see, e.g., Brading and Brown (2003).

References

Anderson, R. L. and Ibragimov, N. H. (1979). *Lie-Bäcklund Transformations in Applications*. SIAM Studies in Applied Mathematics. Philadelphia: SIAM.

Bessel-Hagen, E. (1921). Über die Erhaltungssätze der Elektrodynamik. *Mathematische Annalen*, **84**, 258–76.

Brading, K. A. and Brown, H. R. (2003). Symmetries and Noether's theorems. In K. Brading and E. Castellani, eds., *Symmetries in Physics, Philosophical Reflections*, Cambridge: Cambridge University Press, pp. 89–109.

Brewer, J. W. and Smith, M. K. eds. (1981). *Emmy Noether, a Tribute to Her Life and Work*. Monographs and Textbooks in Pure and Applied Mathematics, vol. 69. New York, Basel: Marcel Dekker.

Deligne, P. and Freed, D. S. (1999). Classical field theory. In P. Deligne, P. Etingof, D. S. Freed, et al., eds., *Quantum Fields and Strings: A Course for Mathematicians*. Vol. 1. Providence, RI: American Mathematical Society/Institute for Advanced Study, pp. 137–225.

Deser, S. (2019). Energy in gravitation and Noether's theorems. *Journal of Physics A: Mathematical and Theoretical*, **52**, 381001.

Dick, A. (1970). *Emmy Noether 1882–1935*, Beihefte zur Zeitschrift 'Elemente der Mathematik', no. 13, Basel: Birkhäuser. English translation by Heidi I. Blocher, Boston, Basel, Stuttgart: Birkhäuser, 1981.

Euler, L. (1744). *Methodus inveniendi lineas curvas, maximi minimive proprietate gaudentes*. Lausanne and Geneva, p. 245.

Fulp, R. O., Lada, T. J. and Stasheff, J. D. (2003). Noether's variational theorem II and the BV formalism. (Proceedings of the 22nd Winter School 'Geometry and Physics', Srní, 2002). *Rendiconti del Circolo Matematico di Palermo*, Series II, Supplement no. 71, pp. 115–26.

Gel'fand, I. M. and Fomin, S. V. (1961). *Variacionnoe ischislenie* [Russian]. Moscow: Gosudarstvennoe Izdatelstsvo fizikomatematicheskoj literatury. English translation by Richard A. Silverman, *Calculus of Variations*, Englewood Cliffs, NJ: Prentice-Hall, 1963.

Hydon, P. and Mansfield, E. (2011). Extensions of Noether's second theorem: from continuous to discrete systems. *Proceedings of the Royal Society, Series A, Mathematical, Physical and Engineering Sciences*, **467**(2135), 3206–21.

Kastrup, H. A. (1987). The contributions of Emmy Noether, Felix Klein and Sophus Lie to the modern concept of symmetries in physical systems. In M. G. Doncel, A. Hermann, L. Michel, and A. Pais, eds., *Symmetries in Physics (1600–1980)*, 1st International Meeting on the History of Scientific Ideas, Sant Feliu de Guíxols, Catalonia, Spain (1983). Bellaterra (Barcelona): Seminari d'Història de les Ciències, Universitat Autònoma de Barcelona, pp. 113–63.

Klein, F. (1910). Über die geometrischen Grundlagen der Lorentzgruppe. *Jahresbericht der Deutschen Mathematiker-Vereinigung*, **19**, 281–300. Reprint, *Physikalische Zeitschrift*, **12**, 17–27; *Gesammelte mathematische Abhandlungen*, vol. 1, pp. 533–52.

Klein, F. (1918). Zu Hilberts erster Note über die Grundlagen der Physik. *Nachrichten von der Königlichen Gesellschaft der Wissenschaften zu Göttingen, Mathematisch-physikalische Klasse* (1917), pp. 469–82 (meeting of 25 January 1918); *Gesammelte mathematische Abhandlungen*, vol. 1, pp. 553–67.

Klein, F. (1921–23). *Gesammelte mathematische Abhandlungen*, 3 vols. Berlin: J. Springer. Reprint, Berlin, Heidelberg, New York: Springer, 1973.

Kosmann-Schwarzbach, Y. (1987). Sur les théorèmes de Noether. In Y. Choquet-Bruhat, B. Coll, R. Kerner and A. Lichnerowicz, eds., *Géométrie et physique* (Journées relativistes de Marseille-Luminy, 1985). Travaux en Cours, 21, Paris: Hermann, pp. 147–60.

Kosmann-Schwarzbach, Y. (2004). *Les théorèmes de Noether, Invariance et lois de conservation au XXe siècle*, Éditions de l'École polytechnique. Revised edition 2006.

Kosmann-Schwarzbach, Y. (2010). *The Noether Theorems, Invariance and Conservation Laws in the Twentieth Century*. Translated by Bertram E. Schwarzbach. New York: Springer (Corrected publication 2018).

Krasil'shchik, J. S. and Vinogradov, A. M. eds. (1997). *Simmetrii i zakony sokhranenija uravnenij matematicheskoj fiziki* [Russian] Moscow: Factorial. English translation by A. M. Verbovetsky and I. S. Krasil'shchik, *Symmetries and Conservation Laws for Differential Equations of Mathematical Physics*. Providence, RI: American Mathematical Society, 1999.

Kupershmidt, B. A. (1976). Lagrangian formalism in the calculus of variations [Russian]. *Funkcional. Anal. i Priložen.* **10**(2), 77–8.

Kupershmidt, B. A. (1980). Geometry of jet bundles and the structure of Lagrangian and Hamiltonian formalisms. In G. Kaiser and J. E. Marsden, eds., *Geometric Methods in Mathematical Physics* (Lowell, MA, 1979), Lecture Notes in Mathematics, 775. Berlin, Heidelberg, New York: Springer-Verlag, pp. 162–257.

Lagrange, J. L. (1788). *Méchanique analitique*. Paris: Chez la Veuve Desaint.

Lagrange, J. L. (1811). *Mécanique analytique*, 2nd ed., Vol. 1. Paris: Mme Veuve Courcier.

Logan, J. D. (1973). First integrals in the discrete variational calculus. *Aequationes Mathematicae*, **9**, 210–20.

Magri, F. (1978). Sul legame tra simmetrie e leggi di conservazione nelle teorie di campo classiche. Fourth National Congress of Theoretical and Applied Mechanics, AIMETA, Florence, 25–8 October.

Manin, Y. I. (1978). Algebraic theory of nonlinear differential equations [Russian]. *Itogi Nauki i Tekhniki, Sovremennye Problemy Matematiki*, **11**, 5–152. English translation, *Journal of Soviet Mathematics*, **11**, 1–122.

Martínez-Alonso, L. (1979). On the Noether map. *Letters in Mathematical Physics*, **3**, 419–24.

Ne'eman, Y. (1999). The impact of Emmy Noether's theorems on XXIst century physics. In M. Teicher, ed., *The Heritage of Emmy Noether*. Israel Mathematical Conference Proceedings, vol. 12. [Ramat-Gan]: Bar-Ilan University, pp. 83–101.

Noether, E. (1907). Über die Bildung des Formensystems der ternären biquadratischen Form. *Sitzungsberichte der Physikalisch-medizinischen Societät zu Erlangen*, **39**, 176–9; *Gesammelte Abhandlungen*, pp. 27–30.

Noether, E. (1908). Über die Bildung des Formensystems der ternären biquadratischen Form. *Journal für die reine und angewandte Mathematik*, **134**, 23–90, with two tables; *Gesammelte Abhandlungen*, pp. 31–99.

Noether, E. (1918). Invariante Variationsprobleme. *Nachrichten von der Königlichen Gesellschaft der Wissenschaften zu Göttingen, Mathematisch-physikalische Klasse*, pp. 235–57; *Gesammelte Abhandlungen*, pp. 248–70. English translation, 'Invariant variation problems', by M. A. Tavel, *Transport Theory and Statistical Physics*, **1**(3) (1971), 186–207. English translation, 'Invariant variational problems', in Kosmann-Schwarzbach (2010), pp. 1–22.

Noether, E. (1983). *Gesammelte Abhandlungen/Collected Papers*. Edited by N. Jacobson. Berlin, Heidelberg, New York: Springer-Verlag.

Olver, P. (1986a). *Aplications of Lie Groups to Differential Equations*, Graduate Texts in Mathematics, vol. 107, New York: Springer-Verlag (2nd ed., revised, 1993).

Olver, P. (1986b). Noether's theorems and systems of Cauchy-Kovalevskaya type. In B. Nicolaenko, D. D. Holm and J. M. Hyman, eds., *Nonlinear Systems of Partial Differential Equations in Applied Mathematics*. Lectures in Applied Mathematics, Vol. 23, Part 2, Providence, RI: American Mathematical Society, pp. 81–104.

Olver, P. (2018). Emmy Noether's enduring legacy in symmetry. *Symmetry: Culture and Science*, **29**, 475–85.

O'Raifeartaich, L. (1997). *The Dawning of Gauge Theory*. Princeton: Princeton University Press.

Pauli, W. (1921). Relativitätstheorie. *Encyklopädie der mathematischen Wissenschaften*, Vol. 2, Leipzig: B. G. Teubner, 1921, pp. 539–775. Reprinted as a monograph, with a preface by Arnold Sommerfeld, Leipzig: Teubner, 1921. *Collected Scientific Papers*, vol. 1, pp. 1–237.

Rowe, D. E. (1999). The Göttingen response to general relativity and Emmy Noether's theorems. In J. J. Gray, ed., *The Symbolic Universe, Geometry and Physics 1890–1930* (Milton Keynes, 1996). Oxford and New York: Oxford University Press, pp. 189–233.

Smale, S. (1970). Topology and mechanics. *Inventiones Mathematicae*, **10**, 305–331.

Souriau, J.-M. (1964). *Géométrie et relativité*. Paris: Hermann.

Souriau, J.-M. (1970). *Structure des systèmes dynamiques*. Paris: Dunod. English translation by C. H. Cushman-de Vries, *Structure of Dynamical Systems: A Symplectic View of Physics*, Springer, 1997.

Srinivasan, B. and Sally, J. D. eds. (1983). *Emmy Noether in Bryn Mawr*. New York, Berlin, Tokyo: Springer-Verlag.

Stasheff, J. D. (1997). Deformation theory and the Batalin-Vilkovisky master equation. In D. Sternheimer, J. Rawnsley and S. Gutt, eds., *Deformation Theory and Symplectic Geometry* (Ascona, 1996). Mathematical Physical Studies. 20. Dordrecht: Kluwer Academic Publishers, pp. 271–84.

Trautman, A. (1967). Noether equations and conservation laws. *Communications in Mathematical Physics*, **6**, 248–61.

Trautman, A. (1972). Invariance of Lagrangian systems. In L. O'Raifeartaigh, ed., *General Relativity*, papers in honor of J. L. Synge. Oxford: Clarendon Press, pp. 85–99.

Utiyama, R. (1947). On the interaction of mesons with the gravitational field. I. *Progress of Theoretical Physics*, **2**(2), 38–62.

Utiyama, R. (1956). Invariant theoretical interpretation of interaction. *Physical Review*, (2), **101**(5), 1597–1607.

Vinogradov, A. M. (1977). On the algebro-geometric foundations of Lagrangian field theory. *Doklady Akad. Nauk SSSR*, **236**, 284–7; English translation, *Soviet Mathematics Doklady*, **18**, 1200–4.

Vizgin, V. P. (1972). *Razvitije vzaimosvyazi principov invariantnosti s zakonami sokhranenija v klassicheskoj fizike* [Russian] (Development of the Interconnection between Invariance Principles and Conservation Laws in Classical Physics), Moscow: Nauka.

Weyl, H. (1918). *Raum, Zeit, Materie*. Berlin: J. Springer; 2nd ed., 1919; 3rd ed., revised, 1919, and later editions.

Weyl, H. (1935). Emmy Noether (Memorial address delivered in Goodhart Hall, Bryn Mawr College, on April 26, 1935). *Scripta Mathematica*, **3**, 201–20; *Gesammelte Abhandlungen*, vol. 3, pp. 425–44.

Yang, C. N. and Mills, R. L. (1954). Conservation of isotopic spin and isotopic gauge invariance. *Physical Review*, **96**, 191–5.

Zuckerman, G. J. (1987). Action principles and global geometry. In S. T. Yau, ed., *Mathematical Aspects of String Theory* (University of California, San Diego, 1986), Advanced Series in Mathematical Physics, Vol. 1. Singapore, New Jersey, Hong Kong: World Scientific, pp. 259–84.

2

Felix Klein and Emmy Noether on Invariant Theory and Variational Principles

DAVID E. ROWE

2.1 Introduction

The title of this chapter calls for some preliminary remarks and clarification. One of its aims is to shed new light on Emmy Noether's paper 'Invariante Variationsprobleme' (Noether 1918b), regarded today as one of her most important works because of its relevance for mathematical physics.[1] Its importance stems from the significance now attached to Noether's First Theorem, which accounts for the relationship between symmetries in physical systems and their related conservation laws. Although less well-known in physics today, Noether's Second Theorem was arguably the more significant result historically, as it helped to resolve much of the confusion surrounding the role of variational principles in general relativity.

The contents of Noether's paper have been analyzed elsewhere[2] and will only be touched upon here. Instead, I want to highlight various connections between this famous work and some of Noether's lesser-known contributions to invariant theory and related formal variational methods. At the same time, I shall underscore the importance of Felix Klein's contemporaneous lecture courses and research program for appreciating the context from which Noether's work arose. Indeed, to appreciate the broader motivation behind her work from the period 1916 to 1918, it should be placed against the background of Göttingen responses to Einstein's general theory of relativity.[3]

Variational analysis gradually arose from physics (the brachistochrone problem) or even metaphysical interests, dating back to Maupertuis's principle of least action. Alongside multivariable calculus, the subject gradually spawned formal operations that mimicked the rules of the Leibnizian calculus, the most important being the δ operator, which commutes with d, thus $\delta dy = d\delta y$. The simplest type of variational problem involves finding a function $y(x)$ with given boundary values $y(a), y(b)$ for which

$$\int_a^b F\big(y, y', x\big)dx \tag{2.1}$$

[1] See Uhlenbeck (1983) and Olver (1986/1993).

[2] For a detailed analysis of Noether (1918b) and its reception by the mathematics and physics communities, see Kosmann-Schwarzbach (2006/2011); in this volume, see Kosmann-Schwarzbach, Chapter 1.

[3] This larger background is the topic of Rowe (1999); for an account of Noether's career and influence, see Rowe (2021).

yields a stationary value. Formulated in the classical δ symbolism, this reads:

$$\delta \int_a^b F(y, y', x) dx = 0. \tag{2.2}$$

To solve this, one calculates $\delta F(y, y', x) = \epsilon \left(\frac{\partial F}{\partial y}\phi + \frac{\partial F}{\partial y'}\phi' \right)$, and finds

$$\delta \int_a^b F(y, y', x) dx = \int_a^b \delta F(y, y', x) dx = \epsilon \int_a^b \left(\frac{\partial F}{\partial y}\phi + \frac{\partial F}{\partial y'}\phi' \right) dx. \tag{2.3}$$

The standard trick at this point is to rewrite the second term on the right by using partial integration:

$$\int_a^b \frac{\partial F}{\partial y'}\phi' dx = \left[\frac{\partial F}{\partial y'}\phi \right]_a^b - \int_a^b \frac{d}{dx}\left(\frac{\partial F}{\partial y'} \right) \phi dx. \tag{2.4}$$

Since ϕ vanishes on the boundary, only the second term remains, and substituting in (2.2) yields

$$\delta \int_a^b F(y, y', x) dx = \int_a^b \left(\frac{\partial F}{\partial y} - \frac{d}{dx}\frac{\partial F}{\partial y'} \right) \phi dx = 0. \tag{2.5}$$

Since this equation holds for arbitrary $\phi(x)$, the integrand itself must vanish, which yields the classical Euler-Lagrange equation:

$$\psi(x) = \frac{\partial F}{\partial y} - \frac{d}{dx}\frac{\partial F}{\partial y'} = 0. \tag{2.6}$$

This well-known argument lies in the background of the formal methods employed in 'Invariante Variationsprobleme' (Noether 1918b). In Noether's setting, however, the function F can be much more complicated, and instead of a single variable x, one integrates over an n-dimensional region. The corresponding solution (2.27) then involves n Lagrangian expressions ψ_i, which arise as differential invariants from the corresponding variational problem. These mathematical underpinnings are familiar from analytical mechanics, where one takes $F = L = T - V$. The Lagrangian $L(q_1, \ldots, q_n, \dot{q}_1, \ldots, \dot{q}_n, t)$ uses generalized coordinates $q_i(t)$ and their derivatives $\dot{q}_i(t)$ to describe the position and velocity of a physical system over a time interval $t \in [a, b]$. The equations of motion are then given by the n equations

$$\psi_i(t) = \frac{d}{dt}\frac{\partial L}{\partial \dot{q}_i(t)} - \frac{\partial L}{\partial q_i(t)} = 0. \tag{2.7}$$

As Noether remarked at the beginning of 'Invariante Variationsprobleme', her results combined methods from the formal calculus of variations with those from Lie's theory of continuous groups. The latter take hold when one requires that the integral remain invariant under the infinitesimal action of the generators of a Lie group, whereas the former hold for *any* variational problem of the type Noether considered. These variational methods happen to be intimately connected with the generation of differential invariants, which will be the primary focal point of interest in this paper. Indeed, this connection was a central theme in

Felix Klein's lectures, beginning with his modernized interpretation of the derivation of the Lagrangian equations of motion (see equation (2.31) below).

Emmy Noether was a true virtuoso in wielding formal variational methods as a tool for systematically generating differential invariants. That this part of her mathematical legacy has been almost entirely overlooked can largely be explained by the fact that, until fairly recently, the formal calculus of variations had fallen out of fashion. This point was made by E. J. McShane, who wrote that her works 'constitute a major contribution to a highly formalistic aspect of the calculus of variations that received much attention in the nineteenth century, but was already becoming less interesting to analysts when her theorems appeared' (McShane 1981, p. 130).[4] As was pointed out in Kosmann-Schwarzbach (2006/2011, pp. 84, 138–9), however, by the mid-1970s there was a resurgence of interest in formal calculus of variations. It should also be added that Noether was well aware of the need to supplement and sharpen formal methods when solving concrete problems in the calculus of variations; see her remarks in Noether (1923, p. 442).

In his lecture courses on relativity, Klein took a retrospective approach. He presented various topics genetically rather than systematically, pulling together threads from earlier publications in order to reveal their underlying connections. Provisional texts from these lectures circulated for a time, either in handwritten or mimeographed form, but these differed markedly from the version prepared by Stephan Cohn-Vossen and published posthumously (Klein 1927). Klein's earlier lectures on the development of mathematics in the nineteenth century, by contrast, were not altered greatly in Klein (1926).[5] Both sets of lectures had ambitious goals that were never completely realized.[6] In spirit, Klein's lectures on relativity were akin to those he delivered earlier on the development of mathematics in the nineteenth century, except that he now had a more specific agenda in mind that required closer attention to technical aspects. Two central topics were invariant theory and variational methods, both major research interests of Emmy Noether at this time. Her famous paper 'Invariante Variationsprobleme' was preceded by a related study on differential invariants (Noether 1918a), which exploited what she called formal variational methods. The problem she attacked there amounted to a far-reaching generalization of E.B. Christoffel's reduction theorem (Christoffel 1869), a key result taken up in Klein's lectures (Klein 1927, pp. 195–9).

An important publication for present purposes is Noether's (1922) rarely mentioned note that appeared in the German *Encyclopedia of the Mathematical Sciences*. Although brief and somewhat opaque, this text will prove highly suggestive for contextualizing her earlier work. One year later, Noether (1923) elaborated on some of the same ideas by sketching recent progress in algebraic and differential invariant theory. Both of these notes refer to Noether's (1918a) generalization of Christoffel's reduction theorem for homogeneous quadratic differential forms. The immediate motivation for Noether's (1918a) work came from Klein's lectures on the mathematical foundations of general relativity, held during the

[4] The pitfalls and shortcomings of formal calculus of variations are discussed in Ewing (1985, chap. 4), where the author emphasized that these methods do not appear in the other chapters of his book.

[5] For details concerning Klein's lecture courses during this period, see Tobies (2019, pp. 455–63).

[6] Richard Courant nevertheless decided to publish edited versions of both posthumously; whether or not Klein ever assented to this plan remains unclear.

summer semester of 1917.[7] In these, Klein highlighted the methodological advantages of the approach taken earlier by Bernhard Riemann and Rudolf Lipschitz for attacking the problem of equivalence for differential forms, a problem first posed and solved by Christoffel (see Klein (1927, pp. 187–8, 195–9)). Noether also worked closely with Hermann Vermeil, who also attended Klein's lectures and contributed to the latter's research program, above all in a paper in which he cited her assistance (Vermeil 1919). Vermeil became Klein's last assistant in 1919 and played a major role in preparing all three volumes of his collected works (Klein 1921–3).

Although none of these investigations can be discussed in any detail here, they clearly point to Felix Klein's strong interest in older mathematical ideas as well as his desire to develop these further. Apart from this, his collaboration with Noether reveals how quickly she mastered the methodological tools needed to pursue the program he had in mind. It was Klein who submitted Noether's 'Invariante Variationsprobleme' (Noether 1918b) as well as her earlier note (Noether 1918a) for publication in the *Nachrichten* of the Göttingen Scientific Society. Both of these papers employed invariant theory and variational principles, and both arose from their mutual research interests at that time. These joint efforts of Klein and Noether, however, were fairly quickly forgotten, in part because they were superseded by subsequent developments and results.[8] The contributions of Göttingen mathematicians to the new physics, on the other hand, were to some extent canonized in Richard Courant's classic textbook *Methoden der mathematischen Physik* (Courant and Hilbert 1924), which went through several editions and was later translated into English.[9] As noted by Kosmann-Schwarzbach, this was one of the few early texts that provided an account of both Noether theorems (Kosmann-Schwarzbach 2006/2011, 2011, p. 96).

2.2 On Klein's Research Agenda, 1916–1918

Most discussions of general relativity in Göttingen have focused somewhat narrowly on David Hilbert's efforts to wed Einstein's theory of gravitation with Gustav Mie's electromagnetic theory of matter. Seen in retrospect, this was the first in a series of abortive attempts to establish a unified field theory. Einstein, who later explored numerous approaches for constructing a viable UFT, dismissed Hilbert's initial program out of hand as naive. Hilbert's own initial enthusiasm for this project seems to have waned as well (see Renn and Stachel 2007). Very few sources exist that might shed some light on Emmy Noether's work with him, but allusions in Hilbert's lecture course from 1916 to 1917 suggest that he hoped to exploit insights from invariant theory to advance his physical program

[7] During the previous two semesters, Klein offered lectures on 'Invariant Theory of Linear transformations' (SS 1916) and 'Invariant Foundations of Special Relativity' (WS 1916/17). His lectures during the SS 1917 were entitled 'Invariant Theory of General Point Transformations.' He began the WS 1917/18 with lectures on 'Invariant Foundations of General Relativity,' but broke these off after the Christmas vacation. These final lectures were never published, but the others appeared ten years later in a posthumously published edition prepared by Stephen Cohn-Vossen (Klein 1927). Some of Klein's contemporaries read earlier *Ausarbeitungen*, which already circulated during the war years. Arnold Sommerfeld, who worked closely with Klein and who edited Band 5 of the *Encyklopädie*, helped to distribute these texts from Klein's lectures.

[8] One of the puzzles surrounding the differential form for energy-momentum conservation in general relativity disappeared once it was realized that this was an immediate consequence of the contracted Bianchi identities; see Rowe (2018, pp. 263–72).

[9] This was Volume I, but the second volume did not appear until 1937 when Courant was in the United States.

(Sauer and Majer 2009, pp. 287–9). If so, these hopes came to naught, nor does it seem likely that Noether's expertise could have been of any help in this venture, given her limited knowledge of physics. In fact, she was already pursuing her own ideas in algebra and invariant theory, inspired by her reading and still-ongoing collaboration with Ernst Fischer.[10] By the final year of the Great War, Hilbert's research shifted away from mathematical physics to the foundations of mathematics. Nevertheless, his work opened an array of mathematical problems that served as a catalyst for Klein's initial publications on general relativity. The latter then prompted Noether to analyze the underpinnings of Hilbert's Theorem 1, which he had stated without proof (Hilbert 1915).

Notwithstanding the central importance of these matters, they represent only one particular aspect within the larger complex of problems Klein pursued during the years 1917–19. One can gain a clear idea of Klein's interests from that time by reading Klein (1927) while comparing it with Part II of Wolfgang Pauli's *Theory of Relativity*, the part on 'Mathematical Tools' in his report for the *Encyklopädie* (Pauli 1958, pp. 21–70). Doing so also reveals that Pauli had carefully studied the *Ausarbeitungen* from Klein's lectures as well as the latter's published works, which he highlighted in the final section of Part II on variational principles (Pauli 1958, Sec. 23). Klein read the proofs of Pauli's report and sent him several letters in response, some published (Pauli 1979), though for some mysterious reason Pauli failed to mention Noether's "Invariante Variationsprobleme", which was written in order to clarify the status of conservation laws and energy principles in general relativity. Instead, he cited Klein's papers from the final section of Volume 1 of his collected works (Klein 1921–3), entitled 'Zum Erlanger Programm.' The final three papers under that heading reflect Klein's research agenda during the period he worked closely with Noether on topics in general relativity.

In general terms, Klein aimed to link spacetime physics with the approach to geometry he outlined in his 'Erlanger Programm' (Klein 1872). Originally, he conceived of this as a method for unifying geometrical research by studying transformation groups and their associated invariants. Already decades earlier, global transformations had begun to assume a central place in geometrical investigations, though it was only around 1870 that geometers associated these infinite families of transformations with the finite groups studied by algebraists. Klein's 'Erlanger Programm' marked the culmination of his collaboration with Sophus Lie during the period 1869–72. As Thomas Hawkins pointed out (Hawkins 1984), its ideas at first met with a slow reception, though this changed by the the 1890s, the decade during which Lie's novel theory of continuous groups drew widespread international attention.[11]

The original conception outlined by Klein (1872) amounts to the following idea. Klein identified a geometry with a coordinatized space (or a 'manifold') M and a group G of coordinate transformations that acted on it. Given a geometry (M, G), one could then induce an equivalent geometry on a manifold M' by means of a bijective mapping $f: M \longrightarrow M'$, since the group G acting on M will by means of f also act on M'. Furthermore, the properties

[10] For an idea of the scope of her research program in invariant theory, see Noether (1923).
[11] On the history of Lie groups and Lie algebras, see Hawkins (2000).

of a geometry (M, G) could be studied systematically by developing the invariant theory for G. Historically, this had been done only for projective groups, once projective geometry emerged as a major field of research in the nineteenth century. Within the context of Klein's 'Erlanger Programm,' projective and Euclidean geometry were the two paradigmatic cases, although he described several other geometries as well.

In his old age, Felix Klein tried to picture this scheme as marking the beginning of a grand narrative that culminated with Einstein's special and general theories of relativity, both of which inspired a great deal of interest among mathematicians. While working closely with Noether, he published three papers on general relativity in the *Göttinger Nachrichten* (Klein 1918a, 1918b, 1919). He then incorporated these in Volume I of his collected works, along with commentaries in a special section entitled 'Zum Erlanger Programm' (Klein 1921–3, I, pp. 411–612). Emmy Noether supported this undertaking from beginning to end, when she read the page proofs for this section of the volume.[12]

Klein had no difficulty interpreting special relativity within the context of his 'Erlanger Programm,' since the Poincaré group acts globally on Minkowski space. General relativity, on the other hand, was based on the principle of general covariance. In this setting, one takes the coordinate transformations to be defined by bijective analytic functions with a non-vanishing functional determinant; this ensures the existence of local bijective inverses, and thus a local transformation group. Sophus Lie had begun to develop such a theory in the 1880s by studying local differential invariants (Lie 1884), whereas Klein's research after 1872 dealt with applications of finite and infinite discrete groups. Eduard Study, a leading expert on algebraic invariant theory, later exposed several weaknesses in Lie's approach, which differs in many respects from modern Lie theory (Study 1908).

During the mid-1870s, Klein continued working closely with his former Erlangen colleague, Paul Gordan, a leading expert on formal methods in invariant theory. Gordan was joined in Erlangen by Max Noether in 1875, the year Klein departed for Munich, and it was Gordan who later guided the doctoral research of his colleague's daughter.[13] Emmy Noether completed her dissertation in 1907, after which time she assisted her father and other Erlangen mathematicians. After 1911, she struck up an intense collaboration with Ernst Fischer, an Austrian mathematician who came to Erlangen as Gordan's successor. Soon thereafter, she began churning out papers closely related to Hilbert's famous invariant-theoretic works from the late 1880s and early 1890s. These eventually caught the latter's attention, and so it happened that both Hilbert and Klein became well aware of Emmy Noether's expertise in invariant theory. As the war dragged on and the usual reservoir of mathematical talent in Göttingen began to dissipate, both were keen to gain her assistance for their ongoing work. This turning point was the one and only time in her career when she could actually profit from not being a man.[14]

12 Klein (1921–3, 1, p. v). Hubert Goenner has pointed out that modern spacetime theories involve fields attached to the geometry. Physicists typically study the symmetries of Lie groups or Lie algebras associated with such fields rather than groups that act on the spacetime itself; see Goenner (2015).

13 On mathematics in Erlangen during this period, see Rowe (2021, pp. 1–37).

14 Her petition to habilitate in Göttingen initially failed, however, and she only gained the right to teach courses in 1919; see Rowe (2021, pp. 39–61).

2.3 Einstein on the Ricci Calculus

Noether arrived in Göttingen in the spring of 1915, shortly before Einstein came to deliver six lectures on his new approach to gravitation based on general relativity.[15] On this occasion, he presented the so-called *Entwurf* theory, which he had developed with the help of his friend Marcel Grossmann, the mathematician who first introduced him to the Ricci calculus. Conceptually, this theory differed very little from Einstein's mature theory, as set forth by Einstein (1916). However, it lacked generally covariant field equations, which Einstein first found in November 1915.

Late in his life, Einstein wrote about how he had come to the realization that a scalar theory of gravitation was inadequate. He therefore sought a theory based directly on the differential form $g_{\mu\nu}dx^{\mu}dx^{\nu}$ in which the coefficients $g_{\mu\nu}$ transform in accordance with general coordinate transformations. Einstein's problem would then be reduced to finding generally covariant differential equations for the $g_{\mu\nu}$, whose solutions would then determine the geodesic curves in spacetime for test bodies in a gravitational field. After explaining this to Grossmann, the latter undertook a search of the literature. Soon thereafter, he reported to Einstein that the mathematical theory for which he was looking already existed and could be found in the works of Riemann, Gregorio Ricci, and Tullio Levi-Civita (Goodstein 2018). Einstein summarized what Grossmann told him as follows:

> This entire development was connected with the Gaussian theory of surface curvature, in which systematic use of curved coordinates appeared for the first time. Riemann's accomplishment was the greatest. He showed how tensors with second derivatives can be constructed from the $g_{\mu\nu}$. From that, one could see how the differential equations of gravitation must appear – if one required invariance with respect to the group of all continuous coordinate transformations. To see that this requirement was justified took some doing, as I had found reasons for believing otherwise. These mistaken misgivings led to the circumstance that not until 1916 did the theory appear in its final form. (Seelig 1956, p. 16, my translation)

In what follows, we will briefly consider the extent to which Grossmann's reading of the classic paper (Ricci and Levi-Civita 1901) may have given Einstein a faulty impression of Riemann's approach to differential invariants. Einstein's attention was naturally focused on generalizing the theory of relativity to arbitrary reference frames. The case of a flat spacetime thus held special interest, whereas the general form problem first addressed by Christoffel involved a much more elaborate treatment of differential invariants.[16] The immediate relevance of the Riemann–Christoffel tensor for relativity, as referenced in a footnote by Einstein and Grossmann (1913, p. 337), was that it provided necessary and sufficient conditions for the possibility of obtaining the Minkowski metric in a finite region of spacetime (this can always be done in an infinitesimal region). Following Einstein's physical interpretation, the vanishing of the 20 components of the Riemann–Christoffel

[15] Until recently it was uncertain whether Emmy Noether was present during Einstein's lectures, as she had left Göttingen in May 1915 around the time of her mother's death. Thanks to research undertaken by Cordula Tollmien, we now know that she returned to Göttingen well before Einstein's arrival.

[16] Pauli referred briefly to this equivalence problem for differential forms, but noted that it had thus far found no relevant application for physics (Pauli 1958, p. 42).

tensor meant that one could transform away the gravitational field (Einstein 1916, p. 318). Einstein and Grossmann also tried to make use of its contracted form, the Ricci tensor with 10 components, for finding gravitational field equations as conditions on the metric tensor $g_{\mu\nu}$. Their efforts floundered, however, because they mistakenly thought that this approach was incompatible with the Poisson equations for a static gravitational field.[17]

Einstein's classic paper (Einstein 1916) opens with a discussion of special relativity focused on the need to extend the original theory beyond inertial reference frames. He showed how this can be done by introducing general coordinate transformations. By means of these, the Minkowski metric passes over to a quadratic differential form, which served as the starting point for Riemann's investigations of n-dimensional differentiable manifolds. Einstein briefly alludes to this background in his introduction – naming Gauss, Riemann, and Christoffel – but only insofar as their work led to the Ricci calculus (Ricci and Levi-Civita 1901). 'All the necessary mathematical tools for the general theory of relativity,' he wrote, 'lay ready in the "absolute differential calculus", ...which Ricci and Levi-Civita had systematized and already had applied to problems in theoretical physics' (Einstein 1916, p. 284). Immediately after writing this, Einstein expressed his thanks to his 'friend, the mathematician Marcel Grossmann' for sparing him the task of having to study the relevant mathematical literature.[18] That Einstein felt relieved that Grossmann provided him with these ready-made mathematical tools is further confirmed by the fact that he cited none of the relevant literature in his paper.[19]

As Klein pointed out (Klein 1927, pp. 189, 195), Einstein's derivation of the Riemann–Christoffel tensor in (Einstein 1916, pp. 314–15) was identical to that given in (Ricci and Levi-Civita, 1901, p. 143)). Thus, the relevant chain of events began with Christoffel, who found the Riemann–Christoffel tensor independently. Grossmann, working with Einstein, wrote this down in the form found in Christoffel (1869, p. 54), which he cited in (Einstein and Grossmann 1913, p. 336). On the same page, he repeated the incorrect citation (Riemann 1876, p. 270), a reference he obviously just copied from Ricci and Levi-Civita (1901). What should be emphasized is that Marcel Grossmann, who studied geometry under Wilhelm Fiedler at the Zurich Polytechnic and eventually succeeded his mentor there, lacked a strong background in differential geometry. He was therefore by no means an expert on the mathematical methods he conveyed to Einstein. This alone would account for why he probably had little knowledge of Riemann's innovations; nor was such knowledge really needed for the purpose Einstein had in mind.

[17] Einstein and Grossmann (1913, p. 337); for a detailed account of how Einstein eventually found his way back to general covariant equations based on the Ricci tensor, see Janssen and Renn (2007).

[18] The English translation of Einstein's paper published in the classic collection (Sommerfeld 1952) omits this introduction entirely.

[19] Nor did he do so in his earlier paper (Einstein 1914), whereas Grossmann did cite some mathematical literature in Part II of Einstein and Grossmann (1913). It would appear doubtful, however, that Grossmann had any familiarity with the literature beyond Ricci and Levi-Civita (1901), Christoffel (1869), and Bianchi (1896). Moreover, there is little reason to believe that Ricci was deeply influenced by Riemann's 'Commentatio mathematica' (Riemann 1876a) at the time he and his former student wrote their classic paper (Ricci and Levi-Civita 1901). They cite it, to be sure (Ricci and Levi-Civita 1901, p. 142), but refer incorrectly to page 270 instead of Riemann (1876, p. 370), which is where the text begins in the first edition of the *Werke*. This was probably just a typographical error, but if the authors wanted to give a substantive reference, they ought to have indicated that the relevant passages came from Riemann (1876, pp. 380–2 or 1892, pp. 401–3). Their purpose was clearly only to acknowledge Riemann's priority.

2.4 On Klein's Göttingen Lectures, 1916–1917

As is well known, Riemannian geometry and the theory of differential forms took on new life in the wake of Einstein's general theory of relativity, which highlighted the importance of Gregorio Ricci's absolute differential calculus, dubbed tensor analysis by Einstein. In the 1880s, Ricci was the first to systematize the theory of differential invariants in the form of a calculus. During the last years of the war, Klein turned to these topics in his historical lectures, delivered in part to celebrate the legacy of the Göttingen mathematical tradition. He described an analytical tradition – associated with the names of Christoffel, Bianchi, and Ricci – setting this against a second tradition, which he identified with the names of Riemann and Rudolf Lipschitz; the latter happened to have been his formal doctoral supervisor in Bonn, following the sudden death in 1868 of his actual mentor, Julius Plücker.

As Abraham Pais pointed out in his scientific biography of Einstein (Pais 1982/2005, p. 217), Felix Klein situated Einstein's work in the analytic tradition of Christoffel and Ricci. Pais warmly recommended Klein's posthumously published lectures (Klein 1927),[20] but he apparently spent little time reading these himself. Had he done so, he surely would have noticed that Klein sharply contrasted between Riemann's methodological approach and the methods Einstein and Grossmann adopted from Ricci and Levi-Civita. To put the matter plainly, Klein's interpretation placed Einstein's work outside the mainstream of the Riemannian tradition. This latter direction drew on variational methods and employed so-called normal coordinates, methods developed further by Lipschitz, who explicitly cited Riemann's work (Lipschitz 1869, 1870). Christoffel's methods for classifying differential forms, on the other hand, were entirely formal and algebraic. Ricci, in fact, saw these as major virtues, and he went out of his way to criticize the use of the calculus of variations as a method for finding differential invariants, a technique employed in earlier works of Jacobi and Beltrami (Ricci and Levi-Civita 1901, p. 127). Klein mentioned Ricci's dismissive attitude toward Beltrami's work, before he went on to say:

> The Christoffel-Ricci representation has found widespread transmission. It takes the place of honor in Edmund Wright's monograph *Invariants of Quadratic Differential Forms* (Cambridge Tracts 1908), whereas Riemann is only dealt with in passing and Lipschitz not at all. Einstein, too, grew up in this tradition. (Klein 1927, p. 189)[21]

In short, Klein's principal goal in his lectures on differential invariants was to resurrect this Riemannian approach. His presentation culminated with a fairly detailed account of its methods and results, tools that provided the background and motivation for Noether (1918a). As will be described below, these ideas formed a major role in Klein's overall agenda, and Noether pursued them with remarkable success. She also attended some of his earlier and better-known courses on the development of mathematics in the nineteenth century. He taught these over a span of three semesters, starting in the winter of 1914–15. Although she was then still in Erlangen, Noether evidently received a copy of the manuscript from

20 The third chapter stems from his SS 1917 lectures, which were supplemented by the editor, Stephan Cohn-Vossen. This chapter is entitled, 'Gruppen analytischer Punkttransformationen bei Zugrundelegung einer quadratischen Differentialform.'

21 Wright's text (Wright 1908), published roughly a decade before Klein's lectures, summarized the state of the art prior to the advent of general relativity.

those initial lectures that Klein's daughter, Elisabeth Staiger, wrote up for him. Klein made a note of this in preparation for a meeting with Noether on 28 April 1915, when he began his second series of lectures.[22]

Klein's lectures on the mathematics of the past century were to have ended with a fourth and final series, held during the summer semester of 1916, in which he originally planned to discuss the works of Lie and Henri Poincaré. Spurred by Einstein's and Hilbert's recent publications, however, he opted instead to lecture on mathematical developments related to the theory of relativity. His assistant, Walter Baade, wrote up these lectures in a 100-page manuscript to which Klein appended a provisional table of contents.[23] Nevertheless, he conceived of these as a direct continuation of his earlier courses; he thus labeled these lectures Teil IV, Kap. 9, so as to follow directly on the eight chapters in the earlier *Ausarbeitungen*.[24] Judging by the many topics touched upon, these lectures represent Klein's first attempt to gain an overview of the literature. Some of the material he covered was probably reworked into subsequent *Ausarbeitungen*, which he circulated through Sommerfeld and Einstein.[25]

Klein's interest in special relativity largely focused on the invariant theory of the Lorentz group, starting with Maxwell equations, which he wrote in a manifestly covariant form. Some years before this, Klein (1910) gave a projective interpretation of Minkowski space, which placed it within the larger context of metric geometries of constant curvature. Like Euclidean geometry, it corresponds to a flat space, but one in which the metric is indefinite rather than positive definite. Klein had been the first to exploit the possibility of attaching different types of quadrics to a projective space in order to introduce a metric, an idea inspired by Arthur Cayley's realization of Euclidean geometry by means of the so-called Cayley metric.[26] In this special case, the quadric is a degenerate imaginary figure, whereas the Minkowski metric corresponds to a real degenerate quadric. These possibilities for deriving different metrical geometries served as a major inspiration for Klein's 'Erlanger Programm,' especially since the same general approach could be applied in many different settings. After Minkowski geometrized what later came to be called 'special' relativity, Klein proposed calling 'invariant theory relative to a group of transformations the relativity theory of a group' (Klein 1910, p. 539). Emmy Noether underscored this viewpoint at the very end of "Invariante Variationsprobleme" (Noether 1918b, p. 257) in order to emphasize how her various results fully accorded with this position.

22 Nachlass Klein 21J, SUB, Göttingen. Noether was enrolled as one of the 28 auditors for these lectures, which Klein held in Carathódory's seminar. She was also one of the 20 who attended his third series of lectures during the winter semester of 1915–16 (Nachlass Klein 21K).

23 Nachlass Klein 21N, SUB, Göttingen.

24 These lectures were entitled 'Die Infinitesimalgeometrie bei Gauß und Riemann und ihre Bedeutung für die neueste mathematische Physik'.

25 Einstein reacted in a letter to Klein from 15 December 1917, in which he criticized the latter's tendency to overestimate the importance of formal methods (Einstein 1998a, pp. 569–70). Pais later commented about the irony of this pronouncement in light of Einstein's later views about the guiding role of mathematical ideas for physical theories (Pais 1982/2005, p. 325).

26 In a metric geometry, the distance between two points is an elementary invariant under congruence transformations (isometries); whereas four collinear points determine an invariant, the cross ratio, in projective geometry. By attaching a quadric to the space as an 'absolute' figure, one can introduce a metric, owing to the fact that the line joining any two points will meet the quadric in two more. If the transformation group is then restricted to automorphisms that leave the 'absolute' figure fixed, the invariance of the cross ratio ensures that the metric thus defined will remain invariant under these transformations.

A prime goal of Klein's lectures on the mathematical foundations of relativity was to describe how the relevant ideas first arose. Invariant theory and the calculus of variations both had a long history; so did group theory, including the theory of transformation groups first launched by Klein and Lie in the early 1870s. In 1904, Henri Poincaré noted that the Lorentz transformations form a group, and four years later Minkowski geometrized special relativity by inaugurating spacetime physics.[27] One of the major strands in Klein's lectures dealt with the posthumous reception of Bernhard Riemann's famous ideas on what came to be called Riemannian geometry. In his younger years, Klein had been strongly influenced by Riemann's geometric theory of complex functions, in particular the notion of so-called Riemann surfaces. He also read the habilitation lecture Riemann delivered in 1854 on a topic chosen by Carl Friedrich Gauss (Riemann 1868). Hermann von Helmholtz first became aware of Riemann's reflections after reading a lecture delivered by Ernst Schering in December 1866, the year of Riemann's death (Schering 1867). As he emphasized soon thereafter in his celebrated essay (Helmholtz 1868), his own reflections on the empirical roots of human space perception ran along somewhat parallel lines as those set out by Riemann. Klein recalled how deep and mysterious Riemann's ideas seemed to him and his generation at the time they first appeared (Klein 1927, p. 165).

One reason for this had to do with the circumstance that Riemann was speaking to the entire philosophical faculty. His allusions to mathematical concepts were thus exceedingly condensed, though sufficient for Gauss to follow and appreciate, especially since Riemann indicated clearly how his concept of curvature for n-dimensional manifolds was related to Gaussian curvature for surfaces. Richard Dedekind reported on Gauss's praise for this accomplishment in the biographical essay he wrote for the first edition of Riemann's *Werke* (Riemann 1876, pp. 507–26). This volume also contained Riemann's 'Commentatio Mathematica' (Riemann 1876a), a paper he composed in 1861 in response to a prize problem set by the Paris Academy. Since his submission failed to win a prize, the manuscript languished among Riemann's papers until it was rescued by Heinrich Weber, who wrote a fairly lengthy commentary on it (Riemann 1876, pp. 384–99, and 1892, pp. 405–23). About half of his commentary was devoted to a single page of the text, which contained the analytic formulas that Riemann suppressed in his lecture from 1854.[28] Klein's 'Erlanger Programm' was written long before the advent of the modern notion of differential manifolds. In fact, it appeared even before Heinrich Weber published Riemann's 'Commentatio' (Riemann 1876a) in the first edition of his *Werke*. Klein's longstanding interest in Riemann's work eventually led to the discovery of new material from his lectures and manuscripts, which were then edited by Max Noether and Wilhelm Wirtinger for publication in 1902.[29] Despite such efforts,

[27] Einstein initially considered this a largely superfluous formalism, but after Arnold Sommerfeld and Max von Laue developed it further, he soon came to appreciate the virtues of 4-vectors (Walter 2007). In the mathematical section of Einstein and Grossmann (1913, p. 328), Grossmann cited works by these authors for readers who wished to study the tensor calculus used in special relativity.

[28] Weber elucidated these in Riemann (1876, pp. 384–91), but to meet criticisms of that commentary he gave a revised version in the second edition (Riemann 1892, pp. 405–15). Olivier Darrigol recently offered a somewhat different interpretation based on a study of fragments found in Riemann's literary estate; see Darrigol (2015).

[29] This undertaking coincided with another, much larger project, namely the Gauss edition, which Klein took over after Schering's death in 1897.

the thread of sources tracing back to Riemann that might shed light on his approach to Riemannian geometry is exceedingly thin.

Riemann's habilitation lecture presents only the idea behind normal coordinates in qualitative terms (Riemann 1892, pp. 276–8), whereas Weber was the first to work out the analytic form (Riemann 1892, pp. 405–11). Starting with an arbitrary point O as origin and the bundle of vectors v_O emanating from it, one coordinatizes the points P at a distance d from O in the direction of v_O by the lengths along the geodesic curve determined by the given vector. Since these geodesic curves are invariants of the metric attached to a differential manifold, Riemannian normal coordinates are part of its intrinsic structure.[30]

Klein traced these ideas through Gaussian surface theory in his lectures, introducing the Lagrangian differential equations for geodesics on a surface. His starting point was the variational principle

$$\delta \int ds = \delta \int \sqrt{Edu^2 + 2Fdudv + Gdv^2} = 0, \tag{2.8}$$

which leads to two differential equations when u and v are varied independently. In order to connect this with classical physics, Klein introduced a time variable t so that the geodesic curves correspond to the paths of inertial motion of test particles on the surface once launched with a given initial velocity. This well-known fact evidently provided Einstein with an important clue in 1912, the year he began searching for a non-scalar theory of gravitation. As he later recalled in his Kyoto address from 1922:

> I suddenly realized that there was good reason to believe that the Gaussian theory of surfaces might be the key to unlock the mystery. I realized at that point the great importance of Gaussian surface coordinates. However, I was still unaware of the fact that Riemann had given an even more profound discussion of the foundations of geometry. I happened to remember that Gauss's theory had been covered in a course I had taken during my student days with a professor of mathematics named Geiser.[31] From this I developed my ideas, and I arrived at the notion that geometry must have physical significance. (Einstein 2012, p. 638, my translation)

Following Lipschitz (1870), Klein rewrote (2.8) for a general quadratic differential form as

$$\delta \int \frac{\sum g_{ik} dx^i dx^k}{dt^2} dt = 0,$$

which implies that for arbitrary dx^r the invariant

$$2d\left(\sum g_{ik} dx^i \delta x^k\right) - \delta\left(\sum g_{ik} dx^i dx^k\right) = 2\sum \Psi_r(d,d)\delta x^r = 0. \tag{2.9}$$

This implies that $\Psi_r(d,d)$ is a covariant vector, and the equation describes the geodesics for an n-dimensional differential manifold R^n. When written out, it amounts to the usual form written with Christoffel symbols. Lipschitz obtained $3n$ differential equations in all:

$$\Psi_r(d,d) = 0, \qquad \Psi_r(d,\delta) = 0, \qquad \Psi_r(\delta,\delta) = 0, \tag{2.10}$$

30 For an overview of the uses of normal coordinates, see Pauli (1958, pp. 44–52).

31 The geometer Carl Geiser was a professor at the ETH Zürich from 1869 to 1913.

where the first and third are special cases of the second, which can be written as

$$\Psi_r(d,\delta) = \sum_i g_{ir} d\delta x^i + \sum_{i,k} \frac{1}{2}\left(\frac{\partial g_{ir}}{\partial x_k} + \frac{\partial g_{kr}}{\partial x_i} - \frac{\partial g_{ik}}{\partial x_r}\right) dx^i dx^k = 0. \tag{2.11}$$

Lipschitz applied this formalism to derive Riemann's curvature form $[\Omega]$ using methods adapted by Emmy Noether (1918a). Starting from the differential form $f = \sum g_{ik} dx^i dx^k$, Riemann first introduced

$$\Omega = \delta\delta \sum g_{ik} dx^i dx^k - 2d\delta \sum g_{ik} dx^i \delta x^k + dd \sum g_{ik} \delta x^i \delta x^k. \tag{2.12}$$

The operators d and δ commute and have the same formal properties as in ordinary calculus. Thus,

$$\delta\delta \sum g_{ik} dx^i dx^k = \delta\left(\sum \delta g_{ik} dx^i dx^k + \sum g_{ik}\left(\delta dx^i dx^k + dx^i \delta dx^k\right)\right),$$

where $\delta g_{ik} = \sum_r \frac{\partial g_{ik}}{\partial x_r} \delta x_r$. In connection with Riemann and Lipschitz, Noether referred to (2.12) as the 'normal form of the second variation.' She showed how similar methods could be used to derive corresponding normal forms of higher degree (Noether 1918a).

Riemann derived $[\Omega]$ by using the geodesic differential equations (2.10) to find $d^2, d\delta, \delta\delta,$ whereas Lipschitz eliminated the higher-order differentials directly and found:

$$[\Omega] = \Omega - 2\left\{\sum g^{rs} \Psi_r(d,\delta) \Psi_s(d,\delta) - \sum g^{rs} \Psi_r(d,d) \Psi_s(\delta,\delta)\right\}. \tag{2.13}$$

Klein used Grassmannian coordinates to express the numerator and denominator of the Riemannian curvature tensor K_R. Starting with $f(d,d) = \sum g_{ik} dx^i dx^k$, he formed its polar $f(d,\delta) = \sum g_{ik} dx^i \delta x^k$, and then the determinant

$$F = \begin{vmatrix} f(d,d) & f(d,\delta) \\ f(\delta,d) & f(\delta,\delta) \end{vmatrix},$$

writing this in Grassmannian notation as

$$F = \sum (g_{ir} g_{ks} - g_{kr} g_{is}) p_{ik} p_{rs}. \tag{2.14}$$

He did the same with the numerator, writing $[\Omega] = \sum (ik,rs) p_{ik} p_{rs}$, and then

$$K_R = -\frac{[\Omega]}{2F}. \tag{2.15}$$

Klein had long been familiar with Grassmannian coordinates, which in the case $n = 4$ provide an elegant method for representing the lines in projective 3-space. Taking two points with coordinates (x_1, x_2, x_3, x_4) and (y_1, y_2, y_3, y_4), the six determinants $p_{ik} = x_i y_k - x_k y_i$ yield homogeneous coordinates for the line joining them. These line coordinates satisfy a quadratic identity, namely $P = p_{12}p_{34} + p_{13}p_{42} + p_{14}p_{23} = 0$. The same relations hold for arbitrary n, where for any two vectors d, δ, the $p_{ik} = dx_i \delta x_k - dx_k \delta x_i$ form an anti-symmetric tensor. Klein was later astonished to learn that these were connected with Riemannian differential geometry, an insight Lipschitz never conveyed to him in the late 1860s when Klein was immersed in his dissertation topic, a systematic investigation of quadratic line complexes in line geometry. Thus, in his 'Erlanger Programm,' he noted the significance of Grassmann's work and related ideas, but he was then still unaware (at age 23) of this striking relationship between line coordinates and differential invariants (Klein 1927, p. 168).

In his lectures, Klein thus used invariant-theoretic constructions in an attempt to retrace Riemann's path for generalizing Gaussian curvature, the key differential invariant in surface theory. Klein could now turn to the next step in approaching this goal, namely to explore the geometrical links between Riemannian and Gaussian curvature. Following Riemann's habilitation lecture, he used normal coordinates to write the metric for a surface in the form

$$ds^2 = (dx^2 + dy^2) + \alpha(ydx - xdy)^2. \tag{2.16}$$

The two expressions on the right are invariant under orthogonal transformations, whereas α evidently measures the deviation from the Euclidean metric in the neighborhood of the origin. In fact, the Gaussian curvature is $K = -3\alpha$, where K here arises intrinsically (one can also define K extrinsically as the product of the two principal curvatures at a point).

Riemann's approach for an n-dimensional manifold R^n involved taking pairs of vectors d, δ at a point P, forming a local surface element $S_{d\delta}$ by taking geodesics in the directions of the vectors in $\chi d + \lambda\delta$, and then finding the Gaussian curvature at P, which represents a directional invariant called the sectional curvature $K_{d\delta}$. In 1916 Gustav Herglotz, who was then in Leipzig, used the $K_{d\delta}$ to give a geometrical interpretation of the Ricci tensor R_{ik} and the Riemannian curvature scalar R (see Pauli 1958, p. 47). Taking n orthogonal vectors in R^n in pairs numbered (rs), he showed that

$$R = 2\sum_{(rs)} K_{rs}, \tag{2.17}$$

which can be interpreted as the mean curvature at a point P. If the vector 1 is ξ^i, then Herglotz furthermore showed that

$$\frac{R_{ik}\xi^i\xi^k}{\xi_i\xi^i} = \sum_{r=2}^{n} K(1r). \tag{2.18}$$

From (2.17) and (2.18), he then deduced the mean curvature of the subspace R^{n-1} perpendicular to the direction 1 as given by ξ^i:

$$\sum_{r\neq 1, s\neq 1} K(rs) = \frac{1}{2}R - \frac{R_{ik}\xi^i\xi^k}{\xi_i\xi^i} = -\frac{G_{ik}\xi^i\xi^k}{\xi_i\xi^i}, \tag{2.19}$$

where

$$G_{ik} = R_{ik} - \frac{1}{2}Rg_{ik} \tag{2.20}$$

is the Einstein tensor.

Klein was obviously eager to explain these recent developments in order to bring out the differential geometry underlying Einstein's theory of gravitation (Klein 1927, pp. 180–4).

He was also particularly interested in the case of constant Riemannian curvature, where the metric in normal coordinates has the form

$$ds^2 = \sum dy_i^2 - \frac{1}{3a^2} \sum (y_i dy_k - y_k dy_i)^2. \tag{2.21}$$

Riemann had noted in his habilitation lecture that the vanishing of the curvature tensor characterizes the Euclidean case, i.e. a differential manifold with curvature zero for which $ds^2 = \sum dy_i^2$. This metric remains invariant under the orthogonal group, which has dimension $n(n+1)/2$, but the same holds for the metric 2.21 in normal coordinates. From this, Klein concluded that the sectional curvature $K_{d\delta}$ will be independent of the surface direction $S_{d\delta}$. He sketched a general proof of this that he had suggested to Noether and Vermeil (see Klein 1927, pp. 187–8).

In the general case, using Riemannian normal coordinates, the constant on the right is replaced by a power series $\mathfrak{P}_{ik,rs}$ in the y_i, and the metric takes the form

$$ds^2 = \sum dy_i^2 + \sum \mathfrak{P}_{ik,rs}(y_i dy_k - y_k dy_i)(y_r dy_s - y_s dy_r). \tag{2.22}$$

Here (2.22) is defined for arbitrary convergent power series $\mathfrak{P}_{ik,rs}$, and Klein related this metric expression to the Riemann curvature tensor, which he wrote in the form $K_R = -[\Omega]/2F$. He then claimed that if the $\mathfrak{P}_{ik,rs}$ are appropriately normed,[32] the metric (2.22) will be uniquely determined by K_R; a detailed proof was published (Vermeil 1919). This, Klein proposed, was what Riemann had in mind when he alluded to such a fact in his habilitation lecture.

As a first approximation, Klein took only the leading terms of the power series $\mathfrak{P}_{ik,rs}$, and writing these as $\alpha_{ik,rs}$ he obtained in place of (2.22)

$$ds^2 = \sum dy_i^2 + \sum \alpha_{ik,rs}(y_i dy_k - y_k dy_i)(y_r dy_s - y_s dy_r). \tag{2.23}$$

Using this, the Riemannian curvature tensor simplifies to

$$K_R = -3\frac{\sum \alpha_{ik,rs} p_{ik} p_{rs}}{\sum p_{ik}^2}. \tag{2.24}$$

Klein introduced what he called the principle of movability for normal coordinate systems. This amounted to generalizing the construction of geodesic surfaces $S_{d\delta}$ to higher dimensions. He thus noted that one could coordinatize lower-dimensional geodesic subspaces simply by letting a certain number of the coordinates be zero. In particular, one could take all but two coordinates to be zero, the case of geodesic surfaces considered above. Here he recovered the Gaussian curvature K from K_R by letting all but the first two coordinates $y_i = 0$. The metric in normal coordinates then reads:

$$ds^2 = dy_1^2 + dy_2^2 + \alpha_{12,12}(y_1 dy_2 - y_2 dy_1)^2, \tag{2.25}$$

and from (2.24) we have $K_R = -3\alpha_{12,12} = K$, in agreement with what was claimed above.

32 This involved adopting the same condition $P = 0$ on the coefficients, thus $\mathfrak{P}_{12,34} + \mathfrak{P}_{13,42} + \mathfrak{P}_{14,23} = 0$, as must hold for the expressions $y_i dy_k - y_k dy_i$.

2.5 On Klein's Collaboration with Noether

The collaboration between Klein and Noether began in 1917 when she was attending his private lectures on the mathematical foundations of relativity theory. The fact that they never co-authored a single paper should come as no surprise. Co-authorship was still relatively uncommon during this period, and German professors were famous for exploiting work done by their assistants. In the case of Klein and Noether, she was obviously in a subservient role; not until 1919 was she allowed to join the faculty as an unpaid lecturer (*Privatdozent*). Nevertheless, Klein's papers contain many references to her work as well as allusions to her vital assistance, whereas she included similar comments in her papers, indicating how her work was closely tied with his. Emmy Noether was, in fact, a key member in a small team of researchers who were engaged in Klein's large-scale research program aimed at elaborating mathematical methods in modern physics. Others who assisted him included Hermann Vermeil, Vsevolod Frederiks, and Walter Baade.

It appears that Noether was no longer working closely with Hilbert by the summer of 1917, since he had ended his lectures on relativity theory and was now teaching set theory. Her work with Klein, on the other hand, began to build momentum around this time. On 22 August 1917, she wrote Fischer to announce that she had finally solved a problem that had occupied her attention since spring, namely the extension of a theorem proved by E. B. Christoffel and G. Ricci for quadratic differential forms to arbitrary forms of any degree (Dick 1981, p. 33). Noether presented a lecture on her general 'Reduction Theorem' on 15 January 1918 at a meeting of the Göttingen Mathematical Society, and 10 days later, Klein submitted her paper (Noether 1918a) for publication. Drawing on methods in the calculus of variations going back to Lagrange, she showed how the problem of finding a complete system of differential invariants could be reduced to classical invariant theory, i.e. finding all algebraic invariants of a corresponding projective group. One begins with one or more differential expressions $f(x,dx) = f(x_1,\ldots,x_n;dx_1,\ldots,dx_n)$ and the group of all analytic transformations $x_i = x_i(y_1,\ldots,y_n)$, where $f(x,dx)$ is mapped to $g(y,dy)$. On the level of the differentials, this general group corresponds to the group of all linear transformations:

$$dx_i = \sum \frac{\partial x_i}{\partial y_k} dy_k; \qquad \delta x_i = \sum \frac{\partial x_i}{\partial y_k} \delta y_k.$$

A general differential invariant is an analytic function J, defined for any f and any number of its partial derivatives with respect to either the variables, the differentials, or combinations of both, that remains unchanged under the general transformation group. If, however, none of the derivatives involves the variables x_i, and only first-order differentials appear in J, then these are called projective invariants, as they are restricted to the general linear group. Noether's argument thus aimed to show how to generate a system of differential invariants that can then be reduced to projective invariants by systematically removing higher-order differentials.

Noether (1922, p. 407) briefly described how her theorem was related to classical variational problems (2.2), thus leading to the introduction of Riemannian normal coordinates, which transform the geodesic curves in a differential manifold into lines. The corresponding coordinate transformations will consequently be linear mappings, and the system of differential invariants can then be transformed to another equivalent system whose underlying transformation group consists of linear transformations. How all this can be carried out is only sketched by Noether (1918a, pp. 245–6), as she originally planned to publish a detailed analysis in *Mathematische Annalen*, a plan that never materialized. She noted, however, that the case for quadratic forms was worked out in detail by Vermeil (1919), who closely followed her general ideas. Above all, Noether emphasized that her note demonstrated the scope and power of formal variational methods as opposed to the far less transparent methods of Christoffel using elimination theory (Noether 1922: 407). Noether (1922) refers explicitly to Klein's seminar lectures from the summer semester of 1917 on mathematical methods in relativity, which clearly served to launch her work on variational methods in the theory of differential invariants. Still, it seems very doubtful that Klein imagined one could extend Christoffel's 'Reduction Theorem' to all possible differential forms. Only Noether had the full grasp of technical resources to take on such a difficult problem.

Einstein also read her paper and wrote to Hilbert on 24 May 1918:

> Yesterday I received from Frl. Noether her very interesting paper on the construction of invariants. It impresses me that one can view these things from such a general standpoint. It would have done no harm to the troops returning to Göttingen from the field if they had been sent to school under Frl. Noether. She appears to know her métier well! (Einstein 1998b, p. 774)

Klein had submitted Noether's (1918a) paper in January, and in the meantime his collaboration with Noether had intensified. It was during this time that their attention shifted to understanding the status of the various formulations of energy-momentum conservation in general relativity. Unlike Einstein, Klein distinguished sharply between these new findings and traditional conservation laws in classical mechanics. This soon became part of his program for promoting the 'Erlangen Programm' (Klein 1872) as a framework for the new physics. In Klein's view, relativity theory was best conceived as a broad approach linking mathematics and physics. He thus saw it not in terms of two groups – the Lorentz group and the group of general point transformations – but rather as the invariant theory *relative to some given group* bearing on a particular physical theory. He thus had this general context in mind when he emphasized the distinction between conservation laws in classical mechanics, special relativity, and the general theory of relativity.

These issues surfaced when Klein and Hilbert exchanged their views regarding the status of the various conservation laws in Klein (1918a). Hilbert not only agreed with Klein's general position, he went further by claiming there was no analogue for classical energy conservation in general relativity. He even asserted that *one could prove a theorem* effectively ruling out conservation laws for general transformations analogous to those that hold for the transformations of the orthogonal group. That remark caught Klein's attention, and he replied: 'It would interest me very much to see the mathematical proof carried out that you alluded to in your answer' (Klein 1918a, p. 565). Emmy Noether was clearly very interested, too.

She wrote Klein from Erlangen on 29 February 1918: 'I thank you very much for sending me your note Klein (1918a) and today's letter, and I'm very excited about your second note Klein (1918b); the notes will certainly contribute much to the understanding of the Einstein–Hilbert theory.'[33] Noether then proceeded to explain her progress on the problem of distinguishing between classical and relativistic conservation laws. Twelve days later, she wrote Klein again about some key ideas that she would later publish in "Invariante Variationsprobleme" (see Rowe 2021, p. 81). The next day, 13 March, her father, the eminent algebraic geometer Max Noether, was pleased to report on her progress in a letter to Klein. Alluding to her collaboration with Klein, he wrote, 'it has been a great source of satisfaction for me that our mutual relations have been revived again through the activities of my daughter; I see every day how her creative powers grow and hope that these will lead to many new results' (Nachlass Klein 12, SUB Göttingen). Four months later, on 23 July, Emmy Noether presented her main results to the Göttingen Mathematical Society. Klein then submitted a preliminary version of her findings to the Göttingen Scientific Society three days later. After receiving proofs, she made the final revisions in late September. Noether (1918b) gave a precise answer to Hilbert's conjecture, though he apparently never acknowledged this. In "Grundlagen der Physik" (Hilbert 1924, p. 5), he cited her paper in a footnote, though only because her second theorem provided a proof for Hilbert's Theorem I in "Die Grundlagen der Physik I" (Hilbert 1915).

As emphasized above, Noether's methods for constructing differential invariants were closely related to those of Riemann and Lipschitz. Moreover, as can be seen from her brief survey article (Noether 1922), the same invariant-theoretic methods she used in Noether (1918a) were central for setting forth the results in her far more famous paper (Noether 1918b). In the latter paper, Noether began by noting that one can derive identities involving the Lagrangian expressions ψ_i for any general variational problem, defined for a function f that depends on n independent variables $x_1, \ldots, x_n$, a set of functions of these variables, $u_1(x), \ldots, u_\mu(x)$, and their derivatives. One varies the $u(x)$ so that δu and all derivatives vanish on the boundary ∂D of the domain D, from which follows:

$$\delta I = \delta \int \cdots \int f\left(x, u, \frac{\partial u}{\partial x}\right) dx = \int \cdots \int \delta f \, dx. \tag{2.26}$$

This leads to

$$\int \cdots \int \sum \left(\psi_i \left(x, u, \frac{\partial u}{\partial x}\right) \delta u_i \right) dx, \tag{2.27}$$

where ψ_i are the usual Lagrangian expressions that arise in solving $\delta I = 0$.[34] Since the δu are varied arbitrarily, the integrands must be equal, and applying integration by parts, Noether obtains the general identity:

$$\sum \psi_i \delta u_i = \delta f + Div\, A = \delta f + \frac{\partial A_1}{\partial x_1} + \cdots + \frac{\partial A_n}{\partial x_n}, \tag{2.28}$$

[33] E. Noether to F. Klein, 29 February 1918, Nachlass Klein 22B, SUB, Göttingen.

[34] To compute such an n-fold integral, one takes D to be a Cartesian product of closed intervals in R^n and then iterates the calculation of the n single integrals.

where *Div A* arises from the boundary terms. She noted that for the simplest case, a single integral and a function f that depends on no derivatives of the u higher than the first, the corresponding identity takes the form:

$$\sum \psi_i \delta u_i = \delta f - \frac{d}{dx}\left(\sum \frac{\partial f}{\partial u_i'} \delta u_i\right), \tag{2.29}$$

where $u_i' = \frac{du_i}{dx}$.

Noether calls this Heun's central Lagrangian equation, but without giving any explanation or reference. It appears that this offhand remark has never received much notice, but, in fact, this was one of the most important threads connecting Klein's lectures with Noether's publications from this time. To recognize what this says in a more familiar setting, let us look again at the Lagrangian equations (2.7), where $f = L, u = q_i$, and $u' = \dot{q}_i$. Then (2.29) reads[35]

$$\begin{aligned}\sum \psi_i \delta q_i &= \sum \left[\frac{\partial L}{\partial q_i(t)} - \frac{d}{dt}\frac{\partial L}{\partial \dot{q}_i(t)}\right] \delta q_i \\ &= \sum \frac{\partial L}{\partial q_i(t)} \delta q_i - \frac{d}{dt}\left(\sum \frac{\partial L}{\partial \dot{q}_i(t)}\right) \delta q_i \\ &= \delta L - \frac{d}{dt}\left(\sum \frac{\partial L}{\partial \dot{q}_i(t)} \delta q_i\right).\end{aligned} \tag{2.30}$$

Equation (2.29) also occupies a central place in Noether's note on 'Formal Variational Calculus and Differential Invariants' (Noether 1922), which appeared at the end of the report on algebraic and differential invariants in the *Encyklopädie der mathematischen Wissenschaften*. The report itself was written by Roland Weitzenböck, a leading expert on invariant theory who was then working in Graz. He submitted the final manuscript in 1921, around the time that Felix Klein was giving Wolfgang Pauli fairly detailed feedback on the proofs for his far better-known report of relativity (Pauli 1921), which contains numerous references to related works by Klein and his young collaborators. Extant correspondence probably no longer survives that would confirm how Noether's note came to be attached to Weitzenböck's report, but the circumstances alone strongly suggest that Klein asked her to write about a special topic that had been at the center of their collaboration during the last two years of the war. Noether references a page in Heun's lengthy report on 'Ansätze und allgemeine Methoden der Systemmechanik' in Band IV of the *Encyklopädie*, which was edited by Klein. Since her brother Fritz was Karl Heun's assistant in Karlsruhe when this article appeared in 1913, she could have learned about the so-called central Lagrangian equation from him. Far more likely, though, she picked up on the mathematical importance of this equation from Klein's lectures.

In those lectures, which Klein delivered in the summer of 1917, his very first topic was Lagrangian mechanics interpreted by way of the invariant theory of a general group of point transformations (or, as one would say today, a local diffeomorphism group). Introducing Lagrangian coordinates q_α, Klein derived the equations of motion in the usual way by

35 The final step follows from $\frac{d\delta q_i}{dt} = \delta \frac{dq_i}{dt} = \delta \dot{q}_i$.

varying the q_α over the time interval while leaving them fixed on the boundary. Writing T for the kinetic energy and $\sum P_\alpha \delta q_\alpha$ for the work associated with a virtual displacement, Klein inquires as to why the equations of motion remain invariant under general coordinate transformations. His simple answer was that the integrand in the variational formulation

$$\delta \int \left[\sum \left(\frac{\partial T}{\partial q_\alpha} - \frac{d\left(\frac{\partial T}{\partial \dot{q}_\alpha}\right)}{dt} + P_\alpha \right) \delta q_\alpha \right] dt = 0 \tag{2.31}$$

is an invariant. He then noted that Lagrange had avoided using variational analysis and instead employed the principle of virtual displacement to derive the same result using what Heun called the central Lagrangian equation (Klein 1927, pp. 141–2).

In Noether (1922) as well as in Noether (1918b), the physics behind Heun's equation has been stripped away. Instead it becomes a tool for showing how the Lagrangian expressions ψ_i enter as differential invariants, which can be used to elucidate several important relations with other differential invariants. For quadratic forms, Noether writes the equation in the form:

$$\delta \sum g_{ik} dx_i dx_k - 2d \sum g_{ik} dx_i \delta x_k = -2 \sum \psi_\mu(d,d) \delta x_\mu . \tag{2.32}$$

The notation on the right is suggestive for exploiting other identities. Indeed, one can derive a more general identity using d and δ that contains the above as a special case:

$$\begin{aligned} D \sum g_{ik} dx_i \delta x_k - \delta \sum g_{ik} dx_i D x_k - \\ d \sum g_{ik} \delta x_i D x_k = -2 \sum \psi_\mu(d,\delta) D x_\mu . \end{aligned} \tag{2.33}$$

Setting $\delta = d$, we recover the identity above, where D now appears in place of δ. Noether then observes that $\psi_\mu(d,\delta)$ is a covariant vector, which leads to a contravariant vector $p(d,\delta)$, where $p(d,d) = 0$ are equations for the geodesic curves. By means of $p(d,\delta)$, one can carry out covariant differentiation, etc. The condition $p(d,\delta) = 0$ was used by Levi-Civita to interpret curvature in terms of parallel displacement of vectors.

Emmy Noether hardly considered her work on differential invariants the last word on the subject, though she herself wrote about it only retrospectively. In a report on 'Algebraic and Differential Invariants' (Noether 1923) that she delivered in September 1922 in Leipzig at the annual meeting of the German Mathematical Society, she summarized recent progress but also alluded to a number of still-outstanding problems. In her opening remarks, she recalled Hilbert's earlier synopsis of developments in algebraic invariants as having unfolded in three phases: a naive period, followed by a formal, and then a critical stage (his own work). Noether remarked that she would only speak with reference to this final phase, noting that Hilbert's work had reformed algebraic invariant theory by exploiting arithmetical methods in algebra. For differential invariants, she identified the critical developments with Riemann's name, as he was the first to treat this branch of research with formal variational methods. Her report ended with some brief remarks about the role these methods play in her reduction theorem (proved in Noether 1918a). Pointing to recent work by Weyl and J. A. Schouten, which was no longer based on variational principles, Noether threw open the question whether higher-order differentials could nevertheless be eliminated in analogy with the results she had found for Lagrangian expressions. In this regard, she mentioned that Weitzenböck had proved a reduction theorem for invariants of low degree, but she added that a general Ansatz for attacking this problem was still lacking.

2.6 On Rehabilitating Riemannian Normal Coordinates

Felix Klein's picture of earlier research on differential invariants clearly clashed with Hilbert's notion of a development in three stages. Klein saw the contemporaneous tradition of Riemann and Lipschitz as overshadowed by the dominant approach of Christoffel and Ricci (as reflected in Wright 1908). There seem, however, to have been several competing methods for studying differential invariants, as J. Edmund Wright dealt briefly with two others: the first based on Sophus Lie's group-theoretic approach to differential equations, the second on Heinrich Maschke's adaptation of the Clebsch–Aronhold formalism from algebraic invariant theory. Christoffel's approach to differential invariants arose from the problem of classifying quadratic differential forms. A natural challenge – taken up by both Riemann and Christoffel – was to find necessary and sufficient conditions to determine when two such forms were equivalent and thus could be transformed into one another. Since several fundamental concepts in modern differential geometry would be developed only much later, this classical equivalence problem was somewhat vaguely formulated. Still, it served as a major stimulus for research on the theory of differential invariants quite apart from potential applications in physics.

One can gain several insights into how the field of differential invariants changed during the period from 1908 to 1927 by comparing Wright's text, which was No. 9 in Cambridge Tracts in Mathematics and Mathematical Physics, with the one Oswald Veblen published two decades later under the same title as Cambridge Tract No. 24 (Veblen 1927). Both are short works of about 100 pages, but the former discussed only applications to differential geometry and analysis, whereas the latter was meant as an introduction suitable for those who wished to study the new approaches to spacetime physics following the lead of Weyl, Élie Cartan, and the Princeton school.[36] It remains unclear whether Veblen was aware of Klein's lectures, but be that as it may, in his closing historical remarks he gave a short synopsis of the Riemann–Lipschitz–Noether tradition while citing Noether (1918a) (Veblen 1927, pp. 100–1).

As just one example of Veblen's modern approach, when discussing surface theory, he introduced the Euclidean metric in tensor form as

$$g_{ij}dx^i dx^j = Edu^2 + 2Fdudv + Gdv^2. \tag{2.34}$$

He then noted that the non-zero components of the Riemann–Christoffel tensor are equal up to their sign, so that

$$\frac{1}{4}\epsilon^{ij}\epsilon^{ab}R_{ijab} \tag{2.35}$$

[36] Luther Eisenhart was Princeton's leading differential geometer. Along with Veblen and their students, he inaugurated the study of path spaces; on Veblen's place in this tradition, see Ritter (2011).

is a relative scalar of weight 2 and

$$K = \frac{\epsilon^{ij}\epsilon^{ab}R_{ijab}}{4(EG - F^2)} \tag{2.36}$$

an absolute scalar, namely the Gaussian curvature.

Veblen's final chapter dealt with affine geometry, where one generalizes the differential equations for geodesic curves by replacing the Christoffel symbols with an an affine connection satisfying

$$\frac{d^2x^i}{dt^2} + \Gamma^i_{jk}\frac{dx^j}{dt}\frac{dx^k}{dt} = 0. \tag{2.37}$$

Veblen then proved that the curves that satisfy (2.37) form a local path space, so that any two points in a given neighborhood are connected by a unique curve in the system. His proof was effected by showing that one can find a system of coordinates for which the solutions of (2.37) are linear. These are called affine normal coordinates, as they are obviously a direct generalization of the normal coordinates for Riemannian spaces. In his historical remarks, Veblen refers to their role in the works of Riemann, Lipschitz, and Noether. He also added that G. D. Birkhoff used Riemannian normal coordinates systematically (Birkhoff 1923), but that otherwise they had been neglected by most authors who wrote about general relativity (Veblen 1927, p. 101). This suggests that Klein's general assessment from 10 years earlier was quite correct; it seems that little had changed in the meantime, at least in this respect.

2.7 Hermann Weyl Recalling Klein and Noether

Although Felix Klein and Emmy Noether are remembered today as two famous Göttingen mathematicians, their names evoke radically different associations. His long career was nearing its end, of course, whereas hers was just beginning. Since their mathematical tastes differed very sharply, it might seem difficult to imagine them working together, although Klein's earlier collaboration with Paul Gordan reveals the same kind of striking clash of styles. During the twilight of Klein's career, Emmy Noether and the now-forgotten Hermann Vermeil were his two most important assistants.[37] Hermann Weyl knew both Klein and Noether very well. He had studied under Klein and Hilbert in Göttingen before joining the faculty as a *Privatdozent* in 1910. After teaching at the ETH in Zurich from 1913 to 1930, he returned to Göttingen as Hilbert's successor at a difficult time during which he became quite close with Emmy Noether (Rowe 2021, pp. 189–211). Their friendship continued as fellow exiles in the United States. It was at Bryn Mawr College, shortly after her death in April 1935, that Weyl delivered his tribute to Emmy Noether in the memorial lecture (Weyl 1935). Five years earlier, when they were both together in Göttingen, Weyl had spoken about the mathematical legacy of Felix Klein during the opening ceremonies for the new Mathematics Institute building (Weyl 1930).

[37] Hermann Vermeil (1889–1959) took his doctorate under Otto Hölder in Leipzig in 1913. He began attending Klein's lectures in 1917–18 and became his assistant in 1919; afterward, he served as an editor of Klein's Collected Works (Klein 1921–3).

As might be expected, these two lectures confirm that Klein and Noether were very different types of mathematicians whose respective styles practically defy comparison. Weyl noted that Noether shared with several other major figures – Dedekind, Kronecker, and even Weierstrass – a strong inclination to algebraicize the broad terrain of analysis, whereas Klein's general tendency was to invoke continuity arguments, or as Weyl expressed it, topological reasoning, which gave his analytical and algebraic works a strong geometric flavor. Since Weyl's own mathematical tastes lay closer to Klein's, he waxed more than a little enthusiastically when speaking of the latter's 'Erlanger Programm' from 1872, which he clearly saw as the centerpiece of Klein's mathematical legacy. Weyl made the particularly striking claim – one that would have greatly pleased Klein – that the understanding of geometry in the latter's 'Erlanger Programm' was 'nothing other than *relativity theory* in its general, mathematically formulated form' (Weyl 1930, p. 299).

Weyl's lecture in honor of Emmy Noether has been cited far more often and for good reason. This was a dramatic occasion, the capstone of truly harrowing times that both of them had gone through together (Rowe 2021, pp. 199–265). In the fall of 1933, Hermann Weyl abandoned his chair in Göttingen for a research professorship at Princeton's Institute for Advanced Study (IAS). Beginning in February 1934, he again saw Emmy Noether on a regular basis, as Oswald Veblen had arranged for her to give weekly lectures at the IAS. Most commentators who have subsequently surveyed Noether's career have followed Weyl in viewing it as composed of three periods: (1) 1907–19, period of relative dependence; (2) 1920–26, ideal theory; (3) 1927–35, non-commutative algebras with applications to commutative number fields (Weyl 1935, p. 439). As a general scheme, this tripartite division is certainly useful, even though the boundaries between these periods should not be drawn too sharply. More problematic is the label 'relative dependence' for (1), when Emmy Noether emerged as a leading authority in the field of *invariant theory*, producing work that carried over into the second period.

Weyl treated the first period rather dismissively, though he was well aware of her two papers (Noether 1918a, 1918b) written in collaboration with Klein. He recalled those times, when Hilbert was 'over head and ears in the general theory of relativity' and Klein saw 'its connection with his old ideas of the Erlangen program [that] brought the last flareup of his . . . mathematical production' (Weyl 1935, pp. 430–1). He noted how the posthumously published lectures (Klein 1927) bore witness to this, but also how Emmy Noether's expertise in invariant theory had supported Hilbert's research as well as Klein's. Commenting on her two papers, he wrote:

> For two of the most significant sides of the theory of general relativity she gave at that time the genuine and universal mathematical formulation: First, the reduction of the problem of differential invariants to a purely algebraic one by the use of 'normal coordinates'; second, the identities between the left sides of Euler's equation of a problem of variation which occur when the (multiple) integral is invariant with respect to a group of transformations involving arbitrary functions (identities that contain the conservation theorem of energy and momentum in the case of invariance with respect to arbitrary transformations of the four world coordinates). (Weyl 1935, p. 431)

It is worth noting that Weyl's brief allusion to 'Invariante Variationsprobleme' addresses only Noether's second theorem, not the first. When physicists write about her paper today, they often refer to *the* Noether theorem, meaning the first of her two main results, thereby overlooking that she proved two theorems. Seen in historical context, her second theorem was the central finding, since it clarified the status of identities arising from invariant variational problems; these were first noticed by Hilbert and played a major role in his 1915 publication. Hermann Weyl had, of course, an excellent vantage point from which to judge the motivation behind these two studies, but by 1935, the year of her death, only a small number of experts knew what was in these papers.

Around the centenary of her birth, Emmy Noether's career was celebrated in several new publications and with the publication of her *Collected Papers* (Noether 1983), edited by Nathan Jacobson. Noether's reputation at that time rested squarely on her contributions to modern algebra, an area of research that had expanded greatly during the half-century since Noether's death. By this time, some mathematical physicists had come to appreciate the importance of her "Invariante Variationsprobleme", but few seem to have appreciated the larger context of interests that motivated her work, despite the fact that she cited several other related papers in it. In "Formale Variationsrechnung und Differentialinvarianten" (Noether 1922, p. 408), she repeated the citations of these six earlier works, noting that they were motivated by physical problems.[38] Noether characterized her own paper as motivated by questions of principle.

References

Bianchi, L. (1896). *Vorlesungen über Differentialgeometrie*. Max Lukat, Über. Leipzig: Teubner.

Birkhoff, G. D. (1923). *Relativity and Modern Physics*. Cambridge, MA: Harvard University Press.

Brewer, J. W. and Smith, M. K. (1981). *Emmy Noether. A Tribute to Her Life and Work*. New York: Marcel Dekker.

Christoffel, E. B. (1869). Ueber die Transformation der homogenen Differentialausdrücke zweiten Grades. *Journal für die reine und angewandte Mathematik* **70**, 46–70.

Courant, R., and Hilbert, D. (1924). *Methoden der mathematischen Physik*, Bd. 1. Berlin: Springer.

Darrigol, O. (2015). The mystery of Riemann's Curvature, *Historia Mathematica*, **42**, 47–83.

Dick, A. (1981). *Emmy Noether*. H. I. Blocher, trans. Boston: Birkhäuser.

Einstein, A. (1914). Die formale Grundlage der allgemeinen Relativitätstheorie. *Sitzungsberichte der Königlichen Preußischen Akademie der Wissenschaften*, 1030–85. Reprinted in Einstein (1996: 72–129).

Einstein, A. (1916). Grundlagen der allgemeinen Relativitätstheorie. *Annalen der Physik*, **49**, 769–822. Reprinted in Einstein (1996: 283–339); English trans. in Sommerfeld (1952: 109–64).

Einstein, A. (1996). *The Collected Papers of Albert Einstein, vol. 6. The Berlin Years: Writings, 1914–1917*. A. J. Kox, M. J., Klein and R. Schulmann, eds. Princeton: Princeton University Press.

[38] The authors were George Hamel, Gustav Herglotz, H.A. Lorentz, Adriaan Fokker, Weyl, and Klein.

Einstein, A. (1998a). *Collected Papers of Albert Einstein, Vol. 8A. The Berlin Years: Correspondence, 1914–1917*. R. Schulmann, A. J. Kox, M. Janssen, and J. Illy, eds. Princeton: Princeton University Press.

Einstein, A. (1998b). *Collected Papers of Albert Einstein, Vol. 8B: The Berlin Years: Correspondence, 1918*. R. Schulmann, A. J. Kox, M. Janssen and J. Illy, eds. Princeton: Princeton University Press.

Einstein, A. (2012). *Collected Papers of Albert Einstein, Vol. 13. The Berlin Years: Writings and Correspondence, January 1922–March 1923*. D. Buchwald, J. Illy, Z. Rosenkrnz, and T. Sauer, eds. Princeton: Princeton University Press.

Einstein, A. and Grossmann, M. (1913). Entwurf einer verallgemeinerten Relativitätstheorie und einer Theorie der Gravitation. Leipzig: Teubner.

Ewing, G. M. (1985). *Calculus of Variations with Applications*. New York: Dover.

Goenner, H. (2015). Klein's 'Erlanger Programm': Do traces of it exist in physical theories? In L. Ji and A. Papadopoulos, eds., *Sophus Lie and Felix Klein: The Erlangen Program and Its Impact in Mathematics and Physics*. Strasbourg: European Mathematical Society, pp. 77–90.

Goodstein, J. R. (2018). *Einstein's Italian Mathematicians: Ricci, Levi-Civita, and the Birth of General Relativity*. Providence, RI: American Mathematical Society.

Gray, J., ed. (1999). *The Symbolic Universe. Geometry and Physics, 1890–1930*. Oxford: Oxford University Press.

Hawkins, T. (1984). The Erlanger Programm of Felix Klein: Reflections on its place in the history of mathematics, *Historia Mathematics*, **11**(4), 442–70.

Hawkins, T. (2000). *Emergence of the Theory of Lie Groups: An Essay in the History of Mathematics, 1869–1926*. New York: Springer.

Helmholtz, H. von. (1868). Ueber die Thatsachen, die der Geometrie zu Grunde liegen, *Nachrichten der Göttingen Gesellschaft der Wissenschaften*. 1868: 193–221.

Hilbert, D. (1915). Die Grundlagen der Physik I. *Nachrichten der Göttingen Gesellschaft der Wissenschaften*, pp. 395–407. Reprinted in (Sauer and Majer 2009, 28–46); English translation in Renn (2007, Vol. IV: 1003–16).

Hilbert, D. (1924). Grundlagen der Physik, *Mathematische Annalen*, **92**, 1–32.

Janssen, M. and Renn, J. (2007). Untying the knot: How Einstein found his way back to field equations discarded in the Zurich Notebook. In Renn (2007, Vol. II: 839–925).

Klein, F. (1872). *Vergleichende Betrachtungen über neuere geometrische Forschungen*. Erlangen: A. Deichert. Reprinted in Klein (1921–3, Vol. I: 460–96).

Klein, F. (1910). Über die geometrische Grundlage der Lorentzgruppe, *Jahresbericht der Deutschen Mathematiker-Vereinigung*, **19**, 281–99. Reprinted in Klein (1921–3, 1: 533–52).

Klein, F. (1918a). Zu Hilberts erster Note über die Grundlagen der Physik, *Nachrichten der Königlichen Gesellschaft der Wissenschaften zu Göttingen*, Mathematisch-Physikalische Klasse, 1918: 469–82. Reprinted in Klein (1921–3, 1: 553–65).

Klein, F. (1918b). Über die Differentialgesetze für die Erhaltung von Impuls und Energie in der Einsteinschen Gravitationstheorie, *Nachrichten der Königlichen Gesellschaft der Wissenschaften zu Göttingen*, Mathematisch-Physikalische Klasse. Reprinted in Klein (1921–3, 1: 568–84).

Klein, F. (1919). Über die Integralform der Erhaltungssätze und die Theorie der räumlich geschlossenen Welt, *Nachrichten der Königlichen Gesellschaft der Wissenschaften zu Göttingen*, Mathematisch-Physikalische Klasse. Reprinted in Klein (1921–3, 1: 586–612).

Klein, F. (1921–23). *Gesammelte Mathematische Abhandlungen*, 3 Bde., Berlin: Julius Springer.

Klein, F. (1926). *Vorlesungen über die Entwicklung der Mathematik im 19. Jahrhundert*, vol. 1. Berlin: Julius Springer.

Klein, F. (1927). *Vorlesungen über die Entwicklung der Mathematik im 19. Jahrhundert*, vol. 2. Berlin: Julius Springer.

Kosmann-Schwarzbach, Y. (2006/2011). *Les Théorèmes de Noether. Invariance et lois de conservation au XXe siècle, avec une traduction de l'article original Invariante Variationsprobleme*. Éditions de l'École Polytechnique, deuxième édition, révisée et augmentée. English trans.: *The Noether Theorems: Invariance and Conservation Laws in the Twentieth Century*, Sources and Studies in the History of Mathematics and Physical Sciences, Springer.

Lie, S. (1884). Ueber Differentialinvarianten. *Mathematische Annalen*, **24**, 537–78.

Lipschitz, R. (1869). Untersuchungen in Betreff der ganzen homogenen Functionen von n Differentialen. *Journal für die reine und angewandte Mathematik*, **70**, 71–102.

Lipschitz, R. (1870). Fortgesetzte Untersuchungen in Betreff der ganzen homogenen Functionen von n Differentialen. *Journal für die reine und angewandte Mathematik*, **72**, 1–56.

McShane, E. J. (1981). The calculus of variations. In Brewer and Smith (1981: 125–30).

Noether, Emmy. (1918a). Invarianten beliebiger Differentialausdrücke. *Nachrichten der Königlichen Gesellschaft der Wissenschaften zu Göttingen*, Mathematisch-Physikalische Klasse, pp. 37–44. Reprinted in Noether (1983: 240–7).

Noether, E. (1918b). Invariante Variationsprobleme. *Nachrichten der Königlichen Gesellschaft der Wissenschaften zu Göttingen, Mathematisch-Physikalische Klasse*, pp. 235–57. Reprinted in Noether (1983, 248–70). French trans. in Kosmann-Schwarzbach (2006/2011, 2006: 1–28); English trans. in Kosmann-Schwarzbach (2006/2011, 2011: 3–22).

Noether, E. (1922). Formale Variationsrechnung und Differentialinvarianten. In *Encyklopädie der mathematischen Wissenschaften mit Einschluss ihrer Anwendungen*, Band III (Geometrie), 3 D10, pp. 68–71. Reprinted in Noether (1983: 405–8).

Noether, E. (1923). Algebraische und Differentialinvarianten, *Jahresbericht der Deutschen Mathematiker-Vereinigung*, **32**, 177–184. Reprinted in Noether (1983, 436–43).

Noether, E. (1983). *Gesammelte Abhandlungen*. Heidelberg: Springer.

Olver, P. J. (1986/1993). *Applications of Lie Groups to Differential Equations*. New York: Springer.

Pais, A. (1982/2005). *'Subtle Is the Lord...' The Science and the Life of Albert Einstein*. Oxford: Clarendon Press.

Pauli, W. (1921). Relativitätstheorie, *Encyklopädie der mathematischen Wissenschaften*, Vol. V19. Leipzig: Teubner.

Pauli, W. (1958). *Theory of Relativity*. English translation of Pauli (1921) by G. Field. Oxford: Pergamon Press.

Pauli, Wolfgang. (1979). *Wolfgang Pauli. Wissenschaftlicher Briefwechsel mit Bohr, Einstein, Heisenberg u.a.*, vol. 1, A. Hermann, K. v. Mayenn, and V. F. Weisskopf, eds., New York: Springer–Verlag.

Renn, J., ed. (2007). *The Genesis of General Relativity*, 4 vols. Dordrecht: Springer.

Renn, J., and Stachel, J. (2007). Hilbert's foundation of physics: From a theory of everything to a constituent of general relativity. In Renn (2007, Vol. IV: 857–974).

Ricci, G., and Levi-Civita, T. (1901). Méthodes de calcul différentiel absolu et leurs applications. *Mathematische Annalen* **54**, 125–201.

Riemann, B. (1868). Ueber die Hypothesen, welche der Geometrie zu Grunde liegen, *Abhandlungen der Königlichen Gesellschaft der Wissenschaften zu Göttingen*, Bd. 13, (Riemann 1876: 254–69).

Riemann, B. (1876a). Commentatio mathematica, qua respondere tentatur quaestioni ab Ill Academia Parisiensi propositae. Reprinted in Riemann (1876: 370–83; 1892: 391–404).

Riemann, B. (1876). *Gesammelte Mathematische Werke*, 1. Aufl. Heinrich Weber, ed. Leipzig: Teubner.

Riemann, B. (1892). *Gesammelte Mathematische Werke*, 2. Aufl. Heinrich Weber, ed. Leipzig: Teubner.

Ritter, J. (2011). Geometry as physics: Oswald Veblen and the Princeton School. In K.-H. Schlote and M. Schneider, eds., *Mathematics Meets Physics: A Contribution to Their Interaction in the 19th and the First Half of the 20th Century*, Frankfurt: Verlag Harry Deutsch, pp. 145–79.

Rowe, D. E. (1999). The Göttingen response to general relativity and Emmy Noether's theorems. In Gray (1999: 189–233).

Rowe, D. E. (2018). *A Richer Picture of Mathematics: The Göttingen Tradition and Beyond*. New York: Springer.

Rowe, D. E. (2021). *Emmy Noether – Mathematician Extraordinaire*. Basel: Springer International Publishing.

Sauer, T. and Majer, U. eds. (2009). *David Hilbert's Lectures on the Foundations of Physics, 1915–1927*. Heidelberg: Springer.

Schering, E. (1867). Zum Gedächtniss an B. Riemann, *Nachrichten der Göttingen Gesellschaft der Wissenschaften*, pp. 305–14.

Seelig, C., ed. (1956). *Helle Zeit – Dunkle Zeit: In Memoriam Albert Einstein*. Zurich: Europa Verlag.

Sommerfeld, A., ed. (1952). *The Principle of Relativity: A Collection of Original Memoirs by H. A. Lorentz, A. Einstein, H. Minkowski, and H. Weyl*. New York: Dover.

Srinivasan, B. and Sally, J. D., eds. (1983). *Emmy Noether in Bryn Mawr. Proceedings of a Symposium Sponsered by the Association for Women in Mathematics in Honor of Emmy Noether's 100th Birthday*. New York: Springer-Verlag.

Study, E. (1908). Kritische Betrachtungen über Lies Invariantentheorie der endlichen kontinuierlichen Gruppen. *Jahresbericht der Deutschen Mathematiker-Vereinigung*, **17**, 125–42.

Tobies, R. (2019). *Felix Klein. Visionen für Mathematik, Anwendungen und Unterricht*. Heidelberg: Springer.

Uhlenbeck, K. (1983). Conservation laws and their applications in global differential geometry, in Srinivasan and Sally (1983, pp. 103–16).

Veblen, O. (1927). *Invariants of Quadratic Differential Forms*. Cambridge: Cambridge University Press.

Vermeil, H. (1919). Bestimmung einer quadratischen Differentialform aus der Riemannschen und den Christoffelschen Differentialinvarianten mit Hilfe von Normalkoordinaten, *Mathematische Annalen*, **79**, 289–312.

Walter, S. (2007). Breaking in the 4-Vectors: the four-dimensional movement in gravitation, 1905–1910. In Renn (2007, Vol. III: 193–252).

Weyl, H. (1930). Felix Kleins Stellung in der mathematischen Gegenwart, *Die Naturwissenschaften*, **18**, 4–11. Reprinted in Weyl (1968, Vol. III: 292–300).

Weyl, H. (1935). Emmy Noether (Memorial Address, 26 April 1935). *Scripta Mathematica*, **3**, 201–20. Reprinted in Weyl (1968, Vol. III: 425–44).

Weyl, H. (1968). *Gesammelte Abhandlungen*, 4 vols. K. Chandrasekharan, ed., Berlin: Springer.

Wright, J. E. (1908). *Invariants of Quadratic Differential Forms*. Cambridge: Cambridge University Press.

3

Moscow, Oxford, or Princeton: Emmy Noether's Move from Göttingen (1933)*

TOMOKO L. KITAGAWA

3.1 Introduction

From 1933 onwards, the dismissal of Jewish civil servants by the Nazi regime in Germany created a large number of academic refugees. Emmy Noether (1882–1935) was in the first group of scholars at the University of Göttingen who were forced to leave their teaching positions. By this time, her fame was spread among advanced researchers in mathematics. She had received the prestigious Alfred Ackermann-Teubner Memorial Award for advancement of mathematical science, sharing it with another eminent mathematician, Emil Artin (1898–1962), in 1932. She was also a plenary speaker at the International Congress of Mathematicians (ICM) in Zurich in 1932; giving a plenary talk at ICM was considered prestigious, and she was the first women to do so. By the age of 50, Noether had earned a solid reputation.

In Göttingen, she had been teaching mathematics as *Privatdozent* (a tutor who teaches students independently without having a chair) until 1922 and was given a professorial rank in 1923.[1] After she received the notification of dismissal in April 1933, she had to look outside of Germany for a university where she could continue her mathematical research. The arrangement had to be made quickly, as her dismissal was set to be on 1 October, 1933.

By the end of the year, she moved to Bryn Mawr College in Pennsylvania, the United States, which was a distinguished women's college. Based at Bryn Mawr, she could also visit the Institute for Advanced Study, Princeton, and give guest lectures. But going to Bryn Mawr was not the only option for her: Noether considered moving to Moscow and Oxford. She was enthusiastic about both options, and she in fact accepted an offer from Somerville College, Oxford first. Because she moved to the USA, without going to Oxford or Moscow, the story of the academics in the Soviet Union and the United Kingdom, who wished to support Noether's move out of Germany, has yet to be told in detail. Making an offer to a refugee scholar, especially to a female mathematician, was not easy, and the hosts put significant

* I am grateful to the Principal and Fellows of Somerville College for the use of their archive. The research trip to Göttingen was supported by the British Society for the History of Mathematics. I am grateful for the help of Katarina Habermann (Georg-August-Universität Göttingen) and Kate O'Donnell (Sommerville College Oxford) during my archival research. I also thank Bernard Lightman, Christopher Hollings, Keith Hannabuss, Brigitte Stenhouse, and Pieter Moree for their advice and feedback at various stages of this research.

[1] Her title in German was '*ausserplanmässige (nichtbeamtete ausserordentliche) Professorin*'. This position did not come with any renumeration, thus she was also given a teaching position (*Lehrauftrag*) to support herself financially.

effort into making arrangements for Noether. This chapter recovers the documents written by her supporters, who had a profound impact on Noether's life in the summer of 1933.

3.2 The Notifications of Dismissal

On 25 April 1933, a telegram announcing the discharge of some teaching staff arrived at the University of Göttingen (Figure 3.1). This was because the Civil Service Law came into effect on the 7 April 1933, following the Nazi Party's consolidation of power. The dictatorship of Adolf Hitler led to the removal of non-Aryan descendants from civil service and university posts, and it dismantled the strong intellectual community of mathematics in Göttingen. In essence, all who were of Jewish descent had to leave their jobs.

The telegram listed the names of dismissed teaching staff in mathematics: Richard Honig, Richard Courant, Max Born, Emmy Noether, Felix Bernstein, and Curt Bondi. It declared that they were no longer allowed to teach at the university, and they had to be notified about this decision immediately. This was the first time that the University of Göttingen received such a notice from the Nazi regime, but it was not the last. More than 50 professors and lecturers were to be removed from their positions in the next few years.

Hearing the news of Noether's dismissal, a great number of mathematicians and graduate students wrote petition letters. They testified to Noether's superb ability in maths research, and they commended her unusual commitment to teaching. Her doctoral students said that 'Professor Noether has founded a mathematical school [here in Göttingen]. Despite our

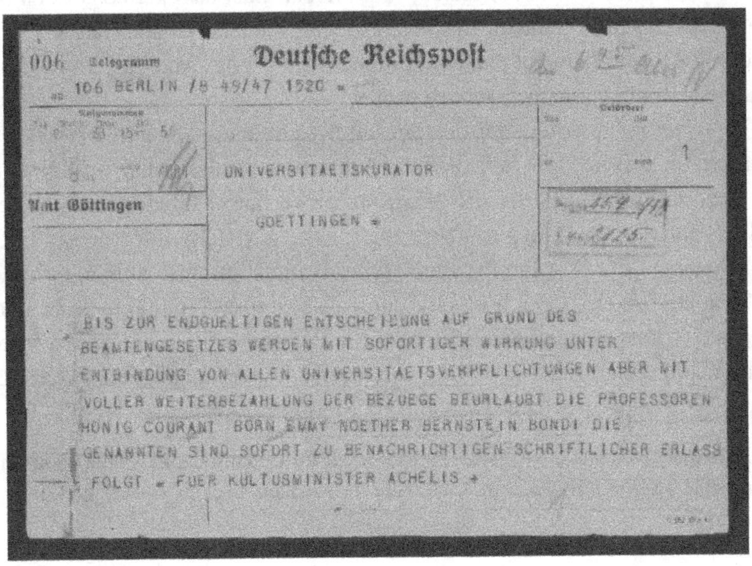

006 Telegramm Deutsche Reichspost
106 BERLIN /8 49/47 1520 =

Amt Göttingen

UNIVERSITAETSKURATOR

GOETTINGEN =

BIS ZUR ENDGUELTIGEN ENTSCHEIDUNG AUF GRUND DES
BEAMTENGESETZES WERDEN MIT SOFORTIGER WIRKUNG UNTER
ENTBINDUNG VON ALLEN UNIVERSITAETSVERPFLICHTUNGEN ABER MIT
VOLLER WEITERBEZAHLUNG DER BEZUEGE BEURLAUBT DIE PROFESSOREN
HONIG COURANT BORN EMMY NOETHER BERNSTEIN BONDI DIE
GENANNTEN SIND SOFORT ZU BENACHRICHTIGEN SCHRIFTLICHER ERLASS
FOLGT = FUER KULTUSMINISTER ACHELIS +

Figure 3.1 Telegram from the Minister of Education to the University Curator Göttingen on 25 April 1933. This is a notification for the first group of professors to be expelled from their teaching positions. Stored at Lower Saxony State and University Library Göttingen. (Public domain)

divergent political views, the personal connections with her are in no way disturbed. . . . We would welcome it if Professor Noether would again be given the opportunity to work as a teacher' (*Dokumente zu Emmy Noether* n.d., 1.1, 7).[2] Noether's collaborator, Helmut Hasse (1898–1979), stated with confidence, 'I am convinced that Miss Noether is one of the leading mathematicians in Germany. Especially for the young generation, it would be a very heavy loss if Miss Noether is forced to move abroad' (*Dokumente zu Emmy Noether*, 1.2, 9).[3] More than a month later, Hasse wrote another letter, saying that 'German mathematics in general would suffer a major moral damage. . . . Noether has always felt like a German. . . . and her mathematics cannot be called "alien" in any way.'[4]

By 31 July 1933, letters came in from all over the world: Copenhagen, Vienna, Cambridge, Bologna, Zurich, Osaka, and Tokyo. While many of her former students emphasised her commitment to teaching, another German mathematician, Oskar Perron (1880–1975), compared Emmy to the outstanding mathematician, the late Sofya Kovalevskaya (1850–91), and said that Noether was a much better mathematician than her because she is 'truly creative' (*Dokumente zu Emmy Noether* n.d., 1.7, 17). Hermann Weyl (1885–1955) also testified that Emmy was more original than the late Sophie Germain (1776–1831) and that she was as important as her contemporary, award-winning physicist Lise Meitner (1878–1968) (*Dokumente zu Emmy Noether* n.d., 1.17, 29). A Japanese mathematician, Teiji Takagi (1875–1960), who had previously studied in Göttingen, was prompt to respond and spread the news to his colleagues in Japan; Noether's former student, Kenjiro Shoda (1902–77), wrote a letter in the hope of saving Noether (*Dokumente zu Emmy Noether* n.d., 1.14, 26). A Danish mathematician, Harald Bohr of Copenhagen (1887–1951), together with G. H. Hardy of Cambridge (1877–1947), stated, 'The effect Emmy has exerted goes far beyond the borders of Germany' (*Dokumente zu Emmy Noether* n.d., 1.15, 27). They knew Noether's talent. They knew her character and commitment to teaching. And they knew the absurdity of this expulsion policy aimed at Jewish professors.

After the telegram, a written notification followed on the 2 September 1933.[5] The Prussian Minister for Science, Art and Popular Education formally deprived Noether of her teaching license. Fortunately, by this time, her move out of Göttingen had already been arranged. As early as the 4 August 1933, the Emergency Committee in Aid of Displaced Foreign Scholars (1933–45) and the Rockefeller Foundation were able to offer a salary of $4,000 (Shen 2019, p. 55), which was the equivalent of over $76,000 in 2019, and her host university was a woman's college, Bryn Mawr in Pennsylvania in the United States.

Noether made up her mind and wrote a letter from Göttingen on 18 August 1933, and sent it to the President of Bryn Mawr College, Marion Edwards Park:[6]

> I would like to thank you and the College, and above all the Department of Mathematics, many times for the surprising offer of a stipendium. I was very pleased, and I accept it with pleasure.

[2] Noether was addressed as 'Frl. Prof. Noether', indicating she was a woman.

[3] From Helmut Hasse to the Curator of the University of Göttingen.

[4] Ihre Mathematik kann man in keiner Weise 'artfremd' nennen. *Dokumente zu Emmy Noether*, 1.5, 14. Here, she was addressed as 'Fräulein Noether'.

[5] The letter arrived at the Mathematical Institute on 21 September.

[6] 18.08.1933, Bryn Mawr Archive. A letter from E. Noether to President Park (English translation is provided by the archive).

> I must, however, request a year's postponement, until the academic year 1934–35. But I hope and assume that this will not create any great difficulties. I have already accepted an invitation for this winter, for the Christmas-Easter term, to give lectures at Somerville College, Oxford, which will also be accessible to mathematicians of the other Colleges. By the end of these lectures the academic year will be almost over.
>
> It is to me personally a very pleasant feeling to be able to make my plans two years in advance; I certainly hope that things will turn out well.

Noether planned to go to Oxford first, and then to Bryn Mawr. The actual move to Oxford did not happen, and the story of Oxford's invitation and Noether's initial acceptance of Somerville's offer has not been discussed in detail. In 'Emmy Noether and Her Influence', Clark Kimberling described the effort of Oxford and quoted a document written by an officer, Professor H. M. Miller, at the Paris Office of the Rockefeller Foundation (RF), saying that 'Miss D [Darbishire, Principal of Somerville College, Oxford] is a classicist and knows nothing about the field of mathematics. She had only recently returned from vacation and was not at all sure what action her College could take in finding funds to supplement the 24 pounds now available for N. [Noether]' (Kimberling 1981, p. 31).[7]

It was not true that Helen Darbishire (1881–1961) did not know what to do. The Paris Office of the RF may not have been informed fully of what she was doing in England.[8] Darbishire was busy contacting possible donors, communicating with the secretary of the Academic Assistance Council, the organisation set up to assist academics who were forced to flee the Nazi regime.[9] Darbishire also listened to what her colleague Dorothy Wrinch (1894–1976), a mathematician, had to say about Noether and her move.[10] The important fact is that Noether did accept Darbishire's offer to teach at Somerville. She ended up not going there as she wished, but this does not mean that the negotiation process, as well as Noether's interest in coming to England, was not important to her. On the contrary, the documents left in the Weston Library in Oxford and in Somerville College, Oxford show that both Noether and members of Somerville were enthusiastic about the possibility that Noether would join the college community.[11] Furthermore, there was one more attractive offer from her colleague in Moscow, which did not come through. Noether considered those two opportunities very seriously. Although she ended up going to the USA, the efforts of both Soviet and British scholars who worked on behalf of Noether to provide her a new academic home had a profound impact on her life during the time of forced expulsion and her subsequent emigration to America.

During this summer, Emmy Noether, her brother Fritz and his wife Regine made a trip to Dierhagen, the beach in northern Germany. The photo is available in Oberwolfach Photo Collection, ID 3116.

7 The letter was written on 25 September 1933.

8 The RF had started to support the refugee scholars since 1933, and they had European offices, including one in Paris.

9 The AAC was founded in 1933, and it was renamed the Society for the Protection of Science and Learning (SPSL) in 1936. It was renamed again in 1999 as the Council for Assisting Refugee Academics, and in 2014 as the Council for At-Risk Academics (Cara). For more about the Academic Assistance Council, see Nossum and Kotůlek (2015).

10 For more about Dorothy Wrinch, see her biography by Senechal (2013).

11 For general information about the series of archives at the Bodleian Library in Oxford, see Nossum (2012).

Figure 3.2 Photo of mathematicians in Nikolausberg (near Göttingen) in July 1933. This was the time when Helen Darbishire approached Noether and offered an invitation to come to Somerville College. From left to right: Ernst Witt; Paul Bernays; Helene Weyl; Hermann Weyl; Joachim Weyl, Emil Artin; Emmy Noether; Ernst Knauf; unidentified person; Chiuntze Tsen, and Erna Bannow (future wife of Ernst Witt). Oberwolfach Photo Collection, Photo ID 9268. Author: Natascha Artin.

Source: Archives of P. Roquette, Heidelberg and C. Kimberling, Evansville and archives of the Mathematisches Forschungsinstitut Oberwolfach.

3.3 Alexandroff in Moscow

The invitation to Moscow was organized by her long-time colleague, Pavel Sergeyevich Alexandroff (1896–1982), at Moscow State University and the Steklov Institute of Mathematics. In the summers of 1926 and 1927, Noether attended the courses on topology given by Alexandroff and Heinz Hopf (1894–1971) in Göttingen. A specialist in abstract algebra, Noether was new to the field of topology. Yet she joined the discussion and made at least one significant contribution, which was mentioned in Hopf's paper *Eine Verallgemeinerung der Euler-Poincaréschen Formel* (1928) Hopf (1964).[12]

Subsequently, Noether visited Moscow and spent the winter of 1928–9 there. Giving a lecture on abstract algebra at Moscow University, she lived in a dormitory. Noether was interested in the lifestyle of the USSR, in particular that of students, and she was very active in making contacts with many Moscow mathematicians (Alexandroff 1981, p. 109). She was influenced by Marxism, and she did not hide it. Upon her return to Germany in 1932, she was condemned by her neighbour as a 'Marxist Jewess' and evicted from her apartment.

[12] In short, she made an expansion of a Betti number by introducing the concept of a Betti group.

According to Alexandroff, when Noether learned that she would be driven out of Germany, she 'seriously considered coming to Moscow for good', as she believed that 'nowhere else would she find the same opportunities to create a splendid new mathematical school that would replace the one that had been taken away from her in Göttingen' (Alexandroff 1981, p. 109). Hearing her desire to live in Moscow for good, Alexandroff negotiated with the Commissariat of Education.

By this time, Alexandroff owed her professionally: Noether had given Alexandroff an opportunity to visit the USA. Recognising his talent, Noether asked her colleague Hermann Weyl to recommend him and Hopf for a Rockefeller scholarship so that they could spend the year 1927–8 working with the leading topologists of their time, Oswald Veblen (1880–1960) and Solomon Lefschetz (1884–1972) at the Institute for Advanced Study (Kimberling 1981, p. 25). Weyl made a recommendation, and they did go to Princeton. Thankful, Alexandroff wished to repay the intellectual debt that he owed to Noether. Alexandroff tried hard to create a chair for Noether at Moscow University. To his disappointment, however, the Commissariat delayed its decision for months and did not respond quickly enough. Time had run out, and Noether had to give up on the idea of going to Moscow to be a part of the mathematical community.

While Emmy Noether did not move to the USSR, Fritz Noether (1884–1941), Emmy's younger brother, did. In response to his dismissal from the University of Breslau, he emigrated and was appointed as mathematics professor at the Institute of Mathematics and Mechanics at the University of Tomsk. He and his family had a difficult time, however. His wife, Regine, went back to Germany and died soon after, and their two sons were sent to Sweden to take asylum. Within five years of his move to the Soviet Union, on 23 October 1938, he was 'found guilty of allegedly spying for Germany and committing acts of sabotage' and sentenced to death 'on the accusation of engaging in anti-Soviet agitation'. He was shot on 10 September 1941.[13]

3.4 Darbishire in Oxford

Although Noether had never visited Oxford, a more concrete offer came from Somerville College, Oxford. There were five women's colleges in Oxford in 1933 – Somerville, Lady Margaret Hall, Society of Home Students (now St Anne's), St. Hugh's, and St Hilda's. Somerville College's Principal was a scholar of English literature, Helen Darbishire. Somerville College was hiring new fellows regularly. The Lady Carlisle Research Fellowship, for example, was the College's most prestigious fellowship that was open 'to unmarried European women, British women, and Jewesses'.[14]

In the letter sent from Darbishire to Sir William Beveridge at the Academic Assistance Council, Darbishire expressed an interest in hosting Noether at Somerville. On 24 June 1933, she wrote, 'we should like to find the person who has the greatest claim by her

[13] Letter from Andrei Parastaev, First Secretary at the USSR Embassy in Washington DC, to Fritz Noether's sons. Dated 12 May 1989. Cited in O'Connor and Robertson (2000).

[14] The 1913 fellowship guidelines are stored in the archives of Somerville College, Oxford.

intellectual qualifications'. She was ready to arrange board in the College, to have Noether elected as a member of the Senior Common Room (SCR), and to provide a room outside the College. One thing that Darbishire added was that she was not expecting Noether to teach: 'we should not expect to use her as a teacher'.[15] In the subsequent letter, sent from Darbishire to Beveridge on 1 July, Darbishire was ready to contact Noether.[16]

It was not Noether who initiated the conversation. Darbishire wrote that Noether had not written to the AAC to receive help. She had learned of Noether's need for assistance from the press, and from Dr. Francis Sowerby Macaulay FRS (1862–1937), one of the donors to Somerville College, who resided in Cambridge. Although he had never met her, he had corresponded with Noether and was worried about her future (Oxford, Bodleian Libraries, MS. S.P.S.L. 53 2/3; 3 July 1933 from Darbishire).[17]

Darbishire decided to get in touch with Noether through Dr. Marie-Elisabeth Lüders. She was the founder and former president of the German Federation of University Women, who was visiting Darbishire in Oxford. She mentioned Noether to Lüders, and asked her to make enquiries upon her return to Germany.[18] On 4 July, Darbishire wrote to the secretary of the AAC, stating that she was prepared to write to Noether.[19] And her attempt to contact Noether continued at least till 11 July.[20] However, either the message through Lüder or the letter to Noether had come through; Darbishire received a letter back from Noether on either 11 or 12 July. She wrote to the Secretary at the Academic Assistance Council, stating that she had a letter from Noether herself 'in answer to an offer which I [Darbishire] made'. She continued to explain what Noether wrote: [21]

> It appears from her letter that she would like to accept the offer of hospitality, provided she had some invitation from the Mathematical profession.
>
> I am trying to find out now whether we could offer her anything in the way of lecturing or teaching in Oxford. I am afraid it would in any case amount to very little. I shall make efforts also in relation with the Women's Colleges in London, where it would be quite easy for her to go up to lecture once a week, supposing she were living with us in Oxford.

Dorothy Wrinch was a mathematics tutor at Somerville, but she was going to take a leave of absence in the 1933–4 academic year. Both Noether and Wrinch attended the International Congress of Mathematicians in Zurich in 1932. At that occasion, they also went on a boat tour together (see Pólya 1987, pp. 82–3, 85). Wrinch must have talked to Noether and knew her well; Darbishire continued in the same letter, explaining what she had heard about Noether from Wrinch:

> Her [Noether's] letter tells nothing about her financial position, but I understand from Dr Wrinch that she was always in poor circumstances. Another point about which I am not clear is whether she is a

[15] Oxford, Bodleian Libraries, MS. S.P.S.L. 53 2/3.
[16] Oxford, Bodleian Libraries, MS. S.P.S.L. 53 2/3.
[17] Oxford, Bodleian Libraries, MS. S.P.S.L. 53 2/3 (3 July 1933, from Darbishire). He was supportive of Darbishire's plan to call Noether; he wrote a letter back. A letter he wrote to an unidentified person on 26 July mentions this. The letter is in the same collection at the Bodleian.
[18] Oxford, Bodleian Libraries, MS. S.P.S.L. 53 2/3 (3 July 1933, from Darbishire).
[19] Oxford, Bodleian Libraries, MS. S.P.S.L. 53 2/3 (4 July 1933, from Darbishire).
[20] Oxford, Bodleian Libraries, MS. S.P.S.L. 53 2/3 (11 July 1933, from Darbishire).
[21] Oxford, Bodleian Libraries, MS. S.P.S.L. 53 2/3 (12 July 1933, from Darbishire).

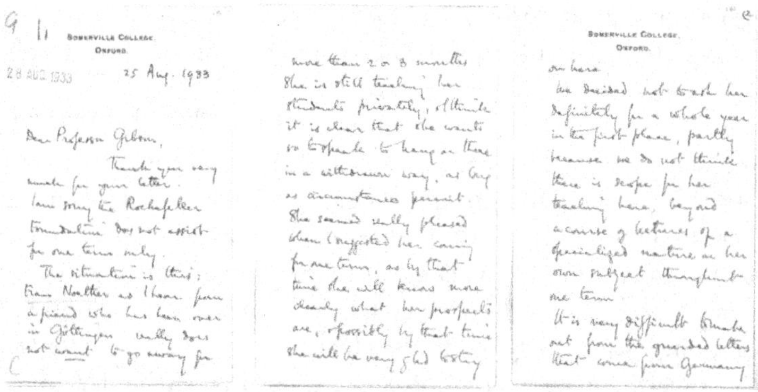

Figure 3.3 A handwritten letter from Helen Darbishire to Gibson at the AAC, explaining what she knows about Noether's situation. Written on 25 August 1933. The material has been reproduced by kind permission of Cara (the Council for At-Risk Academics).

Jewess, or whether she only lost her post owing to the general pressure under which I believe women in the professions are suffering badly.

On 25 July, Darbishire received a letter from the AAC, notifying her that the grant application to cover Noether's maintenance would be sent to Professor Lauder W. Jones, Associate Director of the Rockefeller Foundation in Paris. Professor Gibson at the AAC had already informed Jones about this.[22] By 26 July, Darbishire had contacted Professor Jones to 'put in a tentative plea for a grant towards her [Noether's] maintenance if she comes'.[23] On the same day, the donor, Macauley, had also decided to write to Darbishire regarding the financial assistance he could provide, if she came to Oxford.[24]

During the month of July, Darbishire was active in searching for the grant that could cover the cost of living. After a month had passed, Darbishire wrote to Gibson, stating that she was sorry to hear that 'the Rockefeller Foundation does not assist for one term only'.[25] She explained why Noether could not come to Oxford for more than a term.

The situation is this: Frau Noether as I learn from a friend who has been over in Göttingen really does not want to go away for more than 2 or 3 months. She is still teaching her students privately, and I think it is clear that she wants so to speak to hang on there in a withdrawn way, as long as circumstances permit. She seemed really pleased when I suggested her coming for one term, as by that time she will know more clearly what her prospects are, and possibly by that time, she will be very glad to stay on here.

I've decided not to ask her definitely for a whole year in the first place, partly because we do not think there is scope for her teaching here, beyond a course of lectures of a specialized nature in her own subject throughout one term.

22 Oxford, Bodleian Libraries, MS. S.P.S.L. 53 2/3 (25 July 1933, to Darbishire).
23 Oxford, Bodleian Libraries, MS. S.P.S.L. 53 2/3 (26 July 1933, from Darbishire).
24 Oxford, Bodleian Libraries, MS. S.P.S.L. 53 2/3 (26 July 1933, from Macauley, possibly to Gibson).
25 Oxford, Bodleian Libraries, MS. S.P.S.L. 53 2/3 (25 August 1933, from Darbishire to Gibson).

Darbishire learned that Noether would not be paid after Christmas. More precisely, the final dismissal was expected to be on 1 October 1933. Time was running out, and the arrangement had to be made quickly.

Noether's wish was to stay longer in Göttingen for her students. One of the students who Noether was looking after in Göttingen was from China, Chiungtze Tsen (1898–1940).[26] Noether wrote a letter from Göttingen on 18 August 1933, introduced earlier in this chapter, which was sent to the President of Bryn Mawr College, Marion Edwards Park. She had accepted the offer from Darbishire to come to Oxford for the Christmas-Easter term (i.e., Hilary Term in Oxford). Even in September, her decision had not changed. H. M. Bryant, Principal Darbishire's secretary, wrote that 'she [Noether] is coming to Oxford for the Hilary Term, i.e. about the first three months of next year, and is hoping that Bryn Mawr will postpone their appointment until academic year 1934–35'.[27]

The Academic Assistance Council was also helpful, suggesting a strategy to Darbishire that accommodated Noether's uncertain situation. General Secretary Walter Adams shared his idea:

> I think it would be safer to invite her for a whole year and not merely for one term and then definitely to ask this Council for one of its usual maintenance grants extending over one year. That does not preclude raising the matter again with the Rockefeller Foundation and indeed might facilitate such an approach, since they do not make grants for one term. Then if Professor Noether decided at Christmas that she would prefer to return to Germany, she could do so by formally relinquishing any maintenance grant which she was receiving. It would be far easier to relinquish a grant half way through its duration than to apply for an extension of a grant which had been made on the understanding that it was to extend for only one term.

The Academic Assistant Council tried to be helpful, thus Darbishire replied six days later:[28]

> It was by her [Noether's] own choice that she arranged to come for *one* term, the Hilary Term, 1934. I am not able on my own responsibility to extend an invitation to her for a whole academic year; that would have to wait until our Council meets, but in any case I do not think that this is the best proposition to make, for, as I explained to Professor Gibson, I think that there is little scope for her highly specialised teaching here beyond one course of advanced lectures; frankly, I do not think that we could get an audience for her.

On 13 September 1933, Noether wrote to Richard Brauer (1901–77) who had also looked for a temporary position so he could leave Germany. He himself took up an offer from the University of Kentucky, but Noether told him about her option:

> As for myself, I have been invited to lecture in Oxford for one term, I have chosen the one between Christmas and Easter. Subsequently, I was also offered a research professorship in Bryn-Mawr for 1933/34. I have asked to have postponed for 1934/35 as I have already accepted the Oxford offer. I have no answer yet, but I think it should be alright. . . . Bryn Mawr is a women's college, but Mitchell [Howard Mitchell, professor of mathematics at the University of Pennsylvania] and others are there

26 On 6 December 1933, he defended his dissertation, titled, 'Algebren über Funktionenkörpern' (Algebra over Function Fields), and returned to China in 1935.

27 Oxford, Bodleian Libraries, MS. S.P.S.L. 53 2/3 (11 September 1933, from Bryant).

28 Oxford, Bodleian Libraries, MS. S.P.S.L. 53 2/3. (19 September 1933, from Darbishire to Adams); emphasis in original.

as professors; besides, Veblen has written to me that it is so close to Princeton that they hope I will come over frequently. If through my postponement some funds became available, it could be perhaps somewhere else! By the way, I will live in a women's college in Oxford as well, but I shall lecture to the whole university, which is composed of the different colleges.[29]

At the Meeting of the Education Committee held at Somerville College on 5 October, Darbishire reported the situation to the members. There, Darbishire was in touch not only with Noether, but also with archeologist Margaret Bieber (1879–1978), a Jew. The educational committee minutes recorded the following:

Exiled German Professors.
The Principal reported that she had heard of two eminent German women scholars who had been dismissed from their appointments in Germany owing to their Jewish nationality. It had already been agreed by herself and the Fellows that if possible an endeavour should be made to offer them temporary appointments here.

The Academic Assistance Council was prepared to help in such cases by a grant equal to the amount offered by the College. The Principal hoped to open a fund for subscriptions to augment College payments.

Resolved to recommend to Council:

(1) That Fraulein Noether, formerly Professor of Mathematics in the University of Gottingen should be asked to give a course of 16 lectures in the Hilary Term 1934 with a payment of 25 pounds, on the understanding that this would be made up to 50 pounds from other sources.[30]
(2) That Fraulein Bieber, a distinguished Archaeologist, be offered an Honorary Research Fellowship for the academic year 1933–1934, in the hope that the Women's Colleges in Oxford would help to finance a course of eight lectures in each term.[31]

In the same meeting, they also discussed the status of Dorothy Wrinch. The minutes have a subheading, 'Mathematics', and recorded the resolution that 'Dr. Wrinch should receive the special fee of 30 pounds for the academic year which was paid to her in consideration of special tutorial control'.[32] Wrinch's replacement tutor had already been assigned in June 1933: Miss Margaret Grimshaw of Newnham College, Cambridge.[33] She had worked as a tutor at the University of Göttingen prior to this appointment, and she was going to be paid a salary of 300 pounds for the academic year 1933–4 (*Somerville College Education Committee Meetings Minutes*, 6 June, p. 312).

Three days prior to the education committee meeting at Somerville, on 2 October, Noether had made a decision. President Park at Bryn Mawr received a telegram from Noether, confirming her move there. Delighted, President Park announced her impending arrival at the convocation, held on 3 October. Her term was to be for 2 years, not for 1 year.

29 Letter from Noether to Brauer, Bryn Mawr Special Collections.
30 For a comparison, a student's tuition for mathematics that year was about 32 pounds.
31 Darbishire reported Bierber's work in Oxford, and announced the fund she had created for German refugee scholars (Somerville College 1934, pp. 24–5).
32 The 'tutorial control' meant that Wrinch was taking on consultation from students and other administrative duties while on leave.
33 A tribute to Margaret Eleanor Grimshaw appeared in *Newnham College Roll Letter* 17, 1991.

Financially, Bryn Mawr was able to offer Noether the sum of 2,000 dollars to cover the academic year 1934–5 thanks to the support from the RF (Shen 2019, pp. 59–60).

Oxford's offer to raise the amount to 50 pounds, as stated in the Somerville minutes on 6 October, may not have been delivered to Noether at this point. Subsequently, in the letter, written on 11 October, the College was informed that Emmy Noether 'has accepted the invitation for this year to go to Bryn Mawr in America. This information comes from the Rockefeller Foundation in Paris'.[34] At the Meeting of the Education Committee held at Somerville College on 14 October, Darbishire reported Noether's decision. She said, 'owing to the financial stringency Fraulein Noether had been obliged to accept an offer to lecture in America, and would therefore not come to Oxford during the present academic year'.

Darbishire may not have been aware of the extension of Noether's term at Bryn Mawr. She was willing to help Noether, if she decided to come to Oxford after her appointment at Bryn Mawr for one academic year. Darbishire continued to give assistance to the refugee scholars from European countries. In the *Eighth Annual Supplement to the Report of the College, 1932–1933* (Somerville College 1933), she wrote a paragraph regarding Professor Bieber and the new fund. 'In consultation with the Fellows of the College, I have opened the Fund for the assistance of dismissed German women scholars whom the College resolved, on the ground of the value of their contribution to learning, to invite to work or teach for a period in Oxford' (Somerville College 1934, p. 21). She added, 'I am not making an appeal outside, but if any old students are interested and would like to contribute, I shall be very glad' (Somerville College 1934, p. 21).[35] Noether, however, did not have a chance to use the fund; on 2 February 1935, the college was notified that Bryn Mawr had been gathering funds 'locally' so that Noether could stay there for longer.[36]

3.5 Lecturing at IAS, Princeton

At Bryn Mawr, the department of mathematics was headed by Anna Johnson Pell Wheeler (1883–1966) and the college's president was a classist, President Marion Edward Park (1875–1960). Wheeler, a mathematician, once visited Göttingen to study under Hilbert's guidance as a graduate student. Wheeler chose to conduct research in the field of 'integral equations', on which David Hilbert was a leading scholar; thus she travelled all the way from the USA to Germany. During her study in Göttingen, however, Wheeler had a difficult time; she wrote to Mary Coes (1861–1913), Dean at Radcliffe College, 'In Göttingen I had some trouble with Professor Hilbert and came back to America without a degree' (Grinstein and Campbell 1982, p. 41).[37]

In the USA, she reunited with her husband, Alexander Pell (1857–1921), who had moved to Chicago for his job. (Her husband's real name was Sergei Degaev, an escaped Russian double agent, a fact that would not be revealed during his lifetime; Grinstein and Campbell

[34] Oxford, Bodleian Libraries, MS. S.P.S.L. 53 2/3 (11 October 1933, Assistant Secretary to Darbishire). The underline is in the original.

[35] This fund was successful, and it helped out many refugee scholars up to 1939.

[36] Oxford, Bodleian Libraries, MS. S.P.S.L. 53 2/3 (2 February 1935).

[37] This letter was written in 1910.

Figure 3.4 (left) Noether leaving for the USA. Oberwolfach Photo Collection, photo ID 9265. Author: Natascha Artin.

Source: Archives of P. Roquette, Heidelberg and C. Kimberling, Evansville and archives of the Mathematisches Forschungsinstitut Oberwolfach.

Figure 3.5 (right) The reverse side of this photo states that the photo was taken in Germany. It was shortly before Noether's departure to the USA, or during the summer of 1934 when she made a short return to Germany.

Source: The Bryn Mawr College Archive. (The Archive estimated the year. An inquiry was made to the College Archivist, Ms. Mills.)

1982). Wheeler herself would have known of Noether, through Hilbert. Despite their bitter relationship during her doctoral student year, she kept in touch with Hilbert; she even contributed her portrait to be placed in 1922, for his 60th birthday book, where Noether also put one for herself.[38]

President Park was informed of a possibility of inviting a refugee scholar from the Emergency Committee to Aid Displaced German Scholars. On 11 July 1933, she wrote to the RF in New York, and stated that Bryn Mawr could choose a displaced scholar and invite her with the support of 2,000 dollars. But because the salary she wished to pay was 4,000 dollars, she asked the RF to put in an additional 2,000 dollars. As for the choice of a scholar, she wrote:

> Earlier in the spring Professor Anna Pell Wheeler of our Department of Mathematics was approached by Professor Lefschetz of Princeton with the suggestion that Dr. Emmy Nothe [*sic*] of Göttingen whose resignation from the University had been demanded should be invited to Bryn Mawr College for 1933–34.

Solomon Lefschetz, a mathematics professor at Princeton University, was a keen supporter of Noether. Together with the founder of the IAS, Abraham Flexner (1866–1959), his colleague and the first permanent maths faculty at IAS, Oswald Veblen, and Noether's former colleague at Göttingen, Herman Weyl, they helped to raise funds for Wheeler to extend

[38] The birthday book is stored at the University of Göttingen Archive.

Noether's term at Bryn Mawr. Veblen requested a 'small grant-in-aid' for the academic year 1935–36, and he secured 1,500 dollars and continued fundraising for a 'permanent commitment on the part of the Institute' (Institute for Advanced Study 2017, p. 8). He in fact invited Noether to give weekly lectures at the Institute as a Visitor in the School of Mathematics. This invitation was made at the early stage; recall the letter Noether sent to Brauer. She wrote on 13 September 1933, that 'Veblen has written to me that it [Bryn Mawr] is so close to Princeton that they hope I will come over frequently.'

Noether had sailed to the USA by mid-November Kimberling (1955, 31). She was welcomed by the female academic community at Bryn Mawr. She started to give guest lectures at IAS Princeton in February 1934. Noether had a brand-new start. Professors at many places, near and remote, (such as Swarthmore College, Princeton, Massachusetts Institute of Technology, and Brown University) were willing to help to extend her term in the USA. Wheeler asked the members of the IAS about bringing Noether to Princeton: Solomon Lefschetz responded quickly to the matter, 'she [Noether] is the outstanding refugee German mathematician brought to these shores and if nothing is done for her, it will be a true scandal' (*Dokumente zu Emmy Noether* n.d. 3.2, 80).

References

Archival Documents

18.08.1933, Bryn Mawr Archive.

Birthday Book (photo collection) for David Hilbert at His 60th Birthday. University of Göttingen Archive, 1922.

Dokumente zu Emmy Noether. (n.d.). Compiled by Peter J. Roquette. www.mathi.uni-heidelberg.de/~roquette/Transkriptionen/DOKNOE_070228.pdf.

Emmy Noether, c1935, photograph, Bryn Mawr College Photo Archives, PA_Noether_Emmy_002.

Letter from Noether to Brauer, Bryn Mawr Special Collections.

Newnham College Roll Letter 17, 1991.

Oxford, Bodleian Libraries, MS. S.P.S.L. 53 2/3.

Somerville College Education Committee Meetings Minutes.

Somerville College. (1933). *Eighth Annual Supplement to the Report of the College, 1932–1933.* Oxford: Baxtor Press.

Somerville College. (1934). *Ninth Annual Supplement to the Report of the College, 1933–1934.* Oxford: Baxtor Press.

Secondary Sources

Alexandroff, P. S. (1981). In memory of Emmy Noether. In J. W. Brewer and M. K. Smith, eds., *Emmy Noether: A Tribute to Her Life and Work.* New York: Marcel Dekker, 99–111.

Adams, Pauline. *Somerville for Women: An Oxford College, 1879–1993.* Oxford: Oxford University Press, 1996.

Grinstein, L. S. and Campbell, P. J. (1982). Anna Johnson Pell Wheeler: Her life and work. *Historia Mathematica*, **9**, 37–53.

Hopf H. (1964). Eine Verallgemeinerung der Euler-Poincaréschen Formel. In *Selecta Heinz Hopf.* Berlin, Heidelberg: Springer. https://doi.org/10.1007/978-3-662-25046-4_2

Institute for Advanced Study. (2017). *The Institute Letter.* Spring.

Kimberling, C. (1981). Emmy Noether and her influence. In J. W. Brewer and M. K. Smith, eds., *Emmy Noether: A Tribute to Her Life and Work*. New York: Marcel Dekker, 3–61.
Nossum, R. (2012). Emigration of mathematicians from outside German-speaking academia, 1933–1963, supported by the Society for the Protection of Science and Learning. *Historia Mathematica* **39**(1), 84–104.
Nossum, R. and Kotůlek, J. (2015). The Society for the Protection of Science and Learning as a patron of refugee mathematicians. *BSHM Bulletin: Journal of the British Society for the History of Mathematics* **30**, 153–67.
O'Connor, J. J. and Robertson, E. F. (2000). *MacTutor History of Mathematics*. St Andrews: The University of St Andrews. www-history.mcs.st-and.ac.uk/Biographies/Noether_Fritz.html
Pólya, George. (1987). *The Pólya Picture Album: Encounters of a Mathematician*. Boston: Birkhäuser, 1987.
Senechal, M. (2013). *I Died for Beauty: Dorothy Wrinch and the Cultures of Science*. Oxford: Oxford University Press.
Shen, Q. (2019). A refugee scholar from Nazi Germany: Emmy Noether and Bryn Mawr College. *The Mathematical Intelligencer*, **41**(3), 52–65.

4

Getting to the Bottom of Noether's Theorem*

JOHN C. BAEZ

4.1 Introduction

It is sometimes said Noether showed *symmetries give conservation laws*. But this is only true under some assumptions: for example, that the equations of motion come from a Lagrangian. For which types of physical theories do symmetries give conservation laws? What are we assuming about the world, if we assume it is described by a theories of this type? It is hard to get to the bottom of these questions, but it is worth trying.

We can prove versions of Noether's theorem relating symmetries to conserved quantities in many frameworks. While a differential geometric framework is truer to Noether's original vision, I will study this subject *algebraically*, using ideas from Hamiltonian mechanics.

Atiyah (2002) warned us of the dangers of algebra:

> algebra is to the geometer what you might call the Faustian offer. As you know, Faust in Goethe's story was offered whatever he wanted (in his case the love of a beautiful woman), by the devil, in return for selling his soul. Algebra is the offer made by the devil to the mathematician. The devil says: I will give you this powerful machine, it will answer any question you like. All you need to do is give me your soul: give up geometry and you will have this marvellous machine.

While this is sometimes true, algebra is more than a computational tool: it lets us express concepts in a very clear and distilled way. Further, the geometrical framework developed for classical mechanics is not sufficient for quantum mechanics. An algebraic approach emphasises the similarity between classical and quantum mechanics, clarifying their differences.

Our algebraic study of Noether's theorem relies on an interlocking trio of important concepts used to describe physical systems: 'states', 'observables' and 'generators'. A physical system has a convex set of states. An observable is a real-valued quantity whose value depends – perhaps with some randomness – on the state. More precisely, an observable maps each state to a probability measure on the real line. A generator, on the other hand, is something that gives rise to a one-parameter group of transformations of the set of states – or dually, of the set of observables.

For example, when we say 'the energy of the system is 7 joules', we are treating energy as an observable. When we say 'the Hamiltonian generates time translations', we are treating the Hamiltonian as a generator. In both classical mechanics and ordinary complex quantum

* I thank Javier Prieto for suggesting the connection to the work of Leifer and Spekkens, and Emilio Cobanera for spotting errors.

mechanics, we usually say the Hamiltonian *is* the energy, because we have a way to identify them. But it has repeatedly been noted that observables and generators play distinct roles (Grgin and Petersen 1974; Landsmaan 1997; Zalamea 2018) – and in some theories, such as real or quaternionic quantum mechanics, they are truly different. In all the theories considered here, the set of observables is a Jordan algebra, while the set of generators is a Lie algebra.

When we can identify observables with generators, we can state Noether's theorem as the following equivalence:

The generator a generates transformations that leave the observable b fixed.

$$\Updownarrow$$

The generator b generates transformations that leave the observable a fixed.

In this beautifully symmetrical statement, we switch from thinking of a as the generator and b as the observable in the first part to thinking of b as the generator and a as the observable in the second part. Of course, this statement is *true* only under some conditions, which we explore in detail. But the most fundamental condition, we claim, is the ability to identify observables with generators.

In Section 4.2 we explain how observables and generators are unified in classical mechanics. Here we treat observables as being *the same* as generators, by treating them as elements of a Poisson algebra, which is both a Jordan algebra and a Lie algebra.

In Section 4.3 we turn to quantum mechanics. Here observables are not quite the same as generators. They are both elements of a $*$-algebra: observables are self-adjoint, obeying $a^* = a$, while generators are skew-adjoint, obeying $a^* = -a$. The self-adjoint elements form a Jordan algebra, while the skew-adjoint elements form a Lie algebra.

In ordinary complex quantum mechanics we use a *complex* $*$-algebra. This lets us turn any self-adjoint element into a skew-adjoint one by multiplying it by $\sqrt{-1}$. Thus, *the complex numbers let us identify observables with generators.* In real and quaternionic quantum mechanics, this identification is impossible. Thus, the appearance of complex numbers in quantum mechanics is closely connected to Noether's theorem.

In short, the picture painted by Sections 4.2 and 4.3 is this:

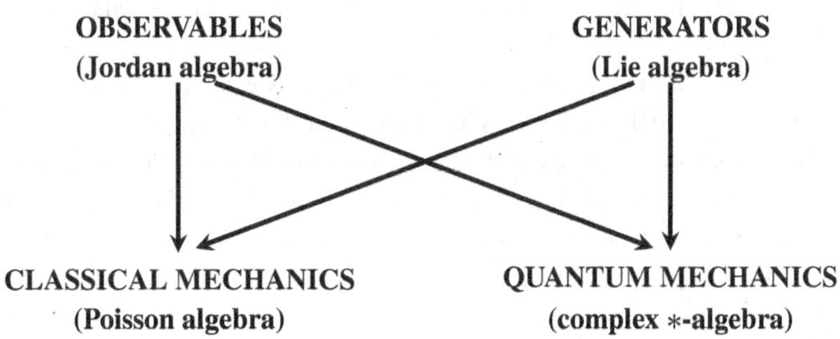

In Section 4.4 we examine generators on their own: that is, Lie algebras. Lie algebras arise very naturally from the concept of 'symmetry'. Any Lie group gives rise to a Lie algebra, and any element of this Lie algebra then generates a one-parameter family of transformations of that very same Lie algebra. This lets us state a version of Noether's theorem solely in terms of generators:

The generator a generates transformations that leave the generator b fixed.

$$\Updownarrow$$

The generator b generates transformations that leave the generator a fixed.

Furthermore, when we translate these statements into equations, their equivalence follows directly from this elementary property of the Lie bracket:

$$[a,b]=0 \iff [b,a]=0.$$

Thus, Noether's theorem is almost automatic *if* we forget about observables and work solely with generators. The only questions left are: why should symmetries be described by Lie groups, and what is the meaning of this property of the Lie bracket? In Section 4.4 we say a bit about both these questions. In particular, the Lie algebra formulation of Noether's theorem follows from a more primitive group formulation, which says that for any two group elements g and h,

$$g \text{ commutes with } h \iff h \text{ commutes with } g.$$

That is, given two ways of transforming a physical system, the first is invariant under the second if and only if the second is invariant under the first.

However, observables are crucial in physics. Working solely with generators in order to make Noether's theorem a tautology would be another sort of Faustian bargain. To really get to the bottom of Noether's theorem, we need to understand the map from observables to generators. In ordinary quantum, this comes from multiplication by i. But this just pushes the mystery back a notch: why should we be using the complex numbers in quantum mechanics?

In Section 4.5 we tackle this issue by examining observables on their own: that is, Jordan algebras. Those of greatest importance in physics are the 'unital JB-algebras', unfortunately named not after the author, but after Jordan and Banach. These allow a unified approach to real, complex and quaternionic quantum mechanics, along with some more exotic theories. Thus, they let us study how the role of complex numbers in quantum mechanics is connected to Noether's theorem.

Any unital JB-algebra O has a partial ordering: that is, we can talk about one observable being greater than or equal to another. With the help of this we can define states on O and prove that any observable maps each state to a probability measure on the real line. More surprisingly, any JB-algebra also gives rise to *two* Lie algebras. The smaller of these, say L, has elements that generate transformations of O that preserve all the structure of this unital JB-algebra. They also act on the set of states. Thus, elements of L truly deserve to be considered 'generators'.

In a unital JB-algebra there is not always a way to reinterpret observables as generators. However, Alfsen and Shultz (1998, 2003) have defined the notion of a 'dynamical correspondence' for such an algebra, which is a well-behaved map $\psi : O \to L$. One of the two conditions they impose on this map implies a version of Noether's theorem. They prove that any JB-algebra with a dynamical correspondence gives a complex $*$-algebra where the observables are self-adjoint elements, the generators are skew-adjoint, and we can convert observables into generators by multiplying them by i.

This result is important, because the definition of JB-algebra does not involve the complex numbers, nor does the concept of dynamical correspondence. Rather, the role of the complex numbers in quantum mechanics *emerges* from a map from observables to generators that obeys conditions including Noether's theorem!

To be a bit more precise, Alfsen and Shultz's first condition on the map $\psi\colon O \to L$ says that every observable $a \in O$ generates transformations that leave a itself fixed. We call this the 'self-conservation principle'. It implies Noether's theorem.

However, in their definition of dynamical correspondence, Alfsen and Shultz also impose a second, more mysterious condition on the map ψ. In Section 4.6 we argue that this condition is best understood in terms of the larger Lie algebra associated to a unital JB-algebra. This is the direct sum $A = O \oplus L$, equipped with a Lie bracket such that

$$[-,-]\colon L \times L \to L \qquad [-,-]\colon L \times O \to O$$

$$[-,-]\colon O \times L \to O \qquad [-,-]\colon O \times O \to L.$$

As mentioned, elements of L generate transformations of O that preserve all the structure on this unital JB-algebra. Elements of O also generate transformations of O, but these only preserve its vector space structure and partial ordering.

In Section 4.6 we argue that these other transformations are connected to *statistical* mechanics. For example, consider ordinary quantum mechanics and let O be the unital JB-algebra of all bounded self-adjoint operators on a complex Hilbert space. Then L is the Lie algebra of all bounded skew-adjoint operators on this Hilbert space. There is a dynamical correpondence sending any observable $H \in O$ to the generator $\psi(H) = iH \in L$, which then generates a one-parameter group of transformations of O defined as follows:

$$a \mapsto e^{itH} a e^{-itH} \qquad \forall t \in \mathbb{R}, a \in O.$$

If H is the Hamiltonian of some system, this is the usual formula for time evolution of observables in the Heisenberg picture. But H also generates a one-parameter group of transformations of O as follows:

$$a \mapsto e^{-\beta H/2} a e^{-\beta H/2} \qquad \forall \beta \in \mathbb{R}, a \in O.$$

Writing $\beta = 1/kT$ where T is temperature and k is Boltzmann's constant, we argue that these are 'thermal transformations'. Acting on a state in thermal equilibrium at some temperature, these transformations produce states in thermal equilibrium at other temperatures (up to normalisation).

The analogy between it and $1/kT$ is often summarised by saying 'inverse temperature is imaginary time'. The second condition in Alfsen and Shultz's definition is a way of capturing this principle in a way that does not explicitly mention the complex numbers. Thus, we may *very roughly* say their result explains the role of complex numbers in quantum mechanics starting from three assumptions:

- observables form a unital JB-algebra,
- the self-conservation principle (and thus Noether's theorem),
- the principle that inverse temperature is imaginary time.

4.1.1 Warning

The concept of 'generating a one-parameter family of transformations' involves a differential equation. This necessarily brings some analysis into our otherwise algebraic discussion. Given a function $f\colon \mathbb{R} \to V$ from the real numbers to a vector space, the meaning of the derivative

$$\frac{df}{dt} = \lim_{h \to 0} \frac{f(t+h) - f(t)}{h}$$

depends on the topology we place on V. In some of our theorems we do not say what topology we are using. Any vector space V has a locally convex topology where a net $v_\alpha \in V$ converges to $v \in V$ if and only if $\ell(v_\alpha) \to \ell(v)$ for all linear functionals $\ell\colon V \to \mathbb{R}$, so we can use this, but in practice we often want to use another. To focus on key ideas, we downplay this issue. In Theorems 4.2.3, 4.3.1, and 4.4.1 we can use any locally convex topology such that the relevant bracket operation is continuous in each argument. In Theorem 4.2.2 we should use the C^∞ topology. In Theorems 4.3.2 and 4.4.3 we should use the norm topology.

All the theorems in this chapter are either easy or already known. Our goal here is to deploy this mathematics to better understand Noether's theorem.

4.2 Classical Mechanics

We begin with a lightning review of classical mechanics. Noether's theorem is often studied in the Lagrangian approach to classical mechanics. Here we take the Hamiltonian approach, and refer the reader elsewhere for the link back to Lagrangian mechanics (Marsden and Ratiu 2013). Our focus is on the dual role that observables play in Hamiltonian mechanics. First, and most obviously, they assign a number to each state. But second, they generate one-parameter groups. This dual role underlies the Hamiltonian formulation of Noether's theorem, Theorem 4.2.3.

In the classical mechanics of a point particle in $\mathbb{R}^n$, the space of states is $\mathbb{R}^{2n}$, with coordinates $q_1, \ldots, q_n$ representing the particle's position, and coordinates $p_1, \ldots, p_n$ representing its momentum. Observables – real-valued quantities depending on the particle's state – are functions on $\mathbb{R}^{2n}$. In order to differentiate these as many times as we want without running into trouble, we take our observables to lie in $C^\infty(\mathbb{R}^{2n})$, the set of smooth real-valued functions on $\mathbb{R}^{2n}$.

The set $C^\infty(\mathbb{R}^{2n})$ is a real vector space where we can add and multiply functions pointwise, and the laws of a commutative algebra hold:

$$f(gh) = (fg)h \qquad fg = gf$$

$$f(\beta g + \gamma h) = \beta fg + \gamma fh$$

for all $f, g, h \in C^\infty(\mathbb{R}^{2n})$ and $\beta, \gamma \in \mathbb{R}$. The multiplication has a clear physical meaning: for example, to measure the observable fg in some state $x \in \mathbb{R}^{2n}$ is the same as to measure f and g separately in this state and then multiply the results. Of course, this becomes problematic in quantum mechanics, but here we assume that both observables can be measured simultaneously with perfect accuracy.

To do interesting physics with our observables, and to state Noether's theorem, we need an additional operation on $C^\infty(\mathbb{R}^{2n})$. This is the *Poisson bracket*, defined by

$$\{f, g\} = \sum_{i=1}^{n} \frac{\partial f}{\partial p_i}\frac{\partial g}{\partial q_i} - \frac{\partial f}{\partial q_i}\frac{\partial g}{\partial p_i}.$$

This gives a way for each observable to define a one-parameter family of transformations of all the rest, at least if a technical condition holds. Most notably, when $f \in C^\infty(\mathbb{R}^{2n})$ is 'energy', these transformations are 'time evolution'.

How does this work? We say $f \in C^\infty(\mathbb{R}^{2n})$ *generates a one-parameter group* it there exist maps

$$F_t^f : C^\infty(\mathbb{R}^{2n}) \to C^\infty(\mathbb{R}^{2n}) \quad \forall t \in \mathbb{R}$$

obeying *Hamilton's equations*:

$$\frac{d}{dt} F_t^f(g) = \{f, F_t^f(g)\} \qquad \forall t \in \mathbb{R}$$

and the initial conditions

$$F_0^f(g) = g$$

for all $g \in C^\infty(\mathbb{R}^{2n})$. Such a family of maps may not exist, but conditions are known under which it does: for example, when the partial derivatives of f are bounded. When such a family exists, it is unique, and it obeys the equations

$$\begin{aligned} F_0^t &= \mathrm{id} \\ F_{s+t}^f &= F_s^f \circ F_t^f \qquad \forall s, t \in \mathbb{R}, \end{aligned}$$

which are the usual definition of a one-parameter group. It follows from these equations that each transformation F_t^f is invertible.

Example 4.2.1 *Take $\mathbb{R}^2$ to be the space of states for a particle on a line, with coordinates (q, p). Let $H \in \mathbb{C}^\infty(\mathbb{R}^2)$ be the energy of a particle of mass $m > 0$ in a smooth potential $V: \mathbb{R} \to \mathbb{R}$, namely*

$$H(q, p) = \frac{p^2}{2m} + V(q).$$

The position and momentum of the particle at time t, if they are defined, are observables given by

$$q_t = F_t^H(q), \qquad p_t = F_t^H(p).$$

Hamilton's equations say

$$\frac{dq_t}{dt} = \{H, q_t\}, \qquad \frac{dp_t}{dt} = \{H, p_t\}.$$

Working out the Poisson brackets, these give

$$\frac{dq_t}{dt} = \frac{p_t}{m}, \qquad \frac{dp_t}{dt} = -V' \circ q_t$$

as one would expect. However, these equations may not have global-in-time solutions: for example, when $V(q) = -q^2$, the particle will typically run off to infinity in a finite amount of time. But if V is well-behaved – for example, when it has bounded derivative or is bounded below – Hamilton's equations have global-in-time solutions and H generates a one-parameter group.

Suppose that $f \in C^\infty(\mathbb{R}^{2n})$ generates a one-parameter group. Then the transformations F_t^f preserve all the structure on observables we have mentioned so far. First, these transformations are linear. One can prove this with a well-known calculation using the linearity of the Poisson bracket in its second argument:

$$\{f, \beta g + \gamma h\} = \beta\{f, g\} + \gamma\{f, h\}.$$

Second, these transformations preserve multiplication, which one can prove using the *Leibniz law*:

$$\{f, gh\} = \{f, g\}h + g\{f, h\}.$$

And third, they preserve the Poisson bracket, which one can prove using the *Jacobi identity*:

$$\{f, \{g, h\}\} = \{\{f, g\}, h\} + \{g, \{f, h\}\}.$$

We can summarise all this by saying the transformations F_t^f are Poisson algebra automorphisms.

Finally, we have Noether's theorem relating symmetries and conserved quantities. Suppose $f \in C^\infty(\mathbb{R}^{2n})$ generates a one-parameter group. We say that f generates *symmetries* of $g \in C^\infty(\mathbb{R}^{2n})$ if

$$F_t^f(g) = g \qquad \forall t \in \mathbb{R}.$$

In this situation we also say g is *conserved* by the group generated by f. Both are ways of saying that the one-parameter group generated by f leaves g unchanged.

Noether's theorem says that if both f and g generate one-parameter groups, then

$$f \text{ generates symmetries of } g \iff g \text{ generates symmetries of } f.$$

In other words, g is conserved by the group generated by f if and only if g generates symmetries of f. The proof, which we give more generally below, relies crucially on one extra law obeyed by the Poisson bracket, its *antisymmetry*:

$$\{f, g\} = -\{g, f\}.$$

We shall see that $\{f, g\} = 0$ iff f generates symmetries of g, while $\{g, f\} = 0$ iff g generates symmetries of f.

Everything so far can be abstracted and generalised, giving an algebraic approach to classical mechanics where we assume observables form a 'Poisson algebra'. A *Poisson algebra* is a real vector space A equipped with a multiplication making A into a *commutative algebra*:

$$a(bc) = (ab)c \qquad ab = ba$$

$$a(\beta b + \gamma c) = \beta ab + \gamma ac$$

together with a bracket making A into a *Lie algebra*:

$$\{a, \{b, c\}\} = \{\{a, b\}, c\} + \{b, \{a, c\}\} \qquad \{a, b\} = -\{b, a\}$$

$$\{a, \beta b + \gamma c\} = \beta\{a, b\} + \gamma\{a, c\}$$

and obeying the *Leibniz law*:

$$\{a, bc\} = \{a, b\}c + b\{a, c\}$$

for all $a, b, c \in A$ and $\beta, \gamma \in \mathbb{R}$.

A typical example of a Poisson algebra is the algebra of smooth functions on a symplectic manifold. More generally, a *Poisson manifold* is a smooth manifold M such that $C^\infty(M)$, made into a commutative algebra in the usual way, is also equipped with a bracket making it a Poisson algebra. For an introduction to symplectic and Poisson manifolds in physics, see Guillemin and Sternberg (1990), Marsden and Ratiu (2013), and Vaisman (1994).

Suppose A is a Poisson algebra and $a \in A$. Let us say that *a generates a one-parameter family* if for each $b \in A$ there exists a unique solution $F_t^a(b)$ to Hamilton's equations

$$\frac{d}{dt} F_t^a(b) = \left\{a, F_t^a(b)\right\} \qquad \forall t \in \mathbb{R},$$

obeying the initial conditions

$$F_0^a(b) = b.$$

Note that we do not demand the one-parameter group law

$$F_s^a(F_t^a(b)) = F_{s+t}^a(b) \qquad \forall s, t \in \mathbb{R}.$$

In well-behaved situations this follows. For example, we have:

Theorem 4.2.2 *Let M be a compact Poisson manifold and $A = C^\infty(M)$ the resulting Poisson algebra. Then every element $a \in A$ generates a one-parameter family $F_t^a \colon A \to A$. Furthermore, F_t^a is a one-parameter group of linear transformations preserving the multiplication and the Poisson bracket.*

Proof Given $a \in A$, the operation $\{a, -\}$ is a derivation of A and thus corresponds to a smooth vector field v on M. Any smooth vector field on a compact manifold is integrable, so there is for any $x \in M$ a unique solution of

$$\frac{d}{dt}\phi_t(x) = v(\phi_t(x)) \qquad \forall t \in \mathbb{R}$$

obeying the initial conditions

$$\phi_0(x) = x \qquad \forall x \in X.$$

This defines a 'flow', that is, a smooth one-parameter family of diffeomorphisms $\phi_t : M \to M$ obeying $\phi_s \circ \phi_t = \phi_{s+t}$ for all $s, t \in \mathbb{R}$. We can define F_t^a by

$$F_t^a(b) = b \circ \phi_t$$

and this is a one-parameter group of linear transformations preserving the multiplication in A. These transformations also preserve the Poisson bracket (Marsden and Ratiu (2013): Prop. 10.3.1). The uniqueness of this one-parameter group can be shown by expressing the Poisson bracket in terms of the Poisson tensor Π, which in local coordinates gives

$$\{a,b\}(x) = \sum_{i,j} \Pi^{ij}(x)\,(\partial_i a)(x)(\partial_j b)(x),$$

and then noting that

$$\frac{d}{dt}F_t^a(b)(x) = \sum_{i,j} \Pi^{ij}(x)(\partial_i a)(x)\,(\partial_j F_t^a(b))(x)$$

has a unique local-in-time solution by standard methods in differential equations. □

Thus, the Poisson algebra arising from a compact Poisson manifold is a structure with a wonderful self-referential property: each element generates a one-parameter group of transformations, preserving all the operations in this structure.

We are now ready to prove a version of Noether's theorem. Suppose that A is a Poisson algebra and $a, b \in A$ generate one-parameter families. We say that a generates *symmetries* of b, or b is *conserved* by the family generated by a, if

$$F_t^a(b) = b \qquad \forall t \in \mathbb{R}.$$

Noether's theorem, formulated in terms of Poisson algebras, says the following:

Theorem 4.2.3 *Let A be a Poisson algebra and suppose $a, b \in A$ generate one-parameter families. Then a generates symmetries of b if and only if b generates symmetries of a.*

Proof The proof proceeds by a chain of equivalences. Note that a generates symmetries of b:

$$F_t^a(b) = b \qquad \forall t \in \mathbb{R}$$

if and only if

$$\frac{d}{dt}F_t^a(b) = 0 \qquad \forall t \in \mathbb{R}$$

which in turn holds if and only if

$$\{a, F_t^a(b)\} = 0 \qquad \forall t \in \mathbb{R}$$

which in turn holds if and only if

$$\{a,b\} = 0.$$

By the antisymmetry of the bracket, i.e. $\{a,b\} = -\{b,a\}$, the last statement is true if and only if

$$\{b,a\} = 0,$$

so, running the chain of implications backwards with a and b switched, we get

$$F_t^b(a) = a \qquad \forall t \in \mathbb{R},$$

meaning that b generates symmetries of a.

The only step above that is not immediate is that

$$\{a,b\} = 0 \quad \Longrightarrow \quad \{a, F_t^a(b)\} = 0 \qquad \forall t \in \mathbb{R}.$$

To show this, assume $\{a,b\} = 0$. Recall that $F_t^a(b)$ is the *unique* solution of Hamilton's equations

$$\frac{d}{dt}F_t^a(b) = \{a, F_t^a(b)\} \qquad \forall t \in \mathbb{R}$$

obeying the initial conditions

$$F_0^a(b) = b.$$

Note that if we set $G(t) = b$ for all t, then $G(t)$ is another solution of Poisson equations:

$$\frac{d}{dt}G(t) = 0 = \{a,b\} = \{a, G(t)\} \qquad \forall t \in \mathbb{R}$$

obeying the same initial conditions. By uniqueness we must have

$$F_t^a(b) = b$$

for all t, and thus

$$\{a, F_t^a(b)\} = \{a,b\} = 0 \qquad \forall t \in \mathbb{R},$$

as desired. □

The proof above has an arch-like structure, where the keystone of the arch is the antisymmetry of the bracket:

$$\{a,b\} = 0 \quad (a \text{ generates symmetries of } b)$$
$$\Updownarrow$$
$$\{b,a\} = 0 \quad (b \text{ generates symmetries of } a)$$

So, in some sense, the antisymmetry of the bracket contains the essence of Noether's theorem. What, however, is the meaning of the bracket's antisymmetry? So far we have simply *posited* it in the definition of Poisson algebra, thus essentially building Noether's theorem in from the start.

One possible answer is as follows. Our original definition of Poisson algebra assumes the bracket is linear in the second argument and antisymmetric. If we instead assume it is bilinear – that is, linear in each argument – antisymmetry follows from assuming $\{a,a\} = 0$ for all a. The proof is easy:

$$\{a,b\} + \{b,a\} = \{a,a\} + \{a,b\} + \{b,a\} + \{b,b\} = \{a+b, a+b\} = 0.$$

If a generates a one-parameter group, the equation $\{a,a\} = 0$ has a simple meaning: a is conserved by the one-parameter group it itself generates. We call this the *self-conservation principle*: each observable generates symmetries of itself. Most famously, energy is conserved by time translations. Momentum in any direction is conserved by spatial translations in that direction. Angular momentum about any axis is conserved by rotations around that same axis (but not necessarily other axes), and so on.

In Theorem 4.3.1 we derive a very general form of Noether's theorem from the self-conservation principle. First, we broaden the scope of investigation by bringing in quantum mechanics.

4.3 Quantum Mechanics

As we have seen, elements of a Poisson algebra play a dual role: they are both observables and generators of one-parameter families of transformations. To accomodate this, a Poisson algebra is a hybrid structure, with the commutative algebra of observables and the Lie algebra of generators tied together by the Leibniz law:

$$\{a,bc\} = \{a,b\}c + b\{a,c\}.$$

The mathematics becomes more tightly unified in quantum mechanics. Now we drop the commutativity of multiplication and describe observables using a *real associative algebra*: a real vector space A with a multiplication obeying

$$a(bc) = (ab)c$$

$$a(\beta b + \gamma c) = \beta ab + \gamma ac \qquad (\alpha a + \beta b)c = \alpha ac + \beta bc$$

for all $a,b,c \in A$, $\alpha,\beta,\gamma \in \mathbb{R}$. If we define the *commutator* $[-,-]\colon A \times A \to A$ by

$$[a,b] = ab - ba,$$

then we get the Lie algebra laws

$$[a,[b,c]] = [[a,b],c] + [b,[a,c]] \qquad [a,b] = -[b,a]$$

$$[a, \beta b + \gamma c] = \beta[a,b] + \gamma[a,c]$$

and the Leibniz law

$$[a, bc] = [a,b]c + b[a,c]$$

for free. They follow from the associative algebra laws!

But there's a catch: not all elements of A are observables, and not all are generators. To accomodate these two roles in one structure we typically require that A be a *real* $*$*-algebra*, meaning an associative algebra with a real-linear operation $*\colon A \to A$ that is an *anti-homomorphism*:

$$(ab)^* = b^* a^* \qquad \forall a, b \in A$$

and **involutory**:

$$(a^*)^* = a \qquad \forall a \in A.$$

We then decree that

- generators are *skew-adjoint* elements of A, obeying $a^* = -a$;
- observables are *self-adjoint* elements of A, obeying $a^* = a$.

To justify this, note that in a real $*$-algebra A, the set of skew-adjoint elements

$$L = \{a \in A\colon\ a^* = -a\}$$

is closed under the commutator $[a,b] = ab - ba$, and it is a Lie algebra, as one would expect of generators. The set of self-adjoint elements

$$O = \{a \in A\colon\ a^* = a\}$$

is closed under the *Jordan product*

$$a \circ b = \frac{1}{2}(ab + ba)$$

and it forms a *Jordan algebra*, meaning that

$$(a \circ a) \circ (a \circ b) = a \circ ((a \circ a) \circ b) \qquad a \circ b = b \circ a$$

$$a \circ (\beta b + \gamma c) = \beta(a \circ b) + \gamma(a \circ c)$$

for all $a, b, c \in A$, $\alpha, \beta, \gamma \in \mathbb{R}$. We say more about why observables should form a Jordan algebra in Section 4.5. The factor of $\frac{1}{2}$ is traditional here: it ensures that if A has a multiplicative unit of 1, then $1 \circ a = a$.

Whenever A is a real $*$-algebra, we have

$$A = O \oplus L$$

because we can write any element $a \in A$ uniquely as the sum of self-adjoint and skew-adjoint parts:

$$a = \tfrac{1}{2}(a + a^*) + \tfrac{1}{2}(a - a^*).$$

This is an interesting situation: while observables and generators are packaged into a single algebraic structure, they are distinct.

To probe this in more detail, let us compare complex, real, and quaternionic quantum mechanics. Quantum mechanics can be done using Hilbert spaces over all three of the associative normed division algebras $\mathbb{R}$, $\mathbb{C}$, and $\mathbb{H}$, and all three forms of quantum mechanics play a role in known physics (Baez 2012). However, they behave quite differently when it comes to the relation between observables and generators.

In complex quantum mechanics, the distinction between observables and generators is a mere formality. Here we commonly take A to be the algebra of bounded linear operators on a complex Hilbert space. In this case, we can turn any self-adjoint bounded linear operator into a skew-adjoint one simply by multiplying it by i. This sets up an isomorphism of real vector spaces

$$\begin{array}{rcl} \psi\colon O & \to & L \\ a & \mapsto & ia \end{array}$$

The distinction between observables and generators becomes significant in real quantum mechanics, where A is the algebra of bounded linear operators on a *real* Hilbert space. There is no way to multiply an observable by i and get a generator. Indeed, suppose A is the algebra of linear transformations of $\mathbb{R}^n$, treated as real Hilbert space with its usual inner product. Then O has dimension $n(n+1)/2$, while L has dimension $n(n-1)/2$. These spaces are not isomorphic when $n > 0$: there are 'more observables than generators'. In fact, there is no nonzero linear map from O to L, or vice versa, that is invariant under the relevant group of symmetries, namely the orthogonal group $\mathrm{O}(n)$.

Alternatively, consider the case where A is the algebra of bounded linear operators on a quaternionic Hilbert space. Now we can multiply an observable by i, but the result is not a generator. Indeed, if the quaternionic Hilbert space is n-dimensional, O has dimension $2n - n^2$ while L has dimension $2n^2 + n$. These spaces are again not isomorphic when $n > 0$, but now there are 'more generators than observables'. And again, there is no nonzero linear map between O and L that is invariant under the relevant symmetry group, namely the quaternionic unitary group $\mathrm{Sp}(n)$.

All this illustrates a clear superiority of complex quantum mechanics over the real and quaternionic versions. Only in the complex case can we identify observables with generators of one-parameter groups! Thus, only in the complex case can we state Noether's theorem in this form:

The one-parameter group generated by a preserves the observable b.

$$\Updownarrow$$

The one-parameter group generated by b preserves the observable a.

It is the number $i \in \mathbb{C}$ that lets us turn observables into generators, and vice versa.

In what follows, we make this idea precise in a more general context. First we prove an easy result, showing that the complex numbers are *sufficient* to unify generators and observables in a framework where Noether's theorem holds. In Section 4.5 we discuss results showing that the complex numbers are also *necessary*, given some technical side-conditions.

In complex quantum mechanics we can treat generators and observables in a unified fashion using a 'complex $*$-algebra'. A *complex associative algebra* is a complex vector space A with a multiplication such that the associative algebra laws hold

$$a(bc) = (ab)c$$

$$a(\beta b + \gamma c) = \beta ab + \gamma ac \qquad (\alpha a + \beta b)c = \alpha ac + \beta bc$$

for all $a, b, c \in A$ and $\alpha, \beta, \gamma \in \mathbb{C}$. A *complex $*$-algebra* is a complex associative algebra with an operation $*\colon A \to A$ that is an involutory antihomomorphism and also *antilinear*:

$$(\alpha a)^* = \overline{\alpha} a^*$$

for all $\alpha \in \mathbb{C}, a \in A$.

Any complex $*$-algebra A has an underlying real $*$-algebra, so we can define a real vector space of skew-adjoint elements:

$$L = \{a \in A\colon\ a^* = -a\}$$

which forms a Lie algebra under the commutator, and a real vector space of self-adjoint elements:

$$O = \{a \in A\colon\ a^* = a\}$$

which forms a Jordan algebra under the Jordan product. But now we also have an isomorphism of real vector spaces:

$$\begin{array}{rcl} \psi\colon O & \to & L \\ a & \mapsto & ia. \end{array}$$

Given a complex $*$-algebra A, we say *$a \in O$ generates a one-parameter family* if for each $b \in O$ there exists a unique solution $F_t^a(b)$ to *Heisenberg's equation*

$$\frac{d}{dt} F_t^a(b) = [ia, F_t^a(b)] \qquad \forall b \in B$$

obeing the initial conditions

$$F_0^a(b) = b.$$

As usual, we say a generates *symmetries* of b if

$$F_t^a(b) = b \qquad \forall t \in \mathbb{R}.$$

We then have this version of Noether's theorem:

Theorem 4.3.1 *Let A be a complex $*$-algebra and suppose $a, b \in O$ generate one-parameter families. Then a generates symmetries of b if and only if b generates symmetries of a.*

Proof The proof is formally just like that of Theorem 4.2.3 for Poisson algebras. Note that a generates symmetries of b if and only if

$$\frac{d}{dt}F_t^a(b) = 0 \qquad \forall t \in \mathbb{R},$$

which holds if and only if

$$[a, F_t^a(b)] = 0 \qquad \forall t \in \mathbb{R},$$

which in turn holds if and only if $[a,b] = 0$. This is true if and only if $[b,a] = 0$, so running the argument in reverse, we see that b generates symmetries of a.

The only non-trivial step is that

$$[a,b] = 0 \qquad \Longrightarrow \qquad [a, F_t^a(b)] = 0 \qquad \forall t \in \mathbb{R}.$$

However, if we assume $[a,b] = 0$, the fact that $F_t^a(b)$ is the unique solution of the Heisenberg equations

$$\frac{d}{dt}F_t^a(b) = [ia, F_t^a(b)] \qquad \forall t \in \mathbb{R}$$

obeying the initial conditions $F_0^a(b) = b$, together with the fact that the constant function $G(t) = b$ is another solution with the same initial conditions, implies that $F_t^a(b) = b$. We thus obtain

$$[a, F_t^a(b)] = [a,b] = 0 \qquad \forall t \in \mathbb{R}$$

as desired. □

The theorem above leaves open the question of which observables generate one-parameter families. This is a question about existence and uniqueness of solutions of differential equations, so we need a bit of analysis to solve it. The nicest complex $*$-algebras are the C*-algebras. In a C*-algebra, *every* observable generates a one-pameter family.

A *C*-algebra* is a complex $*$-algebra A equipped with a norm making it into a Banach space (i.e. a complete normed vector space) and obeying

$$\|ab\| \le \|a\|\,\|b\| \qquad \|aa^*\| = \|a\|^2.$$

for all $a, b \in A$. The following result is well-known, so we merely sketch the proof:

Theorem 4.3.2 *If A is a C*-algebra then every element of $O = \{a \in A\colon\ a^* = a\}$ generates a one-parameter group of transformations that preserve all the C*-algebra structure.*

Proof If $a \in O$, then for each $t \in \mathbb{R}$, we can define

$$e^{ita} = \sum_{n=0}^{\infty} \frac{(ita)^n}{n!} \in A$$

where the power series converges in the norm topology because

$$\left\| \frac{(ita)^n}{n!} \right\| \leq \frac{t^n \|a\|^n}{n!}.$$

For any $b \in O$ we can define

$$F_t^a(b) = e^{ita} b e^{-ita}.$$

One can show by manipulating convergent formal power series that $F_t^a(b) \in O$ for all $t \in \mathbb{R}$, that $F_t^a(b)$ obeys Heisenberg's equation with the initial conditions $F_t^a(b) = b$, and that $F_t^a : O \to O$ is a one-parameter group:

$$\begin{aligned} F_0^t &= \mathrm{id} \\ F_{s+t}^f &= F_s^f \circ F_t^f \qquad \forall s,t \in \mathbb{R}. \end{aligned}$$

For each b one can show that $F_t^a(b)$ is the unique solution of Heisenberg's equations with the given initial conditions. Moreover, one can show by calculations that the transformations F_t^a preserve all the C*-algebra structure:

$$\begin{aligned} F_t^a(\beta b + \gamma c) &= \beta F_t^a(b) + \gamma F_t^a(c) \\ F_t^a(bc) &= F_t^a(b) F_t^a(c) \\ F_t^a(b^*) &= F_t^a(b)^* \\ \|F_t^a(b)\| &= \|b\| \end{aligned}$$

for all $b, c \in A$, $\beta, \gamma \in \mathbb{R}$. □

A huge amount of work on quantum physics has been done using C*-algebras, because they are a very flexible framework that still has close ties to the more traditional complex Hilbert space approach (Arveson 1976; Bratteli and Robinson 1987–1997; Emch 1972; Haag 1992). The so-called C*-axiom $\|a\|^2 = \|a^*a\|$ is very powerful. Among other things, it provides a concrete answer to the question of the physical *meaning* of the norm. This works as follows. Any element a in a unital algebra A has a *spectrum*

$$\sigma(a) = \{z \in \mathbb{C} : \ a - z1 \text{ is not invertible}\}.$$

If A is a C*-algebra and $a \in A$ is self-adjoint, then $\sigma(a) \subseteq \mathbb{R}$, and

$$\|a\| = \sup\{|z| : \ z \in \sigma(a)\}.$$

We think of $\sigma(a)$ as the set of values the observable a can assume. Thus, $\|a\|$ is simply the supremum of $|z|$ where z ranges over all values that a can take on. Furthermore, one can define 'states' on a C*-algebra, and given an observable a and a state ω, we obtain a probability measure on the real line that is supported on the spectrum of a. This gives the probability that a takes on values in any measurable set $S \subseteq \mathbb{R}$ in the state ω. We describe this in more detail in a more general framework in Section 4.5.

To dig deeper into the meaning of Noether's theorem, in the next sections we separately study the two halves of a $*$-algebra: Lie algebras, which describe generators, and Jordan algebras, which describe observables. In the end we shall see that a special kind of Jordan algebra automatically gives rise to a Lie algebra – and under certain conditions connected to Noether's theorem, we can combine the Jordan algebra with this Lie algebra to obtain a C*-algebra.

To lay the groundwork, let us see what happens when we express all the structure of a real or complex $*$-algebra in terms of observables and generators. When we do this for a real $*$-algebra, we get a rather elaborate structure. When we do it for a complex $*$-algebra, we can simplify the answer, since we can use multiplication by i to express generators in terms of observables, or the other way around.

Theorem 4.3.3 *Given a real $*$-algebra A, let*

$$L = \{a \in A\colon\ a^* = -a\}, \qquad O = \{a \in A\colon\ a^* = a\}.$$

Then the commutator and Jordan product on A obey the following conditions: and define

$$\{a,b\} = \frac{1}{2}(ab - ba) \qquad a \circ b = \frac{1}{2}(ab + ba)$$

Then the following conditions hold:

1. *The operations*

$$\{-,-\}\colon L \times L \to L \qquad \{-,-\}\colon L \times O \to O$$

$$\{-,-\}\colon O \times L \to O \qquad \{-,-\}\colon O \times O \to L$$

are bilinear and obey $\{a,b\} = -\{b,a\}$ for all a,b in either L or O.

2. *The operations*

$$\circ\colon L \times L \to O \qquad \circ\colon L \times O \to L$$

$$\circ\colon O \times L \to L \qquad \circ\colon O \times O \to O$$

are bilinear and obey $a \circ b = b \circ a$ for all a,b in either L or O.

3. *For all a,b,c in either L or O,*

$$\{a, b \circ c\} = \{a,b\} \circ c + b \circ \{a,c\}.$$

4. *For all a,b,c in either L or O,*

$$(a \circ b) \circ c - a \circ (b \circ c) = \{a,\{b,c\}\} - \{\{a,b\},c\}.$$

Conversely, given any pair of vector spaces L, O and any operations obeying conditions 1–4, $A = L \oplus O$ becomes a real $*$-algebra if we define

$$ab = a \circ b + \frac{1}{2}\{a,b\}$$

and define a linear map $*\colon A \to A$ by $a^* = a$ for $a \in O$, $a^* = -a$ for $a \in L$.

Proof The proof is a calculation. In the forward direction, we simply check that given a $*$-algebra and defining $[-,-]$ to be the commutator and $a \circ b$ to be the Jordan product, conditions 1–4 hold. In the reverse direction, to check the associative law

$$(ab)c = a(bc)$$

for all $a,b,c \in O \oplus L$, we expand it using $ab = a \circ b + \{a,b\}$ to obtain

$$(a \circ b) \circ c + \{a \circ b, c\} + \{a,b\} \circ c + \{\{a,b\},c\} =$$
$$a \circ (b \circ c) + \{a, b \circ c\} + a \circ \{b,c\} + \{a,\{b,c\}\}.$$

To check this equation we use item 3, the anticommutative law for $\{-,-\}$, and the commutative law for $\circ$ to simplify it to

$$(a \circ b) \circ c - a \circ (b \circ c) = \{a,\{b,c\}\} - \{\{a,b\},c\},$$

which is item 4. The law $(ab)^* = b^*a^*$ follows from the definitions, the anticommutative law for $\{-,-\}$, and the commutative law for $\circ$. The other $*$-algebra laws are easier to check. □

Remarkably, in this theorem we do not assume from the start that O is a Jordan algebra or L is a Lie algebra to prove that $O \oplus L$ is a real $*$-algebra. The most complicated of the Jordan and Lie algebra laws, namely the *Jordan identity*

$$(a \circ a) \circ (a \circ b) = a \circ ((a \circ a) \circ b)$$

and the Jacobi identity

$$\{a,\{b,c\}\} = \{\{a,b\},c\} + \{b,\{a,c\}\},$$

actually *follow* from conditions 1–4. Moreover, conditions 3 and 4 cannot be derived from the Jordan and Lie algebra laws alone. This raises a question: what do these conditions mean? We address this in Section 4.5.

Condition 1 in Theorem 4.3.4 implies this purely algebraic version of Noether's theorem:

$$\{a,b\} = 0 \iff \{b,a\} = 0 \qquad \forall a,b \in O.$$

Thus, we can say that a bracket operation $[-,-]\colon O \times O \to L$ obeying this version of Noether's theorem is a prerequisite for O to become the space of self-adjoint elements of a real $*$-algebra.

The result for complex $*$-algebras is simpler:

Theorem 4.3.4 *Given a complex $*$-algebra A, let $O = \{a \in A : a^* = a\}$ and define*

$$\{a,b\} = \tfrac{i}{2}[a,b] \qquad a \circ b = \tfrac{1}{2}(ab + ba)$$

for all $a,b \in O$. Then

1. *The operation $\{-,-\}\colon O \times O \to O$ is bilinear and obeys*

$$\{a,b\} = -\{b,a\} \qquad \forall a,b \in O.$$

2. *The operation $\circ\colon O \times O \to O$ is bilinear and obeys*

$$a \circ b = b \circ a \qquad \forall a,b \in O.$$

3. *For all $a,b,c \in O$,*

$$\{a, b \circ c\} = \{a,b\} \circ c + b \circ \{a,c\}.$$

4. For all a,b,c in O,

$$(a \circ b) \circ c - a \circ (b \circ c) = \{\{a,b\},c\} - \{a,\{b,c\}\}.$$

Conversely, given a real vector space O and operations obeying conditions 1–4, the complex vector space $A = \mathbb{C} \otimes O$ becomes a complex $*$-algebra in a unique way such that

$$ab = a \circ b - i\{a,b\}$$

and

$$(a + ib)^* = a - ib$$

for all $a,b \in O$.

Proof This follows from Theorem 4.3.3 by noting that for a complex $*$-algebra we can use the linear bijection

$$\begin{array}{rcl} \psi: O & \to & L \\ a & \mapsto & ia \end{array}$$

to express all the structure in terms of O, and rewrite $A = L \oplus O$ as $\mathbb{C} \otimes O = \{a + ib : a,b \in O\}$. □

The above result is implicit in Alfsen and Shultz's (1998) theorem on 'dynamical correspondences', and in Section 4.5 we use it to sketch a proof of their theorem. The reader should also compare work on Jordan–Lie–Banach algebras (Emch 1984; Grgin and Petersen 1974; Halvorson and Clifton 1999; Landsmaan 1997).

4.4 Lie Algebras

In Section 4.2 we described a link between Noether's theorem and the 'self-conservation principle', namely that each observable generates symmetries of itself. Using this idea we can present a version of Noether's theorem with minimal hypotheses. Looking at the proof of Theorems 4.2.3 and 4.3.1, we see they use nothing about the multiplication of observables or the Jacobi identity for the bracket. Thus, we can eliminate these, and suppose we merely have vector space with a bracket operation obeying a few conditions:

Theorem 4.4.1 *Suppose L is a vector space with a bilinear map $\{-,-\}: L \times L \to L$ such that:*

- *For each $a,b \in L$ there is a unique one-parameter family of elements $F_t^a(b) \in L$ obeying*

$$\frac{d}{dt} F_t^a(b) = \left\{a, F_t^a(b)\right\} \quad \forall t \in \mathbb{R}$$

with initial conditions

$$F_0^a(b) = b.$$

- *Each $a \in L$ generates symmetries of itself:*

$$F_t^a(a) = a \qquad \forall t \in \mathbb{R}.$$

Then for any pair $a, b \in L$, this version of Noether's theorem holds:

$$\begin{array}{rcl} F_t^a(b) & = & b \quad \forall t \in \mathbb{R} \\ & \Updownarrow & \\ F_t^b(a) & = & a \quad \forall t \in \mathbb{R}. \end{array}$$

That is, a generates symmetries of b if and only if b generates symmetries of a.

Proof First note that the hypotheses imply

$$\{a,a\} = \frac{d}{dt} F_t^a(a)\Big|_{t=0} = \frac{da}{dt} = 0$$

for all $a \in A$. Since the bracket is bilinear, this implies that the bracket is antisymmetric. From this point on, the proof is exactly like that of Theorem 4.2.3. □

This theorem amounts to dropping *observables* entirely and stating Noether's theorem purely in terms of *generators*. In practice, the vector spaces obeying the conditions of Theorem 4.4.1 are almost invariably Lie algebras. Thus, to dig deeper, we need to ask why Lie algebras are important in physics.

In one sense the answer to this question is simple enough: Lie algebras tend to come from Lie groups, and symmetries in physics tend to form Lie groups. Hilbert's fifth problem goes further and asks which topological groups are Lie groups. After decades of work, mathematicians arrived at a nice answer.

Theorem 4.4.2 *Suppose G is a connected locally compact topological group for which the identity has a neighborhood containing no subgroup except the trivial group. Then G is a Lie group. If G is a Lie group, then its tangent space at the identity has the structure of a finite-dimensional Lie algebra. Conversely, any finite-dimensional Lie algebra arises in this way from some topological group G of this form, which is unique up to isomorphism if we also require that it is simply connected.*

Proof The first statement follows from results of Gleason and Yamabe building on a long line of previous work (Tao 2014). The rest is a combination of Lie's first, second, and third theorems. □

Note that the assumptions on the group G above do not even mention the real numbers, and yet we conclude that G is a manifold! If we assume the symmetries of a physical system form a topological group with these properties, we get a finite-dimensional Lie algebra and – we shall see – a version of Noether's theorem for free.

But infinite-dimensional Lie algebras are also important in physics, so to state this version of Noether's theorem, let us start with a larger class of Lie algebras. A *Banach–Lie algebra* is a Banach space L that is also a Lie algebra where the bracket obeys

$$\|[a,b]\| \leq C\|a\|\,\|b\| \qquad a, b \in L$$

for some $C > 0$. By rescaling the norm we can assume $C = 1$. Any finite-dimensional Lie algebra can be given a norm making it a Banach–Lie algebra. Also, the algebra of bounded linear operators on any Banach space V becomes a Banach–Lie algebra using the operator norm

$$\|a\| = \sup_{\psi \in V, \psi \neq 0} \frac{\|a\psi\|}{\|\psi\|}.$$

Theorem 4.4.3 *Suppose L is a Banach–Lie algebra. Then for each $a, b \in L$ there is a unique one-parameter family of elements $F_t^a(b) \in L$ obeying*

$$\frac{d}{dt} F_t^a(b) = \left[a, F_t^a(b)\right] \quad \forall t \in \mathbb{R}$$

with initial conditions

$$F_0^a(b) = b.$$

Moreover, $F_t^a \colon L \to L$ is a one-parameter group of bounded linear transformations preserving the Lie bracket. For any pair $a, b \in L$, this version of Noether's theorem holds:

$$\begin{array}{rcl} F_t^a(b) & = & b \quad \forall t \in \mathbb{R} \\ & \Updownarrow & \\ F_t^b(a) & = & a \quad \forall t \in \mathbb{R}. \end{array}$$

Proof If L is a Banach–Lie algebra and $a \in L$, the linear transformation $T \colon L \to L$ given by

$$T(b) = [a, b]$$

is bounded. By Picard's existence and uniqueness theorem for the solution of ordinary differential equations, whose usual proof also works for Banach-space valued functions, for each $b \in L$ there is a unique solution $F_t^a(b)$ of the differential equation

$$\frac{d}{dt} F_t^a(b) = \left[a, F_t^a(b)\right] = T\left(F_t^a(b)\right)$$

with initial conditions

$$F_0^a(b) = b.$$

An explicit solution is given by $F_t^a(b) = \exp(tT)(b)$ where the exponential is defined by a convergent power series. Thus, $F_t^a \colon L \to L$ is one-parameter group of bounded linear transformations. We can show these transformations preserve the Lie bracket by noting that for all $t \in \mathbb{R}$ and $b, c \in L$, we have

$$\begin{aligned} \frac{d}{dt}[F_t^a(b), F_t^a(c)] &= \left[\left[a, F_t^a(b)\right], F_t^a(b)\right] + \left[F_t^a(b), \left[a, F_t^a(c)\right]\right] \\ &= \left[a, \left[F_t^a(b), F_t^a(c)\right]\right] \end{aligned}$$

while

$$\frac{d}{dt} F_t^a([b, c]) = \left[a, F_t^a([b, c])\right].$$

Thus, if

$$u(t) = F_t^a([b,c]) - \left[F_t^a(b), F_t^a(c)\right]$$

we have

$$\frac{d}{dt}u(t) = [a, u(t)] \qquad \forall t \in \mathbb{R}.$$

Since $u(0) = 0$, the uniqueness result just mentioned implies that $u(t) = 0$ for all $t \in \mathbb{R}$. Thus, the transformations F_t^a preserve the Lie bracket:

$$F_t^a([b,c]) = \left[F_t^a(b), F_t^a(c)\right]$$

for all $t \in \mathbb{R}$, $a, b, c \in L$.

To show that Noether's theorem holds, by Theorem 4.4.1 it suffices to show

$$F_t^a(a) = a \qquad \forall t \in \mathbb{R}, a \in L.$$

This follows from the uniqueness result, since

$$\frac{d}{dt}a = 0 = [a, a]. \qquad \square$$

Banach–Lie algebras arise naturally from a class of groups called Banach–Lie groups (de La Harpe 1972). However, not every Banach–Lie algebra gives rise to such a group (Van Est and Korthagen 1964). Every Banach–Lie algebra gives a *local Banach–Lie group*: a Banach manifold G containing an element 1, with partially defined smooth group operations with open domains of definition, that obey the group laws when defined. Conversely, any local Banach–Lie group has a Banach–Lie algebra.

Since a version of Noether's theorem holds automatically for any Banach–Lie algebra, and any such thing gives a local Banach–Lie group, we would be remiss not to use this fact to shed more light on Noether's theorem.

Suppose L is a Banach–Lie algebra, and let G be the corresponding local Banach–Lie group. (The reader will lose nothing by considering the case where G is a Lie group.) Suppose $a, b \in L$. From the proofs of Theorems 4.4.1 and 4.4.3, we see that in this context, Noether's theorem

a generates transformations that leave b fixed $\Longleftrightarrow$
b generates transformations that leave a fixed

boils down to the statement

$$[a,b] = 0 \iff [b,a] = 0.$$

Given $a, b \in L$, we have smooth curves $\exp(sa)$, $\exp(tb)$ in G, defined for small s and t, and we have

$$[a,b] = \left.\frac{d^2}{dsdt}\exp(sa)\exp(tb)\exp(-sa)\exp(-ta)\right|_{s,t=0}$$

It follows that $[a,b] = 0$ if $\exp(sa)$ and $\exp(tb)$ commute for sufficiently small s, t. By the Baker–Campbell–Hausdorff formula, the converse is true too.

Therefore, Noether's theorem, phrased solely in terms of generators, follows from the fact that

$$g \text{ commutes with } h \iff h \text{ commutes with } g.$$

It is simply an 'infinitesimal' version of this obvious fact! Similarly, the self-conservation principle follows from the fact that

$$g \text{ commutes with } g.$$

4.5 Jordan Algebras

Having reduced Noether's theorem to a triviality by working only with generators, we see that its non-trivial content must involve observables. Thus, we turn to Jordan algebras. These were introduced by Pascual Jordan (1933), who studied them further with von Neumann and Wigner in an attempt to isolate the algebraic properties of observables in quantum mechanics (Jordan, von Neumann and Wigner 1934). By now much is known about them (Chu 2011; Hanche-Olsen and Størmer 1984; Jacobson 1968; McCrimmon 2006; Upmeier 1987). Still, they remain less widely understood than Lie algebras, so before examining their connection to Noether's theorem, we start by explaining them. We shall see that a certain class of Jordan algebras, the 'unital JB-algebras', are a powerful framework for studying the connection between Jordan algebras and the role of the complex numbers in quantum mechanics.

Recall that a Jordan algebra is a real vector space O equipped with an operation $\circ\colon O \times O \to O$ obeying

$$(a \circ a) \circ (a \circ b) = a \circ ((a \circ a) \circ b) \qquad a \circ b = b \circ a$$

$$a \circ (\beta b + \gamma c) = \beta(a \circ b) + \gamma(a \circ c)$$

for all $a, b, c \in O, \alpha, \beta, \gamma \in \mathbb{R}$. These laws were found by listing some of the properties of the Jordan product $a \circ b = \frac{1}{2}(ab + ba)$ of $n \times n$ self-adjoint complex matrices.

The law $(a \circ a) \circ (a \circ b) = a \circ ((a \circ a) \circ b)$ is mysterious: it demands an explanation. If we use the commutative law to rewrite it this way:

$$(a \circ a) \circ (b \circ a) = ((a \circ a) \circ b) \circ a$$

we see it is a special case of the associative law. But there is perhaps a better way to think about it. One can show, through a tiresome inductive argument, that any Jordan algebra is *power-associative*: given any element a, the n-fold product $a \circ \cdots \circ a$ is independent of how we parenthesise it. Thus, we can write this element as a^n without ambiguity. Furthermore, we can prove that

$$a^m \circ (a^n \circ b) = a^n \circ (a^m \circ b) \qquad \forall m, n \geq 1$$

for all a, b in a Jordan algebra (McCrimmon 2006, Theorem 5.2.2). The mysterious law in the definition of Jordan algebra is just the case $m = 2, n = 1$. Thus, we can equivalently define a Jordan algebra to be a vector space with a commutative power-associative bilinear operation $\circ$ having this property: for any element a, all the operations of multiplication by powers a^n commute.

It is difficult to directly explain the physical meaning of $a \circ b$, since there is not a general procedure for measuring $a \circ b$ given a way to measure a and a way to measure b. However, in a Jordan algebra we have

$$a \circ b = \tfrac{1}{2}\big((a+b)\circ(a+b) - a \circ a - b \circ b\big)$$

for all a, b, so we can understand the meaning of $a \circ b$ if we can understand linear combinations of observables and also the *Jordan square* of an observable, $a^2 = a \circ a$.

For this we need to define a concept of 'state' for a Jordan algebra and connect observables to states. This can be done, not for all Jordan algebras, but for a large class. We say a Jordan algebra *is formally real* if

$$a_1^2 + \cdots + a_n^2 = 0 \implies a_1, \ldots, a_n = 0$$

for all $a_1, \ldots, a_n \in O$. This captures our intuition that the sum of squares of real-valued observables can vanish only if each observable vanishes separately.

The condition of being formally real is very powerful. Jordan, von Neumann and Wigner (1934) showed that in the finite-dimensional case, this condition implies the mysterious law $(a \circ a) \circ (a \circ b) = a \circ ((a \circ a) \circ b)$ so long as the operation $\circ$ is also commutative, power-associative, and bilinear. They also classified the finite-dimensional formally real Jordan algebras. They began by proving that any such algebra is a direct sum of 'simple' ones. A Jordan algebra is *simple* when its only ideals are $\{0\}$ and A itself, where an *ideal* is a vector subspace $B \subseteq A$ such that $b \in B$ implies $a \circ b \in B$ for all $a \in A$. Then they proved every simple finite-dimensional formally real Jordan algebra is isomorphic to one on this list:

- The algebra $\mathfrak{h}_n(\mathbb{R})$ of $n \times n$ self-adjoint real matrices with the product $a \circ b = \frac{1}{2}(ab+ba)$.
- The algebra $\mathfrak{h}_n(\mathbb{C})$ of $n \times n$ self-adjoint complex matrices with the product $a \circ b = \frac{1}{2}(ab+ba)$.
- The algebra $\mathfrak{h}_n(\mathbb{H})$ of $n \times n$ self-adjoint quaternionic matrices with the product $a \circ b = \frac{1}{2}(ab+ba)$.
- The algebra $\mathfrak{h}_3(\mathbb{O})$ of 3×3 self-adjoint octonionic matrices with the product $a \circ b = \frac{1}{2}(ab+ba)$.
- The algebras $\mathbb{R}^n \oplus \mathbb{R}$ with product $(x,t) \circ (x',t') = (tx' + t'x, x \cdot x' + tt')$.

All these are *unital*: they have an element 1 that obeys $1 \circ a = a$ for all a. Thus, all finite-dimensional formally real Jordan algebras are unital.

Unital formally real Jordan algebras are important because we can define 'states' of any physical system whose observables form such an algebra. The first step is to note that in a formally real Jordan algebra O, we can define $a \geq b$ to mean that $a - b$ is a sum of Jordan squares. This relation is a *partial order*, since it clearly obeys the laws

$$a \geq a \qquad a \geq b \;\&\; b \geq c \implies a \geq c \qquad a \geq b \;\&\; b \geq a \implies a = b$$

for all $a, b, c \in O$. One can also see that

$$a, b \geq 0 \implies \alpha a + (1-\alpha) b \geq 0 \qquad \forall \alpha \text{ s.t. } 0 \leq \alpha \leq 1$$

$$a \geq 0 \implies \alpha a \geq 0 \qquad \forall \alpha > 0$$

for all $a, b \in O$. We summarise the last two facts by saying that the set

$$O^+ = \{a \in O : a \geq 0\}$$

is a *convex cone*. Conversely, one can construct a formally real Jordan algebra from any convex cone whose interior is 'symmetric' in a certain sense. In the finite-dimensional case, this is nicely explained in the text by Faraut and Korányi (1994); for an infinite-dimensional generalisation, see Chu (2017).

Given a unital formally real Jordan algebra O, a 'state' is a particular kind of linear functional sending each observable $a \in O$ to its expected values. A linear functional $\omega: O \to \mathbb{R}$ is called *positive* if

$$a \geq 0 \implies \omega(a) \geq 0.$$

A *state* is a positive linear functional $\omega: O \to \mathbb{R}$ that is *normalised*:

$$\omega(1) = 1.$$

The states form a convex set:

$$\omega, \nu \text{ are states} \implies \alpha\omega + (1 - \alpha)\nu \text{ is a state} \qquad \forall \alpha \text{ with } 0 \leq \alpha \leq 1.$$

There has been much work on 'generalised probabilistic theories', where one starts with a convex set of states, or perhaps a cone with an 'order unit' playing the role of $1 \in O^+$. There are a number of interesting attempts to justify a focus on generalised probabilistic theories arising from finite-dimensional formally real Jordan algebras (Barnum, Graydon, and Wilce 2020; Barnum and Hilgert 2020; Wilce 2011, 2018). To generalise this work to infinite-dimensional Jordan algebras, we need a bit of analysis, so we use 'JB-algebras'.

A *JB-algebra* is a Jordan algebra O equipped with a norm making it into a Banach space and obeying

$$\|a \circ b\| \leq \|a\|\|b\| \qquad \|a^2\| = \|a\|^2 \qquad \|a^2\| \leq \|a^2 + b^2\|$$

for all $a, b \in O$. The last inequality implies that any JB-algebra is formally real. Conversely, any *finite-dimensional* unital formally real Jordan algebra can be given a norm making it JB-algebra – but this fails in the infinite-dimensional case (Hanche-Olsen and Størmer 1984, Sec. 3.1). So, the concept of JB-algebra really comes into its own in the infinite-dimensional case.

Given any state $\omega: O \to \mathbb{R}$ on a unital JB-algebra, and any element $a \in O$, there is a probability measure on the real line that describes what happens when one measures the observable a in the state ω. This is the unique Borel measure $\mu_{a,\omega}$ such that

$$\omega(a^n) = \int_{\mathbb{R}} x^n \, d\mu_{a,\omega}(x) \qquad \forall n \geq 0.$$

This measure is supported on the *spectrum* of a, which is defined just as for C*-algebras:

$$\sigma(a) = \{z \in \mathbb{C} : a - z1 \text{ is not invertible}\}.$$

The spectrum of $a \in O$ is always a compact subset of the real line, and we have

$$\|a\| = \sup\{|z| : z \in \sigma(a)\}.$$

Thus, we can think of $\sigma(a)$ as the set of values the observable a can assume, and $\|a\|$ as the supremum of $|z|$ over all values z that a can assume. Furthermore, this formula implies that the norm is not really an extra structure: if a unital Jordan algebra has a norm making it a JB-algebra, this norm is unique. For details, see Alfsen, Shultz and Størmer (1978, Prop. 2.4) and the discussion that follows that proposition.

Another nice feature of JB-algebras is that they support a 'functional calculus'. This lets us apply continuous functions to elements of a unital JB-algebra. For any $a \in O$ and any continuous function $f: \sigma(a) \to \mathbb{R}$, we can define $f(a) \in O$ such that

$$\omega(f(a)) = \int_{\mathbb{R}} f(x)\, d\mu_{a,\omega}(x)$$

for every state $\omega: O \to \mathbb{R}$.

Now let us turn to Noether's theorem. The self-adjoint elements of any C*-algebra form a JB-algebra if we set $a \circ b = \frac{1}{2}(ab + ba)$. As we saw in Theorems 4.3.1 and 4.3.2, C*-algebras have Noether's theorem 'built in': every observable generates a one-parameter group of transformations preserving all the C*-algebra structure, and a generates symmetries of b if and only if b generates symmetries of a. Alfsen and Shultz (1998) compared the situation for JB-algebras. They showed that a JB-algebra consists of a self-adjoint elements of a C*-algebra if and only if it obeys some conditions including a version of Noether's theorem. We now discuss this result.

For the rest of this section, let O be a unital JB-algebra. We begin by getting our hands on a Lie algebra L associated to O. In terms of physics, if O is the Jordan algebra of observables of some system, L will be its Lie algebra of generators.

We start by defining a group. The *order automorphisms* of O are the invertible linear transformations $f: O \to O$ that preserve the partial order:

$$f(a) \geq 0 \iff a \geq 0 \qquad \forall a \in O.$$

Order automorphisms are automatically bounded, and they form a Banach–Lie group $\mathrm{Aut}_{\geq}(O)$. A bounded linear map $\delta: O \to O$ is an *order derivation* if

$$\exp(t\delta) \in \mathrm{Aut}_{\geq}(O) \qquad \forall t \in \mathbb{R}.$$

where the exponential is defined using the usual power series. Order derivations naturally form a vector space which we call A, and this is a Lie algebra under commutators (Alfsen and Shultz 1998, Prop. 10). This is the Lie algebra of the Banach–Lie group $\mathrm{Aut}_{\geq}(O)$.

Every order derivation can be written uniquely as a sum of two kinds, called 'skew-adjoint' and 'self-adjoint' (Alfsen and Shultz 1998, Lemma 11). The skew-adjoint order derivations are closed under commutators so they form a Lie algebra L in their own right, a Lie subalgebra of A. The self-adjoint ones correspond to elements of O, and these are not closed under commutators. We thus have

$$A = O \oplus L$$

as vector spaces, but not as Lie algebras.

A *skew-adjoint* order derivation is a bounded linear map $\delta\colon O \to O$ obeying any one of these equivalent conditions:

1. δ is an order derivation and $\delta 1 = 0$.
2. δ is a derivation of the Jordan product:

$$\delta(a \circ b) = \delta a \circ b + a \circ \delta b \qquad \forall a, b \in O.$$

3. For all $t \in \mathbb{R}$, $\exp(t\delta)$ preserves the Jordan product:

$$\exp(t\delta)(a \circ b) = \exp(t\delta)(a) \circ \exp(t\delta)(b) \qquad \forall a, b \in O$$

4. If ω is a state on O, then $\omega \circ e^{t\delta}$ is a state on O for all $t \in \mathbb{R}$.

The equivalence of these conditions follows from (Alfsen and Shultz 1998, Lemma 9). Using condition 2, it is easy to see that skew-adjoint order derivations are closed under commutators. Thus, they form a Lie subalgebra L of the Lie algebra A of all order derivations.

On the other hand, any element $H \in O$ gives an order derivation δ_H defined by

$$\delta_H b = H \circ b \qquad \forall b \in O$$

and such order derivations are called *self-adjoint*. If O consists of all the self-adjoint elements of a C*-algebra, any self-adjoint order derivation δ_H generates a one-parameter group as follows:

$$\exp(s\delta_H)(b) = \exp(sH/2)\, b\, \exp(sH/2) \qquad \forall b \in O, s \in \mathbb{R}.$$

This formula does not make sense in a general unital JB-algebra, since it relies on the associative multiplication in the C*-algebra. However, while expressions of the form aba do not make literal sense in a Jordan algebra, McCrimmon (2006) has emphasised that there is a perfectly good substitute. For any $a \in O$ there is a linear map $U_a\colon O \to O$ given by

$$U_a(b) = 2(a \circ (a \circ b)) - (a \circ a) \circ b \qquad \forall a \in A.$$

When O is *special* – that is, when the Jordan product can be written as $a \circ b = \frac{1}{2}(ab + ba)$ using the multiplication in some possibly larger associative algebra – we have

$$U_a(b) = aba.$$

Otherwise, we can use $U_a(b)$ as a substitute for this expression. Indeed, in any unital JB-algebra, the one-parameter group generated by a self-adjoint derivation δ_H is given as follows:

$$\exp(s\delta_H)(b) = U_{\exp(sH/2)}(b) \qquad \forall b \in O, s \in \mathbb{R}$$

where $\exp(sH/2)$ is defined using the functional calculus. Concretely, we have

$$\exp(sH/2) = \sum_{n=0}^{\infty} \frac{(sH/2)^n}{n!}.$$

Note that the individual terms in this sum make sense because O is unital and power-associative; furthermore, the sum converges in the norm topology.

Let us see what order derivations amount to in some examples. The skew-adjoint derivations are familiar, but the self-adjoint ones are much less so.

Example 4.5.1 *Let $\mathbb{K} = \mathbb{R}, \mathbb{C}$, or $\mathbb{H}$, let $A = M_n(\mathbb{K})$ be the algebra of $n \times n$ matrices with entries in $\mathbb{K}$, and let*

$$L = \{a \in A\colon\ a^* = -a\},$$
$$O = \{a \in A\colon\ a^* = a\}.$$

Then O is a JB-algebra with the Jordan product $a \circ b = \frac{1}{2}(ab + ba)$ and the operator norm

$$\|a\| = \sup_{\psi \in \mathbb{K}^n,\, \psi \neq 0} \frac{\|a\psi\|}{\|\psi\|}.$$

Every state on O is of the form

$$\omega(a) = \mathrm{tr}(ab) \qquad \forall a \in O$$

for some $b \in O$ with $b \geq 0$ and $\mathrm{tr}(b) = 1$. *We call such b a* density matrix.

Every skew-adjoint derivation of O is of the form

$$\lambda_a b = [a,b] = ab - ba \qquad \forall b \in O$$

for some $a \in L$, and we have

$$\exp(t\lambda_a)(b) = \exp(ta)\, b\, \exp(-ta) \qquad \forall t \in \mathbb{R}, b \in O.$$

The transformations $\exp(t\psi_a)$ preserve the Jordan product and thus also the partial ordering on O and unit $1 \in O$. These transformations also act on any state ω to give a state $\omega \circ \exp(t\psi_a)$. If a generates time evolution, this is the usual time evolution of the state ω, though in the complex case we usually write $a = iH$ for some $H \in O$, called the Hamiltonian.

Every self-adjoint derivation of O is of the form

$$\delta_H b = H \circ b = \tfrac{1}{2}(Hb + bH) \qquad \forall b \in O$$

for some $H \in O$, and we have

$$\exp(-\beta\delta_H)(b) = \exp(-\beta H/2)\, b\, \exp(-\beta H/2) \qquad \forall \beta \in \mathbb{R}.$$

The transformations $\exp(-\beta\delta_H)$ preserve the partial ordering on O but typically not the Jordan product or the unit. In Section 4.6 we argue that these transformations are connected to statistical mechanics.

Given a unital JB-algebra O, Alfsen and Shultz (1998) found conditions under which its complexification

$$\mathbb{C} \otimes O = \{a + ib :\ a, b \in O\}$$

becomes a C*-algebra. For this, they define a *dynamical correspondence* to be a linear map

$$\psi : O \to L$$

obeying these conditions:

(A) $\psi_a(a) = 0$ for all $a \in O$;
(B) $[\psi_a, \psi_b] = -[\delta_a, \delta_b]$ for all $a, b \in O$.

Condition (A) is a version of the self-conservation principle, since it implies that a is conserved by the one-parameter group generated by ψ_a:

$$\exp(t\psi_a)(a) = a \qquad \forall t \in \mathbb{R}.$$

It is equivalent to the statement

$$\psi_a(b) = -\psi_b(a) \qquad \forall a, b \in O$$

which implies this version of Noether's theorem:

$$\psi_a(b) = 0 \iff \psi_b(a) \qquad \forall a, b \in O.$$

Condition (B) is a bit more mysterious, but let us state the theorem:

> **Theorem 4.5.1 (Alfsen and Shultz)** *A unital JB-algebra O is isomorphic to the self-adjoint part of a C*-algebra if and only if there exists a dynamical correspondence on O. Each dynamical correspondence ψ on O determines a unique C*-algebra structure on $\mathbb{C} \otimes O$ with multiplication given by*
>
> $$ab = a \circ b - i\psi_a(b)$$
>
> *for all $a, b \in O$, $*$-structure such that elements of O are self-adjoint and elements of L are skew-adjoint, and norm $\|a\| = \sqrt{\|a^*a\|}$ for all $a \in \mathbb{C} \otimes O$, where the norm in the square root is that in O. Conversely, every C*-algebra structure on $\mathbb{C} \otimes O$ with these properties arises from a unique dynamical correspondence on O.*

Proof This is Theorem 23 of Alfsen and Shultz (1998); here we just mention some key steps. The 'only if' direction is easy since if A is a C*-algebra, and we define O to be its Jordan algebra of self-adjoint elements and L to be the Lie algebra of skew-adjoint derivations of O, the linear map

$$\psi : O \to L$$

given by

$$\psi_a(b) = \tfrac{i}{2}[a, b] \qquad \forall a, b \in O$$

obeys conditions (A) and (B) in the definition of dynamical correspondence. For the 'if' direction, we take a dynamical correspondence ψ and define an operation $\{-,-\} : O \times O \to O$ by

$$\{a, b\} = \psi_a(b) \qquad \forall a, b \in O.$$

We use curly brackets here to emphasise that this is not a commutator. Then we use Theorem 4.3.4 to make $\mathbb{C} \otimes O$ into a complex $*$-algebra. Condition 1 of Theorem 4.3.4 follows from condition (A). Condition 2 follows from O being a Jordan algebra. Condition 3 follows from the fact that ψ_a is a skew-adjoint derivation:

$$\begin{aligned}\{a, b \circ c\} &= \psi_a(b \circ c) \\ &= \psi_a(b) \circ c + b \circ \psi_a(c) \\ &= \{a, b\} \circ c + b \circ \{a, c\}.\end{aligned}$$

Condition 4 follows from conditions (A) and (B):

$$\begin{aligned}(a \circ b) \circ c - a \circ (b \circ c) &= c \circ (a \circ b) - a \circ (c \circ b) \\ &= [\delta_c, \delta_a](b) \\ &= -[\psi_c, \psi_a](b) \\ &= -\{c, \{a, b\}\} + \{a, \{c, b\}\} \\ &= \{\{a, b\}, c\} - \{a, \{b, c\}\}\end{aligned}$$

To finish the job, we need to check that $\mathbb{C} \otimes O$ is a C*-algebra, and that any C*-algebra structure on $\mathbb{C} \otimes O$ obeying the three properties listed in the theorem arises from a unique dynamical correspondence. The most technical part of Alfsen and Shultz's proof is checking the C*-axiom $\|a\|^2 = \|a^*a\|$ for all $a \in \mathbb{C} \otimes O$. □

In the finite-dimensional case Barnum and Hilgert proved a similar result for generalised probabilistic theories, where the convex set of states is taken as fundamental (Barnum and Hilgert 2020, Prop. 9.2); this result involves a map in the other direction, from generators to observables. Earlier, Barnum, Mueller and Ududec proved a similar result in terms of convex cones (Barnum, Mueller and Ududec 2014, Sec. VI). It would be interesting to extend such results to the infinite-dimensional case, especially because they do not assume a version of condition (B).

4.6 Statistical Mechanics

Alfsen and Shultz showed that a unital JB-algebra O comes from a complex C*-algebra if and only if it is equipped with a map from observables to generators, $\psi \colon O \to L$, obeying the self-conservation principle:

$$\psi_a(a) = 0 \qquad \forall a \in O$$

and the mysterious condition (B):

$$[\psi_a, \psi_b] = -[\delta_a, \delta_b] \qquad \forall a, b \in O.$$

As we have seen, the self-conservation principle has a clear physical meaning connected to Noether's theorem. We still need to understand condition (B).

When our JB-algebra comes from a C*-algebra, we can take

$$\psi_a(b) = \tfrac{i}{2}[a, b]$$

for all $a, b \in O$, and then condition (B) boils down to

$$[ia, ib] = -[a, b] \qquad \forall a, b \in O.$$

So, there is no mystery from a purely mathematical viewpoint: condition (B) encodes the fact that $i^2 = -1$. The question concerns the *physical* meaning of this condition *before* we take the complex numbers for granted. Why should a map from observables to generators obey this condition?

We do not have a completely satisfactory answer, but to make any progress on this question, we need to better understand the self-adjoint order derivations δ_a. Let us return to Example 4.5.1, and assume that O is the Jordan algebra of self-adjoint real, complex or quaternionic $n \times n$ matrices. Choose $H \in O$ and treat it as the Hamiltonian of some system. Then there is a self-adjoint order derivation δ_H on O defined by

$$\delta_H b = \tfrac{1}{2}(Hb + bH) \qquad \forall b \in O.$$

This generates a one-parameter group of transformations of O such that

$$\exp(-\beta\delta_H)(b) = \exp(-\beta H/2)\, b\, \exp(-\beta H/2) \qquad \forall \beta \in \mathbb{R}, b \in O.$$

If ω is a state on O, composing it with $\exp(-\beta\delta_H)$ typically does not give a state, but it gives a non-zero positive linear functional, which we can normalise to obtain a state ω_β. This is given by

$$\omega_\beta(b) = \frac{1}{Z(\beta)}\, \omega(\exp(-\beta H/2)\, b\, \exp(-\beta H/2)) \qquad \forall b \in O$$

where

$$Z(\beta) = \omega(\exp(-\beta H)).$$

There is a unique state invariant under all Jordan algebra automorphisms of O, given by

$$\omega(b) = \frac{1}{n}\mathrm{tr}(b) \qquad \forall b \in O.$$

In statistical mechanics, its significance is that it maximises the entropy. In this case $Z(\beta)$ is called the *partition function*, and ω_β is the state of thermal equilibrium at temperature T where $\beta = 1/kT$ and k is Boltzmann's constant. Explicitly,

$$\omega_\beta(b) = \frac{\mathrm{tr}(e^{-\beta H} b)}{\mathrm{tr}(e^{-\beta H})} \qquad \forall b \in O.$$

For each $\beta, \gamma \in \mathbb{R}$, we have

$$\exp(-\beta\delta_H)\omega_\gamma = \frac{Z(\beta)}{Z(\gamma)}\omega_{\beta+\gamma}.$$

Thus, up to a factor involving the partition function, $\exp(-\beta\delta_H)$ acts as 'translation in inverse temperature', somewhat as $\exp(-it\psi_H)$ acts as time translation.

The analogy between it and $\beta = 1/kT$ is well known in physics. It has even been enshrined as a principle: 'inverse temperature is imaginary time'. Under the name of 'Wick rotation', this principle plays an important role in quantum field theory (Peeters and Zamaklar 2011). More speculatively, Hawking and many others have used it to study the thermodynamics of black holes (Fulling and Ruijsenaars 1987). Condition (B) is a way of formulating this principle without explicitly mentioning the complex numbers. As we have seen, the minus sign in this condition encodes the fact that $i^2 = -1$.

Of course, the transformations $\exp(-\beta\delta_H)$ also act on states that are not in thermal equilibrium. To understand how, it seems useful to look at Leifer and Spekkens's (2013) work on Bayesian inference in quantum theory. They define a product of non-negative self-adjoint operators as follows:

$$a \star b = a^{\frac{1}{2}}\, b\, a^{\frac{1}{2}}$$

They use this to describe the 'Bayesian updating' of a state with density matrix b. Given the state ω described by a density matrix b,

$$\omega(a) = \mathrm{tr}(ab) \qquad \forall a \in O,$$

we have

$$\omega_\beta(a) = \frac{1}{Z(\beta)}\mathrm{tr}(a(\exp(-\beta H) \star b)) \qquad \forall a \in O.$$

We can roughly say that ω_β is the result of updating ω by multiplying the probability of being in any state of energy E by the Boltzmann factor $\exp(-\beta E)$.

Returning from the example to the case of an arbitrary unital JB-algebra, it is worth noting that the $\star$ product is a special case of McCrimmon's (2006) operation

$$U_a(b) = 2(a \circ (a \circ b)) - (a \circ a) \circ b \qquad \forall a, b \in O,$$

since we can write

$$a \star b = U_{a^{\frac{1}{2}}}(b)$$

where the square root is well-defined using the functional calculus whenever $a \geq 0$. This formula for the $\star$ product reduces to Leifer and Spekkens's definition when O is a special Jordan algebra. If we generalise their ideas to arbitrary unital JB-algebras in this way, we may get a better understanding of the physical meaning of self-adjoint order derivations.

To conclude, it is worth noting that there is a link between self-adjoint and skew-adjoint order derivations even before we have a dynamical correspondence. Recall that there is a Lie algebra of *all* order derivations

$$A = O \oplus L$$

with O consisting of the self-adjoint order derivations and L consisting of skew-adjoint ones. It is easy to check that the bracket in A has

$$[-,-]\colon L \times L \to L \qquad [-,-]\colon L \times O \to O$$

$$[-,-]\colon O \times L \to O \qquad [-,-]\colon O \times O \to L.$$

Mathematically, we summarise this by saying that A is a $\mathbb{Z}/2$-graded Lie algebra with L as its 'even' part and O as its 'odd' part. In particular, the commutator of two self-adjoint order derivations δ_a and δ_b is skew-adjoint. This is what makes it *possible* to demand condition (B).

Since groups are conceptually simpler than Lie algebras, it is good to note that A is the Lie algebra of $\mathrm{Aut}_{\geq}(O)$, the Banach–Lie group of order automorphisms of O, while L is the Lie algebra of $\mathrm{Aut}(O)$, the Banach–Lie group of Jordan algebra automorphisms of O. The smaller group is connected to quantum mechanics, while the larger group is connected to both quantum and statistical mechanics. In the finite-dimensional case, the fact that A is a $\mathbb{Z}/2$-graded Lie algebra makes the quotient space $\mathrm{Aut}_{\geq}(O)/\mathrm{Aut}(O)$ into a 'symmetric space'. This space is just the interior of the convex cone O^+ (Faraut and Korányi 1994). Something similar holds in the infinite-dimensional case, though this is somewhat less studied (Chu 2017).

Thus, in any unital JB-algebra, even before imposing extra conditions, there is a deep interplay between quantum and statistical mechanics. It remains to more deeply understand this interplay and its connection to Noether's theorem and the role of the complex numbers in quantum mechanics.

References

Alfsen, E. M. and Shultz, F. W. (1998). On orientation and dynamics in operator algebras, part I, *Communications in Mathematical Physics*, **194**, 87–108.

Alfsen, E. M. and Shultz, F. W. (2003). *Geometry of State Spaces of Operator Algebras*. Basel: Birkhäuser.

Alfsen, E. M., Shultz, F. W. and Størmer, E. (1978). A Gelfand–Neumark theorem for Jordan algebras. *Advances in Mathematics*, **28**, 11–56.

Arveson, W. (1976). *An Invitation to C*-Algebras*. Berlin: Springer.

Atiyah, M. (2002). Mathematics in the 20th century. *Bulletin of the London Mathematical Society*, **34**, 1–15.

Baez, J. C. (2012). Division algebras in quantum theory. *Foundations of Physics*, **42**, 819–55. arXiv:1101.5690.

Barnum, H., Graydon, M. A. and Wilce, A. (2020). Composites and categories of Euclidean Jordan algebras. *Quantum*, **4**, 359. arXiv:1606.09331.

Barnum, H., Mueller, M. P. and Ududec, C. (2014). Higher-order interference and single-system postulates characterizing quantum theory. *New Journal of Physics*, **16**, 123029. arXiv:1403.4147.

Barnum, H. and Hilgert, J. (2020). Spectral properties of convex bodies. *Journal of Lie Theory*, **30**, 315–44. arXiv:1904.03753.

Bratteli, O. and Robinson, D. W. (1987–97). *Operator Algebras and Quantum Statistical Mechanics*, 2 vols. Berlin: Springer.

Chu, C.-H. (2011). *Jordan Structures in Geometry and Analysis*. Cambridge: Cambridge University Press.

Chu, C.-H. (2017). Infinite dimensional Jordan algebras and symmetric cones. *Journal of Algebra*, **491**, 357–71. arXiv:1707.03610.

De La Harpe, P. (1972). *Classical Banach-Lie Algebras and Banach-Lie Groups of Operators in Hilbert Space*. Springer Lecture Notes in Mathematics, vol. 285. Berlin: Springer.

Emch, G. G. (1972). *Algebraic Methods in Statistical Mechanics and Quantum Field Theory*. New York: Wiley-Interscience.

Emch, G. G. (1984). *Mathematical and Conceptual Foundations of 20th Century Physics*. Amsterdam: North-Holland.

Faraut, J. and Korányi, A. (1994). *Analysis on Symmetric Cones*. Oxford: Oxford University Press.

Fulling, S. A. and Ruijsenaars, S. N. M. (1987). Temperature, periodicity and horizons. *Physics Reports*, **152**, 135–76.

Grgin, E. and Petersen, A. (1974). Duality of observables and generators in classical and quantum mechanics. *Journal of Mathematical Physics*, **15**, 764–9.

Guillemin, V. and Sternberg, S. (1990). *Symplectic Techniques in Physics*. Cambridge: Cambridge University Press.

Haag, R. (1992). *Local Quantum Physics: Fields, Particles, Algebras*. Berlin: Springer.

Halvorson, H. and Clifton, R. (1999). Maximal beable subalgebras of quantum-mechanical observables. *International Journal of Theoretical Physics*, **38**, 2441–84. arXiv:quant-ph/9905042.

Hanche-Olsen, H. and Størmer, E. (1984). *Jordan Operator Algebras*. London: Pitman. https://folk.ntnu.no/hanche/joa/joa-m.pdf

Jacobson, N. (1968). *Structure and Representations of Jordan Algebras*. Providence, RI: American Mathematical Society.

Jordan, P. (1933). Über Verallgemeinerungsmöglichkeiten des Formalismus der Quantenmechanik. *Nachrichten der Akademie der Wissenschaften in Göttingen, Mathematisch-Physikalische Klasse I*, **41**, 209–17.

Jordan, P., von Neumann, J. and Wigner, E. (1934). On an algebraic generalization of the quantum mechanical formalism. *Annals of Mathematics*, **35**, 29–64.

Landsmaan, N. P. (1997). Poisson spaces with a transition probability. *Reviews in Mathematical Physics*, **9**, 29–58. arXiv:quant-ph/9603005.

Leifer, M. and Spekkens, R. (2013). Towards a formulation of quantum theory as a causally neutral theory of Bayesian inference. *Physical Review A*, **88**, 052130. arXiv:1107.5849.

Marsden, J. E. and Ratiu, T. S. (2013). *Introduction to Mechanics and Symmetry: A Basic Exposition of Classical Mechanical Systems*. Berlin: Springer.

McCrimmon, K. (2006). *A Taste of Jordan Algebra*. Berlin: Springer.

Peeters, K. and Zamaklar, M. (2011). *Euclidean Field Theory* [Lecture notes, Durham University]. http://maths.dur.ac.uk/users/kasper.peeters/eft.html.

Tao, T. (2014). *Hilbert's Fifth Problem and Related Topics*. Providence, RI: American Mathematical Society. https://terrytao.files.wordpress.com/2012/03/hilbert-book.pdf.

Upmeier, H. (1987). *Jordan Algebras in Analysis, Operator Theory, and Quantum Mechanics*. Providence, RI: American Mathematical Society.

Vaisman, I. (1994). *Lectures on the Geometry of Poisson Manifolds*. Basel: Birkhäuser.

Van Est, W. and Korthagen, T. (1964). Non-enlargible Lie algebras. *Indagationes Mathematicae (Proceedings)*, **26**, 15–31. https://core.ac.uk/reader/82236448.

Wilce, A. (2011). Symmetry, self-duality, and the Jordan structure of quantum mechanics. arXiv:1110.6607.

Wilce, A. (2018). A royal road to quantum theory (or thereabouts). *Entropy*, **20**(4), 227. arXiv:1606.09306.

Zalamea, F. (2018). The two-fold role of observables in classical and quantum kinematics. *Foundations of Physics*, **48**, 1061–91. arXiv:1711.06914.

5

BV Quantisation in Perturbative Algebraic QFT

Fundamental Concepts and Perspectives

KASIA REJZNER

5.1 Introduction

The rigorous formulation of gauge theories is one of the big open problems in modern mathematical physics. Its importance for mathematics is evidenced by the fact that it is one of the famous Millennium Problems formulated by the Clay Mathematics Institute. Understanding the proper meaning of concepts such as *symmetry* and *duality* in the context of gauge theories (both classical and quantum) is also an important issue from the point of view of philosophy of science. Gauge theories are therefore a perfect ground for physicists, mathematicians and philosophers to work together and develop new concepts to tackle the fundamental questions.

I think one can say that the first fundamental result concerning symmetries in mathematical physics was obtained by Emmy Noether, who made the connection between symmetries and conservation laws (Noether 1918; see also Kosmann-Schwarzbach 2010 for a review and historical perspective). The conserved currents and charges that one obtains using Noether's theorems are not only crucial in classical field theory, but also play important role in quantisation.

In quantum field theory (QFT), in order to quantise models with local symmetries (including gauge theories and gravity), it is convenient to extend the original system (given in terms of a Lagrangian) by adding auxiliary fields (ghosts, antighosts, etc.). The extended system enjoys a rigid symmetry called BRST symmetry (named after its proposers Becchi, Rouet, and Stora (1975); Becchi et al. (1976) and independently Tyutin (1994, p. 640)) that is an extension of the original local symmetry. For this symmetry one can then construct the corresponding Noether charge and quantise this charge to implement the symmetry at the quantum level. This idea is sometimes referred to as the *quantum Noether method* (Hurth and Skenderis 1999).

Soon after the BRST, a generalisation of their method, called *BV formalism*, was proposed by Batalin and Vilkovisky (1981). Later on, it was recognised that this framework for quantisation of gauge theories has deep connections to homological algebra, graded geometry and modern concepts of derived geometry. There is plenty of literature on the subject, so it is hard to give a really comprehensive list. Here I just mention a few titles: Henneaux and Teitelboim (1992) is a standard reference, Henneaux (1995) gives a really comprehensive treatment of Yang–Mills theories, Gomis et al. (1995) provide a physicist-oriented review of the formalism for beginners and Cattaneo and Moshayedi (2019) also include treatment of theories with boundary. For connections to homolofical algebra, homotopy theory and derived geometry, see for example Calaque (2018), Pantev et al. (2013), and Stasheff (1997, 1998).

Nowadays, the BV formalism plays a central role in two mathematically rigorous approaches to perturbative QFT:

- **Perturbative algebraic quantum field theory (pAQFT)**, which is a mathematically rigorous framework for perturbative QFT that combines the main ideas of the axiomatic framework of algebraic QFT (Araki 1999; Haag 1993; Haag and Kastler 1964) with perturbative methods for constructing models. The main contributions focusing on the scalar field are: Boas (2000), Brennecke and Dütsch (2008), Brunetti et al. (2009), Dütsch and Boas (2002), and Dütsch and Fredenhagen (2001a, 2001b, 2003, 2004, 2007). Abelian gauge theories were treated in Dütsch and Fredenhagen (1999), while the Yang–Mills theories are the subject of Hollands (2008). The full incorporation of the BV framework has been done in Fredenhagen and Rejzner (2012a, 2012b) and Rejzner (2011).
- **Factorisation algebras** approach by Costello (2011) and (2017).

For the relations between the two approaches, see Gwilliam and Rejzner (2020) and Benini et al. (2020).

In this chapter, I will focus on the conceptual understanding of the BV formalism from the point of view of pAQFT, and on the definition of the classical and quantum BV operator. After reviewing the basics of the formalism, I will propose a new approach to defining the classical BV operator and show how this naturally leads to quantisation in agreement with Fredenhagen and Rejzner (2012a). In the second part, I will present some new, perhaps speculative, ideas on how one could incorporate the BV formalism into a non-perturbative framework, along the lines of Buchholz and Fredenhagen (2020).

I argue that the BV formalism in pAQFT is the result of applying the following principles to quantisation of gauge theories:

- *Locality*: imposing and preserving locality is crucial for renormalisation, and it allows one to identify the space of observables of interest. More on locality in pAQFT and its generalisations can be found in Rejzner (2019).
- *Deformation*: in pAQFT, quantum models are built from the classical models by means of deformation quantisation. The advantage is that one stays with the same space of observables and what changes is the product. More generally, in Fredenhagen and Rejzner's (2012a) version of the BV formalism, one also deforms differential graded algebras.
- *Homology*: one avoids building explicit quotients or spaces of orbits under a Lie algebra (or group) action. Instead, one uses homological algebra to construct their derived versions. This idea has been very fruitful in mathematics and turns out also to be very successful in understanding the foundations of QFT.

Through the combination of these three principles, one arrives at a formulation that is both mathematically rigorous and physically meaningful. I will now outline the basics of the resulting framework.

5.2 Classical Field Theory

I start with classical theory. I will follow the principles of *locality* and *homology* to construct models, which will then be easy to quantise using *deformation*.

5.2.1 Kinematical Structure

Let M be an oriented, time-oriented globally hyperbolic spacetime (the latter effectively means that it has a Cauchy surface). The conceptual results presented here hold for general theories with local gauge invariance, but as a running example I will work with the self-interacting Yang–Mills theory.

Let $\mathcal{E}$ denote the configuration space of the theory, which in this chapter is always understood as the space of smooth, sections of some vector bundle $E \xrightarrow{\pi} M$ over M. By choosing $\mathcal{E}$, we decide what kind of objects the theory describes (e.g. scalar fields, tensor fields). We will need some notation:

- $\mathcal{E}_c$ denotes the space of smooth, compactly supported sections of E.
- $\mathcal{E}'$, $\mathcal{E}'_c$ denote complexifications of topological duals of $\mathcal{E}$ and $\mathcal{E}_c$ respectively (both equipped with the strong topology).
- $\mathcal{E}^*$ denotes the space of smooth sections of the dual bundle E^*.
- $\mathcal{E}^!$ denotes the complexification of the space of sections of E^* tensored with the bundle of densities over M. By a slight abuse of notation, I use the symbol $\mathcal{E}^!(M^n)$ for the complexified space of sections of the n-fold exterior tensor product of this bundle, seen as a vector bundle over M^n.

In this chapter, I will typically denote the elements of $\mathcal{E}$ by φ, even if they carry indices. This makes the notation simpler. I will invoke the indices again later, when it becomes necessary.

Example 5.2.1 I will focus on two running examples:

- *Scalar field.* Here the configuration space is just $\mathcal{E} = \mathcal{C}^\infty(M, \mathbb{R})$.
- *Yang–Mills theories.* I consider G a semisimple compact Lie group and $\mathfrak{k}$ its Lie algebra. For simplicity, let's take the trivial bundle[1] $P = M \times G$ over M and define the off-shell configuration space of the Yang–Mills theory as $\mathcal{E} = \Omega^1(M, \mathfrak{k})$.

We model classical observables as functionals on $\mathcal{E}$. To make this mathematically precise, let's equip $\mathcal{E}$ with its natural Frechét topology and consider the space of (Bastiani) smooth functionals $\mathcal{C}^\infty(\mathcal{E}, \mathbb{C})$ (see Bastiani 1964) for the details on the appropriate notion of smoothness). These will be our observables. To understand the physical motivation, note that classically, an observable assigns to a given field configuration a number, which corresponds to the value of the measurement of that observable (e.g. energy density at a given point in spacetime). The smoothness requirement assures that all the algebraic structures that we want to introduce on these observables are well defined.

[1] Non-trivial bundles can also be treated, but in this paper I want to focus on the perturbative treatment of Yang–Mills theory, so restricting to trivial bundles is sufficient. For the non-perturbative treatment of the classical configuration space see Benini et al. (2018).

The next important notion is that of a *spacetime support* of a functional. It encodes localisation properties of observables. Another crucial property is *additivity*.

Definition 5.2.2 The spacetime support of a functional is defined by

$$\begin{aligned}\text{supp } F = \{x \in M | \forall \text{ neighbourhoods } U \text{ of } x \ \exists \varphi_1, \varphi_2 \in \mathcal{E}, \text{ supp } \varphi_2 \subset U \\ \text{such that } F(\varphi_1 + \varphi_2) \neq F(\varphi_1)\}.\end{aligned} \tag{5.1}$$

Definition 5.2.3 A functional F is called *additive* if

$$F(\varphi + \chi + \psi) = F(\varphi + \chi) - F(\chi) + F(\chi + \psi), \tag{5.2}$$

for $\varphi + \chi + \psi \in \mathcal{E}$ and $\text{supp } \varphi \cap \text{supp } \psi = \emptyset$.

On one hand this property is as a weaker version of linearity, and on the other hand it encodes *locality*. Recall that, in physics, a functional is called *local* if it can be written in the form:

$$F(\varphi) = \int_M \omega(j_x^k(\varphi))\, d\mu(x),$$

where ω is a function on the jet bundle over M and $j_x^k(\varphi) = (x, \varphi(x), \partial\varphi(x), \dots)$, with derivatives up to order k, is the kth jet of φ at the point x. Brouder et al. (2018) showed (based on ideas presented in Brunetti et al. 2019) that local functionals can be characterised as smooth functionals that obey (5.2) and have smooth first derivatives. More about additivity and its generalisations can be found in Rejzner (2019).

The space of compactly supported smooth local functions on $\mathcal{E}$ is denoted by $\mathcal{F}_{\text{loc}}$. The algebraic completion of $\mathcal{F}_{\text{loc}}$ with respect to the pointwise product

$$F \cdot G(\varphi) = F(\varphi)G(\varphi) \tag{5.3}$$

is the commutative algebra $\mathcal{F}$ of *multilocal functionals*.

We also introduce *regular functionals*. We say that $F \in \mathcal{F}_{\text{reg}}$ if all the derivatives $F^{(n)}(\varphi)$ are smooth, i.e. for all $\varphi \in \mathcal{E}$, $n \in \mathbb{N}$, we have

$$F^{(n)}(\varphi) \in \mathcal{E}^!(M^n).$$

5.2.2 Dynamics and Symmetries

To introduce dynamics, I use a generalisation of the Lagrangian formalism, following Brunetti et al. (2009). Ideally, we would like to derive the equations of motion and symmetries from the action principle. The potential difficulty here is that the manifolds we are working with are non-compact so the integral of a Lagrangian density, for example $\frac{1}{2}(\nabla^\mu\varphi\nabla_\mu\varphi - m^2\varphi^2)$, over the whole manifold M does not converge, if φ is not compactly supported. One could be tempted to restrict attention to compactly supported configurations, but this does not work either, since the equations of motion we want to consider do not have non-trivial compactly supported solutions! To get around this obstruction, we smear the Lagrangian density with a cutoff function $f \in \mathcal{D} \doteq \mathcal{C}_c^\infty(M, \mathbb{R})$ and define all the relevant objects (e.g. the Euler–Lagrange derivative) in a way which is independent of f. To make this more systematic, let's introduce the notion of a *generalised Lagrangian*.

Definition 5.2.4 A *generalised Lagrangian* on a fixed spacetime M is a map $L : \mathcal{D} \to \mathcal{F}_{\text{loc}}$ such that

(i) $L(f+g+h) = L(f+g) - L(g) + L(g+h)$ for $f, g, h \in \mathcal{D}$ with supp $f \cap$ supp $h = \varnothing$ (**Additivity**).
(ii) $\text{supp}(L(f)) \subseteq \text{supp}(f)$ (**Support**).
(iii) Let $\mathcal{G}$ be the isometry group of the spacetime M (for Minkowski spacetime we set $\mathcal{G}$ to be the proper orthochronous Poincaré group $\mathcal{P}_+^\uparrow$). We require that $L(f)(g^*\varphi) = L(g_* f)(\varphi)$ for every $g \in \mathcal{G}$ (**Covariance**).

I denote the space of all generalised Lagrangians by $\mathscr{L}$.

I also assume that the Lagrangians satisfy $L^* = L$ with respect to the involution $*$, which for now is just the complex conjugation, but when we get to graded geometry, the involution will also swap the order of factors.

Now it's time to get rid of the dependence on f.

Definition 5.2.5 (Brunetti et al. 2009) Actions $S(L)$ are defined as equivalence classes of Lagrangians, where two Lagrangians L_1, L_2 are called equivalent $L_1 \sim L_2$ if

$$\text{supp}(L_1 - L_2)(f) \subset \text{supp}\, df, \tag{5.4}$$

for all $f \in \mathcal{D}$.

The idea is, essentially, to identify generalised Lagrangians whose defining Lagrange densities differ by a total derivative. From here on, I will use the notation S rather than $S(L)$ and all objects and constructions that do not depend on the choice of representative in the equivalence class S will be labelled with S rather than L.

Example 5.2.6 The generalised Lagrangian of the free scalar field is

$$L_0(f)[\varphi] = \frac{1}{2}\int_M (\nabla^\mu\varphi\nabla_\mu\varphi - m^2\varphi^2) f\, d\mu_g.$$

For the Yang–Mills theory, we have

$$L_{\text{YM}}(f)[A] = -\frac{1}{2}\int_M f\, \text{tr}(F \wedge *F),$$

where $F = dA + \frac{i\lambda}{2}[A, A]$, $A \in \mathcal{E}$, λ is the coupling constant, $*$ is the Hodge operator and tr is the trace in the adjoint representation, given by the Killing–Cartan metric κ.

Following Buchholz and Fredenhagen (2020), I introduce some further notation.

Definition 5.2.7 Let $L \in \mathscr{L}$, $\varphi \in \mathcal{E}$. Define a functional $\delta L : \mathcal{D} \times \mathcal{E} \to \mathbb{R}$ by

$$\delta L(\psi)[\varphi] \doteq L(f)[\varphi + \psi] - L(f)[\varphi],$$

where $\varphi \in \mathcal{E}$, $\psi \in \mathcal{D}$ and $f \equiv 1$ on supp ψ (the map $\delta L(\psi)[\varphi]$ thus defined does not depend on the particular choice of f).

The above definition can be turned into a difference quotient, and we can use it to introduce the *Euler–Lagrange derivative* of S. The equations of motion are understood in the sense of Brunetti et al. (2009). Concretely, the Euler–Lagrange derivative of S is a 1-form on $\mathcal{E}$, i.e. a map $dS : \mathcal{E} \to \mathcal{E}'_c$ defined by

$$\langle dS(\varphi), \psi \rangle \doteq \lim_{t \to 0} \tfrac{1}{t} \delta L(t\psi)[\varphi] = \int \frac{\delta L(f)}{\delta \varphi(x)} \psi(x), \tag{5.5}$$

with $\psi \in \mathcal{E}_c$ and $f \equiv 1$ on $\operatorname{supp} \psi$. Here $\frac{\delta L(f)}{\delta \varphi}$ is understood as an element of $\mathcal{E}^! \subset \mathcal{E}'_c$. The field equation is now the following condition on φ:

$$dS(\varphi) \equiv 0, \tag{5.6}$$

so geometrically, the solution space is the *zero locus of the 1-form dS*. Note that dS lives on $\mathcal{E}$, rather than M! Let $\mathcal{E}_S \subset \mathcal{E}$ denote the space of solutions to (5.6). We are interested in the space $\mathcal{F}_S$, of functionals on $\mathcal{E}_S$. We will call them *on-shell* functionals.

Example 5.2.8 Examples of equations of motion:

- **Free scalar field**: $dS_0(\varphi) = -(\Box + m^2)\varphi$, where $\Box$ is the wave operator (d'Alembertian).
- **Yang–Mills theory**: $dS_{\mathrm{YM}}(A) = D_A * F$, where D_A is the covariant derivative induced by the connection A.

Remark 5.2.9 For systems with several fields (or components), I will use the notation $\frac{\delta S}{\delta \varphi^\alpha}$ for $\frac{\delta L(f)}{\delta \varphi^\alpha}$ evaluated at $f \equiv 1$ and treated as a component of the form dS. Here α runs from 1 to D, where D is the number of degrees of freedom of the system (for the scalar field it is 1, for the pure Yang–Mills it is equal to 4 times the dimension of $\mathfrak{k}$).

Next, I discuss *symmetries*. These are directions in the configuration space $\mathcal{E}$, around a given point, along which the action is constant. Geometrically, these are vector fields X on $\mathcal{E}$ such that

$$\partial_X S \equiv 0,$$

where

$$\partial_X S \doteq \int \frac{\delta L(f)}{\delta \varphi(x)} X(x), \quad f \equiv 1 \text{ on } \operatorname{supp} X,$$

and $X \in \Gamma(T\mathcal{E})$ is identified with a map from $\mathcal{E}$ to $\mathcal{E}_c^{\mathbb{C}}$. Note that $\partial_X S$ is just the insertion of the 1-form dS into a vector field X.

Formally, we write

$$X = \int X(x) \frac{\delta}{\delta \varphi(x)},$$

and identify the basis on the fiber $T_\varphi \mathcal{E}$ as the antifields $\frac{\delta}{\delta \varphi(x)} \equiv \varphi^\ddagger(x)$.

I will now focus on the situation, where any local symmetry of the system can be expressed as

$$X = \omega \rho(\xi) + I,$$

where I is a symmetry that vanishes identically on $\mathcal{E}_S$, ω is a local function from $\mathcal{E}$ to $\mathcal{D}$ (multiplication with an element of $\Gamma(T\mathcal{E})$ is defined fibrewise) and $\rho\colon \mathfrak{g}_c \to \Gamma(T\mathcal{E})$ is a Lie-algebra morphism, arising from a given local action σ of some Lie algebra $\mathfrak{g}_c$ on $\mathcal{E}$, by means of

$$\rho(\xi)F[\varphi] := \left\langle F^{(1)}(\varphi), \sigma(\xi)\varphi \right\rangle \equiv \int_M \frac{\delta F}{\delta\varphi(x)}\sigma(\xi)\varphi(x).$$

We assume $\mathfrak{g}_c$ to be the space of smooth compactly supported (hence the subscript c) sections of some vector bundle over M and the action σ on $\mathcal{E}$ to be local.

Example 5.2.10 For the Yang–Mills theory, compactly supported local symmetries are given in terms of the Lie algebra $\mathfrak{g}_c = \Gamma_c(M)$, and the local action σ is given by

$$\sigma(\xi)A := d\xi + [A,\xi] = D_A\xi, \quad \xi \in \mathfrak{g}_c.$$

The presence of local symmetries implies that the equations of motion have redundancies, or, in other words, that $\mathcal{E}_S$, the zero locus of dS, consists of *orbits* of the action σ of $\mathfrak{g}$ on $\mathcal{E}$. To see this explicitly, note that, since X is assumed to be local and compactly supported, it can be expressed in terms of some differential operator

$$X^\alpha(x)[\varphi] = Q^\alpha{}_\beta(\varphi)\varphi^\beta(x) = a(x)[\varphi]\varphi(x) + b^\mu(x)[\varphi]\nabla_\mu\varphi(x) + \cdots,$$

so the condition for X to be a symmetry can be expressed as

$$0 = \int \frac{\delta S}{\delta\varphi^\alpha(x)} X^\alpha(x) d\mu(x) = \int \varphi^\beta (Q^\alpha{}_\beta)^* \frac{\delta S}{\delta\varphi^\alpha} d\mu, \tag{5.7}$$

where $*$ denotes the formal adjoint of a differential operator (obtained using integration by parts). This is the second Noether theorem, and it leads to the conclusion that $\frac{\delta S}{\delta\varphi^\alpha}$, the equations of motion of the system, are not all independent. We will come back to this point in Section 5.2.3.1. More on the relation between Noether's second theorem and the BV formalism can be found in Fulp et al. (2003).

Ultimately, we are interested in functionals on the solution space $\mathcal{E}_S$ that are *invariant* under the action ρ of the symmetries. We will denote this space by $\mathcal{F}_S^{\mathrm{inv}}$.

5.2.3 Homological Interpretation

Our goal is to characterise the space $\mathcal{F}_S^{\mathrm{inv}}$ of symmetry-invariant on-shell functionals in a way that will facilitate quantisation. Remember that our aim is not just to construct the classical theory, but rather to use it as a first step towards quantisation.

The conclusion from Noether's second theorem (which we have now rephrased in a slightly different language, following) is that, in the presence of local symmetries, *equations of motion have redundancies*, so the Cauchy problem is not well posed in such systems and one is tempted to remove the redundancy by taking the quotient by the action ρ of infinitesimal symmetries. However, following the guiding idea of *homology*, instead of forming a quotient, we can go to a larger space where the equations of motion are better behaved and we can keep track of relations between equivalent solutions.

From the point of view of deformation quantisation that we want to perform in the end, it is more convenient to work with vector spaces and encode the information about symmetries and equations of motion in maps between these vector spaces. Remarkably, the kind of algebra that one uses in this construction also started with Emmy Noether!

5.2.3.1 Koszul Complex

We start with finding a homological interpretation for $\mathcal{F}_S$. The idea is to describe it as the quotient $\mathcal{F}_S = \mathcal{F}/\mathcal{F}_0$, where $\mathcal{F}_0$ is the space of functionals that vanish on $\mathcal{E}_S$ (redundancy removal by quotienting). There is a nice geometrical way to find such functionals. Recall that the space of solutions $\mathcal{E}_S$ is the space on which the one form dS vanishes. If we take a vector field X, then $\iota_{dS}X$ (the insertion of a one-form into a vector field) vanishes identically on $\mathcal{E}_S$. If we assume X to have appropriate locality properties, then $\iota_{dS}X \in \mathcal{F}_0$. Let's define $\mathcal{V}$ to be the space of multilocal vector fields on $\mathcal{E}$ (for the precise definition, see Rejzner 2016) and introduce a map $\delta_S\colon \mathcal{V} \to \mathcal{F}$ by setting

$$\delta_S(X) := -\iota_{dS}(X).$$

Clearly, the image of δ_S is contained in $\mathcal{F}_0$. One could ask the question whether it is in fact all of $\mathcal{F}_0$. This is less obvious and depends on the system. It can be shown that $\mathcal{F}_0$ is equal to the image of δ_S, provided S satisfies certain *regularity conditions* (see e.g. Henneaux 1990; Henneaux and Teitelboim 1992). One requires that the equations of motion of the system can be split into independent ones,

$$\frac{\delta S}{\delta \varphi^\alpha}(\varphi) = 0, \quad \alpha = 1, \ldots, N, \tag{5.8}$$

and $D - N$ of dependent ones (the relations follow from Noether's second theorem (5.7)), so that the full system of equations $dS(\varphi) = 0$ is fully equivalent to (5.8). Note that $\frac{\delta S}{\delta \varphi^\alpha}$ are local functions from $\mathcal{E}$ to $\mathcal{E}$ (i.e. they depend only on φ and its derivatives at a point), so they can be seen as functions on the jet space (a k-jet of φ at point x is essentially given by $(\varphi(x), \partial\varphi(x), \ldots)$ with derivatives up to order k). Assume that $\frac{\delta S}{\delta \varphi^\alpha} \equiv S_\alpha$, $\alpha = 1, \ldots, N$ can be used (at least locally) as the first N coordinates on the jet space. It is crucial that functionals in $\mathcal{F}_0$ are multilocal, so one can use the standard argument with the fundamental theorem of calculus to show that $\mathcal{F}_0$ is equal to the image of δ_S. I sketch it here for a local functional $F \in \mathcal{F}_0$. I write F as

$$F(\varphi) = \int_M \omega(u_1, \ldots, u_k) d\mu = \int_M \tilde{\omega}(S_1, \ldots, S_N, u_{N+1}, \ldots, u_k) d\mu,$$

where $u_1, \ldots, u_k$ are some arbitrary fixed coordinates on the jet space. Since F vanishes on the solution space, under the regularity assumption, we have

$$\tilde{\omega}(0, \ldots, 0, u_{N+1}, \ldots, u_k) = 0.$$

We can then write

$$F(\varphi) = \sum_{\alpha=1}^{N} \int_M S_\alpha \int_0^1 \frac{\partial \tilde{\omega}}{\partial S_\alpha}(\lambda S_1, \ldots, \lambda S_N, u_{N+1}, \ldots, u_k) \, d\lambda \, d\mu.$$

Next, one shows that the smoothness, locality and the fact that ω is compactly supported on M imply that $\int_0^1 \frac{\partial \tilde{\omega}}{\partial S_\alpha(x)}(\lambda S_1, \dots, \lambda S_N, u_{N+1}, \dots, u_k) d\lambda \equiv X^\alpha(x)$ defines a smooth local compactly supported vector field. I will always assume that the actions we consider satisfy the above regularity conditions.

Note that δ_S also 'knows' about the symmetries, since the kernel of δ_S consists of those vector fields for which

$$\iota_{dS}(X) = \partial_X S \equiv 0,$$

i.e. of *symmetries*. We can summarise all we know up to now in the following chain complex:

$$\begin{array}{ccccccccc} 0 & \to & Sym & \hookrightarrow & \mathcal{V} & \xrightarrow{\delta_S} & \mathcal{F} & \to & 0, \\ & & 2 & & 1 & & 0 & & \end{array}$$

where the numbers below mean *grading*, and they help to keep track of where things belong.

In homological algebra, *homology groups* are defined by taking quotients of the kernel of the map going out of the space by the image of the map going into it. The 0th homology of our complex is $H_0 = \mathcal{F}/\operatorname{Im}(\delta_S) = \mathcal{F}/\mathcal{F}_0$, so it characterises the space of functionals on the solution space! Assume there are no non-trivial (not vanishing on $\mathcal{E}_S$) local symmetries and let $\mathcal{K} \doteq (\Lambda \mathcal{V}, \delta_S)$ (this is the exterior algebra, built out of antisymmetrised tensor products). Then $\mathcal{F}_S = H_0(\mathcal{K})$ and higher homologies vanish. This is called the *Koszul resolution*.

5.2.3.2 Chevalley–Eilenberg Complex

Let's now consider a situation where local symmetries are present. Let $\mathfrak{g}_c$ be the Lie algebra characterising the infinitesimal local symmetries. Since we let them act as derivations on functionals that are themselves compactly supported, we can drop the requirement of compact support for the symmetries and consider $\mathfrak{g}$ instead.

We are interested in the space of symmetry-invariant observables, i.e. functionals F such that

$$\partial_{\rho(\xi)} F = 0,$$

for all $\xi \in \mathfrak{g}$. Algebraically, the space of invariants under the action of a Lie algebra can be characterised using the *Chevalley–Eilenberg complex*.

The underlying graded algebra of the Chevalley–Eilenberg complex is $\mathcal{CE} \doteq \mathcal{C}^\infty_{\mathrm{ml}}(\overline{\mathcal{E}}, \mathbb{C})$, the space of multilocal functionals on the graded manifold $\mathcal{E} \oplus \mathfrak{g}[1] \equiv \overline{\mathcal{E}}$ (see Rejzner 2016 for a precise definition of this space). We identify functionals on $\mathfrak{g}[1]$ with $\Lambda \mathfrak{g}'$, the exterior algebra over $\mathfrak{g}'$. The generators of $\Lambda \mathfrak{g}'$ can be understood as evaluation functionals

$$c^I(x)[\xi] \doteq \xi^I(x),$$

and in physics these are called *ghosts*. The grading of $\mathcal{CE}$ is called the *pure ghost number* #pg. We express $\mathcal{CE}$ as

$$\mathcal{CE} \doteq \left(\Lambda \mathfrak{g}' \widehat{\otimes} \mathcal{F}, \gamma_{\mathrm{ce}}\right),$$

where $\widehat{\otimes}$ is the appropriately completed tensor product.

The Chevalley–Eilenberg differential γ_{ce} is constructed in such a way that it encodes the action ρ of the gauge algebra $\mathfrak{g}$ on $\mathcal{F}$. For $F \in \mathcal{F}$ we define $\gamma_{\text{ce}} F \in \mathcal{C}^{\infty}_{\text{ml}}(\mathcal{E}, \mathfrak{g}')$ as

$$(\gamma_{\text{ce}} F)(\varphi, \xi) \doteq \partial_{\rho(\xi)} F(\varphi), \tag{5.9}$$

where $\xi \in \mathfrak{g}$. In terms of evaluation functionals (i.e. ghosts):

$$\gamma_{\text{ce}} F = \partial_{\rho(c)} F.$$

For a form $\omega \in \mathfrak{g}'$, which doesn't depend on φ we set $\gamma_{\text{ce}}\omega(\xi_1, \xi_2) \doteq -\omega([\xi_1, \xi_2])$ which is an element of $\Lambda^2\mathfrak{g}'$. Again, we can express this using evaluation functionals:

$$\gamma_{\text{ce}} c = -\frac{1}{2}[c, c]$$

If $F \in \mathcal{F}^{\text{inv}}$, then $\gamma_{\text{ce}} F \equiv 0$, so $H^0(\mathcal{CE})$ characterises the gauge-invariant functionals.

5.2.3.3 BV Complex

Now we combine gauge-invariant and on-shell to be able to characterise the space $\mathcal{F}^{\text{inv}}_S$. Note that $\mathcal{CE}$ is the space of multilocal compactly supported functionals on a graded manifold $\overline{\mathcal{E}} = \mathcal{E} \oplus \mathfrak{g}[1]$, so instead of vector fields on $\mathcal{E}$, we consider the vector fields on the *extended configuration space* $\overline{\mathcal{E}}$. This way we obtain the *BV complex*: $\mathcal{BV}$. Its underlying algebra is the algebra of *multilocal polyvector fileds on* $\overline{\mathcal{E}}$, i.e. the space of multilocal compactly supported functionals on the graded manifold

$$\mathcal{E}[0] \oplus \mathfrak{g}[1] \oplus \mathcal{E}^![-1] \oplus \mathfrak{g}^![-2] \equiv T^*[-1]\overline{\mathcal{E}}.$$

More concretely, elements of $\mathcal{BV}$ are multilocal functionals of the field multiplet φ^α and of corresponding antifields $\varphi^\ddagger_\alpha$, where the index α runs through all the physical and ghost indices. For graded functionals, we distinguish between the right $\frac{\delta_r}{\delta\varphi^\alpha}$ and left derivatives $\frac{\delta_l}{\delta\varphi^\alpha}$. We use the convention that the antifields are identified with the right derivatives.

The algebra $\mathcal{BV}$ has two gradings: the ghost number #gh (the main grading) and the antifield number #af (extra grading used later). Functionals of physical fields have both numbers equal to 0. Functionals of ghosts have #af $= 0$ and #gh $=$ #pg (the 'pure ghost' grading, a ghost c, has $\#pg = 1$). All vector fields have a non-zero antifield number given by $\#\text{af}(\varphi^\ddagger_\alpha) = 1 + \#\text{pg}(\varphi^\alpha)$, and #gh $= -$#af.

$\mathcal{BV}$ seen as the space of graded multivector fields is equipped with a graded generalisation of the Schouten bracket, called in this context *the antibracket*, defined by

$$\{X, Y\} \doteq \sum_\alpha \left\langle \frac{\delta_r X}{\delta\varphi^\alpha}, \frac{\delta_l Y}{\delta\varphi^\ddagger_\alpha} \right\rangle - \left\langle \frac{\delta_r X}{\delta\varphi^\ddagger_\alpha}, \frac{\delta_l Y}{\delta\varphi^\alpha} \right\rangle. \tag{5.10}$$

The right derivation δ_S is not inner with respect to $\{.,.\}$, but locally it can be written as:

$$\delta_S X = \{X, L(f)\}, \quad f \equiv 1 \text{ on } \operatorname{supp} X,\ X \in \mathcal{V}.$$

We write this as $\delta_S X = \{X, S\}$. Similarly, one can find an action θ such that $\gamma X = \{X, \theta\}$ and we define the *classical BV differential* as

$$s = \{., S + \theta\} \equiv \{., S^{\text{ext}}\}.$$

We call S^{ext} the *extended action*. The BV differential s has to be nilpotent, i.e. $s^2 = 0$, which leads to the *classical master equation* (CME):

$$\{L^{\text{ext}}(f), L^{\text{ext}}(f)\} = 0, \tag{5.11}$$

modulo terms that vanish in the limit of constant f.

The differential s increases the ghost number by one (i.e. is of order 1 in #gh). It can be expanded with respect to the antifield number as

$$s = \delta + \gamma,$$

where δ is of order -1 in #af and is the extension of δ_S, while γ is of order 0 is the extension of γ_{ce}. In general, there could be higher order terms as well, but I will not discuss this here.

Differential complex $(\mathcal{BV}, \delta)$ is called the *Koszul–Tate complex* and in the simplest case discussed here, it is a resolution (it would not be a resolution if the symmetries were not independent).

Crucially, we have

$$H^0(\mathcal{BV}, s) = \mathcal{F}_S^{\text{inv}} \tag{5.12}$$

which is the reason to work with $\mathcal{BV}$ in the first place, as it contains the same information as $\mathcal{F}_S^{\text{inv}}$ but has a simpler algebraic structure (quotients and spaces of orbits are resolved).

To prove (5.12), one uses the fact that the Koszul–Tate complex $(\mathcal{BV}, \delta)$ is a resolution (the only non-trivial homology is in degree 0), so

$$H^0(\mathcal{BV}, s) = H^0(H_0(\mathcal{BV}, \delta), \gamma).$$

Since $H_0(\mathcal{BV}, \delta)$ is by construction the space of on-shell functionals on $\overline{\mathcal{E}}$ and the 0th cohomology of γ characterises the invariants, we obtain the desired result.

In the next step, we introduce the gauge fixing using an automorphism α_Ψ, defined on generators as

$$\alpha_\Psi(\Phi_\beta^\ddagger(x)) \doteq \frac{\delta\Psi(f)}{\delta\varphi^\beta(x)}, \quad \alpha_\Psi(\Phi^I(x)) = \Phi^I(x)$$

where $f(x) = 1$ and $\Psi_M(f)$ is a fixed generalised Lagrangian of ghost number -1, called *gauge fixing fermion*. The choice of Ψ_M determines the choice of gauge fixing. It can be easily seen that α_Ψ leaves the antibracket invariant, and we choose it in such a way that the #af $= 0$ part of the transformed action gives rise to hyperbolic equations (see Fredenhagen and Rejzner 2012b for details).

Example 5.2.11 To implement a Lorenz-like gauge in Yang–Mills theory, we need to further extend the BV complex with antighosts $\bar{C}$ (in degree -1) and Nakanishi–Lautrup fields B (in degree 0). These form a trivial pair, i.e.:

$$s\bar{C}^I = iB^I, \quad sB^I = 0.$$

The new extended configuration space is written explicitly as

$$\overline{\mathcal{E}} = \mathcal{E} \oplus \mathfrak{g}[1] \oplus \mathfrak{g}[0] \oplus \mathfrak{g}[-1].$$

Since the new generators were introduced as a trivial pair, the cohomology of the resulting complex is the same as the original one, so (5.12) remains true also after this modification. The gauge-fixing fermion is then:

$$\Psi_M(f) = i \int_M f\left(\frac{\alpha}{2}\kappa(\bar{C},B) + \langle\bar{C}, *d*A\rangle\right)d\mu.$$

To talk about the gauge-fixed theory, it is convenient to redefine the gradings. Let #ta denote the *total antifield number*, which is 1 for all the antifield generators and zero for fields. We decompose s with respect to this grading and obtain two terms (which I again denote by δ and γ):

$$s = \gamma + \delta.$$

The total action is still denoted by S^{ext}, and I will denote by S the #ta $= 0$ term in the action. Let $\theta := S^{\text{ext}} - S$. We can express

$$\delta = \{.,S\}, \quad \gamma = \{.,\theta\}.$$

Differential δ acts trivially on fields, and on antifields it gives $\delta\varphi_\alpha^\ddagger = \frac{\delta S}{\delta\varphi_\alpha}$, so the *gauge-fixed* equations of motion are now the equations of motion of S, which are hyperbolic. This implies that the homology of δ is concentrated in degree 0 (there can be no non-trivial local symmetries for hyperbolic equations!), so $(\mathcal{BV},\delta)$ is a resolution, and we again have

$$\mathcal{F}_S^{\text{inv}} = H^0(\mathcal{BV},s) = H^0(H_0(\mathcal{BV},\delta),\gamma).$$

Taking $H_0(\mathcal{BV},\delta)$ is interpreted as 'going on-shell'.

5.2.4 Linearised Theory

We can split the extended action into the term S_0 that is quadratic in fields and antifields and the interaction therm V. S_0 can be written as

$$S_0 = S_{00} + \theta_0,$$

where S_{00} is the term with #ta $= 0$ and θ_0 has #ta $= 1$. Similarly, $V = V_0 + \theta$, and we note that $S = S_{00} + V_0$ is the total antifield independent part of the action.

We define the linearised BRST differential by

$$\gamma_0 F \doteq \{F,\theta_0\},$$

The total linearised BV differential s_0 is

$$s_0 = \delta_0 + \gamma_0,$$

where $\delta_0(\varphi^\ddagger_\alpha) = -\frac{\delta S_{00}}{\delta \varphi^\alpha}$, so the homology of δ_0 describes the space of solutions to the linearised equations of motion. Denote

$$\frac{\delta_l S_{00}}{\delta \varphi^\alpha(x)}(\varphi) \equiv P_{\alpha\beta}(x)(\varphi^\beta(x)),$$

where each component $P_{\alpha\beta}$ is a differential operator. For simplicity, we will often write the equations of motion using the index-free notation: $P\varphi = 0$.

In a similar manner, I will denote

$$\frac{\delta_r \delta_l \theta_0}{\delta \varphi^\sigma(y) \delta \varphi^\ddagger_\alpha(x)} \equiv K^\alpha{}_\sigma(x)\delta(y-x),$$

where each K^α_σ is a differential operator.

The cohomology of s_0 is given by

$$H^0(\mathcal{BV}, s_0) = H^0(H_0(\mathcal{BV}, \delta_0), \gamma_0),$$

since $(\mathcal{BV}, \delta_0)$ is a resolution. Taking $H_0(\mathcal{BV}, \delta_0)$ is again understood as 'going on-shell' (this time for the linearised theory).[2]

Assume that the gauge fixing was done in such a way that P is Green hyperbolic (for gauge theories and gravity this was shown in Fredenhagen and Rejzner 2012b), meaning that there exist unique retarded and advanced Green functions $\Delta^{\mathrm{A/R}}$, i.e. Green functions for the equations of motion operator P such that

$$\mathrm{supp}(\Delta^{\mathrm{R}}(f)) \subset J^+(\mathrm{supp}(f)), \qquad \mathrm{supp}(\Delta^{\mathrm{A}}(f)) \subset J^-(\mathrm{supp}(f)).$$

We define the Pauli–Jordan function by

$$\Delta = \Delta^{\mathrm{R}} - \Delta^{\mathrm{A}}.$$

The CME of the free theory allows one to prove some important properties that hold for $\Delta^{\mathrm{A/R}}$ and Δ (see e.g. Hollands 2008; Rejzner 2014).

Lemma 5.2.12 *Assume that S_{00} is invariant under γ_0 (i.e. the free* cme *holds) and S_{00} induces a normally hyperbolic system of equations of motion: $P\varphi = 0$. Let Δ^* be a retarded, advanced or causal propagator corresponding to P. It follows that Δ^* satisfies the "consistency conditions" (see Hollands 2008):*

$$\sum_\sigma ((-1)^{|\varphi^\alpha|} K^\alpha_\sigma(x')\Delta^*(x',x)^{\sigma\gamma} + K^\gamma_\sigma(x)\Delta^*(x',x)^{\alpha\sigma}) = 0, \tag{5.13}$$

Example 5.2.13 For Yang–Mills, these identities are:

$$K^A_C(x)\Delta^*(x,y)^{C\overline{C}} + K^{\overline{C}}_B(y)\Delta^*(x,y)^{A,B} = 0, \quad -K^{\overline{C}}_B(x)\Delta^*(x,y)^{BA} + K^A_C(y)\Delta^*(x,y)^{\overline{C}C} = 0,$$

[2] Here we note a difference with Hollands (2008), where, in the example of Yang–Mills theory, the terms $\nabla_\mu A^{\ddagger\mu}_I$ in $s_0(c^\ddagger_I)$ and $\overline{C}^\ddagger_I$ in $s_0(B^\ddagger)$ were attributed to the 'δ_0 part' of s_0 rather than to the 'γ_0 part'. We will denote the operators used in Hollands (2008) by $\tilde{\gamma}_0$ and $\tilde{\delta}_0$, with $s_0 = \tilde{\gamma}_0 + \tilde{\delta}_0$. By direct inspection one can see that $\tilde{\delta}_0$ does not respect the total antifield grading (since, for example, $\tilde{\delta}_0(C^\ddagger_I) = id * d\overline{C}_I - dA^\ddagger_I$ has a term with #ta = 0 as well as a term with #ta = 1). Moreover, $\tilde{\delta}_0$ is not nilpotent, and it does not anti-commute with $\tilde{\gamma}_0$. Hence, from the cohomological perspective, using δ_0 and γ_0 is more natural than using $\tilde{\delta}_0$ and $\tilde{\gamma}_0$.

or more explicitly:

$$d_x \Delta_s^*(x,y) + \delta_y \Delta_v^*(x,y) = 0,$$
$$\delta_x \Delta_v(x,y) + d_y \Delta_s^*(x,y) = 0.$$

These relations are an obvious consequence of the fact that d and δ commute with the Hodge Laplacian.

The classical linearised theory is constructed by introducing the Peierls bracket given by:

$$\lfloor F, G \rfloor = \sum_{\alpha,\beta} \left\langle \frac{\delta^r F}{\delta\varphi^\alpha}, \Delta^{\alpha\beta} \frac{\delta^l G}{\delta\varphi^\beta} \right\rangle, \tag{5.14}$$

where $F, G \in \mathcal{BV}$. Unfortunately, $\mathcal{BV}$ is not closed under this bracket and one needs to extend it to a larger space. A good candidate is the space $\mathcal{BV}_{\mu c}$ of *microcausal* functions on $T^*[-1]\overline{\mathcal{E}}$, i.e. functionals that are smooth, compactly supported and their derivatives (with respect to both φ and $\varphi^\ddagger$) satisfy the WF set condition:

$$\mathrm{WF}(F^{(n)}(\varphi, \varphi^\ddagger)) \subset \Xi_n, \quad \forall n \in \mathbb{N}, \ \forall \varphi \in \overline{\mathcal{E}}(M), \tag{5.15}$$

where Ξ_n is an open cone defined as

$$\Xi_n \doteq T^*M^n \setminus \{(x_1, \ldots, x_n; k_1, \ldots, k_n) | (k_1, \ldots, k_n) \in (\overline{V}_+^n \cup \overline{V}_-^n)_{(x_1,\ldots,x_n)}\}, \tag{5.16}$$

5.2.5 Classical BV Operator and the Møller Maps

The key observation of Fredenhagen and Rejzner (2012a) is that one can *define* the interacting quantum BV operator by taking the free one and twisting it with the quantum Møller map. Then, one has to prove that the resulting map is *local*. The advantage of this viewpoint is that one separates the question of definition and existence of the quantum BV operator from the particular technical results one needs to establish its locality. The latter is crucial from the physical viewpoint, but mathematically, one could very well just go ahead with the non-local operator. To understand this idea better, it's good to first have a look at the classical case.

In the classical limit, the interacting classical BV operator should arise from a twist of the free one with the classical Møller map. Here we provide the direct proof that this is indeed the case.

Consider the theory with action $S = S_0 + V$. For simplicity, we will treat the interaction as a local compactly supported functional rather than a generlised Lagrangian. We will also omit the test function in S_0.

For S without non-trivial local symmetries, the inverse Møller map is defined in Dütsch and Fredenhagen (2003; see also Hawkins and Rejzner 2020):

$$r_{\lambda V}^{-1}(F)(\varphi) \doteq F(\mathtt{r}_{\lambda V}^{-1}(\varphi)), \tag{5.17}$$

where

$$\mathtt{r}_{\lambda V}^{-1}(\varphi) = \varphi + \lambda \Delta^{\mathrm{R}} V^{(1)}(\varphi). \tag{5.18}$$

and Δ^{R} is the retarded Green function of the free theory. It can then be inverted as a formal power series to obtain the classical Møller map $r_{\lambda V}$, whereupon

$$\mathtt{r}_{\lambda V}(\varphi) = \varphi - \lambda \Delta_{S_0}^{\mathrm{R}} V^{(1)}(\mathtt{r}_{\lambda V}(\varphi)), \tag{5.19}$$

which is the Yang–Feldmann equation.

In this way of defining things, $r_{\lambda V}$ goes from the interacting to free theory and the image of $r_{\lambda V}$ represents the interacting fields constructed from the free ones. We also have the intertwining relation:

$$\lfloor r_{\lambda V} F, r_{\lambda V} G \rfloor = r_{\lambda V} \lfloor F, G \rfloor_V,$$

where $\lfloor .,. \rfloor$ and $\lfloor .,. \rfloor_V$ are the free and the interacting Poisson bracket respectively. Moreover, it is easily seen that $r_{\lambda V}^{-1}$ maps the ideal generated by the free equations of motion to the ideal generated by the interacting equations of motion, i.e.:

$$r_{\lambda V}^{-1} \frac{\delta S_0}{\delta \varphi} = r_{\lambda V}^{-1}(P\varphi) = P\varphi + \lambda P \circ \Delta^{\mathrm{R}} V^{(1)}(\varphi) = P\varphi + \lambda V^{(1)}(\varphi).$$

It is therefore compatible with taking the quotients by the both ideals.

In the BV-extended version, we set $\mathtt{r}_{\lambda V}$ to act trivial on antifields and the result above about intertwining the ideals implies that

$$r_{\lambda V}^{-1} \circ \delta_0 = \delta \circ r_{\lambda V}^{-1}. \tag{5.20}$$

Remark 5.2.14 Here some caution is required. Since $r_{\lambda V}$ is a non-local map, the statement above *does not imply* that the local cohomologies of δ and δ_0 are the same! When restricted to local functionals, δ_0 and δ yield different cohomologies, as one would expect. In the literature, one always computes the local cohomologies of δ_0 and δ, so the relation (5.20) between these two operators has been apparently overlooked and might seem rather surprising on first sight.

We now move on to the more complicated case, where gauge symmetries are present. First of all, now S_0 has two terms, one of which, S_{00}, does not depend on the antifields, and this is the term that defines P (and hence Δ^{R}).

The formula for the Møller operator is the same as in the scalar case (5.17), but we need to replace (5.18) with

$$\mathtt{r}_V^{-1}(\varphi^\alpha) = \varphi^\alpha + (\Delta^{\mathrm{R}})^{\alpha\beta} \frac{\delta_l V}{\delta \varphi^\beta}(\varphi),$$

and (5.19) with

$$\mathtt{r}_V(\varphi^\alpha) = \varphi^\alpha - (\Delta^{\mathrm{R}})^{\alpha\beta} \frac{\delta_l V}{\delta \varphi^\beta}(\mathtt{r}_V(\varphi)).$$

Theorem 5.2.15 *Let $X \in \mathcal{BV}$ and assume that S_0 satisfies the classical master equation. Then*

$$r_V^{-1}(\{X, S_0\}) = \{r_V^{-1}(X), S_0 + V\} - \int \frac{\delta_r X}{\delta\varphi^\gamma(y)}(r_V^{-1}(\varphi))\Delta^R(y,z)^{\gamma\beta}\frac{\delta_l}{\delta\varphi^\beta(z)}(\mathrm{cme}(S)),$$

where cme(S) *is the classical master equation for the full theory.*

Corollary 5.2.16 *From Theorem 5.2.15 if follows that, assuming* cme, *we can write the classical BV operator of the full theory as*

$$s = r_V^{-1} \circ s_0 \circ r_V,$$

which is the classical analog of the definition of the quantum BV operator proposed in Fredenhagen and Rejzner (2012a).

The proof of the theorem is rather technical, so we present it in the Appendix.

5.3 Quantisation

5.3.1 Free Theory

The quantised algebra of free fields is constructed by means of *deformation quantisation* of the classical algebra $(\mathcal{BV}_{\mu c}, \lfloor .,. \rfloor)$. To this end, we equip the space of formal power series $\mathcal{BV}_{\mu c}[[\hbar]]$ with a noncommutative star product which corresponds to the operator product of quantum observables.

Define the $\star$-product (deformation of the pointwise product) by

$$F \star G \doteq m \circ \exp(i\hbar D_W)(F \otimes G),$$

where m is the multiplication operator, i.e. $m(F \otimes G)(\varphi) = F(\varphi)G(\varphi)$, and D_W is the functional differential operator defined by

$$D_W \doteq \frac{1}{2}\sum_{\alpha,\beta}\left\langle W^{\alpha\beta}, \frac{\delta^l}{\delta\varphi^\alpha} \otimes \frac{\delta^r}{\delta\varphi^\beta}\right\rangle.$$

with W, the *2-point function of a Hadamard state*. W is positive definite and satisfies the appropriate wavefront set condition (Radzikowski 1996) and we have $W = \frac{i}{2}\Delta + H$, where H is a symmetric bisolution for P. In addition to these standard properties, we also need to require the *consistency condition* (Hollands 2008) on the symmetric part:

$$\sum_\sigma((-1)^{|\varphi^\alpha|}K^\alpha_\sigma(x')H(x',x)^{\sigma\gamma} + K^\gamma_\sigma(x)H(x',x)^{\alpha\sigma}) = 0, \tag{5.21}$$

Note that this is automatically fulfilled for Δ (see (5.13)). Under this condition, γ_0 is a right derivation with respect to the star product. Since W is a solution for the linearised equations of motion operator P, δ_0 is also a right derivation with respect to $\star$. We can therefore conclude that

$$s_0(X \star Y) = (-1)^{\#\mathrm{gh}(Y)} s_0 X \star Y + X \star s_0 Y.$$

5.3.2 Interacting Theory

5.3.2.1 Time-Ordered Products

For simplicity, we start our discussion by considering *regular functionals* (functionals whose derivatives at every point are smooth, compactly supported functions) on $T^*[-1]\overline{\mathcal{E}}$. We use the notation $\mathcal{BV}_{\text{reg}}$.

The *time-ordering operator* $\mathcal{T}$ is defined as:

$$\mathcal{T}F(\varphi) \doteq e^{\frac{\hbar}{2}\mathcal{D}_{\Delta^{\text{F}}}},$$

where, for an integral kernel K, we define

$$\mathcal{D}_K \doteq \sum_{\alpha,\beta}\left\langle K^{\alpha\beta}, \frac{\delta^l}{\delta\varphi^\alpha}\frac{\delta^r}{\varphi^\beta}\right\rangle$$

and $\Delta^{\text{F}} = \frac{i}{2}(\Delta^{\text{A}} + \Delta^{\text{R}}) + H$.

Formally, $\mathcal{T}$ corresponds to the operator of convolution with the oscillating Gaussian measure 'with covariance $i\hbar\Delta^{\text{F}}$',

$$\mathcal{T}F(\varphi) \overset{\text{formal}}{=} \int F(\varphi - \phi)\, d\mu_{i\hbar\Delta_F}(\phi).$$

We define the *time-ordered product* $\cdot_{\mathcal{T}}$ on $\mathcal{BV}_{\text{reg}}[[\hbar]]$ by:

$$F \cdot_{\mathcal{T}} G \doteq \mathcal{T}(\mathcal{T}^{-1}F \cdot \mathcal{T}^{-1}G)$$

Remark 5.3.1 Note that $\cdot_{\mathcal{T}}$ is the time-ordered version of $\star$, in the sense that $F \cdot_{\mathcal{T}} G = F \star G$ if the support of F is later than the support of G and $F \cdot_{\mathcal{T}} G = G \star F$, if the support of G is later than the support of F.

5.3.2.2 Peierls Bracket from the Antibracket

Before I continue with the interacting theory, I would like to address one more issue, often omitted in the literature: the precise relation between the antibracket (5.10) and the Peierls bracket (5.14). The example of the scalar field has been discussed in Gwilliam and Rejzner (2020). Here I give the general statement.

First, I need to introduce one more key concept from the AQFT axiomatic framework, which is yet another way to describe the dynamics of the theory. In AQFT, a QFT model is specified by assigning algebras of observables $\mathfrak{A}(\mathcal{O})$ to relatively compact regions $\mathcal{O} \subset \mathcal{M}$ of a given spacetime. In the original work of Haag and Kastler (1964) these algebras were assumed to be C^*-algebra, but to allow for the use of perturbative methods and homological algebra, one has to weaken this assumption. In Gwilliam and Rejzner (2020), together with Gwilliam, we use instead chain complexes in associative, unital $*$-algebras. Being unital means that they have a unit (one can think of it as the identity operator) and $*$ is an involution, the abstract notion of taking the adjoint of an operator. For a more general formulation using homotopical algebra, see Benini et al. (2019).

Example 5.3.2 As an example, consider $\mathfrak{A}(\mathcal{O}) = (\mathcal{BV}(\mathcal{O})[[\hbar]], \star)$, as defined in the previous section, where $\mathcal{BV}(\mathcal{O})$ is obtained by restricting to functionals supported inside a relatively compact $\mathcal{O} \subset M$. It is clearly an associative algebra (the product is $\star$), the unit is the constant functional 1 and the involution is complex conjugation. It is also a chain complex, where the differential is s_0.

Let $\mathcal{O} \mapsto \mathfrak{A}(\mathcal{O})$, with differential d and product $\star$, be an assignment of such chain complexes in algebras to regions. There are two important axioms to impose here:

- **Einstein causality:** for $\mathcal{O}_1, \mathcal{O}_2$ that are spacelike to each other, the commutar $[\mathfrak{A}(\mathcal{O}_1), \mathfrak{A}(\mathcal{O}_2)] = dX$ for some $X \in \mathfrak{A}(\mathcal{O})$ for any $\mathcal{O}$ that contains both $\mathcal{O}$.
- **Time-slice axiom:** for any $\mathcal{N}$, a neighbourhood of a Cauchy surface[3] in the region $\mathcal{O} \subset M$, the map $\mathfrak{A}(\mathcal{N})$ and $\mathfrak{A}(\mathcal{O})$ are quasi-isomorphic, i.e. isomorphic on the level of cohomology groups.

The first axiom is the weak version of causality. The second is the quantum analog of well-posedness of the Cauchy problem.

Recall that $\star$ arises from the deformation of the Peierls bracket $\lfloor .,. \rfloor$ (5.14), so we will start in the quantum theory. Assume we have a theory $\mathfrak{A}$ with the product $\star$ and the differential d that obeys the time-slice axiom and we have a time-ordered product $\cdot_{\mathcal{T}}$ associated with $\star$. Take $F, G \in \mathfrak{A}(\mathcal{O})$ and consider Cauchy surfaces to the future and to the past of $\mathcal{O}$, denoted $\mathcal{N}_+$ and $\mathcal{N}_-$. The time-slice axiom implies that there exist maps β_- and β_+ such that $\beta_+(F)$ is localised in the future of $\mathcal{O}$ and $\beta_-(F)$ in the past. Using the time-slice axiom, modulo the image of d, we can write the $\star$-commutator of F and G as

$$i\hbar[G,F]_\star = G\star F - F\star G = G\star\beta_+(F) - \beta_-(F)\star G = G\cdot_{\mathcal{T}}\beta_+(F) - \beta_-(F)\cdot_{\mathcal{T}} G \mod \mathrm{Im}d.$$

From the time-slice axiom follows also that there exists Ψ such that $\beta_- F - \beta_+ F = s_0\Psi$. Hence, we rewrite the $\star$ commutator as

$$[G,F]_\star = G\star\beta_+(F) - \beta_-(F)\star G = G\cdot_{\mathcal{T}}\beta_+(F) - \beta_-(F)\cdot_{\mathcal{T}} G = G\cdot_{\mathcal{T}}(\beta_- F - \beta_+ F) = G\cdot_{\mathcal{T}} d\Psi,$$

for some Ψ. Therefore, we can express the Peierls bracket as

$$i\hbar\, \lfloor G,F \rfloor = G \cdot_{\mathcal{T}} s_0\Psi \mod \hbar^2, \mathrm{Im}d.$$

Assume that $dG = 0$. We can re-write the right-hand side using the antibracket as follows:

$$i\hbar\, \lfloor G,F \rfloor = s_0(G \cdot_{\mathcal{T}} \Psi) + i\hbar\{G,\Psi\} \mod \hbar^2, \mathrm{Im}d.$$

Hence,

$$\lfloor G,F \rfloor = \{G,\Psi\} \mod \hbar, \mathrm{Im},$$

which can be thought of as the *intrinsic definition of the Peierls bracket, given the antibracket and the time-ordered product in a theory satisfying time-slice axiom.*

5.3.2.3 Interaction

We model interactions as functionals V and for the moment assume $V \in \mathcal{BV}_{\mathrm{reg}}$. We define the quantum observable (of the free theory), associated with V, as $\mathcal{T}V$. In the language of deformation quantisation, we can say that we use $\mathcal{T}$ as the *quantisation map*. By analogy to normal ordering, we use the notation $\mathcal{T}V \equiv :V:$.

[3] A hypersurface in M such that every inextendible causal curve intersects it exactly once. Cauchy surfaces are used for formulating initial-value problems for normally hyperbolic operators, e.g. the wave operator.

We define the *formal S-matrix*, $\mathcal{S}(\lambda :V:) \in \mathcal{BV}_{\text{reg}}((\hbar))[[\lambda]]$ by

$$\mathcal{S}(\lambda V) \doteq e_{\mathcal{T}}^{i\lambda :V:/\hbar} = \mathcal{T}(e^{i\lambda V/\hbar}).$$

Interacting fields are elements of $\mathcal{BV}_{\text{reg}}[[\hbar, \lambda]]$ given by

$$R_{\lambda V}(F) \doteq \left(e_{\mathcal{T}}^{i\lambda :V:/\hbar}\right)^{\star -1} \star (e_{\mathcal{T}}^{i\lambda :V:/\hbar} \cdot_{\mathcal{T}} :F:) = -i\hbar \frac{d}{d\mu} \mathcal{S}(\lambda V)^{-1} \mathcal{S}(\lambda V + \mu F)\big|_{\mu=0}.$$

For $\lambda = 0$, we recover $R_0(F) = :F:$. We define the *interacting star product* as:

$$F \star_{int} G \doteq R_V^{-1}\left(R_V(F) \star R_V(G)\right).$$

5.3.2.4 Renormalisation Problem

The problem that one faces is that interesting interactions and observables are local, but not regular. In fact, polynomial local functionals of order greater than one 2 cannot be regular, as illustrated in the next example.

Example 5.3.3 Consider the free scalar field and the functional

$$F(\varphi) = \int f(x, y)\varphi(x)\varphi(y) d\mu(x) d\mu(y), \quad f \in \mathcal{D}(M^2) := \mathcal{C}_c^\infty(M^2, \mathbb{R}),$$

which is regular. Now contrast it with

$$F(\varphi) = \int f\varphi^2 d\mu = \int f(x)\delta(x - y)\varphi(x)\varphi(y) d\mu(x) d\mu(y),$$

which is local but fails to be regular, since the second derivative is:

$$F^{(2)}(\varphi)(x, y) = f(x)\delta(x - y) d\mu(x) d\mu(y),$$

i.e. it is not smooth.

Because of singularities of Δ^{F}, the time-ordered product $\cdot_{\mathcal{T}}$ is not well defined on local, non-linear functionals, but the *physical interactions are usually local*!

The *renormalisation problem* is then to extend $\mathcal{S}$ to local arguments by extending time-ordered products:

$$\mathcal{S}(V) = \sum_{n=0}^{\infty} \frac{1}{n!} \mathcal{T}_n(V, \ldots, V).$$

We note that the time-ordered product $\mathcal{T}_n(F_1, \ldots, F_n) \doteq F_1 \cdot_{\mathcal{T}} \ldots \cdot_{\mathcal{T}} F_n$ of n local functionals is well defined if their supports are pairwise disjoint. To extend $\mathcal{T}_n$ to arbitrary local functionals, we use the causal approach of Epstein and Glaser (causal perturbation theory). The crucial property one uses in this process is the *causal factorisation property*: if the supports of $F_1 \ldots F_k$ are later than the supports of $F_{k+1}, \ldots F_n$, then

$$\mathcal{T}_n(F_1 \otimes \cdots \otimes F_n) = \mathcal{T}_k(F_1 \otimes \cdots \otimes F_k) \star \mathcal{T}_{n-k}(F_{k+1} \otimes \cdots \otimes F_n), \tag{5.22}$$

5.3.3 QME and the Quantum BV Operator

In the framework of Fredenhagen and Rejzner (2012a), an important role is played by the condition that the S-matrix is invariant under the free classical BV operator:

$$s_0\left(e_{\mathcal{T}}^{i:V:/\hbar}\right) = 0, \tag{5.23}$$

There is a very useful identity satisfied by $\mathcal{T}$, namely:

$$\delta_0(\mathcal{T}F) = \mathcal{T}(\delta_0 F - i\hbar \bigtriangleup F), \tag{5.24}$$

where $\bigtriangleup$ is the *BV Laplacian*, defined by:

$$\bigtriangleup X = (-1)^{(1+\#\mathrm{gh}(X))} \sum_\alpha \int dx \frac{\delta_r^2 X}{\delta\varphi^\alpha(x)\delta\varphi_\alpha^\ddagger(x)}. \tag{5.25}$$

Moreover, from the consistency conditions of (5.21), it follows that

$$\mathcal{T} \circ \gamma_0 = \gamma_0 \circ \mathcal{T}. \tag{5.26}$$

Putting these two together, we note that the left-hand side of (5.23) can be rewritten as:

$$s_0\left(e_{\mathcal{T}}^{i:V:/\hbar}\right) = \mathcal{T}\left(s_0 e^{iV/\hbar} - i\hbar \bigtriangleup e^{iV/\hbar}\right) = \mathcal{T}\left(e^{iV/\hbar}\left(\frac{i}{\hbar}\{V, S_0\} + \frac{i}{2\hbar}\{V, V\} + \bigtriangleup(V)\right)\right).$$

Setting $\bigtriangleup S_0 = 0$ (for symmetry reasons) and using the classical master equation, we can conclude that

$$s_0\left(e_{\mathcal{T}}^{i:V:/\hbar}\right) = \frac{i}{\hbar} e_{\mathcal{T}}^{i:V:/\hbar} \cdot_{\mathcal{T}} \mathcal{T}\left(\frac{1}{2}\{S_0 + V, S_0 + V\} - i\hbar \bigtriangleup (S_0 + V)\right)$$

and we observe that the condition (5.23) is in fact equivalent to the *quantum master equation* (QME):

$$\frac{1}{2}\{S_0 + V, S_0 + V\} = i\hbar \bigtriangleup (S_0 + V),$$

understood as a condition on V, which turns out to be important for the locality of the *quantum BV operator*. In the free theory, we define it as follows:

$$\hat{s}_0 \doteq \mathcal{T}^{-1} \circ s_0 \circ \mathcal{T}, \tag{5.27}$$

so from (5.24) and (5.26), it follows that

$$\hat{s}_0 = s_0 - i\hbar \bigtriangleup.$$

In the interacting theory, the quantum BV operator $\hat{s}$ is defined on regular functionals by:

$$\hat{s} = R_V^{-1} \circ s_0 \circ R_V,$$

so it is the twist of the free classical BV operator by the *(non-local!)* map that intertwines the free and the interacting theory. The classical limit of this definition makes sense, as demonstrated in Theorem 5.2.15.

The 0th cohomology of $\hat{s}$ characterises *quantum gauge invariant observables*. Assuming QME,

$$\hat{s}F = e_{\mathcal{T}}^{-i:V:/\hbar} \cdot_{\mathcal{T}} s_0\left(e_{\mathcal{T}}^{i:V:/\hbar} \cdot_{\mathcal{T}} :F:\right) = \{F, S_0 + V\} - i\hbar \,\triangle(F) = s_0 - i\hbar \,\triangle(F).$$

The second equality is particularly striking, since it shows that $\hat{s}$ is *local*. In contrast to other frameworks, in our approach QME is not necessary for the nilpotency of $\hat{s}$ (this is true by definition), but is crucial for its *locality*.

5.3.4 Renormalised QME and Quantum BV Operator

To extend QME and $\hat{s}$ to local observables, I now replace $\cdot_{\mathcal{T}}$ with the renormalised time-ordered product.

Theorem 5.3.4 (Fredenhagen and Rejzner 2013) *The renormalised time-ordered product $\cdot_{\mathcal{T}_r}$ is an associative product on $\mathcal{T}_r(\mathcal{F})$ given by*

$$F \cdot_{\mathcal{T}_r} G \doteq \mathcal{T}_r(\mathcal{T}_r{}^{-1}F \cdot \mathcal{T}_r{}^{-1}G),$$

where $\mathcal{T}_r : \mathcal{F}[[\hbar]] \to \mathcal{T}_r(\mathcal{F})[[\hbar]]$ is defined as

$$\mathcal{T}_r = (\oplus_n \mathcal{T}_r{}^n) \circ \beta,$$

where $\beta : \mathcal{T}_r : \mathcal{F} \to S^\bullet\mathcal{F}_{\rm loc}^{(0)}$ is the inverse of multiplication m and we set $\mathcal{T}_r|_{\mathcal{F}_{\rm loc}} = \mathrm{id}$ (so $:V: = V$).

Since $\cdot_{\mathcal{T}_r}$ is an associative, commutative product, we can use it in place of $\cdot_{\mathcal{T}}$ and define the renormalised QME and the quantum BV operator using formulas (5.23) and (5.27). These formulas get even simpler if we use the *anomalous Master Ward Identity* (Brennecke and Dütsch 2008; Hollands 2008):

$$s_0(e_{\mathcal{T}_r}^{i:V:/\hbar}) \equiv \left\{e_{\mathcal{T}_r}^{iV/\hbar}, S_0\right\} = \frac{i}{\hbar}e_{\mathcal{T}_r}^{iV/\hbar} \cdot_{\mathcal{T}_r} \left(\tfrac{1}{2}\{V + S_0, V + S_0\}_{\mathcal{T}_r} - i\hbar\Delta_V\right), \tag{5.28}$$

where Δ_V is identified with the *anomaly term*. If S_0 does not depend on antifields, Equation (5.28) reduces to:

$$\int \left(e_{\mathcal{T}_r}^{iV/\hbar} \cdot_{\mathcal{T}_r} \frac{\delta V}{\delta\varphi^\ddagger(x)}\right) \star \frac{\delta S_0}{\delta\varphi(x)} = e_{\mathcal{T}_r}^{iV/\hbar} \cdot_{\mathcal{T}_r} \left(\tfrac{1}{2}\{V + S_0, V + S_0\}_{\mathcal{T}_r} - i\hbar\Delta_V\right), \tag{5.29}$$

Remark 5.3.5 Note that for regular V, one has

$$\begin{aligned}
s_0\left(e_{\mathcal{T}}^{i:V:/\hbar}\right) &= \mathcal{T}(\hat{s}_0 e^{iV/\hbar}) = \mathcal{T}\left(s_0 e^{iV/\hbar} - i\hbar \,\triangle\, e^{iV/\hbar}\right) \\
&= \frac{i}{\hbar}e^{i:V:/\hbar} \cdot_{\mathcal{T}} \mathcal{T}\left(\{V, S_0\} + \frac{1}{2}\{V, V\} - i\hbar \,\triangle\, V\right) \\
&= \frac{i}{\hbar}e^{i:V:/\hbar} \cdot_{\mathcal{T}} \mathcal{T}\left(\frac{1}{2}\{S_0 + V, S_0 + V\} - i\hbar \,\triangle\, V\right),
\end{aligned}$$

so the MWI is the renormalised version of this identity.

The *renormalised quantum master equation* is therefore:

$$\tfrac{1}{2}\{V + S_0, V + S_0\} - i\hbar\Delta_V = 0.$$

Replacing V with $V + \lambda F$ in (5.28) and differentiating with respect to λ leads to the following identity for the classical BV operator:

$$\begin{aligned} s_0(e_{\mathcal{T}_r}^{i:V:/\hbar} \cdot_{\mathcal{T}_r} F) \\ = \frac{i}{\hbar} e_{\mathcal{T}_r}^{iV/\hbar} \cdot_{\mathcal{T}} \mathcal{T}_r \left(\{F, V + S_0\}_{\mathcal{T}_r} - i\hbar\, \Delta_V\, F + \frac{i}{\hbar} F \left(\frac{1}{2}\{S_0 + V, S_0 + V\} - i\hbar\Delta_V \right) \right), \end{aligned}$$

where $\Delta_V(F) \doteq \frac{d}{d\lambda} \Delta_{V+\lambda F}\big|_{\lambda=0}$. Assuming that the renormalised QME holds, this reduces to:

$$s_0 \left(e_{\mathcal{T}_r}^{i:V:/\hbar} \cdot_{\mathcal{T}_r} F \right) = \frac{i}{\hbar} e_{\mathcal{T}_r}^{iV/\hbar} \cdot_{\mathcal{T}_r} \mathcal{T}_r \left(\{F, V + S_0\}_{\mathcal{T}_r} - i\hbar\, \Delta_V\, F \right),$$

so the renormalised BV operator takes the form:

$$\hat{s} F = \{F, V + S_0\} - i\hbar\, \Delta_V\, (F),$$

Hence, by using the renormalised time-ordered product $\cdot_{\mathcal{T}_r}$, we obtained in place of $\Delta(X)$ the interaction-dependent operator $\Delta_V(X)$ (the anomaly). It is of order $\mathcal{O}(\hbar)$ and local. In the renormalised theory, Δ_V is well-defined on local vector fields, in contrast to Δ.

5.4 Towards a Non-Perturbative Formulation

5.4.1 Local S-Matrices

Buchholz and Fredenhagen (2020) have shown that one can formulate interacting quantum theory of the scalar field in terms of local S-matrices $\mathcal{S}$, treated as a family of unitaries labelled by local functionals, generating a C^*-algebra. One then imposes relations that the S-matrices satisfy.

Let F_1, F_2 be local functionals, and let $F_1 \prec F_2$ denote the relation: supp F_1 is not to the future of supp F_2 (i.e. supp F_1 does not intersect $J^+(\text{supp}\, F_2)$). Local S-matrices are required to satisfy:

S1 Identity preserving: $\mathcal{S}(0) = \mathbb{1}$.
S2 Locality: $\mathcal{S}$ satisfies the Hammerstein property, i.e. $F_1 \prec F_2$ implies that

$$\mathcal{S}(F_1 + F + F_2) = \mathcal{S}(F_1 + F)\mathcal{S}(F)^{-1}\mathcal{S}(F + F_2),$$

where $F_1, F, F_2 \in \mathcal{F}_{\text{loc}}$.

Using time-ordered products and star products, one can construct a concrete realisation of local S-matrices. Following Buchholz and Fredenhagen (2020), I will denote by $\mathfrak{A}$ the group algebra over $\mathbb{C}$ of the free group generated by elements $\mathcal{S}(F)$, $F \in \mathcal{F}_{\text{loc}}$, modulo relations **S5.4.1** and **S5.4.1**. Additionally, for a fixed $L \in \mathscr{L}$ (this is interpreted as the Lagrangian of the theory), one defines $\mathfrak{A}_L$ by also quotienting by the following relation proposed by Buchholz and Fredenhagen (2020) that encodes the dynamics:

$$\mathcal{S}(F)\mathcal{S}(\delta L(\varphi)) = \mathcal{S}(F^{\varphi} + \delta L(\varphi)) = \mathcal{S}(\delta L(\varphi))\mathcal{S}(F). \qquad \textbf{(S3)}$$

where $F^{\varphi}(\psi) \doteq F(\varphi + \psi)$, $\varphi, \psi \in \mathcal{E}$ and δL is given in Def. 5.2.7.

Physically, (**S3**) is interpreted as the *Schwinger–Dyson equation* on the level of local S-matrices.

5.4.2 Schwinger-Dyson Equation from Translation Symmetry

In BV formalism in finite dimensions, the Schwinger–Dyson equation is the consequence of the translation invariance of the path integral measure. Using the formal S-matrix language, the condition **S3** should be an expression of symmetry under translations in $\mathcal{E}$.

First, note that the group $(\mathcal{E}_c, +)$ of compactly supported configurations acts on $\mathcal{E}$ by $\sigma_\psi(\varphi) = \varphi + \psi$, where $\varphi \in \mathcal{E}$, $\psi \in \mathcal{E}_c$. This induces the following map:

$$\alpha : \mathcal{E}_c \to \mathrm{Aut}(\mathscr{L})$$
$$\alpha_\psi(L)(f)[\varphi] \doteq L(f)[\sigma_\psi(\varphi)] - L(f)[\varphi],$$

or in shorthand notation,

$$\alpha_\psi(L) \doteq \sigma_\psi^* L - L.$$

Proposition 5.4.1 *The map α defined above is a 1-cocycle of $\mathcal{E}_c$ in* $(\mathrm{Aut}(\mathscr{L}), +)$, *where the addition is the pointwise addition inherited from $\mathscr{L}$.*

Proof We have

$$\begin{aligned}\alpha_{\psi+\chi}(L)[\varphi] &\doteq L[\sigma_{\psi+\chi}(\varphi)] - L[\varphi] = L[\varphi + \psi + \chi] - L[\varphi + \psi] + L[\varphi + \psi] - L[\varphi] \\ &= \alpha_\chi(\sigma_\psi^* L)[\varphi] + \alpha_\psi(L)[\varphi] = (\sigma_\psi^* \alpha_\chi)(L)[\varphi] + \alpha_\psi(L)[\varphi],\end{aligned}$$

where in the last equation, σ^* is a map form $\mathcal{E}$ to automorphisms of $\mathscr{L}$ given by

$$(\sigma_\psi^*(\beta))(L) \doteq \beta(\sigma_\psi^* L),$$

where $\beta \in \mathrm{Aut}(\mathscr{L})$. In a shorthand notation, we have

$$\alpha_{\psi+\chi} = \sigma_\psi^* \alpha_\chi + \alpha_\psi,$$

which is indeed the cocycle condition. The same argument applies if one switches χ with ψ, i.e.

$$\alpha_{\psi+\chi} = \sigma_\chi^* \alpha_\psi + \alpha_\chi.$$

□

I can now re-express definition 5.2.7 as

$$\delta L(\psi) \doteq \alpha_\psi L(f),$$

where $\psi \in \mathcal{D}$ and $f \equiv 1$ on supp ψ. Hence, the co-cycle α encodes the classical dynamics.

The condition (**S3**) can be re-written as

$$\mathcal{S}(F)\mathcal{S}(\alpha_\psi L(f)) = \mathcal{S}(\sigma_\psi^* F + \alpha_\psi L(f)).$$

where $\psi \in \mathcal{D}$ and $f \equiv 1$ on supp ψ. Let

$$\beta_\psi(F) := \delta L(\psi) + \sigma_\psi^* F$$

One can define maps $\hat{\alpha}^{r/\ell} : \mathcal{E}_c \to \mathrm{Aut}(\mathfrak{A})$ by fixing their action on the generators, namely:

$$\hat{\alpha}^r_\psi(\mathcal{S}(F)) \doteq \mathcal{S}(\beta_\psi(F))\mathcal{S}(\alpha_\psi L(f))^{-1} \quad \hat{\alpha}^\ell_\psi(\mathcal{S}(F)) \doteq \mathcal{S}(\alpha_\psi L(f))^{-1}\mathcal{S}(\beta_\psi(F)). \tag{5.30}$$

To simplify the notation, I will write $\hat{\alpha}^r_\psi$ simply as $\hat{\alpha}_\psi$ and use the superscript r only when distinction with $\hat{\alpha}^\ell_\psi$ has to be made.

Proposition 5.4.2 *$\hat{\alpha}$ defines an action of $\mathcal{E}_c$ on $\mathfrak{A}$, i.e. it is a group homomorphism from $\mathcal{E}_c$ to an abelian subgroup of* Aut($\mathfrak{A}$)*:*

$$\hat{\alpha}_{\psi+\chi} = \hat{\alpha}_\psi \circ \hat{\alpha}_\chi = \hat{\alpha}_\chi \circ \hat{\alpha}_\psi .$$

Proof We have

$$\begin{aligned}
\hat{\alpha}_{\psi+\chi}(\mathcal{S}(F)) &= \mathcal{S}(\alpha_{\psi+\chi}L(f) + \sigma^*_{\psi+\chi}F)\mathcal{S}(\alpha_{\psi+\chi}L(f))^{-1} \\
&= \mathcal{S}((\sigma^*_\psi\alpha_\chi + \alpha_\psi)L(f) + \sigma^*_{\psi+\chi}F)\mathcal{S}((\sigma^*_\psi\alpha_\chi + \alpha_\psi)L(f))^{-1} \\
&= \mathcal{S}((\sigma^*_\chi\alpha_\psi + \alpha_\chi)L(f) + \sigma^*_{\psi+\psi}F)\mathcal{S}((\sigma^*_\chi\alpha_\psi + \alpha_\chi)L(f))^{-1},
\end{aligned}$$

where $f \equiv 1$ on the support of $\psi + \chi$. On the other hand:

$$\hat{\alpha}_\psi \circ \hat{\alpha}_\chi(\mathcal{S}(F)) = \mathcal{S}(\alpha_\psi L(f') + \sigma^*_\psi\alpha_\chi L(f'') + \sigma^*_{\psi+\chi}F)\mathcal{S}(\alpha_\psi L(f') + \sigma^*_\psi(\alpha_\chi L(f'')))^{-1},$$

where $f' \equiv 1$ on supp ψ and $f'' \equiv 1$ on supp χ. Hence, taking f such that $f \equiv 1$ on supp $\chi \cup$ supp $\psi \supset$ supp$(\chi + \psi)$, we obtain:

$$\hat{\alpha}_{\psi+\chi} = \hat{\alpha}_\psi \circ \hat{\alpha}_\chi .$$

A similar argument works for $\hat{\alpha}_\chi \circ \hat{\alpha}_\psi$, hence

$$\hat{\alpha}_{\psi+\chi} = \hat{\alpha}_\psi \circ \hat{\alpha}_\chi = \hat{\alpha}_\chi \circ \hat{\alpha}_\psi .$$

□

Using the action $\hat{\alpha}$, one can express (**S3**) as

$$\hat{\alpha}^r_\psi(\mathcal{S}(F)) = \mathcal{S}(F) = \hat{\alpha}^\ell_\psi(\mathcal{S}(F)), \quad \forall\psi \in \mathcal{D}, \tag{5.31}$$

so imposing (**S3**) amounts to quotienting $\mathfrak{A}$ by the action of $\mathcal{E}_c$ (taking the space of the co-invariants), i.e. implementing *translational symmetry*. This makes sense, as (**S3**) is the finite version of the Schwinger–Dyson equation, which, formally, is the consequence of the *translation invariance* of the path integral.

5.4.3 Relation to the BV Perspective

Let's consider a theory with quadratic action L_0. Note that, in the absence of local symmetries other than the translation symmetry, the infinitesimal version of α corresponds to the action of the classical BV operator s_0 and the infinitesimal version of β is the interacting classical BV operator s, as demonstrated below.

In our explicit model with star product and time-ordered product, we have:

$$\begin{aligned}\frac{d}{dt}\hat{\alpha}_{t\psi}(\mathcal{S}(F))\big|_{t=0} &= \frac{d}{dt}(\mathcal{S}(\beta_\psi(F))\mathcal{S}(\alpha_\psi L(f))^{-1})\big|_{t=0}\\ &= -\tfrac{i}{\hbar}\mathcal{S}(F)\star\langle\psi,P\varphi\rangle + \frac{d}{dt}\mathcal{S}(\beta_\psi(F))\big|_{t=0}\\ &= \tfrac{i}{\hbar}\left(-\mathcal{S}(F)\star s_0(X_\psi)+\mathcal{S}(F)\cdot_{\mathcal{T}}\frac{d}{dt}(\beta_{t\psi}F)\big|_{t=0}\right),\end{aligned}$$

where X_ψ is a vector field defined by $X_\psi = \int\psi(x)\frac{\delta}{\delta\varphi(x)}$ and

$$\langle\psi,P\varphi\rangle \doteq \frac{d}{dt}\delta L_0(t\psi)[\varphi]\Big|_{t=0}$$

is the equation of motion term $P\varphi$ smeared with ψ. Assuming F in the kernel of s_0, we obtain

$$\mathcal{S}(F)\star s_0(X_\psi) = s_0(\mathcal{S}(F)\star X_\psi).$$

The right-hand side can also be written as $s_0(\mathcal{S}(F)\cdot_{\mathcal{T}} X_\psi)$, since X_ψ does not depend on fields.

The map β is the finite version of the classical interacting BV operator s, in the sense that

$$\frac{d}{dt}(\beta_{t\psi}F)\big|_{t=0} = \langle\psi,P\varphi\rangle + \partial_{X_\psi}F = s(X_\psi),$$

where s is the classical BV operator for the theory with the interaction S_0+F. Using the antibracket notation, we can write the above formula also as $\{X_\psi, S_0+F\}$.

Putting all these together, we obsere that the infinitesimal version of Eq. (5.31) is

$$\frac{d}{dt}\hat{\alpha}_{t\psi}(\mathcal{S}(F))\big|_{t=0} = -s_0(\mathcal{S}(F)\cdot_{\mathcal{T}} X_\psi)+\mathcal{S}(F)\cdot_{\mathcal{T}} s(X_\psi) = 0.$$

Comparing this with Equation (5.29), the infinitesimal version of MWI, we notice that in this case the term corresponding to the BV Laplacian $\triangle$ vanishes, since X_ψ does not depend on fields. We can therefore write the infinitesimal version of $\hat{\alpha}$ in a conceptually more appropriate form:

$$\frac{d}{dt}\hat{\alpha}_\psi(\mathcal{S}(F))\big|_{t=0} = \mathcal{S}(F)\cdot_{\mathcal{T}}\hat{s}(X_\psi) - s_0(\mathcal{S}(F)\cdot_{\mathcal{T}} X_\psi).$$

Remark 5.4.3 From the conceptual viewpoint, it seems that the Schwinger–Dyson equation relates the classical BV operator to the quantum one, so it relates two cohomology theories. It is not, by itself, a relation that corresponds to computing any of these cohomologies on their own. Compare this with the infinitesimal version MWI (Equation (5.29)), which plays the same role.

5.4.4 Action by Diffeomorphisms

The constructions of the previous sections should generalise from translations to arbitrary compactly supported diffeomorphisms of $\mathcal{E}$, and we restrict ourselves to the diffeomorphisms that arise as exponentiated local, compactly supported vector fields $X \in \mathcal{V}$. I denote this space by $\mathrm{Diff}_{\mathrm{loc}}$. First, we extend α to

$$\alpha : \mathrm{Diff}_{\mathrm{loc}}(\mathcal{E}) \to \mathrm{Aut}(\mathscr{L})$$

by setting

$$\alpha_g(L)(f)[\varphi] \doteq L(f)[g(\varphi)] - L(f)[\varphi].$$

Similarly, define

$$\beta_g(F) \equiv \alpha_g L(f) + \sigma_g^* F,$$

where $f \equiv 1$ on $\operatorname{supp} g$ and $\sigma_g^* F(\varphi) \doteq F(g(\varphi))$. More generally, we can consider the 'anomalous' version of the action β:

$$\beta_g(F) \doteq \alpha_g L(f) + \sigma_g^* F + \mathcal{A}_F(g), \tag{5.32}$$

where again $f \equiv 1$ and $\mathcal{A}_F(X)$ is the 'finite version' of the renormalised BV Laplacian (5.25), aka 'the anomaly term', meaning that

$$\frac{d}{dt}\mathcal{A}_F(e^{tX})\big|_{t=0} = \Delta_F(X).$$

In the path integral language, $\mathcal{A}_F(g)$ should be the logarithm of the Jacobian of the configuration space transformation $g \in G$. If no renormalisation was needed, it would not depend on F. The interpretation of $\mathcal{A}_F(X)$ as the log Jacobian is consistent with the infinitesimal formulation, since the derivative of the Jacobian of g gives the divergence of the vector field X, generating g.

In perturbation theory, the Master Ward Identity is the generalisation of the Schwinger–Dyson equation, where translations are replaced by general compactly supported diffeomorphisms. Similarly, the *unitary Master Ward Identity* should be an appropriate generalisation of the *unitary* Schwinger–Dyson equation (5.31). Note, however, that (5.31) cannot hold for general g (more general than the translations), if we keep the same definition of $\hat{\alpha}^{r/\ell}$. This is because $\mathcal{S}(\alpha_g L(f))$ is no longer central, so multiplying from the right cannot be the same as multiplying from the left in (5.31).

A proposal for the on-shell formulation of the unitary Master Ward Identity is presented in Brunetti et al. (2021), which includes a formulation of the unitary quantum version of Noether's theorem. As for the off-shell version, analogous to the BV formalism, the following geometrical perspective might give a hint as to the correct formulation.

5.4.5 Geometrical Picture

5.4.5.1 Vector Fields

Let $\mathcal{E}^*$ be the space of smooth sections of the dual bundle, and let $\mathcal{E}^! \equiv \mathcal{E}^* \otimes \mathrm{Dens}$ (tensoring with densities). We can identify vector fields with functions on $\bar{\mathcal{E}} \equiv \mathcal{E} \times \mathcal{E}^! \subset T^*\mathcal{E}$ that are linear in the second argument. We will use the notation $X = \int X(x)\frac{\delta}{\delta\varphi(x)}$ and identify q with $\frac{\delta}{\delta\varphi}$ so that

$$X(\varphi, q) = \int X(x)[\varphi]q(x),$$

where $q \in \mathcal{E}^!$.

5.4.5.2 Products

Let $\bar{\mathcal{F}}$ denote the space of smooth functionals on $\bar{\mathcal{E}}$ and let $\bar{\mathcal{F}}_{\mathrm{loc}}$ be its subspace consisting of the local ones. The following products on $\bar{\mathcal{F}}$ will be used to encode how diffeomorphisms e^X are acting on functions on the base:

$$F \underline{*} G = m \circ e^{-i\Delta_\otimes}(F \otimes G),$$
$$F \overline{*} G = m \circ e^{i\Delta_\otimes^T}(F \otimes G),$$

where

$$\Delta_\otimes \doteq \int \frac{\delta}{\delta q(x)} \otimes \frac{\delta}{\delta\varphi(x)}.$$

The two products are related by complex conjugation in the sense that

$$\overline{F * G} = \bar{G} \overline{*} \bar{F}.$$

A useful identity involves the Jacobian of the diffeomorphism e^X. Denote

$$\Delta \doteq \int \frac{\delta^2}{\delta q(x)\delta\varphi(x)}.$$

Using this notation, $\Delta(X)$ is the divergence of the vector field $X \in \mathcal{V}$. The exponential of Δ also allows one to relate the products:

$$e^{i\Delta}(X \underline{*} Y) = (e^{i\Delta}X)\overline{*}(e^{i\Delta}Y).$$

Hence

$$e^{-i\Delta} \circ \exp_{\overline{*}}(iX) = \exp_{\underline{*}}(e^{-i\Delta}(iX)) = e_{\underline{*}}^{iX+\Delta X}.$$

We also have:

$$e_{\underline{*}}^{iX} \underline{*} F \underline{*} e_{\underline{*}}^{-iX} = \sigma_g^* F,$$

as well as:

$$e_{\underline{*}}^{iX+\Delta(X)} \underline{*} F \underline{*} e_{\underline{*}}^{-iX} = \mathrm{Jac}(g)\sigma_g^* F,$$

where $g = e^X$ is the diffeomorphism given by exponentiating X. These equations allow one to encode the action of symmetries on functionals using the above products on $\overline{\mathcal{F}}$.

Let L_0' and L_0'' denote the first and second derivatives of a quadratic Lagrangian L_0, in the sense that

$$\langle L_0', \psi\rangle = \frac{d}{dt}\delta L_0(t\psi)\big|_{t=0},$$

for $\psi \in \mathcal{E}_c$ and

$$\langle L_0'', \psi_1 \otimes \psi_2\rangle = \frac{d^2}{dtds}\delta L(t\psi_1 + s\psi_2)\big|_{t=s=0},$$

where $\psi_1, \psi_2 \in \mathcal{E}_c$, but due to locality, one of the arguments could be non-compactly supported, so L'' induces a differential operator $P\colon \mathcal{E} \to \mathcal{E}$.

Let $g = e^X$, one can then write

$$\alpha_g L = \left\langle L', \frac{\delta}{\delta q} e_{\underline{*}}^X\big|_{q=0}\right\rangle + \frac{1}{2}\left\langle L'', \frac{\delta^2}{\delta q^2} e_{\underline{*}}^X\big|_{q=0}\right\rangle = \left(L + \left\langle L', \frac{\delta}{\delta q}\right\rangle + \frac{1}{2}\left\langle L'', \frac{\delta^2}{\delta q^2}\right\rangle\right) e_{\underline{*}}^X\big|_{q=0}$$

Using the short-hand $\frac{\delta}{\delta q} \equiv \delta_q$ and defining

$$\delta L(\delta_q) \doteq \left\langle L', \frac{\delta}{\delta q}\right\rangle + \frac{1}{2}\left\langle L'', \frac{\delta^2}{\delta q^2}\right\rangle,$$

one obtains for $f \equiv 1$ on supp X:

$$\alpha_g L(f) = \delta L(\delta_q) e_{\underline{*}}^X\big|_{q=0}.$$

The exponential of $\delta L(\delta_q)$ can be thought of as the finite version of δ_0, and hence one can define a map,

$$\tilde{s}_0 = \mathcal{T}^{-1} \circ e^{-i\delta L(i\delta_q)} \circ \mathcal{T},$$

in analogy to Eq. (5.27) in the BV formalism. Using the Baker–Campbell–Hausdorff formula, one finds that

$$e^{-i\delta L(i\delta_q)} \circ \mathcal{T} = \mathcal{T} \circ e^{i\delta L(-i\delta_q)} \circ e^{-i\triangle},$$

which leads to the following heuristic proposal for the finite off-shell anomalous MWI:

$$e^{-i\delta L(i\delta_q)}\mathcal{T}(e_{\underline{*}}^{iX} \overline{*} e^{iF}) = \mathcal{T}\left(\mathrm{Jac}(e^X) e^{i(\sigma_g^* F + \alpha_g L(X)} e_{\underline{*}}^{iX}\right) = \mathcal{T}\left(e^{i(\sigma_g^* F + \alpha_g L(X) - i\mathcal{A}_F(g))} e_{\underline{*}}^{iX}\right), \tag{5.33}$$

which holds identically for regular functionals. Barring the IR problems we also have

$$e^{iL} \overline{*} \mathcal{T}\left(e_{\underline{*}}^{iX} e^{iF}\right) \overline{*} e^{-iL} \doteq e^{-i\delta L(i\delta_q)} \circ \mathcal{T}\left(e_{\underline{*}}^{iX} e^{iF}\right)$$

One can easily imagine that exponentials appearing in the above equation could be identified with some generalisation of S-matrices. However, as it stands, these are not unitary with respect to the star product present in this setting. A similar situation appears if one takes time-ordered exponentials of regular, but not local, functionals. Hence, it is likely that unitarity can be restored as one reverts to local functionals, but then renormalisation has to be performed. This issue and other questions posed in this review will be further addressed in my future works.

Appendix 5.A Proof of Theorem 5.2.15

The proof relies on two lemmas.

Lemma 5.A.1

$$r_V^{-1}(\{X, S_{00}\}) = \{r_V^{-1}X, S_{00} + V\} - \int \frac{\delta_r X}{\delta\varphi^\gamma(y)}(\mathrm{r}_V^{-1}(\varphi))\Delta^{\mathrm{R}}(y,z)^{\gamma\beta}\frac{\delta_l}{\delta\varphi^\beta(z)}\left(\frac{1}{2}\{S_{00} + V, S_{00} + V\}\right)$$

Proof By definition of $\{.,.\}$ and r_V^{-1}, we have:

$$\begin{aligned} r_V^{-1}(\{X, S_{00}\}) &= -\int \frac{\delta_r X}{\delta\varphi_\alpha^\ddagger(x)}(\mathrm{r}_V^{-1}(\varphi))\frac{\delta_l S_{00}}{\delta\varphi^\alpha(x)}(\mathrm{r}_V^{-1}(\varphi)) \\ &= -\int \frac{\delta_r X}{\delta\varphi_\alpha^\ddagger(x)}(\mathrm{r}_V^{-1}(\varphi))\left(\frac{\delta_l S_{00}}{\delta\varphi^\alpha(x)}(\varphi) + \frac{\delta_l V}{\delta\varphi^\alpha(x)}(\varphi)\right), \end{aligned} \tag{5.34}$$

where we used the fact that $\frac{\delta_l S_{00}}{\delta\varphi^\alpha(x)}(\varphi) = P_{\alpha\beta}(x)(\varphi^\beta(x))$, so

$$\frac{\delta_l S_{00}}{\delta\varphi^\alpha(x)}(\mathrm{r}_V^{-1}(\varphi))(\varphi) = \frac{\delta_l S_{00}}{\delta\varphi^\alpha(x)}(\varphi) + \frac{\delta_l V}{\delta\varphi^\alpha(x)}(\varphi),$$

as $P \circ \Delta^{\mathrm{R}}(x,y) = \delta(x-y)$. Next, we notice that applying the chain rule:

$$\frac{\delta_r}{\delta\varphi_\alpha^\ddagger(x)}(r_V^{-1}(X)) = \frac{\delta_r X}{\delta\varphi_\alpha^\ddagger(x)}(\mathrm{r}_V^{-1}(\varphi)) + \int \frac{\delta_r X}{\delta\varphi^\gamma(y)}(\mathrm{r}_V^{-1}(\varphi))\Delta^{\mathrm{R}}(y,z)\frac{\delta_l}{\delta\varphi^\beta(z)}\frac{\delta_r V}{\delta\varphi_\alpha^\ddagger(x)}(\varphi). \tag{5.35}$$

We solve the above equation for $\frac{\delta_r X}{\delta\varphi_\alpha^\ddagger(x)}(\mathrm{r}_V^{-1}(\varphi))$ and insert it into the first term of Equation (5.34) to obtain

$$\begin{aligned} &-\int \frac{\delta_r X}{\delta\varphi_\alpha^\ddagger(x)}(\mathrm{r}_V^{-1}(\varphi))\frac{\delta_l S_{00}}{\delta\varphi^\alpha(x)}(\varphi) \\ &= \int\left(-\frac{\delta_r}{\delta\varphi_\alpha^\ddagger(x)}(r_V^{-1}(X)) + \int \frac{\delta_r X}{\delta\varphi^\gamma(y)}(\mathrm{r}_V^{-1}(\varphi))\Delta^{\mathrm{R}}(y,z)\frac{\delta_l}{\delta\varphi_g^\beta(z)}\frac{\delta_r V}{\delta\varphi_\alpha^\ddagger(x)}(\varphi)\right)\frac{\delta_l S_{00}}{\delta\varphi^\alpha(x)}(\varphi) \end{aligned} \tag{5.36}$$

The first term in (5.36) is what we eventually want to get on the right-hand side of the identity we are trying to prove, and the second term can be rewritten using the Leibniz rule into:

$$\begin{aligned} &\int \frac{\delta_r X}{\delta\varphi^\gamma(y)}(\mathrm{r}_V^{-1}(\varphi))\Delta^{\mathrm{R}}(y,z)^{\gamma\beta}\left(\frac{\delta_l}{\delta\varphi^\beta(z)}\left(\frac{\delta_r V}{\delta\varphi_\alpha^\ddagger(x)}\frac{\delta_l S_{00}}{\delta\varphi^\alpha(x)}\right)\right. \\ &\quad\left. -(-1)^{|\beta|(|\alpha|+1)}\frac{\delta_r V}{\delta\varphi_\alpha^\ddagger(x)}\frac{\delta_l}{\delta\varphi^\beta(z)}\frac{\delta_l S_{00}}{\delta\varphi^\alpha(x)}\right). \end{aligned} \tag{5.37}$$

Now we use the fact that $\frac{\delta_l S_{00}}{\delta\varphi^\alpha(x)} = (-1)^{|\alpha|}\frac{\delta_r S_{00}}{\delta\varphi^\alpha(x)}$ and that composing Δ^{R} with P gives identity. This allows us to rewrite (5.37) as

$$\int \frac{\delta_r X}{\delta\varphi^\gamma(y)}(\mathtt{r}_V^{-1}(\varphi))\Delta^{\mathrm{R}}(y,z)^{\gamma\beta}\frac{\delta_l}{\delta\varphi^\beta(z)}\left(\frac{\delta_r V}{\delta\varphi^\ddagger_\alpha(x)}\frac{\delta_l S_{00}}{\delta\varphi^\alpha(x)}\right)$$
$$-(-1)^{|\alpha|}\int \frac{\delta_r X}{\delta\varphi^\alpha(x)}(\mathtt{r}_V^{-1}(\varphi))\frac{\delta_r V}{\delta\varphi^\ddagger_\alpha(x)} \tag{5.38}$$

There are two terms in (5.38): the first one is in the form we want, but we need to work a bit more on the second one. First, we notice that

$$-(-1)^{|\alpha|}\int \frac{\delta_r X}{\delta\varphi^\alpha(x)}(\mathtt{r}_V^{-1}(\varphi))\frac{\delta_r V}{\delta\varphi^\ddagger_\alpha(x)} = \int \frac{\delta_r X}{\delta\varphi^\alpha(x)}(\mathtt{r}_V^{-1}(\varphi))\frac{\delta_l V}{\delta\varphi^\ddagger_\alpha(x)} \tag{5.39}$$

and, using a calculation analogous to Equation (5.35), we also get

$$\frac{\delta_r X}{\delta\varphi^\alpha(x)}(\mathtt{r}_V^{-1}(\varphi)) = \frac{\delta_r}{\delta\varphi^\alpha(x)}(r_V^{-1}(X)) - \int \frac{\delta_r X}{\delta\varphi^\gamma(y)}(\mathtt{r}_V^{-1}(\varphi))\Delta^{\mathrm{R}}(y,z)\frac{\delta_l}{\delta\varphi^\beta(z)}\frac{\delta_r V}{\delta\varphi^\alpha(x)}(\varphi),$$

which we then insert on the right-hand side of Equation (5.39). Plugging the result back into Equation (5.38) and then into (5.36), we obtain the final expression:

$$\begin{aligned} &r_V^{-1}(\{X,S_{00}\})\\ &= \{r_V^{-1}(X), S_{00}+V\} + \int \frac{\delta_r X}{\delta\varphi^\gamma(y)}(\mathtt{r}_V^{-1}(\varphi))\Delta^{\mathrm{R}}(y,z)^{\gamma\beta}\frac{\delta_l}{\delta\varphi^\beta(z)}\left(-\{S_{00},V\} - \frac{1}{2}\{V,V\}\right)\\ &= \{r_V^{-1}(X), S_{00}+V\} - \int \frac{\delta_r X}{\delta\varphi^\gamma(y)}(\mathtt{r}_V^{-1}(\varphi))\Delta^{\mathrm{R}}(y,z)^{\gamma\beta}\frac{\delta_l}{\delta\varphi^\beta(z)}\left(\frac{1}{2}\{S_{00}+V, S_{00}+V\}\right). \end{aligned}$$

□

Lemma 5.A.2

$$r_V^{-1}(\{X,\theta_0\}) = \{r_V^{-1}X,\theta_0\} - \int \frac{\delta_r X}{\delta\varphi^\gamma(y)}(\mathtt{r}_V^{-1}(\varphi))\Delta^{\mathrm{R}}(y,z)^{\gamma\beta}\frac{\delta_l}{\delta\varphi^\beta(z)}\left(\{V,\theta_0\}_g\right)$$

Proof We have

$$r_V^{-1}(\{X,\theta_0\}) = \int \frac{\delta_r X}{\delta\varphi^\alpha(x)}(\mathtt{r}_V^{-1}(\varphi))\frac{\delta_l\theta_0}{\delta\varphi^\ddagger_\alpha(x)}(\mathtt{r}_V^{-1}(\varphi)) - \int \frac{\delta_r X}{\delta\varphi^\ddagger_\alpha(x)}(\mathtt{r}_V^{-1}(\varphi))\frac{\delta_l\theta_0}{\delta\varphi^\alpha(x)}(\mathtt{r}_V^{-1}(\varphi)). \tag{5.40}$$

Note that $\frac{\delta\theta_0}{\delta\varphi^\alpha(x)}(\mathtt{r}_V^{-1}(\varphi)) = \frac{\delta\theta_0}{\delta\varphi^\alpha(x)}(\varphi)$, as θ_0 is linear in fields and that

$$\begin{aligned} \frac{\delta_l\theta_0}{\delta\varphi^\ddagger_\alpha(x)}(\mathtt{r}_V^{-1}(\varphi)). &= K^\alpha_\sigma\left(\varphi^\sigma(x) + \int_M \Delta^{\mathrm{R}}(x,y)^{\sigma\beta}\frac{\delta_l V}{\delta\varphi^\beta_g(y)}(\varphi)\right)\\ &= K^\alpha_\sigma\varphi^\sigma(x) + \int_M K^\alpha_\sigma(x)\Delta^{\mathrm{R}}(x,y)^{\sigma\beta}\frac{\delta_l V}{\delta\varphi^\beta_g(y)}(\varphi) \end{aligned}$$

So the first term in Equation (5.40) becomes:

$$\int \frac{\delta_r X}{\delta\varphi^\alpha(x)}(\mathtt{r}_V^{-1}(\varphi))\frac{\delta_l\theta_0}{\delta\varphi^\ddagger_\alpha(x)}(\mathtt{r}_V^{-1}(\varphi))$$
$$= \int \frac{\delta_r X}{\delta\varphi^\alpha(x)}(\mathtt{r}_V^{-1}(\varphi))K^\alpha_{\ \sigma}\varphi^\sigma(x) + \int \frac{\delta_r X}{\delta\varphi^\alpha(x)}(\mathtt{r}_V^{-1}(\varphi))K^\alpha_{\ \sigma}(x)\Delta^{\mathrm{R}}(x,y)^{\sigma\beta}\frac{\delta_l V}{\delta\varphi^\beta_g(y)}(\varphi). \tag{5.41}$$

Now we use the fact that

$$\frac{\delta_r X}{\delta\varphi^\alpha(x)}(\mathtt{r}_V^{-1}(\varphi)) = \frac{\delta_r}{\delta\varphi^\alpha(x)}(r_V^{-1}(X)) - \int \frac{\delta_r X}{\delta\varphi^\gamma(y)}(\mathtt{r}_V^{-1}(\varphi))\Delta^{\mathrm{R}}(y,z)\frac{\delta_l}{\delta\varphi^\beta(z)}\frac{\delta_r V}{\delta\varphi^\alpha(x)}(\varphi)$$

to rewrite the first term in Equation (5.41) as

$$\int \frac{\delta_r X}{\delta\varphi^\alpha(x)}(\mathtt{r}_V^{-1}(\varphi))K^\alpha_{\ \sigma}\varphi^\sigma(x) = \int \frac{\delta_r}{\delta\varphi^\alpha(x)}(r_V^{-1}(X))K^\alpha_{\ \sigma}\varphi^\sigma(x)$$
$$-\int \frac{\delta_r X}{\delta\varphi^\gamma(y)}(\mathtt{r}_V^{-1}(\varphi))\Delta^{\mathrm{R}}(y,z)^{\gamma\beta}\frac{\delta_l}{\delta\varphi^\beta(z)}\frac{\delta_r V}{\delta\varphi^\alpha(x)}(\varphi)K^\alpha_{\ \sigma}\varphi^\sigma(x), \tag{5.42}$$

The first term in (5.42) is what we want, and we rewrite the second term, using the Leibniz rule, as:

$$-\int \frac{\delta_r X}{\delta\varphi^\gamma(y)}(\mathtt{r}_V^{-1}(\varphi))\Delta^{\mathrm{R}}(y,z)^{\gamma\beta}\frac{\delta_l}{\delta\varphi^\beta(z)}\left(\frac{\delta_r V}{\delta\varphi^\alpha(x)}(\varphi)\frac{\delta_l\theta_0}{\delta\varphi^\ddagger_\alpha(x)}(\varphi)\right)$$
$$+\int \frac{\delta_r X}{\delta\varphi^\gamma(y)}(\mathtt{r}_V^{-1}(\varphi))\Delta^{\mathrm{R}}(y,z)^{\gamma\beta}\frac{\delta_r V}{\delta\varphi^\alpha(x)}(\varphi)\frac{\delta_l}{\delta\varphi^\beta(z)}\frac{\delta_l\theta_0}{\delta\varphi^\ddagger_\alpha(x)}(\varphi). \tag{5.43}$$

Now we put (5.43) back into (5.42), use the fact that $\frac{\delta_l}{\delta\varphi^\beta(z)}\frac{\delta_l\theta_0}{\delta\varphi^\ddagger_\alpha(x)}(\varphi) = K^\alpha_{\ \beta}(x)\delta(x-z)$ and insert the resulting form of (5.42) into (5.41) to obtain:

$$\int \frac{\delta_r X}{\delta\varphi^\alpha(x)}(\mathtt{r}_V^{-1}(\varphi))\frac{\delta_l\theta_0}{\delta\varphi^\ddagger_\alpha(x)}(\mathtt{r}_V^{-1}(\varphi))$$
$$= \int \frac{\delta_r}{\delta\varphi^\alpha(x)}(r_V^{-1}(X))K^\alpha_{\ \sigma}\varphi^\sigma(x)$$
$$-\int \frac{\delta_r X}{\delta\varphi^\gamma(y)}(\mathtt{r}_V^{-1}(\varphi))\Delta^{\mathrm{R}}(y,z)\frac{\delta_l}{\delta\varphi^\beta(z)}\left(\frac{\delta_r V}{\delta\varphi^\alpha(x)}(\varphi)\frac{\delta_l\theta_0}{\delta\varphi^\ddagger_\alpha(x)}(\varphi)\right)$$
$$-\int \frac{\delta_r X}{\delta\varphi^\gamma(y)}(\mathtt{r}_V^{-1}(\varphi))K^\alpha_{\ \beta}(z)\Delta^{\mathrm{R}}(y,z)^{\gamma\beta}\frac{\delta_r V}{\delta\varphi^\alpha(x)}(\varphi)$$
$$+\int \frac{\delta_r X}{\delta\varphi^\alpha(x)}(\mathtt{r}_V^{-1}(\varphi))K^\alpha_{\ \sigma}(x)\Delta^{\mathrm{R}}(x,y)^{\sigma\beta}\frac{\delta_l V}{\delta\varphi^\beta_g(y)}(\varphi). \tag{5.44}$$

The last two terms in Equation (5.44) combine to give

$$\int \frac{\delta_r X}{\delta\varphi^\gamma(y)}(\mathtt{r}_V^{-1}(\varphi))\left(-K^\alpha_{\ \beta}(z)\Delta^{\mathrm{R}}(y,z)^{\gamma\beta} + (-1)^{|\alpha|}K^\gamma_{\ \beta}(y)\Delta^{\mathrm{R}}(y,z)^{\beta\alpha}\right)\frac{\delta_r V}{\delta\varphi^\alpha(z)}(\varphi),$$

which vanishes due to (5.13). Next, using (5.35), we compute the second term in (5.40), namely:

$$
\begin{aligned}
&-\int \frac{\delta_r X}{\delta\varphi^{\ddagger}_{\alpha}(x)}(\mathtt{r}_V^{-1}(\varphi))\frac{\delta_l\theta_0}{\delta\varphi^{\alpha}(x)}(\mathtt{r}_V^{-1}(\varphi))\\
&=\int\left(-\frac{\delta_r}{\delta\varphi^{\ddagger}_{\alpha}(x)}(r_V^{-1}(X))+\frac{\delta_r X}{\delta\varphi^{\gamma}(y)}(\mathtt{r}_V^{-1}(\varphi))\Delta^{\mathrm{R}}(y,z)\frac{\delta_l}{\delta\varphi^{\beta}_{g}(z)}\frac{\delta_r V}{\delta\varphi^{\ddagger}_{\alpha}(x)}(\varphi)\right)\frac{\delta_l\theta_0}{\delta\varphi^{\alpha}(x)}(\varphi)\\
&=-\int\frac{\delta_r}{\delta\varphi^{\ddagger}_{\alpha}(x)}(r_V^{-1}(X))\frac{\delta_l\theta_0}{\delta\varphi^{\alpha}(x)}(\varphi)\\
&\quad+\int\frac{\delta_r X}{\delta\varphi^{\gamma}(y)}(\mathtt{r}_V^{-1}(\varphi))\Delta^{\mathrm{R}}(y,z)\frac{\delta_l}{\delta\varphi^{\beta}_{g}(z)}\left(\frac{\delta_r V}{\delta\varphi^{\ddagger}_{\alpha}(x)}(\varphi)\frac{\delta_l\theta_0}{\delta\varphi^{\alpha}(x)}(\varphi)\right).
\end{aligned}\tag{5.45}
$$

Inserting (5.44) and (5.45) back into (5.40) we obtain finally:

$$
\begin{aligned}
&r_V^{-1}(\{X,\theta_0\})=\{r_V^{-1}(X),\theta_0\}\\
&-\int\frac{\delta_r X}{\delta\varphi^{\gamma}(y)}(\mathtt{r}_V^{-1}(\varphi))\Delta^{\mathrm{R}}(y,z)\frac{\delta_l}{\delta\varphi^{\beta}(z)}\left(\frac{\delta_r V}{\delta\varphi^{\alpha}(x)}(\varphi)\frac{\delta_l\theta_0}{\delta\varphi^{\ddagger}_{\alpha}(x)}(\varphi)-\frac{\delta_r V}{\delta\varphi^{\ddagger}_{\alpha}(x)}(\varphi)\frac{\delta_l\theta_0}{\delta\varphi^{\alpha}(x)}(\varphi)\right)
\end{aligned}
$$

□

Now we are ready to prove our main result.

Proof (of Theorem 5.2.15). Since $S_0 = \S_{00} + \theta_0$, the lemmas imply that

$$
r_V^{-1}(\{X,S_0\})=\{r_V^{-1}X,S_0+V\}-\int\frac{\delta_r X}{\delta\varphi^{\gamma}(y)}(\mathtt{r}_V^{-1}(\varphi))\Delta^{\mathrm{R}}(y,z)^{\gamma\beta}\frac{\delta_l}{\delta\varphi^{\beta}(z)}(\{S_0,V\}+\tfrac{1}{2}\{V,V\}).
$$

The last term in the brackets can be rewritten as $\frac{1}{2}(\{S,S\}-\{S_0,S_0\})$, so it is a difference between the CME for the full action and the CME of the linearised theory. □

References

Araki, H. (1999). *Mathematical Theory of Quantum Fields*. Oxford: Oxford University Press.

Bastiani, A. (1964). Applications différentiables et variétés différentiables de dimension infinie. *Journal d'Analyse mathématique*, **13**(1), 1–114.

Brennecke, F. and Dütsch, M. (2008). Removal of violations of the Master Ward Identity in perturbative QFT. *Reviews in Mathematical Physics*, **20**(2), 119–51.

Brunetti, R., Dütsch, M. and Fredenhagen, K. (2009). Perturbative algebraic quantum field theory and the renormalization groups. *Advances in Theoretical and Mathematical Physics*, **13**(5), 1541–99.

Brunetti, R., Düetsch, M., Fredenhagen, K. and Rejzner, K. (2021). *The unitary Master Ward Identity: Time-slice axiom, Noether's theorem and anomalies*. arxiv.org/abs/2108.13336.

Brouder, C., Dang, N. V., Laurent-Gengoux, C. and Rejzner, K. (2018). Properties of field functionals and characterization of local functionals. *Journal of Mathematical Physics*, **59**(2). arXiv:math-ph/1705.01937.

Buchholz, D. and Fredenhagen, K. (2020). A C^*-algebraic approach to interacting quantum field theories. *Communications in Mathematical Physics* (2020), **377**, 947–969.

Brunetti, R., Fredenhagen, K. and Ribeiro, P. L. (2019). Algebraic structure of classical field theory: Kinematics and linearized dynamics for real scalar fields, *Communications in Mathematical Physics*, **368**, 519–584.

Boas, F.-M. (2000). *Gauge Theories in Local Causal Perturbation Theory*. DESY-THESIS-1999-032, ISSN 1435-808, 1–84.

Benini, M., Perin, M. and Schenkel, A. (2020). Model-independent comparison between factorization algebras and algebraic quantum field theory on lorentzian manifolds. *Communications in Mathematical Physics*, **377**, 971–997.

Becchi, C., Rouet, A. and Stora, R. (1975). Renormalization of the abelian Higgs-Kibble model. *Communications in Mathematical Physics*, **42**(2), 127–162.

Becchi, C., Rouet, A. and R. Stora. (1976). Renormalization of gauge theories. *Annals of Physics*, **98**, 287–321.

Benini, M., Schenkel, A. and Schreiber, U. (2018). The stack of Yang–Mills fields on Lorentzian manifolds. *Communications in Mathematical Physics*, **359**(2), 765–820.

Benini, M., Schenkel, A. and Woike, L. (2019). Homotopy theory of algebraic quantum field theories, *Letters in Mathematical Physics*, **109**(7), 1487–1532.

Batalin, I. A. and Vilkovisky, G. A. (1981). Gauge algebra and quantization, *Physics Letters B*, **102**(1), 27–31.

Calaque, D. (2018). *Derived stacks in symplectic geometry*. [arXiv preprint arXiv:1802.09643].

Cattaneo, A. S. and Moshayedi, N. (2019). *Introduction to the BV-BFV formalism*. arXiv:1905.08047.

Costello, K. (2011). *Renormalization and Effective Field Theory*. Mathematical Surveys and Monographs. Providence, RI: American Mathematical Society.

Costello, K. and Gwilliam, O. (2017). *Factorization Algebras in Perturbative Quantum Field Theory, Vol. 1*. New Mathematical Monographs, vol. 31. Cambridge: Cambridge University Press.

Dütsch, M. and Boas, F.-M. (2002). The Master Ward Identity. *Reviews in Mathematical Physics*, **14**(9), 977–1049.

Dütsch, M. and Fredenhagen, K. (1999). A local (perturbative) construction of observables in gauge theories: The example of QED. *Communications in Mathematical Physics*, **203** no. (1), 71–105.

Dütsch, M. and Fredenhagen, K. (2001a). Algebraic quantum field theory, perturbation theory, and the loop expansion. *Communications in Mathematical Physics*, **219**(1), 5–30.

Dütsch, M. and Fredenhagen, K. (2001b). Perturbative algebraic field theory, and deformation quantization. *Mathematical Physics in Mathematics and Physics: Quantum and Operator Algebraic Aspects*, **30**, 151–160.

Dütsch, M. and Fredenhagen, K. (2003). The Master Ward Identity and generalized Schwinger-Dyson equation in classical field theory. *Communications in Mathematical Physics*, **243**(2), 275–314.

Dütsch, M. and Fredenhagen, K. (2004). Causal perturbation theory in terms of retarded products, and a proof of the Action Ward Identity. *Reviews in Mathematical Physics*, **16**(10), 1291–1348.

Dütsch, M. and Fredenhagen, K. (2007). Action Ward identity and the Stückelberg-Petermann renormalization group. *Progress in Mathematics*, vol. 251. Basel: Birkhäuser Verlag, pp. 113–23.

Fulp, R., Lada, T. and Stasheff, J. (2003). Noether's variational theorem II and the BV formalism. *Rendiconti del Circolo Matematico di Palermo S*, **71**, 115–26.

Fredenhagen, K. and Rejzner, K. (2012). Batalin-Vilkovisky formalism in perturbative algebraic quantum field theory. *Communications in Mathematical Physics*, **317**(3), 697–725.

Fredenhagen, K. and Rejzner, K. (2012). Batalin-Vilkovisky formalism in the functional approach to classical field theory. *Communications in Mathematical Physics*, **314** (1), 93–127.

Gomis, J., París, J. and Samuel, S. (1995). Antibracket, antifields and gauge-theory quantization. *Physics Reports*, **259**(1–2), 1–145.

Gwilliam, O. and Rejzner, K. (2020). *Relating nets and factorization algebras of observables: free field theories*. Communications in Mathematical Physics, **373**(1), 107–74.

Haag, R. (1993). *Local quantum physics*. Berlin: Springer-Verlag.

Henneaux, M. (1990). Lectures on the antifield-brst formalism for gauge theories. Nucl. Phys. Proc. Suppl. **18A**, 47–106.

Henneaux, M. (1995). Homological algebra and Yang–Mills theory. *Journal of Pure and Applied Algebra*, **100**, 1–20.

Haag, R. and Kastler, D. (1964). An algebraic approach to quantum field theory. *Journal of Mathematical Physics*, **5**(7), 848–61.

Hollands, S. (2008). Renormalized quantum Yang-Mills fields in curved spacetime. *Reviews in Mathematical Physics*, **20**, 1033–1172. arXiv:gr-qc/705.3340v3.

Hawkins, E. and Rejzner, K. (2020). The star product in interacting quantum field theory. *Letters in Mathematical Physics*. arXiv:math-ph/1612.09157.

Hurth, T. and Skenderis, K. (1999). Quantum Noether method. *Nuclear Physics B*, **541**(3), 566–614.

Henneaux, M. and Teitelboim, C. (1992). *Quantization of Gauge Systems*. Princeton: Princeton University Press.

Kosmann-Schwarzbach, Y. (2010). *The Noether Theorems: Invariance and Conservation Laws in the Twentieth Century*. Sources and Studies in the History of Mathematics and Physical Sciences. New York: Springer-Verlag.

Noether, E. (1918). Invariant variation problems. *Transport Theory and Statistical Physics*, **1**(3), 186–207.

Pantev, T., Toën, B., Vaquié, M. and Vezzosi, G. (2013). Shifted symplectic structures. *Publications mathématiques de l'IHÉS*, **117**(1), 271–328.

Radzikowski, M. J. (1996). Micro-local approach to the Hadamard condition in quantum field theory on curved space-time. *Communications in Mathematical Physics*, **179**, 529–53.

Rejzner, K. (2011). *Batalin-Vilkovisky formalism in locally covariant field theory*, Ph.D. thesis, arXiv:math-ph:1111.5130v1.

Rejzner, K. (2014). Remarks on local symmetry invariance in perturbative algebraic quantum field theory. *Annales Henri Poincaré*, **16**(1), 205–38.

Rejzner, K. (2016). *Perturbative Algebraic Quantum Field Theory: An Introduction for Mathematicians* Mathematical Physics Studies. Cham: Springer.

Rejzner, K. (2019). Locality and causality in perturbative algebraic quantum field theory. *Journal of Mathematical Physics*, **60**(12), 122301.

Stasheff, J. (1997). Homological reduction of constrained Poisson algebras. *Journal of Differential Geometry*, **45**(1), 221–240.

Stasheff, J. (1998). The (secret?) homological algebra of the Batalin-Vilkovisky approach. *Contemporary Mathematics*, **219**, 195–210.

Tyutin, I. (1994). *Generalized Quantization of Gauge Theories*. Rio de Janeiro: Frontiers.

6

Divergence Invariant Variational Problems*

PETER J. OLVER

Harvey R. Brown (Chapter 7, this volume) states:

> One of the most remarkable methodological trends in modern physics has been the *a priori* use of symmetry principles to constrain the action principles of the non-gravitational interactions in quantum electrodynamics (QED) and particle physics.

In other words, the starting point of a modern physical theory is to postulate the relevant physical symmetries, and then construct an invariant variational principle (Lagrangian) whose Euler–Lagrange equations form the field equations whose solutions determine the associated physics. However, the theory of differential invariants, Olver (1993, 1995), implies that there are infinitely many inequivalent invariant Lagrangians that could be employed for this purpose. What is not addressed is the question I pose in (Olver 2018):

> How does this affect the resulting physics governed by the ostensibly different field equations? ... So, either the underlying physics is, in some rather vague and as yet undefined sense, 'the same', in which case choosing the simplest invariant Lagrangian makes sense on purely practical grounds, or, more worryingly, the two physical theories are different, which then begs the question as to which invariant variational problem describes the correct physics – how does one decide among an infinite range of possibilities?

The following note begins with a very preliminary investigation into this issue in the simplest possible case: the one-dimensional classical free particle. It ends with a new result characterising divergence invariant Lagrangians and some speculation as to their role in fundamental physics.

Let us consider the variational principle for the one-dimensional free particle,[1]

$$I[u] = \int \tfrac{1}{2} u_t^2 \, dt, \tag{6.1}$$

where the independent and dependent variables $t, u \in \mathbb{R}$ are scalars, and we use subscripts to denote derivatives. The Euler–Lagrange equation is simply

$$u_{tt} = 0. \tag{6.2}$$

* I thank Harvey R. Brown for an enlightening correspondence concerning Noether's Theorem that inspired me to write this brief note, as well as his comments, questions, and encouragement as it was being composed. I also thank Ian Anderson, Yvette Kosmann-Schwarzbach, Nicholas Teh, and Charles Torre for helpful suggestions and corrections.

[1] The mass of the particle plays no role for now, and will be ignored for simplicity.

Consider the one-parameter group of Galilean symmetries,

$$(t,u) \longmapsto (t, u + \varepsilon\, t), \qquad \varepsilon \in \mathbb{R}. \tag{6.3}$$

The prolonged action is

$$u_t \longmapsto u_t + \varepsilon, \qquad u_{tt} \longmapsto u_{tt}, \qquad \ldots. \tag{6.4}$$

The Lagrangian is not strictly Galilean invariant, since

$$\tfrac{1}{2} u_t^2 \longmapsto \tfrac{1}{2}(u_t + \varepsilon)^2 = \tfrac{1}{2} u_t^2 + \varepsilon\, u_t + \tfrac{1}{2}\varepsilon^2.$$

However, since the additional terms are a total t derivative,

$$\varepsilon\, u_t + \tfrac{1}{2}\varepsilon^2 = D_t\big(\varepsilon\, u + \tfrac{1}{2}\varepsilon^2 t\big),$$

the Lagrangian is divergence invariant (Olver 1993). The corresponding Noether conservation law is

$$D_t(t\, u_t - u) = t\, u_{tt} = 0. \tag{6.5}$$

The fundamental differential invariants (Olver 1993, 1995), of the Galilean group (6.3) are[2]

$$t, \quad v = u_t - \frac{u}{t}, \quad u_{tt}, \quad u_{ttt}, \quad \ldots, \tag{6.6}$$

where, because t is fixed, the invariant differential operator is simply D_t. Note in particular that

$$D_t v = u_{tt} - \frac{u_t}{t} + \frac{u}{t^2} = u_{tt} - \frac{v}{t}.$$

Let us 'invariantise' the Lagrangian in (6.1), and consider the variational principle

$$J[u] = \int \tfrac{1}{2} v^2\, dt = \int \tfrac{1}{2}(u_t - u/t)^2\, dt, \tag{6.7}$$

which, because it is expresed in terms of the differential invariants times the invariant 'volume form' dt, is strictly invariant under the Galilean group (6.3), as can be easily checked directly. Since

$$\frac{1}{2}\left(u_t - \frac{u}{t}\right)^2 = \tfrac{1}{2} u_t^2 - \frac{u\, u_t}{t} + \frac{u^2}{2t^2} = \tfrac{1}{2} u_t^2 - D_t\left(\frac{u^2}{2t}\right),$$

the variational problems (6.1, 6.7) are equivalent and lead to the same free particle Euler–Lagrange equation (6.2). This alternative strictly invariant variational principle can, in fact, be found in Noether's original paper (Noether 1918, p. 245); see also (Brown 2020 eq. (7.9)).

Now, according to Lie's Theorem (Lie 1897; see also Olver 1995, Theorem 7.27), the most general strictly Galilean invariant first-order variational problem has the form

$$K[u] = \int F(t, v, u_{tt}, u_{ttt}, \ldots, u_{nt})\, dt = \int F(t, u_t - u/t, u_{tt}, u_{ttt}, \ldots, u_{nt})\, dt \tag{6.8}$$

[2] These are easy to find directly, but the method of moving frames (Fels and Olver 1999; Kogan and Olver 2003) can be effectively used in more complicated situations.

where u_{nt} denotes the nth derivative of u with respect to t for some finite $n \geq 2$ and F is an arbitrary function, assumed sufficiently smooth for later purposes. Let us study the case when $F = F(v)$ depends only on v. The Euler–Lagrange equation is

$$E = -D_t F'(v) - \frac{F'(v)}{t} = -F''(v)\, v_t - \frac{F'(v)}{t} = -F''(u_t - u/t)\, u_{tt} + \frac{B(u_t - u/t)}{t} = 0, \tag{6.9}$$

where

$$B(v) = v\, F''(v) - F'(v).$$

Since, when written in terms of v, the Euler–Lagrange equation (6.9) is separable, we can integrate once:

$$\int \frac{F''(v)}{F'(v)}\, dv = -\int \frac{dt}{t}, \qquad \text{and hence} \qquad F'(v) = \frac{1}{c\,t},$$

for some integration constant $c \in \mathbb{R}$. On any interval where $F''(v) \neq 0$, we can use the Inverse Function Theorem to solve the latter equation for

$$v = u_t - \frac{u}{t} = G(ct), \qquad \text{where} \qquad G(x) = F'^{-1}(1/x).$$

The resulting first-order ordinary differential equation for u is linear, and its general solution is

$$u(t) = t\big(a + H(ct)\big), \qquad \text{where} \qquad H'(x) = \frac{G(x)}{x}, \tag{6.10}$$

and where a is a second integration constant. In particular, for Noether's variational principle (6.7), $F(v) = \frac{1}{2}\, v^2$, so that $G(x) = 1/x$, $H(x) = -1/x$, and so

$$u(t) = a\,t - 1/c = a\,t + b, \qquad \text{where} \qquad b = -1/c,$$

recovering the standard linear motion of a free particle. The non-linear motion (6.10) induced by the alternative Galilean-invariant Lagrangian is mathematically quite different, and there is no obvious relationship between the two, neither mathematical nor physical.

Of course, (6.3) is just a one-parameter subgroup of the full Galilean group, which also includes the translations in both t and u:

$$(t, u) \longmapsto (t + a, u + \varepsilon\, t + b), \qquad a, b, \varepsilon \in \mathbb{R}. \tag{6.11}$$

The translations do not affect the prolonged action, which coincides with (6.4), and the above calculation shows that the free particle Lagrangian is strictly invariant under the translations, but divergence invariant under the Galilean boost. On the other hand, the alternative Lagrangian (6.7) is strictly invariant under the Galilean boost, but only divergence invariant under the translations. (The latter can be checked either infinitesimally – see below – or directly using the group transformations (6.11).) The conservation laws corresponding to the two independent translation symmetries are

$$D_t(u_t) = u_{tt} = 0, \qquad D_t\big(\tfrac{1}{2}\, u_t^2\big) = u_t u_{tt} = 0, \tag{6.12}$$

the first representing conservation of momentum, and the second conservation of energy.[3] Since $u_t = a$ is constant, the Galilean conservation law (6.5) implies that the motion is linear in time: $u = at + b$.

For the full Galilean group (6.11), the differential invariants are $u_{tt}, u_{ttt}, \ldots$, and hence Lie's Theorem says that the most general strictly Galilean-invariant variational problem takes the form

$$\widetilde{K}[u] = \int F(u_{tt}, u_{ttt}, \ldots, u_{nt})\, dt. \tag{6.13}$$

In particular, *there are no non-constant strictly Galilean-invariant first-order Lagrangians* and the preceding 'invariantisation' trick does not work!

The latter result can be easily proved directly without invoking Lie's Theorem. As in Olver (1993, 1995), a Lagrangian $L(t, u, u_t, \ldots)$ is strictly invariant under a connected transformation group G if and only if

$$\operatorname{pr} \mathbf{v}(L) + L\, D_t \xi = 0 \tag{6.14}$$

for all infinitesimal generators $\mathbf{v} = \xi\, \partial_x + \varphi\, \partial_t$ of G, where pr $\mathbf{v}$ denotes its prolongation to jet space. In the case of the Galilean group (6.11), a basis for the prolonged infinitesimal generators is provided by

$$\operatorname{pr} \mathbf{v}_1 = \partial_t, \qquad \operatorname{pr} \mathbf{v}_2 = \partial_u, \qquad \operatorname{pr} \mathbf{v}_3 = t\, \partial_u + \partial_{u_t}. \tag{6.15}$$

Thus, a first-order Lagrangian $L(t, u, u_t)$ satisfies the infinitesimal invariance condition (6.14) if and only if

$$\frac{\partial L}{\partial t} = \frac{\partial L}{\partial u} = \frac{\partial L}{\partial u_t} = 0,$$

and hence L must be constant.

On the other hand, it is still possible to modify the free particle Lagrangian to obtain a family of higher-order Galilean divergence-invariant Lagrangians of the following form:

$$\widehat{K}[u] = \int \left[\, \tfrac{1}{2} m u_t^2 + f\, u + F(u_{tt}, u_{ttt}, \ldots, u_{nt}) \right] dt, \tag{6.16}$$

where we can regard the constant $m \in \mathbb{R}$ as the mass parameter and, in the case of Newtonian mechanics, the constant $f \in \mathbb{R}$ as a uniform external force. (More generally, f/m represents the constant acceleration a force-free body experiences in a Galilean-invariant theory that violates Newton's first law of motion.) The Lagrangian is strictly Galilean invariant if and only if $m = f = 0$. If F depends non-linearly on the nth order derivative u_{nt}, the corresponding Euler–Lagrange equation has order $2n$, and its solutions appear to have very little to do with physical free particle motion.

In particular, there is a two-parameter family of divergence-invariant first order Lagrangians that are not strictly invariant, obtained by setting $F = 0$:

$$\widehat{K}_1[u] = \int \left[\, \tfrac{1}{2} m u_t^2 + f\, u \,\right] dt. \tag{6.17}$$

[3] Recall that we are ignoring the mass parameter here.

One can replace the free particle term by Noether's Lagrangian (6.7), producing the alternative family

$$\widetilde{K}_1[u] = \int \left[\, \tfrac{1}{2} m (u_t - u/t)^2 + f\, u \,\right] dt, \tag{6.18}$$

with the same properties. In both cases, the corresponding Galilean-invariant Euler–Lagrange equations are

$$m\, u_{tt} = f, \qquad \text{with quadratic solutions} \qquad u(t) = \frac{f\, t^2}{2m} + a\, t + b, \tag{6.19}$$

provided $m \neq 0$. The three associated Noether conservation laws are

$$\begin{aligned} D_t(m\, u_t - f\, t) &= m\, u_{tt} - f = 0, \\ D_t\left(\tfrac{1}{2} m\, u_t^2 - f\, u \right) &= u_t (m\, u_{tt} - f) = 0, \\ D_t\left(m\, (t\, u_t - u) - \tfrac{1}{2} f\, t^2 \right) &= t\, (m\, u_{tt} - f) = 0, \end{aligned} \tag{6.20}$$

representing, as before, conservation of momentum and energy, along with the quadratic dependence of the motion on time.

In view of these considerations, an interesting question arises. Given a variational symmetry group, how can one characterise the most general divergence-invariant Lagrangian, and under what conditions does an equivalent strictly invariant Lagrangian exist? These questions are answered using the machinery of the (invariant) variational bicomplex (Anderson 1992; Anderson and Pohjanpelto 1995; Kogan and Olver 2003; Thompson and Valiquette 2011), as follows.

We shall work at the infinitesimal level, using results in Olver (1993). Thus, let $\mathbf{v}$ be a generalised vector field and $\mathbf{v}_Q$ its evolutionary representative. According to Proposition 5.52 and the identity (5.86) in Olver (1993), a Lagrangian L is (infinitesimally) divergence invariant under $\mathbf{v}$ – in the language of Olver (1993), $\mathbf{v}$ is a variational symmetry – if and only if $\mathbf{v}_Q$ is also a variational symmetry and hence satisfies

$$E\left[\, \mathrm{pr}\, \mathbf{v}_Q(L) \,\right] = \mathrm{pr}\, \mathbf{v}_Q\left[\, E(L) \,\right] + D_Q^* E(L) = 0, \tag{6.21}$$

where D_Q^* denotes the adjoint of the Fréchet derivative of Q.

As in Anderson and Pohjanpelto (1995), we will refer to vector fields satisfying (6.21) as *distinguished symmetries* of the Euler–Lagrange equations. The identity (6.21) is used to prove Theorem 5.53 in Olver (1993), which states that every variational symmetry is a symmetry of the Euler–Lagrange equations. (The converse is not valid, the most common counterexamples of non-distinguished symmetries being the generators of groups of scaling transformations.)

We have thereby established a basic lemma:

Lemma 6.1 *A Lagrangian L is divergence invariant under a generalised vector field $\mathbf{v}$ if and only if $\mathbf{v}$ is a distinguished symmetry of the Euler–Lagrange equations $E(L) = 0$.*

The next step is to reformulate this result using the invariant Euler–Lagrange complex (Anderson 1992; Anderson and Pohjanpelto 1995; Kogan and Olver 2003; Thompson and Valiquette 2011). Let $\mathfrak{g}$ be a Lie algebra of prolonged (generalised) vector fields acting on jet space and hence on the differential forms thereon. A differential form ω is $\mathfrak{g}$-invariant if and only if all its Lie derivatives vanish:

$$\mathbf{v}(\omega) = 0 \qquad \text{for all} \qquad \mathbf{v} \in \mathfrak{g}. \tag{6.22}$$

In particular, $\mathfrak{g}$-invariance of a Lagrangian form λ is equivalent to strict invariance of the associated variational problem, while $\mathfrak{g}$-invariance of the associated Euler–Lagrange source form $E(\lambda)$ encodes the fact that the generators of $\mathfrak{g}$ are distinguished symmetries of the Euler–Lagrange equations, and hence, by Lemma 6.1, the Lagrangian form is divergence invariant. The difference between strict and divergence invariance of the Lagrangian is thus prescribed by the cohomology class of the Euler–Lagrange source form.

More precisely, let $H^*(\mathcal{E}_\mathfrak{g})$ denote the cohomology of the $\mathfrak{g}$-invariant Euler–Lagrange complex, denoted by $\mathcal{E}_\mathfrak{g}$. By definition, a $\mathfrak{g}$-invariant form is *closed* if its differential vanishes; it is *exact* if it is locally the differential of a $\mathfrak{g}$-invariant form. In particular, a source form is closed if it is (locally) an Euler–Lagrange source form for some variational problem, and it is exact if it comes from a $\mathfrak{g}$-invariant Lagrangian form. Thus, as a consequence of Lemma 6.1, we deduce our main result.

Theorem 6.2 *Let p denote the number of independent variables. A divergence $\mathfrak{g}$-invariant Lagrangian form λ is equivalent to a strictly $\mathfrak{g}$-invariant Lagrangian form if and only if the cohomology class of its Euler–Lagrange source form vanishes*:

$$0 = \big[\, E(\lambda) \,\big] \in H^{p+1}(\mathcal{E}_\mathfrak{g}). \tag{6.23}$$

A precursor of this theorem can be found in Anderson (1992). Methods for computing the invariant cohomology spaces when the Lie algebra $\mathfrak{g}$ consists of the infinitesimal generators of either a finite-dimensional Lie group G or an infinite-dimensional Lie pseudo-group of projectable[4] transformations can be found in Anderson and Pohjanpelto (1995). In the former case, it is proved that $H_\mathfrak{g}^{p+1}(\mathcal{E})$ is isomorphic to the Lie algebra cohomology space $H^{p+1}(\mathfrak{g})$; cf. Jacobson (1962). This result was generalised to Lie groups of point transformations in Thompson and Valiquette (2011), where algorithmic moving frame methods are applied to implement the isomorphism. This establishes the following result.

Corollary 6.3 *Let G be a connected Lie group of point transformations. Then the condition $H^{p+1}(\mathfrak{g}) = \{0\}$ is necessary and sufficient for every divergence-invariant Lagrangian to be strictly invariant.*

In particular, if $\dim \mathfrak{g} \le p$, then $H^{p+1}(\mathfrak{g}) = \{0\}$ automatically, and hence Corollary 6.3 applies. In the preceding example, there is one independent variable t, and hence $p = 1$. Thus, in this situation, Noether's 'trick' of replacing the free particle Lagrangian with an equivalent strictly invariant Lagrangian can be applied to *any* one-parameter symmetry group, and hence divergence inequivalence is a multi-parameter phenomenon.

[4] Projectable means that the transformations of the independent variables do not depend on the dependent variables; in bundle-theoretic terms, they are fibre-preserving maps. All the group actions considered here, and most of those arising in physics, are projectable.

On the other hand, if $\mathfrak{g}_3$ denotes the three-dimensional Galilean Lie algebra spanned by (6.15), then it can be shown that $H^2(\mathfrak{g}_3) = \mathbb{R}^2$, and hence there is a two-parameter family of non-zero cohomology classes, which correspond to the divergence-invariant Lagrangians (6.17), or, alternatively, (6.18). Theorem 6.2 implies that every divergence invariant Lagrangian under the full Galilean group (6.11) is locally equivalent to one of the above forms (6.16).

Example 6.4 An even simpler example is the abelian Lie algebra $\mathfrak{g}_2$ spanned by the first two vector fields in (6.15), generating the translations $(x,u) \longmapsto (x+a, u+b)$. In this case, the differential-invariants are $u_t, u_{tt}, u_{ttt}, \ldots$, while $H^2(\mathfrak{g}_2) = \mathbb{R}$, and hence there is a one-dimensional space of divergence-invariant Lagrangians that are not strictly translation invariant. Indeed, it is not hard to see that every divergence-invariant variational problem has the forms

$$I[u] = \int \big[\, f\, u + F(u_t, u_{tt}, \ldots, u_{nt}) \big]\, dt, \tag{6.24}$$

where $f \in \mathbb{R}$ and the second term is strictly invariant.

Example 6.5 A more substantial example is provided by combining the Galilean group with the scalings of time and space, producing the five-parameter transformation group

$$(t,u) \longmapsto (\lambda t + a, \mu u + \varepsilon t + b), \qquad a,b,\varepsilon, \in \mathbb{R}, \;\; \lambda,\mu \in \mathbb{R}\setminus\{0\}. \tag{6.25}$$

It is easily checked that this forms a symmetry group of the (unforced) free particle equation (6.2). In addition, the particular scaling symmetries

$$(t,u) \longmapsto (\mu^2 t, \mu u), \qquad \mu \in \mathbb{R}\setminus\{0\} \tag{6.26}$$

form a strict one-parameter variational symmetry group of both the standard Lagrangian (6.1) and the Noether alternative (6.7). The corresponding conservation law is

$$D_t(u\,u_t - t\,u_t^2) = (u - 2t\,u_t)u_{tt} = 0. \tag{6.27}$$

On the other hand, the other scaling symmetries of the Euler–Lagrange equation are not distinguished symmetries of the corresponding source form, and thus are *not* symmetries of either Lagrangian, neither strict nor divergence. They act by rescaling the Lagrangian by a non-unit factor. Thus, they fail to produce a conservation law but instead, according to Steudel (1962); see also Olver (1993, Exercise 5.35) lead to the following divergence identity:

$$D_t(uu_t) = u\,u_{tt} + u_t^2 = u_t^2, \tag{6.28}$$

which is valid on solutions to (6.2). Note that the right-hand side is twice the Lagrangian, i.e., twice the kinetic energy. Such scaling identities are used to establish Clausius' Virial Theorem that governs the behavior of the average kinetic energy of a system; see Goldstein (1980).

Let us finish by determining the structure of variational problems that admit the full group (6.25). The Lie algebra $\mathfrak{g}_5$ is spanned by the three Galilean infinitesimal generators (6.15) along with the two prolonged scaling generators

$$\begin{aligned} \operatorname{pr} \mathbf{v}_4 &= t\,\partial_t - u_t\,\partial_{u_t} - 2u_{tt}\,\partial_{u_{tt}} - 3u_{ttt}\,\partial_{u_{ttt}} - \cdots, \\ \operatorname{pr} \mathbf{v}_5 &= u\,\partial_u + u_t\,\partial_{u_t} + u_{tt}\,\partial_{u_{tt}} + u_{ttt}\,\partial_{u_{ttt}} + \cdots. \end{aligned} \tag{6.29}$$

Using either the method of moving frames (Fels and Olver 1999), or a direct computation, one finds a complete system of differential invariants,

$$I_4 = \frac{u_{tt}\,u_{tttt}}{u_{ttt}^2}, \qquad I_5 = \frac{u_{tt}^2\,u_{ttttt}}{u_{ttt}^3}, \qquad \ldots \qquad I_n = \frac{u_{tt}^{n-3}\,u_{nt}}{u_{ttt}^{n-2}}, \tag{6.30}$$

for $n \geq 4$, along with the fundamental (contact) invariant one-form and dual invariant differentiation operator

$$\omega = \frac{u_{ttt}}{u_{tt}}\,dt, \qquad \mathcal{D} = \frac{u_{tt}}{u_{ttt}}\,D_t, \tag{6.31}$$

the latter mapping differential invariants to higher-order differential invariants. Thus, by Lie's Theorem, the most general strictly invariant variational problem has the form

$$I[u] = \int F(I_4, I_5, \ldots, I_n)\,\omega,$$

where F is an arbitrary function of the basic differential invariants (6.30).

To check for divergence-invariant Lagrangians, we compute the Lie algebra cohomology, finding $H^2(\mathfrak{g}_5) = \mathbb{R}$. This implies that there is a unique, up to constant multiple, divergence-invariant Lagrangian that is not equivalent to a strictly invariant Lagrangian. A moving frame-based computation using the methods in Thompson and Valiquette (2011) is the easiest way to find the following third-order representative:[5]

$$J[u] = \int \frac{u_{ttt}\log u_{ttt} - u_{ttt}}{u_{tt}}\,dt, \tag{6.32}$$

with invariant sixth-order Euler–Lagrange equation

$$D_t^2\left(\frac{u_{tt}\,u_{tttt} - u_{ttt}^2}{u_{tt}^2\,u_{ttt}}\right) = 0. \tag{6.33}$$

One can verify directly that (6.32) is divergence invariant under (6.25) but is not equivalent to a strictly invariant Lagrangian. Of course, this final example is a mere mathematical curiosity, as such higher-order variational principles presumably have very little relevance to physics.

Let us close with a wild speculation that Theorem 6.2 may provide an answer (or at least a partial answer) to the original question. Namely, we propose that variational principles of physical relevance are distinguished by arising from non-zero cohomology classes of the

[5] The Lagrangian is defined only on the subset of jet space where $u_{tt} \neq 0,\ u_{ttt} > 0$, and hence should be restricted to functions $u = f(t)$ whose jet lies in this subset. Interestingly, the Euler–Lagrange equation is less restricted, especially if one clears denominators after evaluating the total derivatives.

underlying physical symmetry group, or, equivalently, are based on divergence-invariant Lagrangians that are not equivalent to any strictly invariant Lagrangian. Of course, this requires using the full physical symmetry group, as restriction to a subgroup, especially if it is just a one-parameter subgroup, may lead to an equivalent strictly invariant variational principle of physical relevance, as we saw in our key example. According to Theorem 6.2 and the identification with Lie algebra cohomology, $H^{p+1}(\mathcal{E}_{\mathfrak{g}}) \simeq H^{p+1}(\mathfrak{g})$, when $\mathfrak{g}$ is a finite-dimensional Lie algebra of infinitesimal point transformations, such variational problems form a finite-dimensional space, whereas the strictly invariant variational problems depend upon an arbitrary function of the infinite family of independent differential invariants which are of arbitrarily high order. In particular, if the relevant cohomology space is one-dimensional, such a variational principle is unique up to a constant multiple. Furthermore, according to computations in Anderson and Pohjanpelto (1995), the relevant cohomology spaces continue to be finite (and low) dimensional for a variety of non-trivial infinite-dimensional pseudo-group actions of physical and mathematical importance. (On the other hand, if the cohomology vanishes, there are no distinguished variational principles, and one must use other criteria to characterise physically relevant invariant variational principles.) Thus, divergence invariance or, equivalently, cohomological considerations may be fundamental to the symmetry-driven formulation of physical theories. In this direction, it would be of great interest to determine the cohomology for the infinite-dimensional group underlying the standard model, although this will be a very challenging computation.

References

Anderson, I. M. (1992). Introduction to the variational bicomplex. *Contemporary Mathematics*, **132**, 51–73.

Anderson, I. M. and Pohjanpelto, J. (1995). The cohomology of invariant variational bicomplexes. *Acta Applicandae Mathematicae*, **41**, 3–19.

Fels, M. and Olver, P. J. (1999). Moving coframes. II. Regularization and theoretical foundations. *Acta Applicandae Mathematicae*, **55**, 127–208.

Goldstein, H. (1980). *Classical Mechanics*, 2nd ed. Reading, MA: Addison–Wesley.

Jacobson, N. (1962). *Lie Algebras*. New York: Interscience Publishers.

Kogan, I. A. and Olver, P. J. (2003). Invariant Euler–Lagrange equations and the invariant variational bicomplex. *Acta Applicandae Mathematicae*, **76**, 137–93.

Lie, S. (1897). Über Integralinvarianten und ihre Verwertung für die Theorie der Differentialgleichungen. *Leipz. Berichte*, **49**, 369–410; also *Gesammelte Abhandlungen*, vol. 6, B. G. Teubner, Leipzig, 1927, pp. 664–701.

Noether, E. (1918). Invariante Variationsprobleme, *Nachr. König. Gesell. Wissen. Göttingen, Math.–Phys. Kl.* 235–57. (See Kosmann-Schwarzbach, Y. (2011), *The Noether Theorems. Invariance and Conservation Laws in the Twentieth Century*, New York: Springer, for an English translation.)

Olver, P. J. (1993). *Applications of Lie Groups to Differential Equations*, 2nd ed, Graduate Texts in Mathematics, vol. 107. New York: Springer–Verlag.

Olver, P. J. (1995). *Equivalence, Invariants, and Symmetry*. Cambridge: Cambridge University Press.

Olver, P. J. (2018). Emmy Noether's enduring legacy in symmetry. *Symmetry: Culture and Science*, **29**, 475–85.

Steudel, H. (1962). Eine Erweiterung des ersten Noetherschen Satzes, *Zeit. fur Naturforsch.* **17A**, 133–5.

Thompson, R. and Valiquette, F. (2011). On the cohomology of the invariant Euler–Lagrange complex. *Acta Applicandae Mathematicae*, **116**, 199–226.

7

Do Symmetries 'Explain' Conservation Laws?

The Modern Converse Noether Theorem *vs* Pragmatism*

HARVEY R. BROWN

7.1 Introduction

In 2004, I argued with Peter Holland (Brown and Holland 2004a) that in her celebrated 1918 paper, Emmy Noether established by way of her (first) theorem[1] a *correlation* between variational symmetries and conservations laws, but not an explanatory one-way arrow from the former to the latter. Our argument was based on the fact that Noether also provided a converse of the theorem in her paper. But as we shall see below, this converse result is not sufficient to make the point: appeal must instead be made to the modern, generalised version of the joint direct and converse theorem due to Luis Martínez-Alonso in 1979 and independently Peter Olver in 1986. A careful scrutiny of the conditions of the theorem will be given below.

Despite this strengthening of the argument, such an egalitarian position in the reading of Noether's theorem is hardly in tune with typical discussions of the theorem in the literature: conservation principles are usually taken to follow from the existence of variational symmetries. Attempts have been made to establish the priority of symmetries on metaphysical grounds.[2] However, this chapter takes a more pragmatic approach, looking at how symmetry principles are actually used by physicists within the Lagrangian framework. This may not strictly establish an explanatory fundamentality for symmetries in relation to conservation principles, but it may help to explain why more emphasis is placed in practice on Noether's direct theorem than on its modern converse.

* I am grateful to James Read, Bryan Roberts and Nicholas Teh for the invitation to contribute to this volume. Thanks go to Adam Caulton, Erik Curiel, and James Read for helpful comments, and to Niels Linnemann for spotting a technical error and making other suggestions for improvements. I am indebted to Adrian Sutton for illuminating discussion of Eshelby's contributions to the theory of elasticity. My greatest debt is to Peter Olver, for a generous and enlightening correspondence about various technical Noether-related matters.

[1] For an outline of a special case of the first theorem, see Section 7.4.1. English translations of Noether (1918) by M. A. Tavel and Y. Kosmann-Schwarzbach are found in Noether (1971) and Kosmann-Schwarzbach (2010, pp. 1–22), respectively. For an account of Noether's second theorem see, e.g., Brading and Brown (2003).

[2] Mark Lange, for instance, has provided an argument to the effect that symmetry principles possess 'a variety of necessity' that is stronger than that of conservation laws (Lange 2007); note that Lange does not base this claim on Noether's theorem *per se*. A hard-hitting critique of Lange's position is found in Smith (2008). A further worry is mentioned in Section 7.5.

7.2 The Standard Account

The following quotations seem to typify discussions of the relationship between symmetries and conservation principles in the literature. We start with Landau and Lifshitz (1960, section 6):

There are some [integrals of the motion] whose constancy is of profound significance, deriving from the fundamental homogeneity and isotropy of space and time. (section 6).

In his 1986 book *Fearful Symmetry*, the theorist and expositor A. Zee wrote:

For years, I did not question where these [energy, linear and angular momentum] conservation laws came from; they seemed so basic that they demanded no explanation. Then, I heard about Noether's insight and I was profoundly impressed. The revelation that these basic conservation laws follow from the assumption that physics is the same yesterday, today, and tomorrow; here, there, and everywhere; east, west, north, and south, was for me, as Einstein put it, essentially spiritual. (Zee 1999, p. 121).[3]

Lewis Ryder, in his 1996 book *Quantum Field Theory*, referred to "the origin of the conservation laws for energy, momentum and angular momentum; they all follow from space-time symmetries via Noether's theorem.' (Ryder 1996, section 3.2). An article (Hanc, Tuleja and Hancova 2004) in the *American Journal of Physics* claimed that

When the German mathematician Emmy Noether proved her theorem, she uncovered the fundamental justification for conservation laws.

In their 2004 book *Classical Mechanics*, Kibble and Berkshire stated that the 'physical reasons' for the existence of conserved quantities are 'expressions of symmetry properties expressed by the system' (Kibble and Berkshire 2004, p. 291) and that 'any symmetry property of the system leads to a corresponding conservation law …' (Kibble and Berkshire 2004, p. 301).

Two prominent voices on symmetries in the philosophical literature are Bas van Fraassen and Mark Lange. Writing in 1989, van Fraassen asserted:

In the twentieth century we have learned to say that every symmetry yields a conservation law (p. 258). (van Fraassen 1989, p. 258).

He went on to say that symmetries 'engender' conservation laws. (p. 433). More recently Lange has referred approvingly to the 'commonly held view' that 'a given symmetry principle explains the associated conservation law' (Lange 2007, p. 462).

Note that neither Landau and Lifshitz nor Kibble and Berkshire mention Noether's theorem explicitly, and it must not be thought that demonstrating the connection between symmetries and conservation laws, in either classical or quantum mechanics, necessarily requires appeal to Lagrangian methods.[4] At any rate, the supposed explanatory priority

[3] This statement is striking because Zee also writes that 'Noether's observation tells them [physicists aware of a conserved quantity] that the action must have a corresponding symmetry' (p. 121).

[4] Nor should it be overlooked that (i) the existence of conservation laws, or integrals of the motion, do not depend in general on the existence of symmetries, and (ii) invariance with respect to part or all of the Galilean group, for example, does not invariably imply in Newtonian mechanics the existence of any of the standard conservation laws. For further details see Havas (1973, pp. 148–50).

of symmetries reflected in all of the remarks cannot be regarded as obvious, for several reasons. First, the real meat in the physics ultimately resides in the equations of motion or field equations (Euler–Lagrange equations in the case of Lagrangian systems) which are the source of *both* the dynamical symmetries and the conservation laws. From this point of view, it is hard to see any distinction between them of the kind we are interested in. Second, in relation to Noether's theorem, there is in general no unique symmetry associated with a given conservation law, or *vice versa*, as we shall see below. Third, even if certain field equations were, as a matter of historical fact, discovered by imposing certain symmetries or conservation laws as *a priori* constraints, this doesn't settle the fundamental explanatory issue. (We return to this point in Section 7.2.) And fourth, as mentioned above, Noether's theorem has a converse, which in its modern form arguably deserves to be better known. But before turning to the converse theorem, it is worth reminding ourselves of the subtleties and limitations of Noether's direct theorem which are not always recognised in conventional statements of its content.

7.3 Subtleties of Noether's Theorem

(i) Not all differential equations in physics have a variational (Lagrangian) formulation; in such cases the application of Noether's theorem is ruled out. Examples are the Navier–Stokes equations (in most cases) and Fourier's heat equation.[5] Furthermore, the very question whether a given dynamical theory has a Lagrangian formulation can depend on how the relevant equations of motion, or field equations, are formulated.[6]

(ii) Noether's theorem does not apply to discrete symmetries of equations of motion. But not every continuous dynamical symmetry is a variational (quasi-)symmetry.[7] Nor is it strictly true that every variational (quasi-)symmetry is a dynamical symmetry, though this is usually the case.[8]

[5] The heat equation features in Section 7.4.3. In Appendix A, it is shown how a 'trick' allows for a variational formulation, analogous to that for the complex version of the (adjoint) heat equation: the time-dependent Schrödinger equation for the free particle.

Wigner (1954) gave the example of a non-Lagrangian mechanical system obeying Newton's first law of motion, but whose force law is associated with velocities rather than accelerations. In the classical case, there are no conservation laws for energy and angular momentum associated with the time translation and spatial isotropy symmetries.

[6] The standard vectorial formulation of Maxwell's theory in terms of **E** and **B** fields does not have a Lagrangian formulation, but that in terms of potentials does; see Olver (1986a, p. 289, exercise 4.6). However, a Lagrangian formulation in terms of **E** and **B** does exist for the partial, time-dependent Maxwell equations; see the independent work by Anderson and Arthurs (1978) and Rosen (1980) as well as Ibragimov (2006a), and the full set of vacuum equations also has an unusual vector Lagrangian formulation in terms of the electromagnetic field tensor $F^{\mu\nu}$ and its dual; see Sudbery (1986).

Another example is the equation for a single particle under the influence of a frictional force: $m\ddot{\mathbf{q}} = -k\dot{\mathbf{q}}$. The equation is symmetric under space and time translations, but has no standard momentum or energy conservation laws. It also has no variational formulation, but an equivalent equation is obtained in the usual fashion from the Lagrangian $(1/2)m\dot{\mathbf{q}}^2 \exp[(k/m)t]$, neither of which is time translation symmetric. So the existence of the conserved energy-like quantity $(1/2)m\dot{\mathbf{q}}^2 \exp[2(k/m)t$ does not follow from Noether's theorem; see Havas (1973, pp. 152–3).

[7] A typical case is that of equations of motion which have a scaling symmetry. See Olver (1986a, pp. 252, 282), and Brown and Holland (2004b, section IV). Other examples in field theory are found in Boyer (1967, p. 448). The definition of a quasi-symmetry is a transformation of the independent and/or dependent variables such that the action is invariant up to a surface term, so the Lagrangian is invariant up to a total divergence. A special case is described in Section 7.4.1, following Eq. (7.3).

[8] A necessary condition is that every dependent variable is dynamical, i.e., that it is subject to Hamilton's variational principle, or equivalently that it satisfies a non-trivial Euler–Lagrange equation; see Brown and Holland (2004a, p. 1138), and Butterfield (2006, p. 56).

(iii) The connection between a given conservation principle and its associated symmetry will in general depend on the choice of Lagrangian. Standardly, conservation of energy and linear and angular momentum, for example, are associated via Noether's theorem with the symmetries of time translation (temporal homogeneity), space translation (spatial homogeneity), and rotation (spatial isotropy), respectively, but these associations are only 'mostly' valid. One can find Lagrangian systems that are counterexamples in the forward sense and others that are counterexamples in the reverse sense.[9] Typically such non-standard relationships in the literature arise because some systems share distinct non-gauge-related Lagrangians, i.e., not related by a total divergence.[10] One of the most striking examples concerns the free Maxwell field. Lagrangian densities can be found such that the conservation of the energy is related to an *internal* symmetry: duality rotations involving the electric and magnetic field strengths ($\mathbf{E}' = \mathbf{E}\cos\theta - \mathbf{B}\sin\theta$; $\mathbf{B}' = \mathbf{B}\cos\theta + \mathbf{E}\sin\theta$)![11]

However, once the true nature of Noether's theorem is revealed by incorporating quasi-symmetries, it is possible to see the degree to which the notion that, for example, conservation of energy is intrinsically linked to temporal homogeneity is misleading – without needing to appeal to the occasional existence of systems sharing 'inequivalent' Lagrangians in the above sense. Hopefully this will be made clear below.

(iv) Not every non-trivial first integral obtained in applications of Noether's theorem is associated with a conservation principle in the usual sense of the word. This is particularly evident in non-conservative systems, which, contrary to widespread belief, may have a Lagrangian formulation. For instance, in the application of Noether's theorem to the linearly damped oscillator, the first integral associated with time translation symmetry is not energy as normally construed.[12] In optical systems, the time translation symmetry can lead, by Noether's theorem, to a trivial mathematical identity.[13]

(v) Noether's theorem normally has to do with temporal evolution, but this is not always the case. We see in Section 7.6 applications in the theory of elastostatics which give rise to 'conservation' laws, defined not with respect to time but to spatial coordinates. In general the independent variables in the Lagrangian need not contain a time variable.

(vi) Finally, in cases where the Euler–Lagrange equations themselves can be expressed as Div $P = 0$, symmetries can sometimes be found for which the Noether conservation law is precisely this equation.[14]

[9] There are systems in which conservation of energy, for example, is not related by the generalised Noether theorem (see below) to temporal homogeneity, and others in which the opposite holds; and similarly for the other cases above of conservation laws. Instructive examples, not always contrived, are found in Santilli (1983, Chart A.11, pp. 340–4, pp. 348–9, and Example A7, pp. 354–6). See Occurrence 2, p. 341, for a counterexample to Lewis Ryder's claim: 'Conservation of energy and momentum, then, holds for any system whose Lagrangian (and therefore action) does not depend on x^μ.' (Ryder 1996, section 3.2).

[10] Such is ubiquitous in two-dimensional systems, though this is not a necessary condition.

[11] See Sudbery (1986) for conservation of the stress energy tensor using a four-component Lagrangian density, and Ibragimov (2006a, section 3), for conservation of energy with a scalar Lagrangian density for the time-dependent Maxwell equations. For further discussion of the role of inequivalent Lagrangians in relation to Noether's first theorem, see Brown and Holland (2004a, pp. 1136–7), Butterfield (2006, p. 56), and Smith (2008, section 2).

[12] See Santilli (1983, Occurrence 2, p. 341), and Smith (2008, section 4).

[13] See Blaker and Tavel (1974) and Olver (1986a, exercise 5.19, p. 372).

[14] For an example in elastostatics, see Olver (1986a, p. 281), and for examples in quantum mechanics and electrodynamics, see Brown and Holland (2004b, sections III and V). Probably the simplest case is the non-relativistic free particle in one

7.4 The Converse Noether Theorem

7.4.1 Noether 1918

Let us start with what Noether's 1918 converse theorem is not. It is *not* the statement that, given a Lagrangian $\mathcal{L}$, and a conservation law emerging from the associated Euler–Lagrange equations, a strict variational symmetry of the action $\int \mathcal{L} dx$ exists which is associated by the direct theorem with this conservation law.

To get a sense of what Noether proved, let us summarise the punchline of the direct theorem. For the sake of simplicity, we consider an action associated with a system of a single degree of freedom of the form $\int \mathcal{L}\,(q(t),\dot{q}(t),t)\,dt$, with q a scalar, where $\dot{q} = \frac{dq}{dt}$. The associated Euler–Lagrange expression E is defined in this case by

$$E \equiv \frac{\partial \mathcal{L}}{\partial q} - \frac{d}{dt}\frac{\partial \mathcal{L}}{\partial \dot{q}}. \tag{7.1}$$

Now suppose there is a group $\mathcal{G}$ of transformations of q and t that depend analytically on a single parameter ϵ, such that the action is invariant under $\mathcal{G}$.[15] It follows that the action will be invariant under the infinitesimal transformations

$$q' = q + Q\epsilon; \quad t' = t + T\epsilon, \tag{7.2}$$

where $Q\epsilon$ and $T\epsilon$ are the terms of lowest order in ϵ in the transformations of q and t, respectively[16] – invariant, that is, up to terms of at least second order in ϵ. If ϵ does not depend on the independent variable t, an exercise in the calculus of variations then leads to the so-called *Noether identity*:

$$E(Q - \dot{q}T) = \frac{d}{dt}\left[-\frac{\partial \mathcal{L}}{\partial \dot{q}}(Q - \dot{q}T) - \mathcal{L}T\right]. \tag{7.3}$$

Note that in the case of a quasi-symmetry, which Noether does not consider in her paper, where the first variation $\delta\mathcal{L} = \left(\frac{d}{dt}C(t,q,\dot{q},\ldots)\right)\epsilon$, the term $\frac{d}{dt}C$ must be added to the RHS of (7.3) (see, e.g., Brown and Holland 2004a, sec. 2).[17] Finally, when the symmetry group has ρ parameters, there will be ρ such Noether identities.

This completes the direct, or first, Noether theorem in the special case of $\mathcal{L} = \mathcal{L}\,(q(t),$ $\dot{q}(t),t)$, and it makes no reference to conservation laws.[18] Of course Noether was aware that when Hamilton's principle is applied, leading to $E = 0$, the term inside the square bracket on the RHS of (7.3) corresponds to the conserved quantity, or first integral, associated with

dimension, where in relation to the standard Lagrangian $\mathcal{L} = \frac{1}{2}m\dot{\mathbf{q}}^2$, the conservation law related to spatial homogeneity is $\frac{d}{dt}(m\dot{\mathbf{q}}) = 0$, which is equivalent to the Euler–Lagrange equation $\ddot{\mathbf{q}} = 0$.

15 This means that $\int \mathcal{L}(q,\dot{q},t) = \int \mathcal{L}(q',\dot{q}',t)$, where the primes represent transformations under $\mathcal{G}$, $\dot{q}' \equiv \partial q'/\partial t'$, and the region of integration on the RHS is the transformation of the arbitrary t-interval for the integral on the LHS.

16 It is assumed that the parameter ϵ can be chosen such that $\epsilon = 0$ corresponds to the identity transformations. Note that since the range of t in the integral $\int \mathcal{L}\,dt$ is arbitrary, invariance of the integral implies that the Lagrangian $\mathcal{L}$ is invariant.

17 For a critical analysis of two different approaches in the literature to defining variational quasi-symmetries, see section 3 of Brading and Brown (2003).

18 Nor does Noether ever use the word 'symmetry' in her 1918 paper; she is concerned only with invariance of (what is normally called) an action functional, and not explicitly with symmetry properties of the Euler–Lagrange equations.

the transformations (7.2).[19] It is worth noting that Noether's theorem does not determine how many independent conservation laws follow from the invariance of the action under a ρ-parameter group of transformations. In relation to the full 10-parameter Galilean group, there are 10 conservation laws, but for the single free particle, only six are independent) see Havas 1973, p. 151).

Note also the role of $Q-\dot{q}T = [q'(t)-q(t)]/\epsilon$ on both sides of the Noether identity. Multiplied by ϵ, it is what Noether denoted by $\overline{\delta q}$, and it is sometimes called the *characteristic* of the symmetry (7.2), We shall see below that defining new transformations (sometimes called *contemporaneous*) with Q replaced by $\tilde{Q} = Q - \dot{q}T$, and T by $\tilde{T} = 0$, represents generally a quasi-symmetry that, assuming Hamilton's principle, leads to the same conserved quantity and obviously preserves the characteristic.[20]

Now Noether's converse theorem, when applied to this simple case, starts from the identity (7.3) – not the conservation law associated with the vanishing of the RHS of (7.3) – and goes through the proof of the direct theorem 'in reverse order'. Noether presupposes a functional $\mathcal{L}(q(t), \dot{q}(t), t)$ along with its Euler–Lagrange expression (7.1), but does not presuppose a physical meaning for Q and T. A series of reverse steps from (7.3) exploit the identity

$$E\delta q = \delta\mathcal{L} - \frac{d}{dt}\left[\frac{\partial\mathcal{L}}{\partial\dot{q}}\delta q\right]; \tag{7.4}$$

which follows from the calculus of variations where the first variation $\delta\mathcal{L}$ arises from the variation $\delta t \equiv t' - t$ in the independent variable and $\delta q \equiv q'(t') - q(t)$ in the dependent variable. After integration, it is confirmed that $\int \mathcal{L}\,dt$ is strictly invariant under the infinitesimal transformations (7.2), the factors T and Q thereby gaining their meaning, and δq and δt are shown to be linear in the parameter ϵ. Thus Noether showed that the existence of a strict variational symmetry associated with $\mathcal{L}$ holds if and only if (7.3) holds, or rather a generalisation thereof.[21] Note that *the assumption that the Euler–Lagrange equations, and hence conservation laws, hold is nowhere used in the proof.*[22] In order to see if a strict variational symmetry can be correlated with a given conserved quantity arising from the Euler–Lagrange equations for the Lagrangian $\mathcal{L}$ of the simple kind we are

[19] In the case of field theory, the counterpart of the RHS is the total divergence of a current, so when Hamilton's principle is applied, (7.3) becomes a continuity equation (which Noether calls a 'law of conservation'), and a constant of the motion (Noether charge) only ensues when suitable boundary conditions are satisfied; see Brown and Holland (2004a, section 4). It is noteworthy that (i) Noether does not mention this widely known last step involving Gauss's divergence theorem (ii) she also considers the case where the Euler–Lagrange expressions do not vanish, but she nonetheless refers to laws of conservation in such cases; see footnote 15 in section 3 of Noether (1918).

[20] When $Q - \dot{q}T = 0$, the transformations will not be a strict symmetry; they will always be a quasi-symmetry associated with the trivial null conserved quantity. This will be made clear in Section 7.4.3: the evolutionary form of these transformations is such that $Q' = T' = 0$.

[21] Noether remarked, however, that in the converse theorem, it is 'left moot' whether in the case of transformations that depend on derivatives of the dependent variables (see below), the corresponding finite transformations form a group; see Noether (1918, section 3).

[22] This point is made particularly clearly in Bessel–Hagen (1921) and Havas (1973, pp. 158, 161). In Chapter 1 of this volume, Kosmann-Schwarzbach correctly states that in her converse theorem Noether started from 'linearly independent divergence relations' (the higher dimensional generalisation of (7.3)). She calls the result of applying the Euler–Lagrange equations to these relations 'conservation laws'. So it is somewhat misleading when she concludes that Noether had shown that

> the existence of ρ linearly independent conservation principles yields the infinitesimal invariance of [the Lagrangian] under a Lie algebra of infinitesimal symmetries of dimension ρ

considering, T and Q must be found such that (7.3) is satisfied and the conservation law corresponds to the vanishing of the RHS of (7.3). As Boyer noted in his careful 1967 study of Noether's theorem in the general case of field theory, the converse theorem 'gives no hint on the crucial question as to when a conserved current can be written in the form of the identity [corresponding to (7.3)]' (see Boyer 1967, p. 457).

However, in a little-known paper Dass (1966) had proved a strengthened version of Noether's converse theorem. He showed that for a certain class of action integrals in field theory and for any divergenceless vector, there exists another such vector that differs from it by terms which vanish when the Euler–Lagrange equations hold, and for which the (generalised) version of (7.3) holds, leading to the existence of (quasi-)symmetry transformations of the action. The restriction on the action is that the Lagrangian does not depend on the independent variables explicitly. As Havas (1973, pp. 158, 162) pointed out, this means that for our 1-dimensional mechanical case of $\mathcal{L} = \mathcal{L}(q(t), \dot{q}(t), t)$, any conservation law is associated with a (quasi-)symmetry of the action as long as $\mathcal{L}$ does not depend explicitly on t[23], although it is not guaranteed in Dass's proof that the symmetry transformations form a group.

Dass's theorem partially anticipates an even stronger general result in the more recent literature that allows for a version of the converse theorem that starts with a conservation law rather than a generalisation of the identity (7.3). In our special case of $\mathcal{L} = \mathcal{L}(q(t), \dot{q}(t), t)$, the quantity P is conserved, i.e., $dP/dt = 0$ holds for all solutions of the Euler–Lagrange equations $E = 0$, if and only if there is an 'equivalent' conserved quantity P' and a function $R = R(q, \dot{q})$ such that $ER = dP'/dt$. Olver calls this the *characteristic form*, and R the *characteristic*, of the conservation law.[24] Now note the similarity between this equation and (7.3) above, in which the analogue of R is the characteristic $Q - \dot{q}T$ of the symmetry (7.2). Indeed, the modern version of Noether's combined forward and converse theorem rests on the key result that a group of transformations defines a variational symmetry group related to a given Lagrangian if and only if its characteristic is the characteristic of a conservation law for the associated Euler–Lagrange equations. A fuller account of the theorem (and the meaning of 'equivalence' above), will be given in Section 7.4.3.

In the meantime, note that in 1970, Candotti, Palmieri, and Vitale demonstrated a solution of a generalised version of (7.3) for strict symmetries in the case of a Lagrangian of the form $\mathcal{L}(q_i, \dot{q}_i, t)$, $i = 1, \ldots, n$ and conserved quantity D. In our case of a single degree of freedom, the solution is

$$Q = -\frac{\partial D}{\partial \dot{q}} H^{-1} + \dot{q}T; \quad T = \frac{1}{\mathcal{L}} \left[\frac{\partial \mathcal{L}}{\partial \dot{q}} \frac{\partial D}{\partial \dot{q}} H^{-1} - D \right], \tag{7.5}$$

where $H = \frac{\partial^2 \mathcal{L}}{\partial \dot{q} \partial \dot{q}}$ is the Hessian of $\mathcal{L}$ (see Candotti, Palmieri and Vitale 1972, sec. IF).[25]

23 The case of a conservation principle for a particle undergoing frictional force in the second paragraph of footnote 6 above is not a counterexample because the Lagrangian depends explicitly on t.

24 For Olver's proof of this result for more general Lagrangians, and for conservation laws of the form Div $P = 0$, see Olver (1986a, p. 270). It seems that the first reference to the characteristic form is due to Steudel (1962). I am grateful to Peter Olver for bringing this paper to my attention.

25 Note that their Euler–Lagrange expression and conserved quantity are the negative of ours. In the same paper, a related result for field theories is derived, for Lagrangian densities that again depend on the spatiotemporal coordinates, the fields and their

In anticipation of the discussion in Section 7.4.3 concerning the generalised Noether theorem and its converse, it can be shown that the analogue of (7.5) in the case of quasi-symmetries is:

$$Q = -\frac{\partial D}{\partial \dot{q}} H^{-1} + \dot{q} T; \quad T = \frac{1}{\mathcal{L}} \left[\frac{\partial \mathcal{L}}{\partial \dot{q}} \frac{\partial D}{\partial \dot{q}} H^{-1} - D - C \right]. \tag{7.6}$$

These transformations correspond to the case of $\delta \mathcal{L} = \frac{dC}{dt} \epsilon$.

7.4.2 The Free Particle

The simplest example of a Lagrangian of the above type is that associated with the non-relativistic free particle in one dimension, $\mathcal{L} = \frac{1}{2}\dot{q}^2$.[26] It can be used to exemplify the points we want to make, by looking at two elements of the Galilean symmetry group, namely boosts and time translations. Related remarks concerning scale symmetry are found in the Appendix.

While not strictly invariant under the one-parameter infinitesimal boost transformations

$$q' = q + \epsilon t; \quad t' = t, \tag{7.7}$$

(where the infinitesimal ϵ has dimension $[qt^{-1}]$) the free particle Lagrangian is quasi-invariant. Following the work of Bessel-Hagen in (1921), it has been widely (though belatedly, as we shall see) recognised that Noether's direct theorem can be generalised to include cases of 'quasi-' or 'divergence' invariance, in the light of the 'gauge' (divergence) freedom associated with Lagrangians.[27] In the present example, the conserved quantity is $q - \dot{q}t$.[28] It follows from the work of Candotti, Palmieri and Vitale (1972) that there is a *strict* variational symmetry (7.5) associated with this conservation law, in relation to the action associated with $\mathcal{L} = \frac{1}{2}\dot{q}^2$.[29] Curiously, Noether herself had already provided the explicit symmetry in her 1918 paper, without any reference to the issue of Galilean boost invariance in particle physics, but as an example of an invariance group $\mathcal{G}$ in which the transformations of the dependent and independent variables may involve derivatives of the dependent variable, called *generalised* symmetries by Olver (1986a, p. 292); the fact that her theorem also applies to such symmetries was long overlooked.[30]

In our notation, Noether's strict infinitesimal symmetry transformations are:

$$q' = q + \left(t - \frac{2q}{\dot{q}} \right) \epsilon; \quad t' = t - \frac{2q}{\dot{q}^2} \epsilon, \tag{7.8}$$

first derivatives. The authors point out that these conditions on the Lagrangians and Lagrangian densities for particles and fields, respectively, can be relaxed 'somewhat', but the derivations become more involved and less conceptually clear. Note finally that (7.5) will have singularities at the isolated points where the Lagrangian vanishes, as will Eqs. (7.6) and (7.13) below.

26 Mass plays no role in Newton's first law of motion and is ignored here.

27 Bessel-Hagen owed the idea of such generalisation to Noether herself; see Bessel–Hagen (1921); Kosmann-Schwarzbach, Chapter 1, this volume.

28 It is common to call the conserved quantities arising in Noether's theorem as first integrals (as Noether did herself), but sometimes a distinction is made between first integrals and constants of the motion (such as $q - \dot{q}t$) which depend on t.

29 In this case, the strict symmetry, if it exists, is unique, but in general in higher dimensions uniqueness holds up only to the addition of a null divergence. See Olver (1986a, exercise 5.23(b), p. 372).

30 Olver (1986a, p. 366) noted that generalized symmetries had been rediscovered many times; see also his Olver (2018) for their importance in soliton theory.

where the infinitesimal parameter ϵ again has dimension $[qt^{-1}]$ (Noether 1918).[31] This result surely *casts doubt on the common claim that conservation of centre of mass motion is intrinsically linked to boost symmetry.*

We might further ask: is there a Lagrangian (expected to be a gauge transformed version of $\frac{1}{2}\dot{q}^2$) for which the boost transformations (7.7) are a strict variational symmetry? Again, Noether gave the answer:

$$\mathcal{L} = \frac{1}{2}\left[\dot{q}^2 - \frac{d}{dt}\left(\frac{q^2}{t}\right)\right], \tag{7.9}$$

and this can be verified using (7.3) with $Q = t$ and $T = 0$, which follow from transformations (7.7). Once more, the conserved quantity can be shown to be $q - \dot{q}t$.[32] This is all consistent with (7.5).[33]

Now the infinitesimal transformations corresponding to time translation

$$q' = q; \quad t' = t + \epsilon, \tag{7.10}$$

constitute both a quasi-symmetry of the Noether Lagrangian (7.9) as well as (famously) a strict variational symmetry of the standard Lagrangian $\mathcal{L} = \frac{1}{2}\dot{q}^2$, the associated conserved quantity in both cases being the Hamiltonian, in this case $\frac{1}{2}\dot{q}^2$.[34] And notice that amongst the quasi-symmetries of the standard Lagrangian are

$$q' = q - \frac{2q}{t}\epsilon; \; t' = t + \left(1 - \frac{2q}{\dot{q}t}\right)\epsilon \tag{7.11}$$

and

$$q' = q - \dot{q}\epsilon; \quad t' = t, \tag{7.12}$$

the associated conserved quantity in both cases being again $\frac{1}{2}\dot{q}^2$. And from the result (7.5) of Candotti et al., we can calculate a strict symmetry associated with Noether's Lagrangian (7.9) and the same conserved quantity. The transformations are:

$$q' = q - \dot{q}(1 - T)\epsilon; \quad t' = t + T\epsilon; \quad T = \frac{1}{\mathcal{L}}\left(\frac{q\dot{q}}{t} - \frac{1}{2}\dot{q}^2\right), \tag{7.13}$$

which can be checked against (7.3), for $\mathcal{L}$ given by (7.9). (In all cases (7.10)–(7.13), ϵ has the dimension $[t]$.) The existence of the quasi-symmetries (7.11)–(7.13) again *flies in the*

31 I have not come across Noether's solution in the physics literature. In Santilli (1983, p. 342) suggests the existence of non-standard symmetries associated with the conservation of centre of mass motion, but leaves the details to the reader.

32 Denman (1966) showed that any Lagrangian for a classical particle moving in one direction which is strictly invariant under the Galilean transformations (7.7) must have the form $\mathcal{L}(w, t)$ where $w = q - \dot{q}t$. Noether's Lagrangian is $(q - \dot{q}t)^2/(2t^2)$.

33 Noether's point in introducing the Lagrangian (7.9) was to show that in going from it with its strict symmetry (7.7), to the gauge-related Lagrangian $\mathcal{L} = \frac{\dot{q}^2}{2}$ with its strict symmetry (7.8), the term $\overline{\delta q} \equiv Q - \dot{q}T$ is preserved, and the new transformations depend on derivatives of q – and that this is generally the case with such gauge-changes to the Lagrangian. We shall see the wider importance of this property of $\overline{\delta q}$ in the next section. For an interesting suggestion as to how Noether discovered generalised symmetries, see Olver (2018).

34 The analogous case of spatial homogeneity hardly needs a Noetherian analysis; the fact that for the standard Lagrangian, q is a cyclic, or ignorable variable, means that the Euler–Lagrange equation takes the form $\frac{d}{dt}\frac{\partial\mathcal{L}}{\partial\dot{q}} = 0$, which implies $\dot{q}$ is a constant of the motion.

face of the common assumption that conservation of energy is intrinsically related to the homogeneity of time.[35]

The suspicion that members of the pair (7.7) and (7.8) are in some sense equivalent, and similarly those of the quadruple (7.10), (7.11), (7.12), and (7.13), is borne out in the next section.

7.4.3 The Modern Noether Map

Being confined to strict symmetries, Noether's 1918 analysis arguably obscured the true nature of both the direct and converse theorems, as they are understood today. The key insight is that a bijective Noether map exists not between variational (quasi-)symmetries and conservation laws, *but between suitably identified equivalence classes of both.* To the best of my knowledge, the first proof of the existence of such a map, allowing for generalised symmetries (see above), is due to Luis Martínez Alonso in a regrettably little-known paper (Martínez Alonso 1979).[36] Then in 1986, Peter Olver independently proved the existence of the Noether map, again allowing for generalised symmetries (see Olver 1986b, p. 293; 1986a). In discussing post-Bessel-Hagen refinements to the direct theorem in his monumental 1986 monograph *Applications of Lie Groups to Differential Equations*, Olver wrote:

> Noether's theorem now provided a complete one-to-one correspondence between one-parameter groups of generalised variational symmetries of some functional and the conservation laws of its associated Euler–Lagrange equations. …Recent results have further crystallised the roles of trivial symmetries and conservation laws in the Noether correspondence for totally nondegenerate systems, with the consequence that each nontrivial variational symmetry group gives rise to a nontrivial conservation law, and conversely. (Olver 1986a, p. 293)

Olver expressed what I shall call the *Martínez Alonso Olver (MAO) theorem* as follows:

> if $\mathcal{L}$ is a nondegenerate variational problem, there is a one-to-one correspondence between equivalence classes of nontrivial conservation laws of the Euler–Lagrange equations and equivalence classes of variational symmetries of the functional [i.e. the action $\int \mathcal{L}\, dx$].[37]

Throughout his book, Olver uses the group theoretical representation of a symmetry of a system of differential equations in terms of a vector field $\mathbf{v}$ over some open subset of the space of independent and dependent variables, where $\mathbf{v}$ incorporates infinitesimal generators of the relevant symmetry; the notion of a variational symmetry relative to some action

35 If an objection is made to the effect that strict symmetries are more fundamental than quasi-symmetries, then what is one to say about the boost symmetry (7.7) in relation to (7.8)?

36 I am grateful to Peter Olver for alerting me to this paper, and for pointing out that it allows for generalised symmetries (private communication).

37 This is part of Theorem 5.42 in Olver (1986a, p. 328). Olver cites (p. 367) related results due to Vinogradov (1984), and to Steudel (1966). In Chapter 1 of this volume, Kosmann-Schwarzbach refers to the 'Noether–Olver–Vinogradov theorem', but I am unable to determine whether Vinogradov's theorem is equivalent to MAO.

It may be worth clarifying that the MAO theorem, being Lagrangian specific, does not imply that knowledge of some standard conservation laws determines a corresponding symmetry group of the dynamical equations of motion. After all, 'Since the conservation laws for a single particle are all a consequence of the constancy of its velocity, they are equally compatible with a law of motion (or indeed an action integral) invariant under the Lorentz group as with one invariant (up to a [surface term] only for the action integral) under the Galilei group' (Havas 1973, pp. 157–8).

$\int \mathcal{L}\, dx$ is likewise defined in these terms. In what follows I will continue to use the more clunky but perhaps more familiar language of explicit transformations of the dependent and independent variables, and for purposes of illustration, I will again use the simple example above of the free particle. The word 'symmetry', when variational in relation to some action, will be taken to mean either a strict or a quasi-symmetry, unless otherwise specified.

In informal terms, Olver's condition of total non-degeneracy for a system of analytic Euler–Lagrange equations ensures local solvability of the system, so that it is neither over- nor under-determined. (The condition rules out systems with 'local' symmetries, the subject of Noether's second theorem.[38]) Two conservation laws are equivalent according to Olver if they differ by a trivial conservation law, which is a linear combination of trivial laws of the first and second kind. An example of equivalence of the first kind for the free Newtonian particle is that between $\frac{d}{dt}\left(\frac{1}{2}\dot{q}^2\right) = 0$ and $\frac{d}{dt}\left(\frac{1}{2}\dot{q}^2 + \ddot{q}\right) = 0$, since $\ddot{q}$ vanishes on-shell. An example of a trivial law of the second kind for higher dimensional systems is Div $P = 0$ when P is a total curl, i.e., the law is a mathematical identity, and holds off-shell (Olver 1986a, pp. 268–70). Finally, two symmetries are equivalent if they differ by a trivial symmetry. In order to understand what Olver means by a trivial symmetry, a short but revealing detour is needed.

A notable feature of any symmetry in relation to a given system of differential equations – whether or not there is a Lagrangian for which they are Euler–Lagrange equations – is that if the associated transformations are not restricted to just the dependent variables, then there is a procedure for generating from them a symmetry that does have this property, associated moreover with the same first integral (for the general case, see, e.g., Boyer 1967, p. 451). Recall that if the transformations (7.2) are a quasi-symmetry in relation to the action $\int \mathcal{L}(q, \dot{q}, t)dt$, then

$$E(Q - \dot{q}T) = \frac{d}{dt}\left[-\frac{\partial \mathcal{L}}{\partial \dot{q}}(Q - \dot{q}T) - \mathcal{L}T + \mathcal{C}\right], \tag{7.14}$$

where $\delta L = \frac{d}{dt}\mathcal{C}\epsilon$. Now consider the transformations defined by

$$q' = q + \tilde{Q}\epsilon; \quad \tilde{Q} = Q - \dot{q}T; \quad t' = t. \tag{7.15}$$

Olver calls these contemporaneous transformations the *evolutionary representative, or form*, of (7.2), preserving the characteristic of (7.2). If Noether's 1918 converse theorem also holds for quasi-symmetries, then it is easy to confirm that the transformations (7.15) are a quasi-symmetry of the same action, with the same conservation law, but with $\mathcal{C}' = \mathcal{C} - \mathcal{L}T$. Indeed, it is a theorem that a dynamical symmetry of a given system of differential equations

38 The corresponding assumption in Martínez Alonso (1979) is his 'normality' condition on a system of partial differential equations. Note that by a symmetry being 'local', I mean one in which the transformations of the independent and dependent variables are spacetime dependent, unlike in (7.2) above, when the parameter ϵ is 'global', i.e., does not depend on t. In the mathematics literature, including Olver (1986), the term 'local' applies to any symmetry if its infinitesimal generator depends only on the independent and dependent variables and their derivatives, i.e., the jet coordinates. According to this definition, both Noether's first and second theorems concern local symmetries.

In order to avoid possible confusion, theories with local (in our sense) symmetries, such as gauge-invariant Maxwellian electrodynamics formulated in terms of 4-potentials, are nonetheless subject to the MAO theorem as long as they also possess symmetries which are global. The first application of Noether's first theorem to electrodynamics was given by Bessel-Hagen (1921).

is a variational symmetry in relation to the action $\int \mathcal{L} dx$ if and only if its evolutionary representative is too (see Olver 1986a, Proposition 5.36, p. 325). Note that it follows immediately from both Eqs. (7.5) and (7.6) that whether the symmetry (7.2) is a strict or a quasi-symmetry, $\tilde{Q}$ takes the form:

$$\tilde{Q} = -\frac{\partial D}{\partial \dot{q}} H^{-1}, \tag{7.16}$$

so apart from the conserved quantity D, $\tilde{Q}$ depends only on the Hessian of the Lagrangian.

Now Olver calls a symmetry of a system of differential equations *trivial* if *its evolutionary form has coefficients which vanish on-shell.* In our case, this means that the symmetry (7.2) is trivial when the coefficient of ϵ in (7.15) vanishes on-shell, such as in the case of $q' = q + \ddot{q}\epsilon'$, $t' = t$. Olver writes:

Two generalized symmetries $\mathbf{v}$ and $\tilde{\mathbf{v}}$ are called *equivalent* if their difference $\mathbf{v} - \tilde{\mathbf{v}}$ is a trivial symmetry of the system. This induces an equivalence relation on the space of generalized symmetries of the given system; moreover, we will classify symmetries up to equivalence so by a *symmetry* of the system we really mean a whole equivalence class of generalized symmetries, each differing from the other by a trivial symmetry (Olver 1986a, p. 298).[39]

Olver gives some examples related to the heat equation (for flow in a one-dimensional rod with unit diffusivity):

$$\frac{\partial u}{\partial t} - \frac{\partial^2 u}{\partial x^2} = 0, \tag{7.17}$$

in terms of the following infinitesimal symmetries:

$$x' = x, \ t' = t + \epsilon, \ u' = u, \tag{7.18}$$

$$x' = x, \ t' = t, \ u' = u - \dot{u}\epsilon', \tag{7.19}$$

and

$$x' = x, \ t' = t, \ u' = u - \frac{\partial^2 u}{\partial x^2}\epsilon'', \tag{7.20}$$

where the infinitesimals ϵ and ϵ' have dimension $[t]$, and ϵ'' has dimension $[x^2]$.[40] As Olver states, (7.19) is the evolutionary form of time translation (7.18), and he claims that all three symmetries are equivalent – and 'for all practical purposes determine the self-same symmetry group' (Olver 1986a).[41] Indeed, subtracting (7.20) from (7.18), as well as (7.20) from (7.19), result in symmetries whose evolutionary forms are trivial on-shell. But subtracting (7.19) from (7.18) results in a symmetry whose evolutionary form is

[39] Note that a dynamical symmetry may be equivalent to a variational quasi-symmetry but not itself be one. Recall the quasi-symmetry (7.7); an equivalent dynamical symmetry obtained by adding $\ddot{q}$ to the q transformation is no longer a variational symmetry. See Olver (1986a, exercise 5.22, p. 372).

[40] Emphasis on the heat equation in this context is perhaps unfortunate, because as a scalar evolution equation it does not have a Lagrangian formulation. But there is a twist in the tale; see Appendix A below.

[41] The notion of equivalence here needs nuancing. The time translation (7.18) has, for an isolated subsystem of the universe, a familiar and straightforward operational rendering that (7.19) and (7.20) do not, related to the homogeneity of time. Olver's notion of equivalence is formal, and meaningful only in relation to a given set of dynamical equations. Similar remarks hold for the 'equivalences' of certain transformations discussed in Section 7.4.2 for the free particle.

trivial *because its coefficients vanish identically*, and thus vanish off-shell. (A treatment of these results in Olver's vectorial notation is given in Appendix 7.A.) In fact, it is easy to prove that the difference between any symmetry and its evolutionary form has this property.

Now it is noteworthy that before Olver gives the above definition of triviality, when introducing the notion of the evolutionary representative of a symmetry he claims that the two are 'essentially the same symmetry' (Olver 1986a, p. 297). Thus there are, as in the case of conservation laws, *two types of triviality* in the case of evolutionary symmetries that Olver is effectively assuming: on-shell and off-shell. Recognition of this point does not, however, change the details of the proof of the MAO theorem.[42]

We return now to the Lagrangian formulation of the dynamics of the free particle. It is easy to confirm that the boost transformations (7.7) are the evolutionary representative of Noether's transformations (7.8), and that the transformations (7.12) are the evolutionary representative of (7.10), (7.11), and (7.13). So (7.7) and (7.8) are equivalent in the off-shell sense, and the same goes for (7.10), (7.11), (7.12), and (7.13) – and note that the formula for $\tilde{Q}$ in (7.16) is consistent with (7.7) and (7.12) given the associated first integrals $q - \dot{q}t$ and $\frac{\dot{q}^2}{2}$, respectively.

Note too that whether two symmetries are equivalent because one is the evolutionary form of the other is insensitive to the gauge chosen for the relevant Lagrangian. Now recall that the MOA theorem above defines equivalence classes of variational symmetries *relative to a given action*. So far we have seen that transformations (7.7), (7.8), and (7.10)–(7.13) are variational symmetries of the action associated with the standard Lagrangian $\mathcal{L} = \frac{1}{2}\dot{q}^2$, and the transformations (7.7) and (7.10) are variational symmetries of the action associated with gauge-related Noether Lagrangian (7.9). Consistency with the MOA theorem and the off-shell definition of equivalence demands that all of these transformations should be variational symmetries of both Lagrangians, with (7.7) and (7.8) having associated conserved quantity $q - \dot{q}t$, and (7.10)–(7.13), having $\frac{\dot{q}^2}{2}$.

Calculations show that this is so, and summary details are provided in Appendix 7.B. It is seen that neither of these Lagrangians is strictly invariant under both boost transformations and time translations. In Chapter 6 of this volumne, Olver (2022) demonstrated that there is no non-constant first-order Lagrangian for the non-relativistic free particle that is strictly invariant with respect to the full Galilean group. In the same chapter he also provided a necessary and sufficient condition, using cohomology theory, for every divergence-invariant Lagrangian related to a connected Lie group of point transformations to be locally equivalent to a strictly invariant Lagrangian.

[42] As mentioned in Section 7.4.1, Olver's proof rests on the key result that a group of transformations defines a variational symmetry group related to a given Lagrangian if and only if its characteristic is the characteristic of a conservation law Div $P = 0$ for the associated Euler–Lagrange equations. In defining an equivalence class of variational symmetries, the only relation between characteristics that is of interest is the on-shell relation: symmetries related by off-shell equivalence automatically have the same characteristic. (A similar situation holds for the proof of Martínez Alonso.) I am grateful to Peter Olver for clarifying this matter (private communication). Note that by admitting off-shell equivalence, an equivalence class of symmetries gains members, but the cardinality of the set of equivalence classes is not altered, and remains the same as that of the equivalence classes of conservation principles.

7.5 Global Spacetime Symmetries

If the original question of explanatory priority cannot be settled in a straightforward way in the light of the modern bijective Noether map, there may be other reasons why symmetries are routinely given privileged status.

The quotations from Landau and Lifshitz, Zee, and Ryder in Section 7.2 refer specifically to space and time symmetries. Properties of space such as homogeneity and isotropy, of time such as homogeneity, and properties of space-time such as boost invariance seem to transcend the physics of any one type of non-gravitational interaction. Such universality might suggest that these properties are primitive attributes of the very spacetime manifold on which different field theories are written, and more fundamental than conservation principles.[43] But note that as late as 1900, H. A. Lorentz would extol two principles as putative universal constraints on theories: the second law of thermodynamics and the conservation of energy (see Frisch 2005, section 3). It was Einstein's special theory of relativity that put the spotlight on the role of symmetries *qua* constraints, and we return to this development in the following section.

At this point it is worth reminding ourselves that some natural-looking properties of space and time have turned out *not* to be universally valid. The weak interactions 'have little respect for symmetries' (Bernabéu and Martínez-Vidal 2015, section 3), at least of the discrete kind. The experimental violation of parity (P, or space-inversion) in 1957 came as a bolt out of the blue; the experimental violation of time inversion symmetry announced in 2012 was less surprising, given the violation of CP symmetry (a combination of P and charge conjugation C) in the weak interactions in 1964 and 2001.[44] These results are hard to reconcile with the claim that spacetime symmetries have a kind of metaphysical necessity: cherry picking is not allowed. It makes sense then that searches still take place for possible violations of Lorentz covariance.[45]

It is also helpful to look at the significance of these global spacetime symmetries from the perspective of Einstein's general theory of relativity (GTR). None of these symmetries emerges from inspection of Einstein's field equations, whose symmetry group is the local diffeomorphism group. They arise from the specific way matter fields couple to the metric field as determined by the strong (Einstein) equivalence principle, which brings the Poincaré group of relativistic boosts, rotations, and translations, and hence the local validity of special relativity, into the story. Space, for instance, is neither intrinsically homogeneous nor isotropic from the perspective of the gravitational (metric) degrees of freedom; these properties emerge from the Euclidean subgroup of the Poincaré group associated with the matter field equations when expressed in the appropriate local freely falling frames. The universality of local Poincaré covariance for the diverse non-gravitational interactions is one of the remarkable features of the strong (or Einstein) equivalence principle in GTR. Such

43 The pertinent issue here is universality, not intuitive appeal. The boost claim is surely less intuitive than the homogeneity claims. As regards the latter, Lewis Ryder (1996, section 3.2) wrote: 'If it were not for these conditions, it is obvious that science itself would be impossible.' More complicated, maybe, but arguably not impossible; physics has accommodated the violation of some space-time symmetries, as we see below.

44 For details see Lees et al. (2012) and Bernabéu and Martínez-Vidal (2015).

45 See Will (2014, section 2.1.2).

covariance is a property of the relevant matter field equations, and in this sense the same conclusions can even be applied to the nature of spacetime in special relativity. Finally, even such continuous, global spacetime symmetries are, from the point of view of GTR, only approximately valid. They hold in circumstances in which the effects of spacetime curvature are negligible.[46]

7.6 A Brief Ode to Symmetries

7.6.1 Pragmatic Considerations

(i) It is striking that, despite its fame in physics, Noether's theorem was largely overlooked by physicists for decades. Olver (1986a) wrote :

> by 1922 [the year following Bessel-Hagen's version of the direct theorem for quasi-symmetries] all the machinery for a detailed, systematic investigation into the symmetry properties and consequent conservation laws of the important equations of mathematical physics was available. Strangely enough, this did not occur until quite recently (p. 288).[47]

As Olver noted, an important event was the publication in 1951 by E. L. Hill of a review of (a special case of) Noether's direct theorem in *Reviews of Modern Physics*. In this paper, Hill lamented that

> Despite the fundamental importance of this theory there seems to be no readily available account of it which is adapted to the needs of the student of mathematical physics, while the original papers [of Klein, Noether and Bessel-Hagen] are not readily accessible. (Hill 1951)

As late as 1964, Tassie and Buchdahl (1964) extended the direct theorem to quasi-symmetries, in apparent ignorance of the 1921 work of Bessel-Hagen. An English translation of Noether's 1918 paper was not published until 1971, at which time applications of her theorem in mathematical physics were still thin on the ground. A translation of Bessel-Hagen's 1921 paper, which applied the generalised Noether theorem to electrodynamics for the first time (including its conformal symmetry) appeared only in 2006. In his comments on the history of the subject, Olver (1986a, p. 288) noted that important applications of conservation principles and Noether-type identities in elasticity, scattering theory, and optics occurred before the connection with Noether's theorem was realised.

It might help to pause for a moment on the case of elasticity. A major advance in the treatment of defects (dislocations, point defects, interfaces, and crack tips) within elastic media emerged in the work of J. D. Eshelby (1951, 1956).[48] Eshelby was able to prove the existence of a stress-energy tensor, and a related conservation law, from which the force on the defect can be calculated, in analogy to the role of the Maxwell stress tensor

[46] This brief discussion of the strong equivalence principle in GTR is over-simplified; for more extensive discussion, see Read et al. (2018).

[47] Kosmann-Schwarzbach (2010) has provided an extensive treatment of the history of the reception of Noether's 1918 paper, a précis of which is found in her Chapter 1 of this volume. For related remarks, see also Olver (2018).

[48] See Eshelby (1951) and Eshelby (1956); Eshelby's contributions to the field extended well beyond these initial papers.

in electrostatics in establishing the force on a charge.[49] The 1968 work of Günther and the independent work by Knowles and Sternberg in 1972 had together established, using Noether's theorem, the relevant conservation law on the basis of spatial translation symmetry of an elastically homogeneous material, a second conservation law that holds in the case of homogeneity and isotropy,[50] and a third law in relation to the assumption of scale invariance.[51] In 1984, Olver found further undetected symmetries of the equations of linear elasticity, associated with new conservation laws.[52]

Noether's theorem played two roles in this development of the theory of defects in elastic media. First, here is Eshelby himself (1975, p. 322):

> The normal theory of elasticity recognizes nothing which corresponds with the force on a defect. ...But in fact the appropriate concept has been to hand ever since the appearance of a paper by Noether ...in 1918, in the form of the energy-momentum tensor which the elastic field possesses in common with every field whose governing equations are derivable from a variational principle, and some for which they are not.

Eshelby had not needed Noether's theorem for the discovery of his stress-energy tensor, but by giving in 1975 a Lagrangian treatment of elasticity, he wanted to show that the theory was amenable 'to the standard results of the classical part of general field theory'.[53]

But more importantly for our purposes, the use of Noether's theorem led to the discovery of new conservation laws for elastic media. And here a discrepancy between symmetries and conservation laws is apparent in practice: it seems that it is easier in general to make progress by using Noether's direct theorem than by using its converse. In the case of elasticity, the symmetries follow directly from the nature of the Lagrangian which in turn reflects the assumed properties of the elastic medium under consideration.

(ii) Mention should be made of the fact that knowledge of first integrals plays a role in the possible solution of equations of motion by quadrature (integration), including numerical integration.[54] For instance in the case of $\mathcal{L}(q,\dot{q},t)$, knowledge of a conserved quantity allows for integration of the Euler–Lagrange equation completely by quadratures. For the

49 A very clear treatment of the application of Eshelby's method in the cases of pressure in an interface and force on a static defect is found in Sutton (2020, chapter 8).

50 The homogeneity and isotropy of an elastic medium are usually defined in terms of the corresponding properties of the elastic constants of the medium, but sometimes the defects themselves are regarded as inhomogeneities. From the point of view of Noether's theorem, what matters is whether the Lagrangian density itself (the negative of the elastic energy density) is symmetric in these ways.

51 See Günther (1962) and Knowles and Sternberg (1972). It is worth noting two features of this application of Noether's theorem. First, in elastostatics, time obviously makes no appearance; the infinitesimal transformations are defined for the spatial coordinates and the displacement vector field, and the conservation laws do not yield constants of the motion in the usual temporal sense of mechanics. Indeed they represent a counterexample to the claim that 'Noether's theorem concerns the way particles behave *under temporal evolution*' (Schroeren 2020, p. 52). (Recall point 5 in Section 7.3.) Second, the theorem holds only in homogeneous regions of space inside the solid body which have no defects, though it allows for forces on defects to be calculated. For subsequent developments involving such conservation laws, see Rice (1985).

52 See Olver (1984a, b). A summary of some of these applications of Noether's theorem is found in Olver (1986a, Example 4.32, pp. 281–3). Federico et al. (2019) provides an explicit relation between Eshelby's inclusion theory and Noether's theorem, and cites several more papers on the role of Noether's theorem in the theory of elasticity published since Olver's work.

53 In this way Eshelby hoped to attract the interest of applied mathematicians in the stress-energy tensor, the lack of which (the work of Gunther and Knowles and Sternberg being exceptions) he attributed tentatively to 'the artificial separation which has grown up between applied mathematics and theoretical physics' (Eshelby 1975, p. 323).

54 For mechanical systems, the original meaning of the term (first) integral was not a constant of the motion but an equation of the form $\frac{d}{dt} f(q_i, \dot{q}_i, \ddot{q}_i, ..., t) = 0$, which can be solved by integration.

more general case involving n-th order variational problems, knowledge of a one-parameter group of variational symmetries allows for reduction of the associated Euler–Lagrange equations of order $2n$ to those of order $2n-2$ (see Olver 1986a, Theorem 4.17, p. 262). Olver (1986a) wrote that "Noether's method is the only really systematic procedure for constructing conservation laws for complicated systems of partial differential equations' (p. 246), although more recently computer algorithms have been developed to generate conservation laws which are independent of variational considerations (for details, see Wolf 2002).

(iii) Finally, following the seminal work of Wigner (1939), there is widespread acceptance by physicists of the claim that the very properties of elementary particles are grounded in continuous symmetry groups. Mass and spin, for example, are represented by Casimir invariants of the Poincaré group. In the words of Steven Weinberg (1993, p. 138n), such properties 'are what they are because of the symmetries of the laws of nature'.[55]

I do not wish to assert that any of the pragmatic considerations in this subsection are likely to be the motivation for the claims quoted in Section 7.2. But these considerations seem to support the notion that Noether's direct theorem has proved more useful in mathematical physics than its converse, as well as to underline the prominent role of symmetries in the articulation of particle physics.

7.6.2 Heuristics

The prominent heuristic role symmetries have played in twentieth-century physics must surely be part of our account. This story is well-known but, to my knowledge has not been stressed in discussions of the explanatory arrow in Noether's theorem, at least in the philosophical literature. So the reader may forgive a brief recap.

Einstein's 1905 derivation of the Lorentz transformations rested on two fundamental symmetry principles: the relativity principle (dynamical equivalence of inertial frames) and the isotropy of space, alongside the postulate governing the constancy of the speed of light with respect to the 'resting' frame.[56] The justification of all these principles did not rest, for Einstein, on any *a priori* notions about the structure of space and time, but was based on 'plenty of experiential knowledge' related to mechanics and electrodynamics.[57] In 1940, Einstein (1940, p. 490) would stress that the theory of special relativity could be summarised in one principle: 'all natural laws must be so conditioned that they are covariant with respect to Lorentz transformations'. This allowed Einstein to say that the theory transcended Maxwell's equations, and what he saw as the awkward emphasis on the role of light in his 1905 formulation.[58] Special relativity is essentially a constraint in the sense that a symmetry is being imposed on the fundamental equations of *all* the non-gravitational interactions.

55 This is also cited in Schroeren (2020), which provides a recent philosophical analysis of such claims.
56 For the role specifically of spatial isotropy in Einstein's derivation, see Brown (2007, section 5.4.3).
57 Einstein to Amiet, 17 December 1947; The Albert Einstein Archives at the Hebrew University of Jerusalem, 25-335.
58 For further discussion, see Brown (2007, chapter 8), and especially Giovanelli, section 4.1.

This amounts to only the second time that a constraint, or set of constraints, on fundamental physics has been given the honorific title of a theory; the first was thermodynamics.[59]

One of the most remarkable methodological trends in modern physics has been the *a priori* use of symmetry principles to constrain the action principles of the non-gravitational interactions in quantum electrodynamics (QED) and particle physics. It is what A. Zee (2003, p. 457) called the new paradigm of 'symmetry → action → experiments' in fundamental physics (see also Zee 1999, chapter 6; Olver 2018). Indeed the use of the Lagrangian formalism has become virtually *de rigeur* in these fields (though less so in axiomatic, algebraic quantum field theory[60]) largely because it is so friendly to the imposition of symmetry constraints in comparison to conservation laws.[61] Symmetries can be said to be worn on the sleeves of the Lagrangian (which is not to say that given a Lagrangian, *all* of its symmetries 'leap out at you'; see Zee 2003, p. 457, for this point). And the issue goes beyond the special relativistic constraint (Lorentz invariance).[62] An equally, if not more, important actor in the story is the Weyl–Yang–Mills *gauge principle* involving internal symmetries.

QED is based on a Lagrangian

$$\mathcal{L}_{\text{QED}} = \mathcal{L}_{\text{Dirac}} + \mathcal{L}_{\text{Maxwell}} + \mathcal{L}_{\text{int}}, \tag{7.21}$$

coupling the Dirac field with the Maxwell field. This Lagrangian can be considered the result of taking the Dirac Lagrangian for the free electron, which is invariant under global phase transformations associated with the abelian U(1) group, and replacing it by one which is invariant under local (spacetime dependent) phase or 'gauge' transformations. Needless to say, the resulting appearance of a 4-vector potential (or connection in the language of fibre bundles), with its appropriate compensating transformations, does not mean anything physically interesting *per se*. It is the further requirements that such a 'gauge' field (a) be dynamical (b) give rise to a locally invariant kinetic energy term in the Lagrangian that depends only on the the 4-potential and its derivatives, and (c) interact with the Dirac field by way of 'minimal coupling', i.e, by replacing derivatives of the Dirac field by a suitable covariant derivative in the interaction term in the Lagrangian,[63] that brings empirical clout

[59] The difference between the 1940 formulation (as distinct from the 1905 one) and thermodynamics is that now the constraint is no longer purely phenomenological, and hence special relativity ceases to be a strict 'principle' theory in Einstein's terminology. See Brown (2007, section 8.4.1).

[60] Orzalesi (1970) argues that it is the converse of Noether's theorem that is valid.

[61] In Pauli's magnificent 1921 monograph on relativity theory, the English translation of which was published in 1958, he wrote that 'it is not at all self-evident, from a physical point of view, that the physical laws should be derivable from an action principle' (Pauli 1958, p. 201). Such qualms are rare today, at least in relation to fundamental physics.

[62] According to Peter Havas, the heuristic role of *universal* Lorentz invariance ultimately finds its justification in terms of belief in the standard conservation principles:

> The fundamental importance of the invariances lies in their role in the development of physical theories, in the help they may provide in our search for 'all the laws of nature'. Thus, once physicists had come to trust the special theory of relativity, the requirement of invariance under the proper orthochronous Lorentz group served as a selection criterion in limiting the number of theories acceptable for further scrutiny as possible candidates for a correct description of parrticular classes of phenomena. If we then further require that all the laws of nature should be derivable from a *single* Lorentz invariant ... variational principle, we are automatically provided with a *universal* set of ten conservation laws which can be readily interpreted in terms of familiar physical concepts.
>
> It is precisely the strength of the physicists' belief in the universal validity of such a set of conservation laws which motivates their preference for a single variational principle invariant under a universal group in the first place. (Havas 1973, p. 160).

[63] Note that the minimal coupling assumption is appropriate only if the free Dirac Lagrangian is chosen to be 'minimal' with respect to Lagrangian gauge transformations; see Sakurai (1964, p. 183).

to the procedure such that the 4-potential gains its familiar electrodynamic currency.[64] Yang and Mills (1954) generalised this 'gauge principle' to a system with global symmetry (isotopic spin rotation for nucleons) associated with a nonabelian group. The rest of the story is encapsulated in the words of Zee (2003, p. 361):

> In the late 1960s and early 1970s, the electromagnetic and weak interactions were unified into an electroweak interaction, described by a nonabelian gauge theory based on the group $SU(2) \otimes U(1)$. Somewhat later, in the early 1970s, it was realised that the strong interaction can be described by a nonabelian gauge theory based on the group $SU(3)$. Nature literally consists of a web of interacting Yang–Mills fields (Zee 2003, p. 361).[65]

It may be unclear why the gauge principle works when it does, and for some commentators the fact that the physics of the non-gravitational interactions is redolent of gauge symmetry seems to represent an awkwardness in the standard model (see, e.g., Zee 2003, p. 456). Be that as it may, the heuristic role of symmetry in the development of post nineteenth-century physics, including string theory, has no comparable historical precedents.[66] It has also led to the use of advanced geometrical and group theoretical methods in particle and condensed matter physics.

It is hard to believe that this development has not, to some extent, influenced the conventional reading of Noether's theorem, at least amongst physicists since the 1970s. While, again, it does not strictly justify the claim that symmetries have explanatory priority in Noether's theorem, recognition of the important heuristic role of symmetries in the standard model may be a contributing factor behind the relative unfamiliarity of the converse Noether theorem and its implications.

[64] It is also necessary in this procedure to assume renormalizability as well as the invariance of the Lagrangian under time reversal or parity; see Peskin and Schroeder (1995, section 15.1). It is noteworthy that in his excellent monograph *The Geometry of Physics* Theodore Frankel (2004, p. 535) gives the misleading impression that step (a) in the gauge principle is a requirement of quantum mechanics, rather than an independent postulate. A more accurate spin on the significance of the 4-vector field is given by Lancaster and Blundell (2015, p. 128):

> It only exists in our description because we've invented it to satisfy our demand for a locally invariant theory, but if such ambitions have any grounding in reality, then the [4-vector] field should have dynamics of its own!

See also Brown (1999, section 2(a)). An amusing article by Heras (1994) shows that applying the gauge principle to the Schrödinger equation for a free electron, without imposing the analogue of (b) above, is consistent with an elliptical variant of Maxwell's equations with Euclidean, rather than Lorentzian, symmetry, and hence possessing no propagating solutions and no invariant speed.

[65] The original problem facing Yang–Mills theory is that it gives rise to massless spin-1 particles which, apart from the photon, are not observed. In the eventual electroweak theory, the Yang–Mills particles acquire mass through the Higgs mechanism, and in quantum chromodynamics, the unobservability of the particles (gluons) is explained through the phenomenon of asymptotic freedom; see, e.g., Zee (2003, chapters VII.2 and VII.3).

[66] The one, isolated pre-Einstein case is that of Christiaan Huygens, who in 1656 used the relativity principle as a postulate in deriving his non-Cartesian theory of collisions; an insightful account of this episode is given by Barbour (2001, section 9.4). Barbour mentions on p. 470 the intriguing possibility, first mooted by Martin Klein, that Einstein may have been influenced by the account of Huygens' theory of collisions in Mach's *Mechanics*, a book which he 'read avidly'. A fable concerning a 1705 anticipation of Einstein's 1905 logic, in which Albert Keinstein uses the relativity principle to derive the Galilean transformations, is found in Brown (2007, chap. 3).

Justification for the claim that Einstein's route to his 1915 theory of gravity, general relativity, was motivated (*inter alia*) by a symmetry principle, namely general covariance, is relatively problematic. Not only did Einstein erroneously think, until at least 1918, that general covariance was a generalisation of the relativity principle, he also abandoned the principle between 1913 and early 1915. For details, see Brown (2007, Appendix A), and especially Janssen (2014, section 3).

Appendix 7.A Heat Equation I

Consider two conserved quantities associated with the heat equation (7.17):

$$Q(u) = \int_{-\infty}^{\infty} u(x,t)dx, \tag{7.22}$$

and

$$\tilde{Q}(u) = \int_{-\infty}^{\infty} xu(x,t)dx, \tag{7.23}$$

which hold under the boundary conditions $u_x|_{-\infty}^{\infty} = 0$ and $(xu_x - u)|_{-\infty}^{\infty} = 0$, respectively (see Steinberg and Wolf 1981, p. 39), where $u_x \equiv \frac{\partial u}{\partial x}$. These conserved quantities cannot be associated in the Noetherian sense with symmetries of (7.17) because the heat equation does not have a variational formulation. However, consider the Lagrangian containing an auxiliary variable $v = v(x,t)$:

$$\mathcal{L} = \frac{1}{2}(vu_t - uv_t) + u_x v_x. \tag{7.24}$$

The Euler–Lagrange equations obtained by varying with respect to v and u are respectively (7.17) and its adjoint:

$$v_t + v_{xx} = 0. \tag{7.25}$$

Noether's first theorem can now be shown to lead to the conservation of Q and $\tilde{Q}$ in relation to the respective symmetries of the Lagrangian (7.22):

$$x' = x,\ \ t' = t,\ \ u' = u + \epsilon;\ \ \ x' = x,\ \ t' = t,\ \ u' = u + x\epsilon'.\text{[67]} \tag{7.26}$$

Now the Schrödinger equation in quantum mechanics for the free particle in one spatial dimension,

$$\psi_t - \frac{i\hbar}{2m}\psi_{xx} = 0, \tag{7.27}$$

is the complex analogue of the heat equation (7.17) with finite (complex) diffusivity. The Lagrangian can be written as

$$\mathcal{L} = \frac{i\hbar}{2}\left(\psi^*\psi_t - \psi_t^*\psi\right) - \frac{\hbar^2}{2m}\psi_x^*\psi_x, \tag{7.28}$$

in analogy to (7.24), where ψ^* is the complex conjugate of ψ. It seems that the first Noether-type demonstration that strict wave-functional analogues exist of the conservation principles associated with the quantities (7.22) and (7.23), from analogues of the transformations (7.26), was published in 2004; the authors were unaware of the analogy with the heat equation (see Brown and Holland 2004b, section III). Note that because the wave function is

[67] The application of Noether's theorem for deriving the conservation of (7.22) is given in Ibragimov and Kolsrud (2004, p. 35), putting $\phi = 1$. The corresponding result for (7.23) follows from the equations (7.3) through (7.9) on p. 30, *op. cit.* A novel non-Noetherian route to (7.22) is found in Ibragimov (2006b, pp. 69, 70), using the Lagrangian $\mathcal{L} = v(u_t - u_{xx})$ and the Galilean symmetry of the heat equation. In all these cases the derivation of a conservation equation of the form $dC_1/dt + dC_2/dx = 0$ must be followed by integration over space and appeal to Gauss's divergence theorem.

complex, so that ψ^* is not an auxiliary function independent of ψ, the Schrödinger equation is time-reversal invariant, unlike the heat equation (7.17).

Appendix 7.B Heat Equation II

In relation to the heat equation (7.17), consider the general infinitesimal transformations

$$x' = x + \xi_x[u]\epsilon, \quad t' = t + \xi_t[u]\epsilon, \quad u' = u + \phi[u]\epsilon. \tag{7.29}$$

In terms of their generators, the combination of these transformations can be represented by a vector field over the space of dependent and independent variables:

$$\mathbf{v} = \xi_x[u]\partial_x + \xi_t[u]\partial_t + \phi[u]\partial_u, \tag{7.30}$$

where $\partial_x \equiv \frac{\partial}{\partial x}$, etc. Its evolutionary form is (Olver 1986a, equation (5.7)).

$$\tilde{\mathbf{v}} = [\phi[u] - (\xi_x[u]\partial_x + \xi_t[u]\partial_t)]\,\partial_u. \tag{7.31}$$

The specific symmetries (7.18), (7.19), and (7.20) of the heat equation are represented by

$$\mathbf{v}_{18} = \partial_t; \quad \mathbf{v}_{19} = -\dot{u}\partial_u; \quad \mathbf{v}_{20} = u_{xx}\partial_u, \tag{7.32}$$

where $u_{xx} \equiv \frac{\partial^2 u}{\partial x^2}$. It is easily seen that (i) $\mathbf{v}_{19}$ is the evolutionary form of $\mathbf{v}_{18}$, (ii) the evolutionary forms of both $\mathbf{v}_{18} - \mathbf{v}_{20}$ and $\mathbf{v}_{19} - \mathbf{v}_{20}$ vanish on-shell, and (iii) the evolutionary form of $\mathbf{v}_{18} - \mathbf{v}_{19}$ is identically zero.

Appendix 7.C Free Particle

(i) In relation to the action defined with respect to the standard action $\mathcal{L} = \frac{1}{2}\dot{q}^2$ for the free particle, the nature of the variational symmetries in Section 7.4.2 is as follows:

- Boost transformations (7.7) are a quasi-symmetry, with divergence term $\dot{q}\epsilon$, and the equivalent (7.8) are a strict symmetry. The corresponding non-trivial conserved quantity in both cases is $q - \dot{q}t$.
- Time translation transformations (7.10) are a strict symmetry; (7.11) are a quasi-symmetry with divergence $\frac{d}{dt}\left(-\frac{q\dot{q}}{t}\right)\epsilon$; (7.12) are a quasi-symmetry with divergence $\frac{d}{dt}\left(-\frac{\dot{q}^2}{2}\right)\epsilon$ and (7.13) are a quasi-symmetry with divergence $\frac{d}{dt}\left(\frac{q\dot{q}}{t} - \frac{\dot{q}^2}{2}\right)\epsilon$. In all these equivalent cases the nontrivial conserved quantity is $\frac{\dot{q}^2}{2}$.

(ii) In relation to the action defined with respect to the Noether Lagrangian (7.9):

- Transformations (7.7) are a strict symmetry and transformations (7.8) are a quasi-symmetry with divergence $\frac{d}{dt}\left(-\frac{2q}{\dot{q}^2}\mathcal{L}\right)\epsilon$, where $\mathcal{L}$ is the Lagrangian (7.9). The conserved quantity as above.
- Time translation transformations (7.10) are a quasi-symmetry with divergence $\frac{d}{dt}\left(\frac{q^2}{2t^2}\right)\epsilon$; (7.11) are a quasi-symmetry with divergence $\frac{d}{dt}\left(\frac{3q^2}{2t^2} - \frac{q\dot{q}}{t} - \frac{q^3}{\dot{q}t^3}\right)\epsilon$;

(7.12) are a quasi-symmetry with divergence $\frac{d}{dt}\left(\frac{q\dot{q}}{t} - \frac{\dot{q}^2}{2}\right)\epsilon$ and (7.13) are a strict symmetry. The conserved quantity as above.

(iii) Consider now the group of rescaling transformations defined by

$$q' = \lambda q;\ t' = \lambda^2 t, \qquad \lambda \in \mathbb{R} \setminus \{0\}. \tag{7.33}$$

Both of the Lagrangians above are strictly invariant with respect to these transformations, and the associated conserved quantity is $\dot{q}(q - \dot{q}t)$, the product of the conserved quantities associated with spatial translations and boosts, respectively.[68] When a potential energy term is included in the standard Lagrangian, the transformations (7.22) are variational symmetries only if the particle moves under the influence of an inverse cube force law. More interesting applications of (approximate) scale invariance have been studied in relativistic quantum field theory (for an introduction, see Jackiw 1972).

References

Anderson, N. and Arthurs, A. M. (1978). A variational principle for Maxwell's equations. *International Journal of Electronics Theoretical and Experimental*, **45**(3), 333–4. https://doi.org/10.1080/00207217808900916.

Barbour, J. (2001). *The Discovery of Dynamics*. New York: Oxford University Press.

Bernabéu, J. and Martínez-Vidal, F. (2015). Colloquium: Time-reversal violation with quantum-entangled *B* mesons. *Reviews of Modern Physics*, **87**(1), 165–82. https://doi.org/10.1103/RevModPhys.87.165

Bessel–Hagen, E. (1921). Über die Erhaltungssätze der Elektrodynamik. *Mathematische Annalen*, **84**, 258–76. English translation by M. Albinus and N. H. Ibragimov: On conservation laws of electrodynamics, in *Archives of ALGA*, Vol. 3 (2006), 33–51. ALGA Publications ISSN 1652-4934. www.researchgate.net/publication/311456950_Archives_of_ALGA_vol_3.

Blaker, J. W. and Tavel, M. A. (1974). The application of Noether's theorem to optical systems. *American Journal of Physics*, **42**, 857–61. https://doi.org/10.1119/1.1987878.

Boyer, T. H. (1967). Continuous symmetries and conserved currents. *Annals of Physics*, **42**, 445–466.

Brading, K. and Brown, H. R. (2003). Symmetries and Noether's theorems. In K. Brading and E. Castellani, eds., *Symmetries in Physics: Philosophical Reflections*. Cambridge: Cambridge University Press, pp. 89–109.

Brown, H. R. (1999). Aspects of objectivity in quantum mechanics. In J. Butterfield and C. Pagonis, eds., *From Physics to Philosophy*. Cambridge: Cambridge University Press, pp. 45–70. http://philsci-archive.pitt.edu/223/1/Objectivity.pdf.

Brown, H. R. (2007). *Physical Relativity. Space-time structure from a dynamical perspective*. Oxford: Clarendon Press.

Brown, H. R. and Holland, P. (2004a), Dynamical vs. variational symmetries: Understanding Noether's first theorem, *Molecular Physics*, **102**, (11–12 Spec. Iss), 1133–9. http://philsci-archive.pitt.edu/2914/1/MolPhys04.pdf.

68 For recent discussions, see Olver (Chapter 6 in this volume), and Fagerholm et al. (2020) where an application to neural systems is given.

Brown, H. R. and Holland, P. R. (2004b). Simple applications of Noether's first theorem in quantum mechanics and electromagnetism. *American Journal of Physics*, **72**, 34–39. http://arxiv.org/quant-ph/0302062.

Butterfield, J. (2006). On symmetry and conserved quantities in classical mechanics. In W. Demopoulos and I. Pitowsky, eds., *Physical Theory and Its Interpretation*. New York: Springer, pp. 43–99.

Candotti, E., Palmieri, C. and Vitale, B. (1972). On the inversion of Noether's theorem in classical dynamical systems, *American Journal of Physics*, **40**, 424–9. https://doi.org/10.1119/1.1986566.

Dass, T. (1966). Conservation laws and symmetries. II, *Physical Review*, **150**(4), 1251–5.

Denman, H. H. (1966). Invariance and conservation laws in classical mechanics. II., *Journal of Mathematical Physics*, **7**(11), 1910–15.

Einstein, A. (1940). Considerations concerning the fundaments of theoretical physics, *Science*, **91**, 487–92.

Eshelby, J. D. (1951). The force on an elastic singularity. *Philosophical Transactions of the Royal Society of London A*, **244**, 87–111. https://doi.org/10.1098/rsta.1951.0016.

Eshelby, J. D. (1956). The continuum theory of lattice defects. *Solid State Physics*, **3**, 79–144. https://doi.org/10.1016/S0081-1947(08)60132-0.

Eshelby, J. D. (1975). The elastic energy-momentum tensor, *Journal of Elasticity*, **5**, 321–35.

Fagerholm, E. D., Foulkes, W. M. C., Gallero-Salas, Y., Helmchen, F., Friston, K. J., Moran, R. J., et al. (2020). Conservation laws by virtue of scale symmetries in neural systems. *PLoS Computational Biology*, **16**(5), e1007865. https://doi.org/10.1371/journal.pcbi.1007865

Federico, S., Alhasadi, M. F. and Grillo, A. (2019). Eshelby's inclusion theory in the light of Noether's theorem. *Mathematics and Mechanics of Complex Systems*, **7**(3), 247–85. https://doi.org/10.2140/memocs.2019.7.247

Frankel, T. (2004). The Geometry of Physics. An Introduction. Cambridge UK: Cambridge University Press.

Frisch, M. (2005). Mechanism, principles, and Lorentz's cautious realism. *Studies in History and Philosophy of Modern Physics*, **36**, 659–79.

Giovanelli, M. (2020). Like Thermodynamics before Boltzmann. On the emergence of Einstein's distinction between constructive and principle theories, Studies in History and Philosophy of Science Part B: Studies in History and Philosophy of Modern Physics, **71**, 118–157. https://doi.org/10.1016/j.shpsb.2020.02.005 http://philsci-archive.pitt.edu/18350

Günther, W. (1962). Über einige Randintegrale der Elastomechanik. *Abhandlungen der Braunschweiger Wissenschaftlichen Gesellschaft*, **14**, 53–72.

Hanc, J., Tuleja, S. and Hancova, M. (2004). Symmetries and conservation laws: Consequences of Noether's theorem. *American Journal of Physics*, **72**, 428–35. https://doi.org/10.1119/1.1591764

Havas, P. (1973). The connection between conservation laws and invariance groups: folklore, fiction and fact. *Acta Phyica Austriaca*, **38**, 145–67.

Heras, J. A. (1994). Electromagnetism in Euclidean four space: A discussion between God and the Devil. *American Journal of Physics*, **62**(10), 914–6.

Hill, E. L. (1951). Hamilton's principle and the conservation theorems of mathematical physics. *Reviews of Modern Physics*, **23**, 253–60.

Ibragimov, N. H. (2006a). Integrating factors, adjoint equations and Lagrangians. *Journal of Mathematical Analysis and Applications*, **318**, 742–57.

Ibragimov, N. H. (2006b). The answer to the question put to me by L.V. Ovsyannikov 33 years ago. *Archives of ALGA*, **3**, 53–80. ALGA Publications ISSN 1652-4934. www.researchgate.net/publication/311456950_Archives_of_ALGA_vol_3.

Ibragimov, N. H. and Kolsrud, T. (2004). Lagrangian approach to evolution equations: symmetries and conservation laws. *Nonlinear Dynamics*, **36**, 29–40.

Jackiw, R. (1972, January). Introducing scale symmetry. *Physics Today*, **1972**, 23–7.

Janssen, M. (2014). 'No success like failure ...': Einstein's quest for general relativity, 1907–1920. In M. Janssen and C. Lehner, eds., *The Cambridge Companion to Einstein*. Cambridge: Cambridge University Press, pp. 167–227.

Kibble, T. W. B. and Berkshire, F. H. (2004). *Classical Mechanics*. London: Imperial College Press.

Knowles, J. K. and Sternberg, E. (1972). On a class of conservation laws in linearized and finite elastostatics. *Archive for Rational Mechanics and Analysis*, **44**, 187–211.

Kosmann-Schwarzbach, Y. (2010). *The Noether Theorems, Invariance and Conservation Laws in the Twentieth Century*. Sources and Studies in the History of Mathematics and Physical Sciences. New York: Springer.

Kosmann-Schwarzbach, Y. (2022). The Noether theorems in context. Chapter 1, this volume, and https://arxiv.org/pdf/2004.09254.pdf.

Lancaster, T. and Blundell, S. J. (2015). Quantum Field Theory for the Gifted Amateur. Oxford: Oxford University Press.

Lange, M. (2007). Laws and meta-laws of nature: Conservation laws and symmetries. *Studies in the History and Philosophy of Modern Physics*, **38**, 457–81.

Landau, L. D. and Lifshitz, E. M. (1960). *Mechanics*. New York: Pergamon Press.

Lees, J. P., et al. (*BABAR* Collaboration). (2012). Observation of time reversal violation in the B^0 meson system. *Physical Review Letters*, **109**, 211801.

Martínez Alonso, L. (1979). On the Noether map. *Letters in Mathematical Physics*, **3**, 419–24.

Noether, E. (1918). Invariante Variationsprobleme. *Göttinger Nachrichten, Mathematisch-physikalische Klasse*, **2**, 235–57.

Olver, P. J. (1984a). Conservation laws in elasticity I: General results. *Archive for Rational Mechanics and Analysis*, **85**, 111–29.

Olver, P. J. (1984b). Conservation laws in elasticity II: Linear homogeneous isotropic elastostatics. *Archive for Rational Mechanics and Analysis*, **85**, 131–60.

Olver, P. J. (1986a). *Applications of Lie Groups to Differential Equations*. New York: Springer-Verlag.

Olver, P. J. (1986b). Noether's theorems and systems of Cauchy-Kovalevskaya type. In B. Nicolaenko, D. D. Holm and J. M. Hyman, eds., *Nonlinear Systems of Partial Differential Equations in Applied Mathematics*, Lectures in Applied Mathematics Vol. 23. Providence, RI: American Mathematical Society, pp. 81–104.

Olver, P. J. (2018). Emmy Noether's enduring legacy in symmetry. *Symmetry: Culture and Science*, **29**(4), 475–85.

Orzalesi, C. A. (1970). Charges and generators of symmetry transformations in quantum field theory. *Reviews of Modern Physics*, **42**(4), 381–408.

Pauli, W. (1921). *Relativitätstheorie, Encyklopädie der mathematischen Wissenschaften*, Vol. 19. Liepsig: B. G. Teubner.

Pauli, W. (1958). The Theory of Relativity. Toronto: Pergamon Press.

Peskin, M. E. and Schroeder, D. V. (1995). *An Introduction to Quantum Field Theory*. Cambridge, MA: Perseus Books.

Read, J., Brown, H. R. and Lehmkuhl, D. (2018). Two miracles in the general theory of relativity. *Studies in History and Philosophy of Modern Physics*, **64**, 14–25. http://philsci-archive.pitt.edu/14589.

Rice, J. R. (1985). Conserved integrals and energetic forces. In B. A. Bilby, K. J. Miller and J. R. Willis, eds., *Fundamentals of Deformation and Fracture (Eshelby Memorial Symposium)*. Cambridge: Cambridge University Press, pp. 43–56.

Rosen, J. (1980). Redundancy and superfluity for electromagnetic fields and potentials. *American Journal of Physics*, **48**, 1071–3. https://doi.org/10.1119/1.12289.

Ryder, L. H. (1996). *Quantum Field Theory*, 2nd ed. Cambridge: Cambridge University Press.

Sakurai, J. J. (1964). *Invariance Principles and Elementary Principles*. Princeton: Princeton University Press.

Santilli, R. M. (1983). *Foundations of Theoretical Mechanics II, Birkhoffian Generalisation of Hamiltonian Mechanics*, New York: Springer-Verlag.

Schroeren, D. (2020). The metaphysics of invariance. *Studies in History and Philosophy of Modern Physics*, **70**, 51–64.

Smith, S. R. (2008). Symmetries and the explanation of conservation laws in the light of the inverse problem in Lagrangian mechanics. *Studies in History and Philosophy of Modern Physics*, **39**, 325–45.

Steinberg, S. and Wolf, K. B. (1981). Symmetry, conserved quantities and moments in diffusive equations. *Journal of Mathematical Analysis and Applications*, **80**, 36–45.

Steudel, H. (1962). Über die Zuordnung zwischen Invarianzeigenschaften und Erhaltungssätzen. *Zeitschrift für Naturforschung*, **17A**, 120–32.

Steudel, H. (1966). Die Struktur der Invarianzgruppe fur lineare Feldtheorien. *Zeitschrift für Naturforschung*, **21A**, 1826–8.

Sudbery, A. (1986). A vector Lagrangian for the electromagnetic field. *Journal of Physics A: Mathematical and General*, **19**, L33–L36.

Sutton, A. P. (2020). *Physics of Elasticity and Crystal Defects*. Oxford: Oxford University Press. ISBN: 9780198860785.

Tassie, L. J. and Buchdahl, H. R. (1964). Gauge-independent theory of symmetry. I. *Australian Journal of Physics*, **17**, 431–9.

Noether, E. (1971). Invariant variation problems. *Transport Theory and Statistical Physics*, **1**(3), 186–207. https://doi.org/10.1080/00411457108231446

van Fraassen, B. C. (1989). *Laws and Symmetry*. Oxford: Clarendon Press.

Vinogradov, A. M. (1984). Local symmetries and conservation laws. *Acta Applicandae Mathematicae*, **2**, 21–78.

Weinberg, S. (1993). *Dreams of a Final Theory*. New York: Pantheon Press.

Wigner, E. (1939). On the unitary representations of the inhomogeneous Lorentz group. *Annals of Mathematics*, **40**, 149–204.

Wigner, E. (1954). Conservation laws in classical and quantum physics. *Progress of Theoretical Physics*, **11**(4–5), 437–40.

Will, C. M. (2014). *The Confrontation between General Relativity and Experiment*. https://arxiv.org/pdf/1403.7377.pdf.

Wolf, T. (2002). A comparison of four approaches to the calculation of conservation laws. *European Journal of Applied Mathematics*, **13**, 129–52. https://doi.org/10.1017/S0956792501004715.

Yang, C. N. and Mills, R. L. (1954). Conservation of isotopic spin and isotopic gauge invariance. *Physical Review*, **96**(1), 191–5.

Zee, A. (1999). *Fearful Symmetry: The Search for Beauty in Modern Physics*. Princeton: Princeton University Press.

Zee, A. (2003). *Quantum Field Theory in a Nutshell*. Princeton: Princeton University Press, Princeton.

8

Noether's First Theorem and the Energy-Momentum Tensor Ambiguity Problem*

MARK ROBERT BAKER, NIELS LINNEMANN, AND CHRIS SMEENK

Dedicated to the late Bessel-Hagen, who when alive had his habilitation thesis thrown into the sea, and even now must feel as if his work was lying somewhere on the seabed.

8.1 Introduction

Physicists have long exploited symmetries to simplify problems. In Lagrangian mechanics, cyclic coordinates (that is, generalised coordinates q_i such that $\partial\mathcal{L}/\partial q_i = 0$ for the Lagrangian $\mathcal{L}$) signal the presence of a symmetry, and the Euler–Lagrange equations imply that the associated conjugate momenta p_i are conserved.[1] It is hard to understate the practical importance of finding conserved quantities, thereby reducing the number of variables and making it much easier to find solutions. Noether's celebrated (1918) paper significantly clarified the mathematical structure underlying earlier results.[2] The rich line of work stemming from her seminal contribution has elucidated three intertwined aspects of physical theories: laws, symmetries, and conservation principles.

Conventional wisdom now holds that Noether's first theorem and its converse universally link a certain kind of continuous symmetry (such as Poincaré translation) to a certain kind of conserved current (such as the energy-momentum tensor).[3] Although based on a kernel of truth, this conventional wisdom reflects an overly simplified picture of the mathematical physics. As a starting point for the discussion below, consider the following specific claim often taken to follow from the Noether machinery: a subset of the variational symmetries of the action, namely spatial and temporal translations, are associated with energy-momentum conservation. Here we encounter an immediate difficulty: applying Noether's first theorem in the context of field theory (as described in Section 8.3), the canonical variational symmetry and 4-parameter translation subgroup of the Poincaré group yields what is called the 'canonical Noether energy-momentum tensor' ($T_C^{\mu\nu}$). For most classical field theories, the

* We are grateful to James Read, Bryan Roberts, and Nicholas Teh for the invitation to contribute to this volume. We thank Harvey R. Brown for encouraging comments and helpful suggestions on the chapter. We also thank an anonymous reviewer for their helpful input.

1 See Butterfield (2006) for a pedagogical presentation.

2 Noether (1918), translated into English by Kosmann-Schwarzbach (2011).

3 The metaphysical work of Lange (2007), for instance, rests on this.

canonical tensor lacks features required for a physically sensible energy-momentum tensor, and differs from known physical energy-momentum tensors established in other ways.[4] Such results raise two related challenges to the conventional wisdom: do the quantities that actually follow from applying Noether's theorem have a clear physical interpretation, and does Noether's theorem need to be supplemented in order to derive physically meaningful conserved quantities? Particularly striking is the existence of inequivalent definitions of the energy-momentum tensor, a central physical quantity in any classical field theory.

Typical textbook presentations leave the impression that Noether's theorem fails to yield the correct energy-momentum tensor. They mention the unappealing features of $T_C^{\mu\nu}$, and then immediately propose a fix. Such fixes amount to variations on a theme going back to Belinfante (1940), who added the divergence of a so-called superpotential to $T_C^{\mu\nu}$ such that a new 'Belinfante' energy-momentum tensor $T_B^{\mu\nu} = T_C^{\mu\nu} + \partial_\alpha b^{[\mu\alpha]\nu}$ recovers the correct answer for electrodynamics[5] if an on-shell condition is imposed (discussed in more detail in Section 8.4). This does not follow directly from Noether's theorem itself, suggesting that some form of 'improvement' is needed to find physically meaningful conserved quantities (as we discuss in Section 8.4.1). But any such improvement approach has the unsavoury air of devising a series of poorly justified steps to arrive at an answer found in the back of the book. What happens when we do not already know, or have independent ways of finding, the correct form for the energy-momentum tensor?

Thankfully, there is another approach, albeit much less common in the literature. Noether's first theorem is often taken to be the result of a general identity applied to a specific type of action symmetry, namely finite symmetries such as the Poincaré group. But we can also consider the consequences of this identity for finite *and* infinite (e.g. gauge) symmetries of the action – although the first theorem is usually assumed to apply only to finite symmetry groups, Noether herself suggested a broader formulation of the theorem.[6]

As we will see below, in the case of electrodynamics this leads directly to the correct energy-momentum tensor. This was Bessel-Hagen's neglected contribution, inspired in part by Noether herself (Bessel-Hagen 1921). By contrast with the 'improvement' approach, we will argue that this approach exploits all of the relevant variational symmetries of the action in applying Noether's theorem. We will argue for the superiority of this approach based on using the converse of Noether's first theorem, which correctly identifies the proper variational symmetries (those derived using Method II in Section 8.4.2) of the Lagrangian from the accepted form of the energy-momentum tensor. The derivation of the canonical stress-energy tensor fails to use the full power of the mathematical machinery that Noether

[4] We will focus on energy momentum in the ensuing discussion, but similar issues arise for other conserved currents, such as angular momentum.

[5] Unless otherwise indicated, the ensuing discussion focuses on sourceless electrodynamics for ease of exposition.

[6] The end of Noether's statement of theorem I holds that it also applies to an infinite-parameter group: 'Dieser Satz gilt auch noch im Grenzfall von unendlich vielen Parametern.' ('This proposition [Satz] also holds in the limit of infinitely many parameters.') To our mind it is ambiguous whether this clarification is to be taken as part of the theorem or as an additional theorem; it is in this sense of Noether's first theorem, however, that Poincaré translation symmetry *and* gauge symmetries of the action can work together. Our main interest is in what Noether's mathematical framework allows one to prove based on symmetries of the action, following from the general identity below (Equation (8.2)); whether one regards a particular result as falling within the scope of the 'first theorem' is partly a terminological issue. Thanks to Harvey R. Brown for pressing us on this point.

has given us by considering only a restricted subset of the variational symmetries. Thus, at least in the context of Lagrangian field theories in flat spacetime, the conventional wisdom of a universal linkage between symmetries and conservation laws can be refined to that of a linkage between a specific variational symmetry – to be introduced below – and the set of physical conservation laws for the theory.

This line of argument does not address Lagrangian field theories in curved spacetime, which lack the global symmetries needed to obtain the energy-momentum tensor via Noether's theorem.[7] Physicists then typically use Hilbert's definition of the energy-momentum tensor $T_H^{\gamma\rho} = \frac{2}{\sqrt{-g}} \frac{\delta \mathcal{L}}{\delta g_{\gamma\rho}}$, which is sometimes referred to as the metric energy-momentum tensor. The Hilbert energy-momentum tensor is by definition symmetric, thereby avoiding one of the major flaws of the canonical Noether tensor. From this expression, the Hilbert energy-momentum tensor in Minkowski spacetime is defined as the curved spacetime Hilbert energy-momentum tensor with all metric tensors $g_{\mu\nu}$ replaced by the Minkowski metric $\eta_{\mu\nu}$. Blaschke et al. (2016) outlines the distinction between the curved spacetime and Minkowski spacetime definitions of the Hilbert tensor.[8] In any case, our discussion will focus on the status of the energy-momentum tensor derived from Noether's first theorem in flat (Minkowski) spacetime.

The plan for the chapter is as follows. This section concludes with a tribute to Bessel-Hagen. We then briefly introduce the energy-momentum tensor, and review properties required for it to be regarded as physically reasonable. In the next two sections, we use classical electrodynamics as a simple case to introduce Noether's first theorem and its subtleties. Section 8.3 introduces Noether's first theorem. Section 8.4 considers the two approaches described above for defining Noether currents, focusing on the energy-momentum tensor: Method I uses translation symmetries to yield the canonical energy-momentum tensor, which requires 'improvement' terms to yield the correct energy-momentum tensor; Method II, by contrast, considers a broader class of variational symmetries and leads directly to the correct energy-momentum tensor. We then argue in favour of the second method based on the converse of Noether's theorem. Section 8.5 considers how this line of argument applies to a variety of other cases, including in particular the challenge of defining an energy-momentum tensor for the gravitational field in linearised gravity. Section 8.6 brings out one of the themes running through the discussion, namely the challenge of tracking the physical significance of these structural properties of field theories. There are several steps in the early sections where it is tempting to describe both symmetries and conserved currents only up to an equivalence class. We aim to identify clearly the limitations of this move, and develop our position by contrast with recent philosophical discussions about how symmetries relate to the representational capacities of our theories (considering, in particular, Brown's Chapter 7 in this volume). Finally we discuss the outlook and conclusions of our work in Section 8.7.

[7] See also Holman (2010), who argues for generalisations of Noether's theorems in Minkowski spacetime from considerations of diffeormorphism invariance, and based on these, he claims to show how to obtain the correct stress-energy tensor.

[8] The Hilbert energy-momentum tensor in Minkowski spacetime ($T_{H,\eta}^{\gamma\rho} = \frac{2}{\sqrt{-g}} \frac{\delta \mathcal{L}}{\delta g_{\gamma\rho}}\big|_{g=\eta}$) is not, in general, equivalent to energy-momentum tensors derived via Noether's theorem from the 4-parameter Poincaré translation (see Baker et al. 2021).

8.1.1 An Ode to Bessel-Hagen

Alongside these systematic aims, we want to use the occasion to clarify the contribution of Erich Bessel-Hagen (1898–1946) to the Noether machinery. On the one hand, Bessel-Hagen seems to be often wrongly treated only as an originator of the generalisation of Noether's theorem to invariance under symmetry transformations of the action *up to a boundary term*.[9, 10]

On the other hand, Bessel-Hagen does not seem to be widely known for his central contribution in that very same paper, namely the introduction of what we call Method II: the application of Noether's first theorem in light of gauge symmetries when deriving the complete set of conformal conservation laws for classical electrodynamics. Bessel-Hagen's work has been independently reproduced – by, among others, Eriksen and Leinaas (1980) and Montesinos and Flores (2006); the reception of his paper in the English-speaking world, however, suffered from the fact that a translation first appeared in 2006 (Albinus and Ibragimov 2006), arguably much too late. Even though the method has resurfaced in some textbooks as well (Burgess 2002; Scheck 2012), it remains relatively unknown in the wider physics literature. The Bessel-Hagen method has recently been applied to a wide class of special relativistic field theories (Baker 2021c). Here we contrast the more common textbook approach with Bessel-Hagen's method, primarily focusing on the simple case of electrodynamics.

8.2 Energy-Momentum Tensors

Einstein took the general formulation of conservation laws in terms of the energy-momentum tensor to be 'the most important new advance in the theory of relativity' (as of 1912). The energy-momentum tensor has a central role in the new conception of mechanics and field theory, as Einstein went on to emphasise:[11]

> To every kind of material process we want to study, we have to assign a symmetric tensor $T_{\mu\nu}\cdots$ The problem to be solved always consists in finding out how $T_{\mu\nu}$ is to be formed from the variables characterizing the processes under consideration. If several processes can be isolated in the energy-momentum balance that take place in the same region, we have to assign to each individual process its own stress-energy tensor ($T^1_{\mu\nu}$, and etc.) and set $T_{\mu\nu}$ equal to the sum of these individual tensors. (CPAE Vol. 4, Doc. 1, p. 63)

[9] Importantly, it is not correct to see Bessel-Hagen as the *sole* originator of this generalisation in any case. As Kosmann-Schwarzbach (2011) (see her section 4.2) notes, Bessel-Hagen himself (ambiguously) acknowledges his debt to Noether herself (to a certain degree, at least):

> Zuerst gebe ich die beiden E. Noetherschen Sätze an, und zwar in einer etwas allgemeineren Fassung als sie in der zitierten Note stehen. Ich verdanke diese einer mündlichen Mitteilung von Fräulein Emmy Noether selbst. (Bessel-Hagen 1921, p. 260) (I first present the two Noetherian propositions, albeit in a slightly more general fashion than they can be found in the cited note. I owe these propositions to an oral communication by Miss Emmy Noether herself. (Own translation))

[10] We do not apply this generalisation in our article. See Sections 8.5 and 8.6 for more discussion.

[11] See Janssen and Mecklenburg (2006) for an insightful discussion of the importance of the energy-momentum tensor in the transition to relativistic mechanics.

Strikingly, Einstein treats all 'material processes,' whether they involve electromagnetic fields or matter as described by continuum mechanics, as on a par: the fundamental dynamical quantity in each case is the energy-momentum tensor. How, then, are we to find an appropriate $T_{\mu\nu}$ for various processes we aim to describe?

Before turning to that question, recall that the energy-momentum tensor (also known as the stress-energy tensor) encodes information regarding energy-momentum densities and fluxes for different kinds of 'material processes'. In relativistic mechanics this is all captured in a single rank-two tensor, $T^{\mu\nu}$: the T^{00} component represents energy density, the T^{0i} and T^{i0} components represent energy and momentum flux, respectively, and the T^{ij} components represent stresses (where $i, j = 1,2,3$).

As an illustration, the energy-momentum tensor for electromagnetism provides a compact summary of familiar facts about the electromagnetic field. Minkowski formulated electromagnetism in terms of the field strength tensor $F_{\mu\nu} = \partial_\mu A_\nu - \partial_\nu A_\mu$, where A_μ is the vector potential. The energy-momentum tensor takes the following form:[12]

$$T^{\mu\nu} = F^{\mu\alpha} F^{\nu}_{\ \alpha} - \frac{1}{4}\eta^{\mu\nu} F_{\alpha\beta} F^{\alpha\beta}. \tag{8.1}$$

The invariance of the field strength tensor $F^{\mu\nu}$ under gauge transformations ($A'_\mu = A_\mu + \partial_\mu \phi$, for a scalar gauge parameter ϕ) implies gauge invariance of $T^{\mu\nu}$. More generally, we require gauge invariance of $T^{\mu\nu}$ because the energy-momentum tensor represents observable quantities directly. In this case, we have constructed the energy-momentum tensor based on what we already know about the relevant field. Historically, von Laue extended this constructive approach, writing down appropriate energy-momentum tensors for extended stressed bodies, relativistic fluids, and other cases, based on prior knowledge about energy and momentum in each case.

To what extent can we determine the form of $T^{\mu\nu}$ for a new classical field ψ (whether scalar, vector, tensor, etc.) based on general principles, or on specific features of ψ's dynamics? There are two main types of constraints a tensor would be expected to satisfy to be plausibly interpreted as representing energy-momentum of the field. The first set of constraints stem from the idea that all matter fields 'carry positive energy-momentum.' More formally, for arbitrary regions of spacetime R, $T^{\mu\nu}$ vanishes on R if and only if the field ψ vanishes.[13] Further constraints can be imposed to capture the idea that the energy-momentum is positive, and that energy-momentum flows respect the causal structure of relativistic spacetime. One fundamental requirement of this kind is that the energy density (the T^{00} component) is bounded from below, so that the field cannot serve as an infinite energy source. Gravity

12 The energy density of electromagnetic fields is given by $U = \frac{1}{2}\left(\epsilon_0 E^2 + \frac{1}{\mu_0} B^2\right)$, the Poynting vector $\vec{S} = \frac{1}{\mu_0} \vec{E} \times \vec{B}$ represents energy flux, and the Maxwell stress tensor σ_{ij} represents stress and momentum fluxes. We can express the energy-momentum tensor in terms of these quantities as follows: $T^{\mu\nu} = \begin{pmatrix} -U & -S_x/c & -S_y/c & -S_z/c \\ -S_x/c & \sigma_{xx} & \sigma_{xy} & \sigma_{xz} \\ -S_y/c & \sigma_{yx} & \sigma_{yy} & \sigma_{yz} \\ -S_z/c & \sigma_{zx} & \sigma_{zy} & \sigma_{zz} \end{pmatrix}$.

13 Here we are setting aside fields with negative energy density, for which $T^{\mu\nu}$ could vanish through a cancellation of positive and negative energy densities. This condition has to be formulated with greater care for quantum fields, which necessarily admit negative expectation values for the energy density at spacetime points, but versions of this condition have been proposed for open regions.

is sensitive to the absolute value of the energy, so that it is meaningful to differentiate positive and negative energies for fields coupled to gravity. Further constraints can then be imposed: The weak energy condition, for example, requires that the energy density (the T^{00} component) is non-negative, as measured by all observers. The dominant energy condition holds if, in addition, momentum fluxes stay within the light cone. There is a long list of other energy conditions that have been used to prove results such as the singularity theorems.[14]

A second set of constraints, and the main focus of the ensuing discussion, regards symmetries and conserved quantities. We will take the satisfaction of an appropriate conservation principle as a defining feature of energy momentum.[15] Given an appropriate $T^{\mu\nu}$, the on-shell conservation principle can be succinctly stated: $\partial_\mu T^{\mu\nu} = k^\nu$. (For a free field, $k^\nu = 0$; otherwise k^ν represents an external force density.[16]) In classical mechanics, the conservation of energy and momentum stem from space-time translation symmetries, so it is plausible to begin by constructing a tensor combining the conserved currents associated with these symmetries. The variational symmetries of an action S consist of the transformations that leave S invariant,[17] and Noether's first theorem associates a conserved quantity with each element of the finite group of transformations. In electrodynamics these symmetry constraints, the variational symmetries of the action for coordinates (conformal symmetries) and fields (gauge symmetries), are required to obtain the known conservation laws, which we detail in Section 8.4.2; the 15 conservation laws are associated to the (finite) 15 parameter conformal group of transformations.

Other properties of the energy-momentum tensor follow from symmetries of the field theory for specific types of fields. For classical field theories with conformal symmetry, for example, the energy-momentum tensor will be trace-free so that the conformal $C^{\rho\alpha}$ and dilatation D^ρ tensors are conserved.[18]

We take these two types of constraints as requirements that a rank-two tensor must satisfy to be a plausible candidatefor an 'energy-momentum tensor' of particular importance in considering energy-momentum tensors proposed for a new field ψ rather than constructed based on prior knowledge. Perusing the physics literature suggests that these two types of constraints do not suffice to determine a unique choice: there are several proposed, apparently inequivalent, candidates for the 'energy-momentum tensor for ψ' (for a variety of different fields). Our overall aim here is to argue against this view. Several of the candidate energy-momentum tensors may not be worthy of the name. Take, for example, the question of whether we should require that the energy-momentum is symmetric under exchange of indices ($T^{\mu\nu} = T^{\nu\mu}$). Failure of this to hold in a mechanical system would lead to torque and

14 See Curiel (2017) for a comprehensive review of energy conditions, their status, and their role in sundry theorems.

15 This leaves open the possibility that there are fields, such as the metric field in general relativity, that lack an energy-momentum tensor in this sense.

16 In the case of electrodynamics sourced by J^α, this is the force density $f^\nu = \partial_\mu T^{\mu\nu} = F^\nu_{\ \alpha} J^\alpha$ which includes the Lorentz force density in the spatial components ($f^i = \partial_\mu T^{\mu i} = J^\rho F^i{}_\rho = \rho\vec{E} + \vec{J} \times \vec{B}$).

17 There are cases where certain transformations leave S invariant only up to a boundary term; see Sections 8.5 and 8.6 for further discussion.

18 In the case of electrodynamics, this statement is directly related to the associated quantum particles are massless (see Garg 2012, p. 563).

the possibility of unlimited angular acceleration.[19] Similar problems arise in field theories. Just as in the case of mechanics, an anti-symmetric energy-momentum tensor would entail failure of angular momentum conservation if the dynamical fields are all scalar quantities. This does not hold for the more general case of tensorial fields, however, which can have non-trivial spin degrees of freedom that also contribute to the total angular momentum (as we will see in more detail).

Although this case is more subtle than the criteria emphasised above, a preference for symmetric rank-two tensors already rules out many candidates discussed in the literature. We will further argue that we need to take into account more than just spacetime translations in building the energy-momentum tensor out of conserved currents, as we will illustrate next by considering the case of electromagnetism in more detail.

8.3 Noether's First Theorem for Classical Electrodynamics

Noether's first theorem, applied to a particular Lagrangian density, yields a relationship between the Euler–Lagrange expressions and Noether current of the theory of the form $E^A\delta\phi_A + \partial_\mu J^\mu = 0$, where E^A is the Euler–Lagrange expression for the rank-A field and J^μ is the Noether current. This relationship is derived by substituting a particular Lagrangian density and action symmetries into a general identity which is sometimes referred to as the Noether identity.[20] For the Lagrangian density of electrodynamics $\mathcal{L} = -\frac{1}{4}F_{\mu\nu}F^{\mu\nu}$, the Noether identity is (e.g., Gelfand and Fomin 2000):

$$\left(\frac{\partial\mathcal{L}}{\partial A_\nu} - \partial_\rho\frac{\partial\mathcal{L}}{\partial(\partial_\rho A_\nu)}\right)\delta\bar{A}_\nu + \partial_\rho\left(\eta^{\rho\beta}\mathcal{L}\delta x_\beta + \frac{\partial\mathcal{L}}{\partial(\partial_\rho A_\nu)}\delta\bar{A}_\nu\right) = 0, \tag{8.2}$$

where

$$\delta\bar{A}_\nu = -\partial^\beta A_\nu\delta x_\beta + \delta A_\nu \tag{8.3}$$

is the complete set of symmetry transformations that are directly obtained upon deriving the Noether identity (8.2).[21]

The two methods we will discuss diverge with regard to the general form of the transformations $\delta\bar{A}_\nu$. Consider first the difference between $\delta\bar{A}_\nu$ and δA_ν: the non-bar transformation of fields is the difference in transformed fields as a function of their respective coordinates,

$$\delta A_\nu = A'_\nu(x') - A_\nu(x). \tag{8.4}$$

[19] An angular momentum tensor $M^{\rho\mu\nu} = x^\mu T^{\rho\nu} - x^\nu T^{\rho\mu}$ is conserved if and only if $T^{\mu\nu}$ is symmetric. The angular momentum relative to a given event chosen as an origin can be obtained by integrating $M^{\rho\mu\nu}$; see chapter 5 of Misner et al. (1973) for further discussion of this point, and regarding the general properties of energy-momentum tensors.

[20] The Noether identity is the starting point for both of Noether's first and second theorems. It is sometimes referred to as the Noether off-shell condition.

[21] Equation (8.2) corresponds to Eq. 12 in Noether's paper (Kosmann-Schwarzbach 2011) for the specific Lagrangian of classical electrodynamics, and Eq. (8.3) corresponds to Eq. 9 in Noether's paper. Our x_β (coordinates) correspond to her independent variables x_n and our A_ν (fields) correspond to her functions of these independent variables u_i.

By contrast, the bar transformations of fields is the difference in transformed fields as a function of the same (non-transformed) coordinates,

$$\delta \bar{A}_\nu = A'_\nu(x) - A_\nu(x), \tag{8.5}$$

where the bar notation is adopted from Noether's paper for this particular transformation in her Eq. 9. The following subsections treat the three types of transformations ($\delta \bar{A}_\nu$): (i) the two terms which correspond to the Lie derivative of the fields (canonical and contragredient transformations) and (ii) the term which corresponds to gauge symmetries of the action.

8.3.1 Transformations Associated to the Lie Derivative

In this subsection, we describe two of the contributions to the transformations of fields $\delta \bar{A}_\nu$ arising from infinitesimal change of coordinates δx^ν. These two transformations follow directly from the Lie derivative of the four potential A_ν with respect to the infinitesimal change in coordinates δx,

$$\pounds_{\delta x} A_\nu = -\delta x^\beta \partial_\beta A_\nu - A_\beta \partial_\nu \delta x^\beta = \delta_C A_\nu + \delta_T A_\nu. \tag{8.6}$$

The Lie derivative represents the coordinate invariant change of a tensor field along the flow of a vector field, which in this case is the infinitesimal change in coordinates δx.

We have denoted the two terms in this expression as $\delta_C A_\nu$ and $\delta_T A_\nu$, respectively. The first term, $\delta_C A_\nu = -\delta x^\beta \partial_\beta A_\nu$, is exactly what is found in the first term of (8.3). This term alone is used to derive the canonical Noether energy-momentum tensor when $\delta x^\beta = a^\beta$ is the 4-parameter Poincaré translation, thus we will refer to this as the canonical transformations. The second term, $\delta_T A_\nu = -A_\beta \partial_\nu \delta x^\beta$, we will refer to as contragredient transformations as Bessel-Hagen did in his article; they are associated to the transformation properties of a tensor. This contribution is zero for $\delta x^\beta = a^\beta$, and thus does not factor into energy-momentum tensor discussion. However, for any non-constant δx^β, this contribution is nonzero and essential for deriving the associated conserved tensors, such as the angular momentum tensor resulting from the remaining parameters of the Poincaré group.

8.3.1.1 Canonical Transformations and the Canonical Noether Energy-Momentum Tensor

If we restrict ourselves to canonical transformations, $-\partial^\beta A_\nu \delta x_\beta$ with $\delta A_\nu = 0$, and no gauge symmetries, we have only

$$\delta \bar{A}_\nu = \delta_C A_\nu = -\partial^\beta A_\nu \delta x_\beta \tag{8.7}$$

to substitute into the Noether identity (8.2). We will use $\delta_C A_\nu$ to indicate these canonical transformations. We then obtain:

$$\left(\frac{\partial \mathcal{L}}{\partial A_\nu} - \partial_\rho \frac{\partial \mathcal{L}}{\partial(\partial_\rho A_\nu)}\right) \delta \bar{A}_\nu + \partial_\rho \left(\left[\eta^{\rho\beta} \mathcal{L} - \frac{\partial \mathcal{L}}{\partial(\partial_\rho A_\nu)} \partial^\beta A_\nu \right] \delta x_\beta \right) = 0 \tag{8.8}$$

The square brackets contain what is known as the 'canonical Noether energy-momentum tensor' $T_C^{\rho\beta}$ for a Lagrangian density of the form $\partial A \partial A$ such as $\mathcal{L} = -\frac{1}{4} F_{\mu\nu} F^{\mu\nu}$ of electrodynamics. In the case of the 4-parameter Poincaré translation $\delta x_\beta = a_\beta$, we can factor out the constant a_β from the divergence yielding $E^\nu \delta \bar{A}_\nu + a_\beta \partial_\rho T_C^{\rho\beta} = 0$, where

$$T_C^{\rho\beta} = \eta^{\rho\beta}\mathcal{L} - \frac{\partial \mathcal{L}}{\partial(\partial_\rho A_\nu)}\partial^\beta A_\nu \tag{8.9}$$

and E^ν is the Euler–Lagrange expression (in the case of $\mathcal{L} = -\frac{1}{4}F_{\mu\nu}F^{\mu\nu}$, E^ν are the non-homogeneous Maxwell's equations).

However, substituting the Lagrangian density for electrodynamics $\mathcal{L} = -\frac{1}{4}F_{\mu\nu}F^{\mu\nu}$, where $F_{\alpha\beta} = \partial_\alpha A_\beta - \partial_\beta A_\alpha$ and $\frac{\partial \mathcal{L}}{\partial(\partial_\rho A_\nu)} = -F^{\rho\nu}$, yields

$$T_C^{\rho\beta} = F^{\rho\nu}\partial^\beta A_\nu - \frac{1}{4}\eta^{\rho\beta}F_{\mu\nu}F^{\mu\nu}, \tag{8.10}$$

the canonical Noether energy-momentum tensor for classical electrodynamics. By contrast, the accepted energy-momentum tensor $T^{\mu\nu}$ for the theory is given by Equation (8.1) above.

This specific case illustrates two distinct problems for the canonical energy-momentum tensor that hold more broadly. First, the result simply does not match with an independently motivated expression for the energy-momentum tensor, based on an understanding of energy and momentum densities and fluxes for the relevant field. Second, the canonical tensor lacks essential properties: In general it is neither symmetric nor gauge invariant, nor trace-free. There are special cases where some of these properties hold. For example, a symmetric tensor follows from (8.9) for a Klein–Gordon scalar field; yet even then, there are alternative tensors which improve on the canonical expression by being trace-free (Callan et al. 1970). In some of these cases, it may not be obvious whether the canonical tensor or some other candidate tensor is to be preferred. We do not claim to have a way to resolve this debate across the board; rather, there are several clear cases (like electromagnetism) where the canonical energy-momentum tensor fails to have the right form.

8.3.1.2 Contragredient Transformations

The non-bar transformation of fields δA_ν (second term in Eq. (8.3)) is referred to by Bessel-Hagen as being associated to the 'contragredient' transformations of the fields, which in current treatments follow simply from the definition of a contravariant tensor (in this case vector):

$$A'^\nu(x') = \frac{\partial x'^\nu}{\partial x^\mu}A^\mu(x) \tag{8.11}$$

Inserting this into the contravariant form of (8.4), we have for δA^ν,

$$\delta A^\nu = \frac{\partial x'^\nu}{\partial x^\mu}A^\mu(x) - A^\nu(x). \tag{8.12}$$

If we consider the transformation of coordinates,

$$x'^{\nu} = x^{\nu} + \delta x^{\nu} \tag{8.13}$$

In particular, then, $\frac{\partial x'^{\nu}}{\partial x^{\mu}} = \delta^{\nu}_{\mu} + \partial_{\mu}(\delta x^{\nu})$. Substituting (8.13) into (8.12), we have $\delta A^{\nu} = A^{\mu}\partial_{\mu}\delta x^{\nu}$. To determine the covariant form of this expression, we can consider the identity $A^{\mu}A_{\mu} = A'^{\mu}A'_{\mu}$ as a function of their respective coordinates, and solve for the transformation $\delta_T A_{\nu}$, which is exactly the contragredient transformation presented by Bessel-Hagen in his Eq. 18,

$$\delta_T A_{\nu} = -A_{\mu}\partial_{\nu}\delta x^{\mu} \tag{8.14}$$

where $\delta_T A_{\nu}$ indicates that this is the transformation based on the definition of a tensor T. Note that we require the covariant form of this transformation due to our presentation of the Noether identity in (8.2).

For higher-rank tensors this contribution can easily become quite complicated. However, in the case of energy-momentum tensor derivation, when we have the 4-parameter Poincaré translation $\delta x^{\mu} = a^{\mu}$ – regardless of Method I or Method II – $\delta_T A_{\nu} = 0$ since a^{μ} is a constant. For this reason, since most of the discussions of conservation laws focus solely on the energy-momentum tensor at the expense of other conserved quantities such as angular momentum, this contribution usually drops out of the picture. (Yet we need the contragredient transformations for the derivation of conservation laws linked to non-constant coordinate symmetries δx^{μ}.)

8.3.2 Gauge (Field) Symmetries of the Action

There is also the possibility of gauge (field) symmetries of the action, often overlooked from the perspective of Noether's first theorem because they are thought to be relevant only to Noether's second theorem. The Bessel-Hagen et al. approach uses these symmetries as well to derive the known conservation laws of electrodynamics directly from Noether's general identity.

Mixing the complete set of coordinate and field symmetries is essential to obtaining the accepted energy-momentum tensor of electrodynamics. To highlight this point we briefly touch on Noether's second theorem, again with a focus on electrodynamics.

Noether's second theorem is another application of the Noether identity in (8.2). The basic idea behind Noether's second theorem is that field (gauge) symmetries that leave the action invariant (in her case, the 'infinite continuous group' of transformations of the functions u) can be integrated by parts to remove the derivatives on the field transformations; neglecting boundary terms (instead of keeping them as in the case of the Noether current in the first theorem) results in an identity in terms of the Euler–Lagrange expressions.[22]

In the case of electrodynamics, discarding boundary terms (all terms under a total divergence) leaves the standard Euler–Lagrange expression, Maxwell's $\partial_{\rho}F^{\rho\nu}$, as,

$$\partial_{\rho}F^{\rho\nu}\delta\bar{A}_{\nu} = 0 \tag{8.15}$$

[22] See Eq. 16 in Noether (1918) and associated discussion for statements on Noether's second theorem. Notably, the converse also holds, namely that the existence of such identity implies invariance of the action under an infinite continuous group.

Now taking the gauge transformation $A'_\nu = A_\nu + \partial_\nu \phi$ (where ϕ is a scalar), we have,

$$\delta \bar{A}_\nu = \delta_g A_\nu = \partial_\nu \phi \tag{8.16}$$

where we denote $\delta_g A_\nu$ to emphasise the transformation associated to the gauge symmetry of the action. From (8.15) and (8.16), we therefore have $\partial_\rho F^{\rho\nu} \partial_\nu \phi = 0$. Integrating by parts and discarding the resulting boundary term, we are left with the well-known identity for Noether's second theorem in electrodynamics,

$$\partial_\rho \partial_\nu F^{\rho\nu} \phi = 0 \tag{8.17}$$

and thus $\partial_\rho \partial_\nu F^{\rho\nu} = 0$. It is the incorporation of this transformation (8.16) that is then also essential for directly deriving the complete set of conservation laws from Noether's first theorem, including the accepted energy-momentum tensor (8.1). By use of the converse of Noether's second theorem we have a concrete methodology for obtaining the variational gauge symmetry $\delta \bar{A}_\nu$ that is required for application of Method II.

8.3.3 Summary

In summary, Noether's identity (8.2) can be used to obtain a relationship between the Euler–Lagrange expressions and conservation laws for field theories such as electrodynamics.

The Noether current depends on the coordinate symmetry transformation δx_β and field symmetry transformations $\delta \bar{A}_\nu$; i.e. any symmetry transformation of the action must be introduced through these contributions in order to derive corresponding on-shell conserved currents. We distinguished three main types of transformations of fields, which can be simultaneously applied to (8.2) in the form,

$$\delta \bar{A}_\nu = \delta_C A_\nu + \delta_T A_\nu + \delta_g A_\nu \tag{8.18}$$

where we have the canonical transformations (8.7), contragredient transformations (8.14) and gauge transformations (8.16). In the case of electrodynamics, this gives

$$\delta \bar{A}_\nu = -\partial^\mu A_\nu \delta x_\mu - A_\mu \partial_\nu \delta x^\mu + \partial_\nu \phi \tag{8.19}$$

These transformations are the complete set required to derive conservation laws in standard field theories such as electrodynamics. Bessel-Hagen derived all 15 conservation laws of electrodynamics which are associated to the 15 parameter conformal group of infinitesimal coordinate transformations,

$$\delta x_\alpha = a_\alpha + \omega_{\alpha\beta} x^\beta + S x_\alpha + 2\xi_\nu x_\alpha x^\nu - \xi_\alpha x_\nu x^\nu \tag{8.20}$$

In the case of the four-parameter Poincaré translation $\delta x_\beta = a_\beta$, the coordinate symmetry associated to energy-momentum tensor derivation, we have $\delta_T A_\nu = 0$, leaving only two contributions to the transformation of fields $\delta \bar{A}_\nu$.[23]

[23] The term $\omega_{\alpha\beta} x^\beta$, associated to the angular momentum tensor $M^{\rho\alpha\beta} = x^\alpha T^{\rho\beta} - x^\beta T^{\rho\alpha}$, consists of the remaining six parameters in the Poincaré group through the antisymmetric parameter $\omega_{\alpha\beta}$. Terms $S x_\alpha$ and $2\xi_\nu x_\alpha x^\nu - \xi_\alpha x_\nu x^\nu$ correspond to the dilatation tensor D^ρ and conformal tensor $C^{\rho\alpha}$. Direct substitution of the various terms in (8.20) into (8.19) give the $\delta \bar{A}_\nu$ which can be directly substituted into (8.2) to derive known physical conservation laws in, e.g., electrodynamics.

Notably, the presence of mixed coordinate and field transformations should be no surprise to anyone who has actually read Noether's paper, as she explicitly admits that her first theorem holds for a combination of the symmetries:

> In the case of a 'mixed group,' if one assumes similarly that Δx and Δu are linear in the ε and the $p(x)$, one sees that, by setting the $p(x)$ and the ε successively equal to zero, divergence relations … as well as identities … are satisfied. (translated from Noether 1918, p. 243)[24]

where the Eq. 13 she refers to is the Noether identity associated to her first theorem that we consider in this chapter. It is exactly this freedom that Bessel-Hagen, in fact in consultation with Noether herself, used to apply the first theorem successfully to electrodynamics; this is the topic of the section on Method II.

In the section on Method I, we will discuss the case when $\delta_g A_\nu = 0$, i.e. when we use only the canonical transformations, yielding the canonical Noether energy-momentum tensor $T_C^{\mu\nu}$. Since this is not the correct energy-momentum tensor $T^{\mu\nu}$ of electrodynamics, various ad-hoc 'improvements' have been considered in the literature that add terms to $T_C^{\mu\nu}$ in order to obtain the desired result. In the section on Method II, we will discuss the less common method in the literature, which does not make restrictions on $\delta \bar{A}_\nu$ and keeps the most general (8.18). In this case, the well-known $T^{\mu\nu}$ of electrodynamics is directly derived with no ad-hoc 'improvements' needed.

8.4 A Tale of Two Methods

8.4.1 Method I: Canonical Tensor Plus 'Improvements'

In this section, we consider the most common method in the literature for deriving the generally accepted energy-momentum tensor 'from' Noether's first theorem. As emphasised above, the canonical Noether tensor $T_C^{\mu\nu}$ obtained on this approach (8.10) differs from the accepted energy-momentum tensor (8.1). Hence, Noether's theorem apparently fails to properly identify the conserved quantities associated with symmetries even in the most familiar case. On our view, this 'canonical' result reflects a basic mistake: it does not take into account all of the relevant variational symmetries needed to build an energy-momentum tensor. We will see how to employ Noether's theorem more effectively to do so through what we call Method II in Section 8.4.2.

Usually the the canonical energy-momentum tensor is 'improved' by adding specific terms, such as the divergence of a superpotential and terms proportional to the equations of motion.[25] We give an example of this for electrodynamics in this section. One could put the task – a bit provocatively – as follows: Given that $T_C^{\rho\beta}$ is not the result we wanted (or expected), what terms can we add to get the correct answer? Of course this is an ad-hoc

Transformations (8.20) can be found by solving the conformal Killing's equation. See Rosen (1972) for a self-contained derivation of the conformal invariance of electrodynamics.

24 'Setzt man entsprechend einer ,gemischten Gruppe' Δx und Δu linear in den ε und den $p(x)$ an, so sieht man, indem man einmal die $p(x)$, einmal die ε Null setzt, daß sowohl Divergenzrelationen …, wie Abhängigkeiten … bestehen.'

25 There are a number of proposals regarding how to 'improve' the energy-momentum tensor in the literature; see Forger and Römer (2004) and Blaschke et al. (2016) for recent surveys.

approach to fixing the problem, but if it is the best available method we have to obtain the accepted $T^{\rho\beta}$, one might just bite the bullet.[26]

The required 'improvement' term in the case of electrodynamics is simply the difference between (8.1) and the canonical expression (8.10),

$$T^{\rho\beta} - T_C^{\rho\beta} = -F^{\rho\nu}\partial_\nu A^\beta. \tag{8.21}$$

All of the various improvements for electrodynamics in the literature ultimately need to give us this term on the right-hand side, at minimum after imposing on-shell conditions (Blaschke et al. 2016; Forger and Römer 2004). The challenge is how to get the correct tensor, by starting from the canonical expression, and adding 'improvement' terms through a well-defined procedure. We will briefly discuss the Belinfante improvement procedure since it is by far the most commonly adopted in the literature.[27]

But before doing so, it is worth discussing the general idea of improvement by superpotentials and terms proportional to the equations of motion; together these form the bulk of possible 'improvement' terms. Superpotentials have the form $\Psi^{[\rho\alpha]\sigma}$, the divergence of which $\partial_\alpha\Psi^{[\rho\alpha]\sigma}$ can be added to an energy-momentum tensor without affecting on-shell conservation. This is because indices $[\rho\alpha]$ are anti-symmetric; the divergence of the divergence of a superpotential $\partial_\rho\partial_\alpha\Psi^{[\rho\alpha]\sigma}$ is identically zero off-shell. Adding a superpotential to a Noether energy-momentum tensor for a specific Lagrangian does not spoil conservation (the superpotential is conserved as a mathematical identity on its own), yet doing so may lead to an energy-momentum tensor with the required properties. Terms that vanish on-shell, i.e. terms proportional to the equations of motion, can also be added while preserving on-shell equivalence; in practice terms of this type often must also be added to obtain the accepted form of the energy-momentum tensor.

For electrodynamics, the difference between the accepted and canonical energy-momentum tensors is given by (8.21) above. Writing the extra term as the divergence of a superpotential, we have $-F^{\rho\nu}\partial_\nu A^\beta = \partial_\alpha[-F^{\rho\alpha}A^\beta] + A^\beta\partial_\nu F^{\rho\nu}$, where $\Psi^{[\rho\alpha]\sigma} = -F^{\rho\alpha}A^\beta$, and Maxwell's equations $E^\rho = \partial_\nu F^{\nu\rho}$. Thus, we have,

$$T^{\rho\beta} = T_C^{\rho\beta} + \partial_\alpha\Psi^{[\rho\alpha]\beta} - A^\beta E^\rho. \tag{8.22}$$

The Belinfante improvement procedure yields exactly the same superpotential; the divergence of this superpotential as well as a term proportional to the equation of motion can be used to recover the accepted energy-momentum tensor in cases such as electrodynamics. Therefore just by knowing (8.1), we know the form of the required additional terms. The Belinfante procedure provides a derivation of this superpotential, which we will detail in the following subsection.

8.4.1.1 The Belinfante Symmetrisation Procedure

Turning an arbitrary tensor into a symmetric tensor is in principle straightforward: decompose the tensor into a symmetric and antisymmetric part, and then add a new contribution

[26] It is worth noting that proponents of Method I are usually unaware of the Bessel-Hagen et al. approach.

[27] Belinfante (1940) is commonly cited as the origin of the Belinfante improvement procedure, also known as the Belinfante symmetrisation procedure, but it is not clear to what extent he shared the motivations of later work regarding the 'improvement' of energy-momentum tensors..

to cancel out the antisymmetric part (in this case, from a superpotential). But more interestingly, arguably, Belinfante (1940)[28] showed that a suitable superpotential of this kind can be derived – and he argued that it is related to the spin angular-momentum of the model.[29] We wish to add the divergence of the Belinfante superpotential $\partial_\alpha b^{[\rho\alpha]\sigma}$ to the canonical expression (8.10) to form the Belinfante tensor $T_B^{\rho\sigma}$,[30]

$$T_B^{\rho\sigma} = T_C^{\rho\sigma} + \partial_\alpha b^{[\rho\alpha]\sigma} \tag{8.23}$$

where the superpotential $b^{[\rho\gamma]\sigma}$ is defined by a combination of the spin angular momentum tensor $S^{\rho[\sigma\gamma]}$ of the form (Belinfante 1940):

$$b^{[\rho\gamma]\sigma} = \frac{1}{2}(-S^{\rho[\sigma\gamma]} + S^{\gamma[\sigma\rho]} + S^{\sigma[\gamma\rho]}). \tag{8.24}$$

In electrodynamics this contribution is defined as $S^{\gamma[\alpha\beta]} = \frac{\partial \mathcal{L}}{\partial\partial_\gamma A^\mu}[\eta^{\alpha\mu}A^\beta - \eta^{\beta\mu}A^\alpha]$. Therefore, we have

$$S^{\gamma[\alpha\beta]} = -F^{\gamma\mu}[\delta^\alpha_\mu A^\beta - \delta^\beta_\mu A^\alpha]. \tag{8.25}$$

Inserting (8.25) into the Belinfante superpotential (8.24), we have $b^{[\rho\gamma]\sigma} = -F^{\rho\gamma}A^\sigma$. But this is the same superpotential found from (8.22)! Thus, for the Belinfante procedure applied to electrodynamics we have,

$$T_B^{\mu\nu} = T_C^{\mu\nu} + \partial_\alpha(F^{\alpha\mu}A^\nu). \tag{8.26}$$

This differs from the accepted energy-momentum tensor (8.1), according to (8.22), by on-shell terms.[31]

Therefore the accepted energy-momentum tensor (8.1) is related to the Belinfante tensor (8.26) as follows:

$$T^{\mu\nu} = T_B^{\mu\nu} - A^\nu E^\mu. \tag{8.27}$$

The Belinfante prescription alone does not yield the correct expression without adding this additional term ($-A^\nu E^\mu$) proportional to the equations of motion; equivalence to $T_B^{\mu\nu}$ alone can be established only after imposing the on-shell condition $E^\mu = 0$. Note that requiring such an on-shell condition for just formulating the energy-momentum tensor is a severe restriction; in contrast, the Noether energy-momentum tensor directly obtained in Method II can be defined *without* any on-shell condition – only conservation requires imposition of the equations of motion.

[28] Through the help of an uncited Dr. Podolansky (presumably theoretical physicist Dr. Julius Podolanski); see Belinfante (1940).

[29] Since the 'total' angular momentum tensor $M^{\rho\alpha\beta} = x^\alpha T^{\rho\beta} - x^\beta T^{\rho\alpha}$ is based on the symmetric energy-momentum tensor $T^{\mu\nu}$, decomposing it into parts (including 'spin' angular momentum), one can obtain terms missing in the canonical Noether tensor $T_C^{\mu\nu}$. The missing terms have the appropriate form for spin angular momentum; if this tensor vanishes the canonical tensor is symmetric, otherwise we have to follow the procedure described in the text.

[30] We note that the Belinfante tensor $T_B^{\rho\sigma}$ is sometimes referred to as the Belinfante–Rosenfeld tensor, since Rosenfeld (1940) independently came to some of these results and published them shortly after Belinfante (1939) first presented them.

[31] On-shell equivalence of the Belinfante and Hilbert tensors is a well-established result (Forger and Römer 2004; Pons 2011). The Hilbert tensor in Minkowski spacetime is the accepted energy-momentum tensor in cases such as electrodynamics (but not in general; see Baker et al. 2021), thus for electrodynamics the Belinfante–Hilbert relationship can be used to obtain the Belinfante superpotential and associated on-shell terms in (8.26) which are required to correctly improve the canonical Noether tensor.

So we see that, when interested in symmetric energy-momentum tensors, the Belinfante symmetrisation procedure does provide an on-shell procedural fix in, e.g., the case of electrodynamics. But it is just a symmetrisation procedure; it is not clear how it would, for instance, help to obtain the physical (i.e. also tracefree and gauge-invariant and not just symmetric) energy-momentum tensor in general.[32] Furthermore, it is limited in scope: it focuses only on the energy-momentum tensor, at the expense of other conserved quantities. By contrast, Method II treats all conserved currents on the same footing; there is no privileging of any specific quantity as most fundamental – all conserved currents follow directly and uniquely from Noether's first theorem.

If we take the most charitable possible view of Method I, that the ad-hoc improvements are entirely physically justified and a necessary correction after applying Noether's first theorem, one unavoidable fact remains: the improvement procedure still requires on-shell conditions to equate the accepted $T^{\mu\nu}$ to the improved tensor. In Method II, no on-shell conditions are required to obtain the complete set of conservation laws.

8.4.2 Method II: Including Gauge Symmetries

There is another method for deriving the energy-momentum tensor, such that we directly obtain it from Noether's first theorem.

Instead of using the restrictive condition of Method I where we only consider the canonical transformations (8.7), we instead use the most general picture of all possible field transformations, such as outlined in (8.18). Considering the transformations of the action

$$\delta \bar{A}_\alpha = -F^{\nu}_{\ \alpha}\delta x_\nu, \tag{8.28}$$

Bessel-Hagen (1921) – as well as Eriksen and Leinaas (1980); Montesinos and Flores (2006); among others – derived all 15 accepted conservation laws of electrodynamics directly from Noether's first theorem using (8.2), (8.28), and the sourcefree Lagrangian density $\mathcal{L} = -\frac{1}{4}F_{\mu\nu}F^{\mu\nu}$. In other words, (8.28) are what we call proper transformations,

$$E^\nu \delta \bar{A}_\nu + \partial_\rho\left(\left[F^{\rho\nu}F^{\beta}_{\nu} - \frac{1}{4}\eta^{\rho\beta}F_{\mu\nu}F^{\mu\nu}\right]\delta x_\beta\right) = 0, \tag{8.29}$$

where E^ν is the Euler–Lagrange expression. Immediately, Noether's first theorem leads to the physical energy-momentum tensor (8.1) in square brackets.[33] More precisely, for the case of the 4-parameter Poincaré translation, $\delta x_\beta = a_\beta$ is a constant that can be pulled out of the total divergence leaving the elegant identity,

$$E^\nu \delta \bar{A}_\nu + a_\beta \partial_\rho T^{\rho\beta} = 0 \tag{8.30}$$

[32] Blaschke et al. (2016) provide an improvement procedure based on requiring gauge invariance instead of just symmetry.

[33] Using (8.29), the 15 conformal conservation laws of electrodynamics are immediately obtained by inserting δx_β from (8.20): four from the divergence of the energy-momentum tensor $T^{\rho\beta} = F^{\rho\nu}F^{\beta}_{\nu} - \frac{1}{4}\eta^{\rho\beta}F_{\mu\nu}F^{\mu\nu}$, six from the divergence of the angular momentum tensor $M^{\rho\mu\nu} = x^\mu T^{\rho\nu} - x^\nu T^{\rho\mu}$, one from the divergence of the dilatation tensor $D^\rho = T^{\rho\beta}x_\beta$, and four from the divergence of the conformal tensor $C^{\rho\alpha} = T^{\rho\beta}(2x_\beta x^\alpha - \delta^\alpha_\beta x_\lambda x^\lambda)$.

Immediately we have a compact identity relating the Euler–Lagrange expression E^{ν} and energy-momentum tensor $T^{\rho\beta}$ of electrodynamic theory. The Lorentz force law and Poynting's theorem are compactly derived alongside Maxwell's equations. This compact identity makes it easy to appreciate the celebrated elegance of Noether's theorems.

8.4.2.1 Deriving the Proper Transformations from the Bessel-Hagen Method

How can one obtain the proper transformations (8.28) that lead to the physical conservation laws? The various authors Bessel-Hagen (1921), Eriksen and Leinaas (1980), and Montesinos and Flores (2006) that independently came to this conclusion used slightly different rationales, largely to do with requiring gauge invariance of the Noether current or requiring gauge invariance of the transformations themselves. We will follow the Bessel-Hagen approach because he was first to present this result, and he took advice from Noether herself on his paper. More explicit application of the Bessel-Hagen method to electrodynamics can be found in Baker (2021c). Starting from the general transformations of fields (8.19) in the case of electrodynamics we have,

$$\delta\bar{A}_{\nu} = -\partial^{\beta}A_{\nu}\delta x_{\beta} - A_{\mu}\partial_{\nu}\delta x^{\mu} + \partial_{\nu}\phi \tag{8.31}$$

The question Bessel-Hagen asked is how to derive the parameter ϕ such that we have the unique gauge invariant energy-momentum tensor of (sourcefree) electrodynamics. To do this we substitute $\delta\bar{A}_{\nu}$ into the Noether current (8.2) and solve for ϕ to obtain a current which is gauge invariant. The ϕ must depend on both the vector potential A_{α} (in order to obtain a gauge-invariant current) and the infinitesimal transformations of coordinates δx^{α} (in order to factor out the 4-parameter Poincaré translation from the current and obtain the energy-momentum tensor). Bessel-Hagen solved for ϕ, obtaining for the gauge parameter $A_{\mu}\delta x^{\mu}$, which is the most trivial scalar combination of the required components. Inserting this ϕ into (8.31) and differentiating the third term, we have

$$\delta\bar{A}_{\nu} = -\delta x_{\beta}\partial^{\beta}A_{\nu} - A_{\mu}\partial_{\nu}\delta x^{\mu} + \delta x_{\beta}\partial_{\nu}A^{\beta} + A_{\mu}\partial_{\nu}\delta x^{\mu}. \tag{8.32}$$

Remarkably, the second and last terms on the right-hand side cancel (those associated to the contragredient transformations), and we are left with exactly $\delta\bar{A}_{\nu} = -F^{\beta}_{\ \nu}\delta x_{\beta}$ as in (8.28)! Therefore, the proper transformations that directly yield the physical conservation laws can be thought of as a mixing of the various symmetries of the action, as opposed to an independent application of symmetries as in the case of the canonical Noether energy-momentum tensor or Noether's second theorem.

We note that the selection of the gauge parameter, while aided by knowledge of the unique gauge invariant energy-momentum tensor in the case of electrodynamics, can be obtained from the Noether current. For this reason, the method applies more generally to models where the energy-momentum tensor is not already known. The more general application of Bessel-Hagen to exactly gauge invariant actions in this way is the subject of Baker (2021c), in which the Bessel-Hagen method has successfully been applied to several field theories such as Yang–Mills, Kalb–Ramond, third-rank antisymmetric fields, and linearised Gauss–Bonnet gravity.

The proper form of the transformation was noticed for the Yang–Mills theory by Jackiw (1978), without deriving this from a procedure such as Method II. While the vast majority of textbooks give the canonical picture alone, some, such as Burgess (2002) and Scheck (2012), have noticed the proper transformations and avoided the restrictive canonical presentation. One of our goals in the following is to settle this ambiguity in favor of the proper transformations through appeal to the converse of Noether's first theorem.

8.4.2.2 *Proper Transformation as Gauge-Invariant Transformations*

We now know that the appropriate choice of $\delta \bar{A}_\alpha = \delta_C A_\alpha + \delta_g A_\alpha$ in the Noether identity for classical electrodynamics directly leads to the accepted energy-momentum tensor, and that the proper transformation can be chosen by solving for a $\delta_g A_\alpha$ that makes the current invariant. We will explore how to justify the specific choice of $\delta_g A_\alpha$ a posteriori via the converse of Noether's first theorem in the next section, i.e. by starting from the accepted energy-momentum tensor.

Before doing so, we want to explore how to motivate the choice of $\delta_g A_\alpha$ other than by solving the Noether identity for $\delta_g A_\alpha$ while requiring that the energy momentum is gauge-invariant. To this end, we will consider Eriksen and Leinaas (1980), who argued that gauge invariance of the transformation $\delta \bar{A}_\alpha$ is the property one can use to determine the proper transformation $\delta \bar{A}_\alpha = -F^{\nu}_{\ \alpha} \delta x_\nu$ as in (8.28). Eriksen and Leinaas (1980) starts from the gauge condition in the case of sourcefree electrodynamics, $\delta_g A_\alpha = \partial \chi$ with $\chi = \chi(A)$. The parameter χ is taken to be an arbitrary gauge parameter we must solve for based on the condition that $\delta_g A_\alpha$ must be gauge invariant. By combining the $\delta_T A_\alpha$ (contragredient) and $\delta_g A_\alpha$ (gauge) transformations, the authors find an equation for χ which does not uniquely determine χ; however, they choose the 'simple' solution that leaves $\delta_g A_\alpha$ as a whole gauge invariant, which is identically $\chi = \delta x^\nu A_\nu$ – exactly what was found by Bessel-Hagen!

One could now note how intuitive the requirement of a gauge-invariant transformation is: as long as all expressions in the assumptions of the Noether theorem are gauge-invariant, the resulting energy-momentum tensor should come out as gauge-invariant too. However, there exist cases where the proper transformations that are used to derive the unique energy-momentum tensor for a theory are not themselves gauge invariant, as we will discuss in Section 8.5. This indicates limits in the scope of application of Eriksen and Leinaas's (1980) method. We note that the Bessel-Hagen method works more broadly because it treats both the cases where the transformations themselves are gauge invariant, as well as cases where they are not.

8.4.3 *Converse of Noether's First Theorem as a Test for Noetherian Currents*

We now use the converse of Noether's first theorem relative to the Lagrangian density of electrodynamics and the accepted energy-momentum, tensor (8.1) in order to arrive at the relevant variational symmetry linked to this $T^{\mu\nu}$. As we will see, the converse can generally be used to decide whether an energy-momentum tensor can be directly derived from Noether's first theorem – and thus from Method II.

We can derive the form of the transformations $\delta \bar{A}_\nu$ using the converse of Noether's first theorem based on the accepted energy momentum tensor, (8.1), as follows: We start with

$$E^\nu \delta \bar{A}_\nu + a_\nu \partial_\mu T^{\mu\nu} = 0 \tag{8.33}$$

and the Noether identity (8.2). Since the 4-parameter Poincaré translation $\delta x_\beta = a_\beta$ is associated to the energy-momentum tensor, it follows that $\delta \bar{A}_\nu = U_\nu^\beta a_\beta$, namely the transformation of fields, must be proportional to the 4-parameter a_β. Therefore, we must solve for U_ν^β,

$$E^\nu \delta \bar{A}_\nu + a_\beta \partial_\rho \left(-F^{\rho\nu} U_\nu^\beta - \frac{1}{4} \eta^{\rho\beta} F_{\mu\nu} F^{\mu\nu} \right) = 0 \tag{8.34}$$

Subtracting Eqs. (8.33) and (8.34), we have,

$$a_\beta \partial_\rho \left(-F^{\rho\nu} U_\nu^\beta - \frac{1}{4} \eta^{\rho\beta} F_{\mu\nu} F^{\mu\nu} \right) = a_\beta \partial_\rho \left(F^{\rho\nu} F_\nu^\beta - \frac{1}{4} \eta^{\rho\beta} F_{\mu\nu} F^{\mu\nu} \right) \tag{8.35}$$

We have $U_\nu^\beta = -F_\nu^\beta$, and thereby recover the proper transformations for the Poincaré translation $\delta \bar{A}_\nu = -F_\nu^\beta a_\beta$. More generally, we can solve for δx_ν from the Noether identity, and again we have the requirement $\delta \bar{A}_\alpha = -F_\alpha^\nu \delta x_\nu$. Thus, if we consider the converse of Noether's first theorem on the accepted $T^{\mu\nu}$ (8.1), the canonical transformation (8.7) associated to the canonical Noether energy-momentum tensor (8.10) never appears in isolation! In other words, in the case of electrodynamics, the converse of Noether's first theorem supports Method II. At the same time, we can now see that Method I uses the wrong symmetry to begin with. The failure to recognise the properly adapted symmetry transformations leads to the need to introduce – and justify, if possible – ad hoc 'improvements'.

The lesson from electrodynamics generalises: Given a proposed energy-momentum tensor, we learn through Noether's converse which (if any) symmetries are linked to it; if there are none linked to it, then the energy-momentum tensor cannot be derived directly from Noether's first theorem.[34]

So, in cases like electrodynamics, for which the canonical energy momentum tensor lacks essential properties, we find that the improvements can be avoided by using Noether's first theorem properly (that is, by exploiting the complete set of variational symmetries of the action) – and thus that there is nothing wrong with the Noether method to begin with. In cases where the converse does not give symmetries linked to an energy-momentum tensor, we at least learn that this energy-momentum tensor cannot be derived from Noether's first theorem.

We acknowledge that the converse approach we are advocating is not as straightforward if we do not already know the appropriate energy-momentum tensor for the relevant fields. We have criticised the improvement approach in Method I because it relies on knowing the proper form of the energy-momentum tensor in order to find the appropriate improvement terms; yet, we also need to know the proper form of the energy-momentum tensor to obtain the proper variational symmetries from the converse of Noether's first theorem. If the proper

[34] This raises the question whether such an object earns the title of energy-momentum tensor in the first place.

energy-momentum tensor is not uniquely known (several candidates exist), this can lead to a kind of reflective equilibrium in assessing candidates for energy-momentum tensors and the associated symmetries. We will turn to just such a case in the next section, namely linearised spin-2 fields where numerous different energy-momentum tensors have been proposed.

8.5 Beyond Electrodynamics

Up to this point we have used electrodynamics to explicate Method II – but it has much broader scope. A recent series of papers has shown how Method II applies to several classical, relativistic field theories, such as:

- *(Sourcefree) Yang–Mills* Baker (2021c), with the Lagrangian $\mathcal{L}_{YM} = -\frac{1}{4}F^a_{\mu\nu}F^{\mu\nu}_a$. Applying Method II with the 'mixed' variational symmetry $\bar{\delta}A^a_\mu = -F^a_{\mu\nu}\delta x^\nu$ (with $\delta x^\nu = a^\nu$) leads to the energy-momentum tensor $T^{\mu\nu} = F^{\mu\lambda}_a F^{a\nu}{}_\lambda - \frac{1}{4}\eta^{\mu\nu}F^a_{\lambda\rho}F^{\lambda\rho}_a$.[35] The energy-momentum tensor is invariant under the gauge transformation $\delta_g\bar{A}_{a\mu} = \partial_\mu\theta_a + C_{abc}A^b_\mu\theta^c$.
- *Linearised Gauss–Bonnet gravity* Baker (2021a); Baker and Kuzmin (2019), with the Lagrangian $\mathcal{L} = \frac{1}{4}(R_{\mu\nu\alpha\beta}R^{\mu\nu\alpha\beta} - 4R_{\mu\nu}R^{\mu\nu} + R^2)$.[36] The variational symmetry $\delta\bar{h}_{\rho\sigma} = -2\Gamma^\nu{}_{\rho\sigma}\delta x_\nu$ (with $\delta x^\nu = a^\nu$) leads to the generally accepted energy-momentum tensor,[37] which is gauge-invariant under the spin-2 gauge transformation $\delta_g\bar{h}_{\mu\nu} = \partial_\mu\xi_\nu + \partial_\nu\xi_\mu$.

Just as with electrodynamics, applying the mixed variational symmetry in each of these cases leads directly to the accepted energy-momentum tensor. Baker (2021c) discusses several other cases as well.

What can then be said about the scope of the method? For a given gauge-invariant Lagrangian density, an exact variational symmetry can be found such that the Noether current associated to the Poincaré translation will be the physical energy-momentum tensor. Notably, it is *not* a necessary criterion that the total symmetry transformation is gauge-invariant: Recalling Section 8.4.2.2, the decisive symmetry transformation in electrodynamics (as given by the standard Lagrangian) is gauge-invariant, but the symmetry transformation is not always itself gauge-invariant (e.g. linearised Gauss–Bonnet gravity). This means that the proper variational symmetries can not always be systematically obtained from requiring gauge-invariance of the proper transformation, which showcases the restrictions of the procedure presented in Section 8.4.2.2. In other words, an exactly gauge-invariant symmetry transformation will only be sufficient but not necessary for obtaining a gauge-invariant conserved current; Method II à la Bessel-Hagen has a much wider scope than the method of Eriksen and Leinaas (1980).

Put the other way around, problems arise for Method II when (i) we have a model that does not have an exactly gauge invariant action, so that solving for the right

[35] The field strength tensor is given by $F_{a\mu\nu} = \partial_\mu A_{a\nu} - \partial_\nu A_{a\mu} + C_{abc}A^b_\mu A^c_\nu$ where C_{abc} is the totally antisymmetric structure constant.

[36] Appearing in the Lagrangian are the linearised Riemannian tensor, defined as $R^{\mu\nu\alpha\beta} = \frac{1}{2}(\partial^\mu\partial^\beta h^{\nu\alpha} + \partial^\nu\partial^\alpha h^{\mu\beta} - \partial^\mu\partial^\alpha h^{\nu\beta} - \partial^\nu\partial^\beta h^{\mu\alpha})$, and contractions of it.

[37] Explicitly, $T^{\omega\nu} = -R^{\omega\rho\lambda\sigma}R^\nu{}_{\rho\lambda\sigma} + 2R_{\rho\sigma}R^{\omega\rho\nu\sigma} + 2R^{\omega\lambda}R^\nu_\lambda - RR^{\nu\omega} + \frac{1}{4}\eta^{\omega\nu}(R_{\mu\lambda\alpha\beta}R^{\mu\lambda\alpha\beta} - 4R_{\mu\nu}R^{\mu\nu} + R^2)$

gauge-transformation becomes problematic or (ii) when the energy-momentum tensor is questionable but the BH method is not applicable because there is just no gauge symmetry to begin with. We will outline now how in both cases, at least the Noetherian[38] nature of the energy-momentum tensor can be checked upon application of Noether's converse.

With respect to (8.2), an interesting application of Noether's converse is to reveal the improved Callan–Coleman–Jackiw (CCJ) traceless energy-momentum tensor of the Klein–Gordon theory (see Callan et al. 1970) as non-Noetherian relative to the standard Lagrangian.[39]

However, an energy-momentum tensor that is non-Noetherian relative to some Lagrangian may be Noetherian to the same Lagrangian up to a divergence term; this is exactly the case of the CCJ-energy-momentum tensor (see Kuzmin and McKeon 2001).

Turning to (8.1), Noether's (original) theorem involves the invariant action condition $S = S'$, namely that an action does not change after simultaneous transformation of the independent variables (e.g. coordinates) and dependent variables (e.g. fields). This condition requires strict invariance, namely the exact symmetries which we have referred to throughout the chapter. In this case Noether's theorem can be applied as in her original paper, in a straightforward manner which we consider for both Method I and Method II. There are generalisations of this theorem to transformations which do not satisfy this action condition, and instead after simultaneous transformation of the independent and dependent variables differ by a boundary term, e.g. $S' - S = \int \partial_\mu B^\mu dV$. Therefore, in cases where an action has an inexact symmetry (transforming the action results in a boundary term), we cannot trivially apply Noether's first theorem as in her article, and must turn to one of the generalisations of invariance up to a boundary term. These 'modern' generalisations originate with the other result in Bessel-Hagen's paper, invariance up to boundary terms, but extend to the so-called MAO theorem, which we will discuss in Section 8.6. We do not apply these generalisations in our article because our main concern is to compare the conventional application of Noether's first theorem (Method I) to the lesser-known method (Method II), both in the context of the action condition $S = S'$. However, we will now discuss a possible valuable application of these generalisations, which we could not find any studies of in the literature.

An appealing possible application with respect to (8.1) is the case of the spin-2 Fierz–Pauli action where the gauge symmetry of the equation of motion (linearised diffeomorphisms) is not an exact symmetry of the action; the action is only invariant up to a boundary term (see Baker and Kuzmin 2019). Furthermore, there is no gauge-invariant energy-momentum tensor for spin-2 gravity (see Magnano and Sokolowski 2002) to begin with, so we cannot use gauge invariance to help pick out a unique expression.[40] If an action is not exactly invariant, such as in the case of spin-2, generalisations of Noether's theorem to symmetries up to boundary terms (i.e. the inexact symmetries method in Bessel-Hagen 1921) must be applied; the application of these methods to spin-2 Fierz–Pauli theory is the subject of future work.

[38] See Section 8.6 for discussion of Noetherian vs. non-Noetherian currents.

[39] We call an energy-momentum tensor Noetherian relative to $\mathcal{L}$ if it is directly derivable from a variational symmetry of $\mathcal{L}$ via Noether's identity.

[40] Arguably, this jeopardises the application of the BH method which is centrally about achieving a gauge-invariant current.

To elaborate a bit on the issue: For linearised gravity (massless spin-2 gravity), there are numerous proposals for $T^{\mu\nu}$ (see Bičák and Schmidt 2016 for an overview). This ambiguity cannot be avoided in, for example, attempts to derive general relativity from a spin-2 field theory that proceed by taking the spin-2 field $h_{\mu\nu}$ to be self-coupled. Which $T^{\mu\nu}$ should be added to the action to represent this self-coupling? Here authors disagree on whether the Einstein field equations can be derived from spin-2 Fierz–Pauli theory, to a large degree based on their choice of which $T^{\mu\nu}$ to select (if they even grant that it is physically well-defined despite its inevitable gauge-dependent nature). See Padmanabhan (2008) for a criticism, and Barceló et al. (2014) for a defense, of conventional wisdom on this issue.

Linearised (massless) spin-2 gravity is given by the (massless) Fierz–Pauli Lagrangian density

$$\mathcal{L}_{FP} = \frac{1}{4}\left(\partial_\alpha h^\beta_\beta \partial^\alpha h^\gamma_\gamma - \partial_\alpha h_{\beta\gamma}\partial^\alpha h^{\beta\gamma} + 2\partial_\alpha h_{\beta\gamma}\partial^\gamma h^{\beta\alpha} - 2\partial^\alpha h^\beta_\beta \partial^\gamma h_{\gamma\alpha}\right)$$

with canonical Noether energy-momentum tensor $T^{\mu\nu}_C = \eta^{\mu\nu}\mathcal{L} - \frac{\partial\mathcal{L}}{\partial(\partial_\mu h_{\alpha\beta})}\partial^\nu h_{\alpha\beta}$ – the linearised Einstein energy-momentum tensor gives exactly this canonical expression (Szabados 1992). Strikingly, this tensor is neither gauge-invariant nor symmetric nor traceless. As we know from Magnano and Sokolowski (2002), there is no energy-momentum tensor for spin-2 Fierz–Pauli theory that is gauge-invariant. Candidates for such improved energy-momentum tensors for linearised gravity usually presented include, for example, the linearised Hilbert and Landau–Lifshitz expressions – both of which can be obtained by adding the appropriate divergence of superpotential and terms proportional to the equations of motion to $T^{\mu\nu}_C$ (see Baker 2021b). Strong adherents to Method I might then suggest that since any such energy-momentum tensor (conserved on-shell using the spin-2 Fierz–Pauli equation of motion) follows from the addition of improvement terms, that all such linearised gravity energy-momentum tensors are in some sense connected to Noether's first theorem. Baker (2021b) shows that there are infinitely many such improved energy-momentum tensors for linearised gravity.

A much more straightforward approach for spin-2 linearised gravity, then, is to apply the converse of Noether's first theorem to the various expressions in the literature as we did for electrodynamics in the previous section. This would concretely determine which (if any) can yield δX_β and $\delta\bar{h}_{\mu\nu}$ symmetry transformations to prove a direct and meaningful connection to Noether's first theorem. Concretely, one would have to use the Noether identity

$$\left(\frac{\partial\mathcal{L}}{\partial h_{\mu\nu}} - \partial_\rho\frac{\partial\mathcal{L}}{\partial(\partial_\rho h_{\mu\nu})}\right)\delta\bar{h}_{\mu\nu} + \partial_\rho\left(\eta^{\rho\beta}\mathcal{L}\delta X_\beta + \frac{\partial\mathcal{L}}{\partial(\partial_\rho h_{\mu\nu})}\delta\bar{h}_{\mu\nu}\right) = 0 \qquad (8.36)$$

for each energy-momentum tensor and solve for $\delta\bar{h}_{\mu\nu}$ in the same way we solved for the variational symmetries of electrodynamics in Section 8.4.3, where $\delta\bar{h}_{\mu\nu} = -\partial^\beta h_{\mu\nu}\delta X_\beta + \delta h_{\mu\nu}$. As in the case of electrodynamics, the term proportional to the Lagrangian density ($\eta^{\rho\beta}\mathcal{L}$) must have a Lagrangian density $\mathcal{L}$ which yields the spin-2 equation of motion in the Euler–Lagrange equation – otherwise, the energy-momentum tensor in question will not be associated to spin-2 Fierz–Pauli theory in the context of Noether's first theorem, regardless of the transformations we consider. Regardless of the outcome of this calculation, we will have a strong statement about the energy-momentum tensors for spin-2 theory in the literature: either that Noether transformations can select a preferred expression, confirm numerous expressions can be obtained from the Noether approach, or show that there are problems with applying Noether's first theorem to this model as a whole. If transformations can uniquely be solved for by the converse of Noether's first theorem, then we can say that a given expression can be directly derived. If there is no solution, then a given expression cannot be claimed to be associated to Noether's first theorem for a particular Lagrangian density. Whether or not the various published expressions can be directly obtained, regardless of outcome, will provide clear insight into the relationship between the linearised gravity energy-momentum tensors in the literature and Noether's first theorem. The treatment of energy in linearised gravity may have some bearing on active disputes regarding energetic quantities in GR (see, for instance, Dürr 2019; Read 2020). This application of the converse of Noether's first theorem to spin-2 is the subject of future work.

8.6 Equivalence Classes

Mathematics often draws finer distinctions than physics requires. Physicists typically treat a unique definition of a fundamental quantity as necessary for understanding its physical significance. This can conflict with the embarrassment of riches resulting from new frameworks produced by mathematicians, particularly when they lead to quantities with an ambiguous physical status. A natural response is to regard some range of mathematically distinguished possibilities as falling within an equivalence class, such that physical interpretations need not draw distinctions among its members. Recent philosophical discussions by Brown (Chapter 7 in this volume) have pointed out that modern Noether-*like* theorems by Martínez, Alonso, and Olver (dubbed the MAO theorem by Brown) should be read as relating an equivalence class of symmetries to an equivalence class of conservation laws.[41] Perhaps we could follow a similar strategy and elucidate only an equivalence class of energy-momentum tensors, setting aside the search for a unique energy-momentum tensor we have been pursuing as misguided or unnecessary? In this section we aim to adjudicate these questions regarding uniqueness and the appropriate criteria of equivalence, or (perhaps more accurately) at least to survey some of the considerations that bear on them.

To make this concrete, suppose we treat two candidate energy-momentum tensors as equivalent iff they differ by superpotential terms and linear combinations of the equations of motion. In the case we have focused on in Method II, this equivalence class consists of tensors generated from the standard energy-momentum tensor (8.1) by adding the divergence of a superpotential and terms proportional to the equations of motion:[42]

[41] To be more precise, Olver's (2000) statement of Noether's theorem (Theorem 5.58, p. 334) introduces the following notion of equivalence between variational symmetries – transformations on the space of dependent and independent variables of the differential equations, in the form of a (generalised) vector field, that preserve the action. Two variational symmetries are equivalent if they differ by a trivial symmetry; a trivial symmetry is a transformation on the space of solutions that is just the generator of the dynamical flow (up to parameterisation). The generalised notion of a vector field includes 'evolutionary vector fields'; a trivial symmetry transformation corresponds to an evolutionary vector field that vanishes on the solutions of the differential equation under consideration. The need to 'quotient out' by trivial symmetries arises as a result of the shift to a different mathematical setting.

[42] Conventionally these 'improvements' (the divergence of a superpotential and terms proportional to the equations of motion) are applied to the canonical Noether tensor (8.10) as in (8.22) for Method I, but they can be applied to any Noether current (such as in (8.37) using Method II), satisfying the Noether identity (8.2) on-shell.

$$T_G^{\mu\nu} = F^{\mu\alpha}F^{\nu}_{\ \alpha} - \frac{1}{4}\eta^{\mu\nu}F_{\alpha\beta}F^{\alpha\beta} + C_1\partial_\alpha\Psi^{[\mu\alpha]\nu} + C_2A^\nu E^\mu + C_3A^\mu E^\nu + C_4\eta^{\mu\nu}A_\alpha E^\alpha \tag{8.37}$$

where $\Psi^{[\rho\alpha]\sigma}$ is the most general rank-three tensor defined in terms of the potential and derivatives of the potential, with the required symmetry properties. The subscript indicates that this is the general form for tensors in this equivalence class, with the coefficients C_n taking arbitrary values. We trivially recover the standard expression (8.1) by setting $C_n = 0$ (for all n) because Method II yields (8.1) without requiring the C_n terms. But even allowing C_n to take arbitrary values, (8.37) will still satisfy the Noether identity (8.2) on-shell.[43] Therefore, despite the mathematical freedom to allow for an on-shell equivalence class, we see that the equivalence class is not at all needed to yield the unique $T^{\mu\nu}$ of the theory.

The point generalises to other conserved currents. Based on our analysis above, we see more generally that Noether's theorem and its converse specify a relationship between a properly 'Noetherian' current J and the variational symmetries of the action that hold off-shell, regardless of whether the Euler–Lagrange equations are satisfied. But just as we can add the terms with coefficients $\{C_1, \ldots, C_4\}$ to T_G above, we can add terms to other Noether currents: schematically, $J_G = J + J_1 + J_2$, for currents J_1 that can be expressed as the divergence of a superpotential and terms J_2 proportional to the equations of motion. The divergence of the J_1 terms vanish in the Noether identity by construction, regardless of the features of the Lagrangian; the J_2 terms vanish on-shell. We classify these further terms as 'non-Noetherian' because applying the converse of Noether's theorem does not yield any relationship to variational symmetries of the Lagrangian. The proposed definition of equivalence then amounts to generating an equivalence class of conserved currents by adding 'non-Noetherian' terms.

There are plausible empiricist motivations in favour of this definition of equivalence. We have only empirical access to quantities defined on-shell: all empirical data is in the solutions, so to speak. Alternatively, this data can be seen as codified into the dynamical equations and the space of possible initial conditions (provided the problem is well-posed). Since the collection $\{T_G\}$ (or $\{J_G\}$) and the Noetherian currents agree on-shell, by construction, we cannot measure the differences among them, and it makes sense to treat them as belonging to an equivalence class.

But *should* we indeed treat the symmetries as determining only an equivalence class of energy-momentum tensors, or currents, in this empiricist sense? A positive answer would contrast sharply with the common practice in physics of taking (8.1) as the 'correct' expression for the energy-momentum tensor, and of delineating properly Noetherian currents. One possibility is that physicists simply choose one element of the equivalence class by convention or as a matter of convenience. (If the elements of the equivalence class truly 'represent the same physical situation', it would be a mistake to demand physical justification of the choice.) But this is not the position one finds. Instead, there are active debates regarding, for example, what is the uniquely correct physical expression for the energy-momentum tensor for various classical field theories (Gotay and Marsden 1992; Forger and Römer 2004).

43 In the case of spin-2 Fierz–Pauli theory, any of the published energy-momentum tensors for linearised gravity can be obtained from the canonical Noether tensor by adding superpotential terms and on-shell contributions (Baker 2021b); there are infinitely many energy-momentum tensors in this equivalence class.

There seems to be a more insightful explanation than that of a mere conventionalist, or of (unfounded) convenience, as to why practitioners do not adopt such empiricist equivalence classes. To see this, we first note that this empiricist definition of equivalence builds on a sort of provincialism on what to count as empirically relevant. This can be challenged when we consider a given classical field theory in a somewhat broader theoretical context, such as how it relates to other theories. While off-shell differences may not be empirically relevant classically, they do, for instance, become empirically relevant upon quantisation (think of the Feynman path integral picture). Furthermore, regarding the energy-momentum tensor in particular, coupling to gravity does render the absolute value of matter energy-momentum empirically meaningful, suggesting that we should care about the absolute value even outside of gravitational theories (in other words, we should not be indifferent towards shifts in superpotential either).

But we need not turn to the relationships to other theories for reasons to reconsider what should qualify as an empirically meaningful quantity, as Rovelli's (2014) take on gauge symmetry illustrates. Conventional wisdom treats gauge-dependent quantities as redundant, 'descriptive fluff'; the collection of states related by gauge transformations are treated as elements of an equivalence class. We should then characterise the state of a system entirely in terms of gauge-invariant observables, even if gauge-dependent quantities appear as a useful bookkeeping device in some calculations. Rovelli (2014) argues that this view betrays a limited conception of how the theory represents systems. Briefly put, it makes sense if we take the theory to represent isolated systems, but when we use gauge theory to describe interacting subsystems, the gauge degrees of freedom are essential to describing coupling among these systems. Without going further into the details, we hope that the main point relevant to our discussion is clear: interpretative questions, e.g. whether the theory represents isolated systems or coupled systems, have to be settled prior to deciding what quantities count as empirically meaningful. In conjunction with our point in the previous paragraph, namely that relations among theories may also require more fine-grained distinctions, this tells against accepting an equivalence class of energy-momentum tensors.

Both considerations support a general methodological point of view, namely a preference for individuating physical quantities in terms of deep principles of the theory whenever that is possible. Doing so makes it more likely that the physical quantities will be well-defined in a broader range of contexts, rather than being applicable only to the case at hand, and will continue to be useful as we modify or extend current theory. We can justify a strong preference for a definition of Noetherian currents, and the preferred form of the energy-momentum tensor, from this point of view as follows. The expressions $\{J_G\}, \{T_G\}$ are obtained by adding on 'non-Noetherian' terms that lack any clear connection with the Lagrangian, and only hold on-shell. By contrast, Noether's first theorem and its converse, as described in detail above, establish a direct relationship that holds off-shell between currents and the variational symmetries of the Lagrangian. The considerations above support, in line with this methodological principle, a much more stringent criterion of equivalence than the empiricist endorses, and supports the preference for a unique expression for the energy-momentum tensor and other conserved currents.

Before closing this section, we should acknowledge that these interpretative questions depend sensitively on the formal context in which they are pursued, as well as on the general methodological issues we have just highlighted. Above we have used the Lagrangian formulation of classical field theories, such that solutions to the relevant field equations are found via variational methods, and focused on the relationship between variational symmetrics and conserved quantities. Noether's theorem can be applied in a variety of formalisms, including to Hamiltonian systems (see Butterfield 2006) and directly to the solutions of differential equations (whether or not they have been derived using variational techniques). For the latter case, there are several distinct notions of symmetry, defined in terms of transformations on the space of solutions of the differential equations whose infinitesimal generators have specific properties (e.g. Belot 2013). (One can think of the differential equation as introducing a structure that these symmetries preserve, where this is more general than a Lagrangian.) The subtle relationship between these different contexts in which we can discuss symmetries and conservation laws is nicely illustrated by Smith (2008). In classical particle mechanics, there are several examples of non-standard Lagrangians that are dynamically equivalent to a 'standard' Lagrangian, of the form '$L = T - V$', yet with a different variational symmetry. Noether's theorem then yields different conservation laws for the two Lagrangians, even though they yield the same equations of motion. What grounds would there be for choosing among the class of Lagrangians $\{\mathcal{L}_1, \mathcal{L}_2, ...\}$ in order to determine the 'correct' conserved quantity? Whether one admits physical grounds to choose among the candidate Lagrangians, and so recover a clear connection between symmetry and conserved quantities, depends on whether we can go above and beyond the dynamical equations themselves.

8.7 Conclusions

Noether's first theorem is one of the most celebrated results in physics. Yet, standard textbook and literature presentation gives the picture that this method fails to derive standard physical conservation laws: the canonical Noether energy-momentum tensor, which is derived using a restricted condition placed on Noether's first theorem, does not give the known physical energy-momentum tensor in foundational models such as electrodynamics and Yang–Mills theory. All of this creates the impression that Noether's first theorem, despite frequent praise in the scientific community, is in some sense not working in practice for our most significant theories. We hope that our presentation of the Bessel-Hagen method (Method II) has let the reader regain confidence in the power of Noether's first theorem when applied to exactly gauge-invariant field theories: Using the complete set of (mixed) symmetries of the action (both gauge and coordinate symmetries), one obtains transformations that directly yield the known physical energy-momentum tensor of electrodynamics and theories with a gauge-invariant Lagrangian density more generally. No 'improvement' of the energy-momentum tensor is needed to supplement (nor actually advised for by) the Noether machinery.

In showcasing the proper application of Noether's theorem in the context of exactly gauge-invariant theories, we have, moreover, learned that the conventional wisdom that a specific variational symmetry (namely the canonical variational symmetry) is linked to a specific conservation law by the Noether machinery after all remains true within the bulk of classical field theory in practice. As we had already said in the introduction, one is free to question the linkage by more theoretical counterexamples – but this is a question for another day. Yet another interesting insight was gained along the way: contra common characterisations in the literature, Noether's first theorem is not solely concerned with what we called canonical variational symmetries exclusively but rather the complete set of symmetries of the action (this includes gauge symmetries which are sometimes portrayed as being only associated to Noether's second theorem). We have used the converse of Noether's first theorem as a method for emphasising this fact, as the canonical variational symmetries do not follow from the converse theorem for the majority of accepted physical energy-momentum tensors in the literature.

Finally, there is a sense in which our overall message in favour of Method II could be made even more strongly: Throughout the article, we had tacitly accepted the common theme in the literature to pay special attention to the energy-momentum tensor over and above other conserved currents in special relativistic field theory. This is important to note, as it is quite possible that many of the non-uniqueness and ambiguity problems associated with tensor conservation laws are a result of limiting oneself to the case of energy-momentum tensor specifically, and that the large variety of methods for energy-momentum construction compared to the other tensors is rather an issue that may not be solved by treating the energy-momentum tensor as a privileged standalone object. From the point of view of Noether's first theorem and Method II, none of the standard conserved tensors (energy-momentum, angular momentum, conformal and dilatation) are privileged compared to each other. Thus, if we agreed to consider only methodology which links all of the conserved tensors of a theory to variational symmetries simultaneously, Noether's first theorem in the sense of Method II may give a much-needed uniqueness in methodology akin to the Euler–Lagrange equation for an equation of motion. Such a view, if adopted, has promise to end the various ambiguity and non-uniqueness problems associated to the energy-momentum tensor once and for all.

References

Albinus, M. and Ibragimov, N. H. (2006). *On Conservation Laws of Electrodynamics*. ALGA Publications, **11**. ISBN ISSN 1652-4934.

Baker, M. R. (2021a). A connection between linearized Gauss–Bonnet gravity and classical electrodynamics II: Complete dual formulation. *International Journal of Modern Physics D*, **30**(4), 2150030.

Baker, M. R. (2021b). Canonical Noether and the energy-momentum non-uniqueness problem in linearized gravity. *Classical and Quantum Gravity*, **38**, 095007.

Baker, M. R. (2021c). Field theories from physical requirements: Noether's first theorem, energy-momentum tensors and the question of uniqueness. Electronic Thesis and Dissertation Repository, 7802. https://ir.lib.uwo.ca/etd/7802.

Baker, M. R. and Kuzmin, S. (2019). A connection between linearized Gauss–Bonnet gravity and classical electrodynamics. *International Journal of Modern Physics D*, **28**(07), 1950092.

Baker, M. N., Kiriushcheva, N. and Kuzmin, S. (2021). Noether and Hilbert (metric) energy-momentum tensors are not, in general, equivalent. *Nuclear Physics B*, **962**, 115240.

Barceló, C., Carballo-Rubio, R. and Garay, L. J. (2014). Unimodular gravity and general relativity from graviton self-interactions. *Physical Review D*, **89**(12), 124019.

Belinfante, F. J. (1939). On the spin angular momentum of mesons. *Physica*, **6**(7–12), 887–98.

Belinfante, F. J. (1940). On the current and the density of the electric charge, the energy, the linear momentum and the angular momentum of arbitrary fields. *Physica*, **7**(5), 449–74.

Belot, G. (2013). Symmetry and equivalence. In R. Batterman, ed., *Oxford Handbook for the Philosophy of Physics*. Oxford: Oxford University Press, pp. 318–39.

Bessel-Hagen, E. (1921). Über die Erhaltungssätze der Elektrodynamik. *Mathematische Annalen*, **84**(3–4), 258–76.

Bičák, J. and Schmidt, J. (2016). Energy-momentum tensors in linearized Einstein's theory and massive gravity: The question of uniqueness. *Physical Review D*, **93**(2), 024009.

Blaschke, D. N., Gieres, F., Reboud, M. and Schweda, M. (2016). The energy–momentum tensor(s) in classical gauge theories. *Nuclear Physics B*, **912**, 192–223.

Burgess, M. (2002). *Classical Covariant Fields*. Cambridge Monographs on Mathematical Physics, Cambridge: Cambridge University Press. ISBN 9781139432979.

Butterfield, J. (2006). On symmetry and conserved quantities in classical mechanics. In W. Demopoulos and I. Pitowsky, eds., *Physical Theory and Its Interpretation*. Dordrecht: Springer, pp. 43–100.

Callan, C. G., Jr, Coleman, S. and Jackiw, R. (1970). A new improved energy-momentum tensor. *Annals of Physics*, **59**(1), 42–73.

Curiel, E. (2017). A primer on energy conditions. In D. Lehmkuhl, G. Schlemann and E. Scholz, eds., *Towards a Theory of Spacetime Theories*. Dordrecht: Springer, pp. 43–104.

Dürr, P. M. (2019). Against 'functional gravitational energy': a critical note on functionalism, selective realism, and geometric objects and gravitational energy. *Synthese*, **199**, 299–33.

Eriksen, E. and Leinaas, J. M. (1980). Gauge invariance and the transformation properties of the electromagnetic four-potential. *Physica Scripta*, **22**(3), 199.

Forger, M. and Römer, H. (2004). Currents and the energy-momentum tensor in classical field theory: a fresh look at an old problem. *Annals of Physics*, **309**(2), 306–89.

Garg, A. (2012). *Classical Electromagnetism in a Nutshell*. Princeton, NJ: Princeton University Press.

Gelfand, I. M. and Fomin, S. V. (2000). *Calculus of variations*. Translated and edited by R. A. Silverman. Chelmsford, MA: Courier Corporation.

Gotay, M. J. and Marsden, J. E. (1992). Stress-energy-momentum tensors and the Belinfante–Rosenfeld formula. *Contemporary Mathematics*, *132*, 367–92.

Holman, M. (2010). Generalized Noether theorems for field theories formulated in Minkowski spacetime. *arXiv preprint arXiv:1009.1803*.

Jackiw, R. (1978). Gauge-covariant conformal transformations. *Physical Review Letters*, **41**(24), 1635.

Janssen, M. and Mecklenburg, M. (2006). From classical to relativistic mechanics: Electromagnetic models of the electron. In V. F. Hendricks, K. F. Jørgensen, J. Lützen and S. A. Pedersen, eds., *Interactions: Mathematics, Physics and Philosophy, 1860–1930*. Boston Studies in the Philosophy of Science, Vol. 251. Dordrecht: Springer, pp. 65–134.

Kosmann-Schwarzbach, Y. (2011). *The Noether Theorems: Invariance and Conservation Laws in the Twentieth Century*. Translated by B. E. Schwarzbach. New York: Springer.

Kuzmin, S. V. and McKeon, D. G. C. (2001). Automatic derivation of improved symmetry currents in the context of the Wess–Zumino model. *Physical Review D*, **64**(8), 085009.

Lange, M. (2007). Laws and meta-laws of nature: Conservation laws and symmetries. *Studies in History and Philosophy of Science Part B: Studies in History and Philosophy of Modern Physics*, **38**(3), 457–81.

Magnano, G. and Sokolowski, L. M. (2002). Symmetry properties under arbitrary field redefinitions of the metric energy–momentum tensor in classical field theories and gravity. *Classical and Quantum Gravity*, **19**(2), 223.

Misner, C. W., Thorne, K. S. and Wheeler, J. A. (1973). *Gravitation*. New York: Macmillan.

Montesinos, M. and Flores, E. (2006). Symmetric energy-momentum tensor in Maxwell, Yang-Mills, and Proca theories obtained using only Noether's theorem. *Revista Mexicana de física*, **52**(1), 29–36.

Noether, E. (1918). Invariante variationsprobleme. *Nachrichten von der Gesellschaft der Wissenschaften zu Göttingen, Mathematisch-Physikalische Klasse*, pp. 235–57.

Olver, P. J. (2000). *Applications of Lie Groups to Differential Equations*. Volume 107. New York: Springer Science & Business Media, 2000.

Padmanabhan, T. (2008). From gravitons to gravity: Myths and reality. *International Journal of Modern Physics D*, **17**, 367–98.

Pons, J. M. (2011). Noether symmetries, energy-momentum tensors, and conformal invariance in classical field theory. *Journal of Mathematical Physics*, **52**(1), 012904.

Read, J. (2020). Functional gravitational energy. *The British Journal for the Philosophy of Science*, **71**(1), 205–32.

Rosen, G. (1972). Conformal invariance of Maxwell's equations. *American Journal of Physics*, **40**(7), 1023–7.

Rosenfeld, L. J. H. C. (1940). *Sur le tenseur d'impulsion-énergie*. Brussels: Palais des Académies.

Rovelli, C. (2014). Why gauge? *Foundations of Physics*, **44**(1), 91–104.

Scheck, F. (2012). *Classical Field Theory*. New York: Springer.

Smith, S. R. (2008). Symmetries and the explanation of conservation laws in the light of the inverse problem in Lagrangian mechanics. *Studies in History and Philosophy of Science Part B: Studies in History and Philosophy of Modern Physics*, **39**(2), 325–45.

Szabados, L. B. (1992). On canonical pseudotensors, Sparling's form and Noether currents. *Classical and Quantum Gravity*, **9**(11), 2521.

9

Noether's Theorems and Energy in General Relativity*

SEBASTIAN DE HARO

9.1 Introduction

This chapter has three main aims: first, to give a pedagogical introduction to Noether's two theorems: and in particular to the role that her second theorem played in important discussions between Hilbert, Klein, Noether, and Einstein about the conservation of energy and momentum in general relativity (GR). These discussions have been the subject of recent scholarship, and are still valuable today.[1]

Second, I introduce for a philosophical audience two proposals for the gravitational energy-momentum tensor: namely, Einstein's pseudo-tensor (and the quasi-local expressions derived from it[2]) and the quasi-local definition by Brown and York, especially as modified in the recent literature on the holographic principle. I argue that arguments from gauge-gravity duality support the quasi-local quantities as correctly expressing the energy and momentum of general relativity.

Third, I discuss and critically assess the various proposals in light of a recent philosophical discussion of energy and momentum in general relativity, by Hoefer (2000), Pitts (2010), Lam (2011), Dürr (2019a), and Read (2020).

General relativity suffers from a number of problems regarding its local conservation laws for energy and momentum. This was the subject of a crucial discussion between Hilbert, Klein, Noether, and Einstein between 1915 and 1918. What is perhaps less well-known is that the problem of the definition and conservation of energy[3] played a crucial role in both: the genesis of general relativity and Noether's reasons for proving her two theorems.

To define gravitational energy and momentum, Einstein introduced a gravitational *pseudo-tensor,* i.e. a quantity that is non-covariant, and is in some ways analogous to

* I thank Bryan Roberts and Nic Teh for organising the conference 'The Philosophy and Physics of Noether's Theorems' at the University of Notre Dame, London Global Gateway, and James Read, Bryan Roberts, and Nic Teh, for their invitation to contribute to this volume. I thank James Nester for detailed correspondence and for his patience answering my questions, and James Read, Henrique Gomes, Harvey R. Brown, Jeremy Butterfield, and Jeroen van Dongen for comments. I also thank audiences at the University of Notre Dame, London Global Gateway, the Hong Kong University of Science and Technology, and the Tsinghua Sanya International Mathematics Forum in Hainan, China.

1 See for example Kastrup (1987, pp. 116–25), Brading (2005, pp. 125–7), Brading and Brown (2003, p. 91), Kosmann-Schwarzbach (2011, pp. 37–46), Rowe (1999, pp. 209–21; 2019a. pp. 7–11, 29-034; 2019b).

2 Especially in recent work by Chang et al. (1999) and Chen et al. (2017, 2018a).

3 For brevity, I will sometimes follow the literature in using 'energy' or 'energy-momentum' for the longer 'energy-momentum and angular momentum', especially in expressions like 'conservation of energy-momentum and angular momentum'.

the Christoffel symbols. However, unlike the Christoffel symbols, pseudo-tensors are not geometric objects. The question of energy and momentum, and the pseudo-tensor proposal, are discussed in virtually all of Einstein's papers on the subject.[4] Part of the reason it played such a prominent role, including in his November (1915a, 1915b, 1915c) papers, is because the pseudo-tensor is intimately linked to Einstein's field equations and to the contracted Bianchi identities.

Noether's own work was directly motivated by the question of energy-momentum conservation, and prior to publishing her two theorems she had worked on comparing Hilbert's and Einstein's approach to it. According to Hilbert, Klein, and Noether, the conservation law in general relativity is *improper*, because it is given in terms of a superpotential whose conservation law is an identity. On Klein's interpretation, the conservation law does not make direct use of the equations of motion, and so it has no physical content.

Einstein's pseudo-tensor, and thus the energy, are known to have a number of other problems: the pseudo-tensor is non-covariant, even non-geometric, and there appears to be no unique expression for it. For many different expressions have appeared in the literature. This non-uniqueness stems from the ability to add to the pseudo-tensor the derivative of an arbitrary 'superpotential'. The superpotential is such that its divergence is identically zero, and so an arbitrary superpotential can be added into the conservation law without modifying it. Thus, the very definition of the pseudo-tensor appears to be ambiguous.

The philosophical interest in energy and momentum in general relativity has so far been moderate.[5] For example, Hoefer (2000), Lam (2011), and Dürr (2019a, 2019b) have criticised, and Read (2020) has defended, local definitions of energy and momentum, in particular those involving the energy pseudo-tensor. I will give considerable attention to this recent philosophical discussion in Sections 9.5.2 and 9.6.

There are two aspects that these recent discussions have not focussed on, and which I will address in this chapter: namely, the alleged non-uniqueness of the superpotential, and the quasi-local (as opposed to local) definitions of energy and momentum.

I will show in Section 9.2 that, once the Lagrangian and its variation are fixed, Noether's second theorem fixes the superpotential. This holds not only for general relativity, but for any theory with a local symmetry. In general relativity, the gravitational pseudo-tensor, which is the Noether current corresponding to the local translation symmetry, is thus fixed, through Noether's theorem, by the choice of the action. In particular, the freedom to add an arbitrary superpotential corresponds to the freedom to add boundary terms to the action without changing the equations of motion.[6] Thus, the superpotential is not arbitrary but is fixed, through Noether's theorem, by the choice of boundary terms in the action. Although this is known to the experts, I have not seen these consequences of Noether's theorem

[4] See for example Janssen and Renn (2007, p. 844).

[5] For a detailed analysis of Hilbert's (1916, 1917) two papers on the 'Foundations of Physics', see Renn and Stachel (2007, pp. 878–902, 914–57). Hilbert's first paper, including the changes that he introduced into the proofs, is discussed, and compared with Einstein's four November 1915 papers, in Sauer (1999, pp. 539–57), Renn and Stachel (2007, pp. 872–933), and Corry et al. (1997, pp. 1271–3). A brief account of the interactions between Einstein and Hilbert in 1915 is in Corry (1999, pp. 177–9).

[6] In a different context, Chang et al. (1999, pp. 3–4) and Chen et al. (2018a, p. 4) have also made a related connection between the superpotential and boundary conditions in the Hamiltonian, although they do not mention Noether's theorem in that connection.

discussed in the literature: in fact, the 'problem of non-uniqueness' is still frequently cited without a discussion of the implications of the choice of boundary conditions and boundary terms. For example, even Brading and Brown (2003, pp. 101–2), whose formulation of Noether's second theorem I largely follow, have the superpotential implicit in their formulas but do not seem to recognise it as such, and they do not comment on the definition and alleged non-uniqueness of the pseudo-tensor in connection with Noether's theorems.[7]

The detailed consideration of Noether's second theorem also allows one to address Klein's worry about the alleged 'lack of physical content' of the local energy conservation law of general relativity. Indeed, I will argue that the distinction between 'proper' and 'improper' conservation laws, in the context of comparing general relativity with classical mechanics, is a red herring: and that the claim that local energy conservation laws lack physical content is misleading.

I will also introduce two different quasi-local expressions: one, due to Chang et al. (1999), Nester (2004), and Chen et al. (2017, 2018a, 2018b), directly obtained from the total energy-momentum (i.e. the energy-momentum tensor plus the gravitational pseudo-tensor) suitably integrated over a volume. These authors show that, under assumptions that I will discuss in Sections 9.5 and 9.6, this expression can be obtained from the covariant Hamiltonian formalism of general relativity. Indeed, part of their aim appears to be to justify the use of the pseudo-tensor from the Hamiltonian formalism.

The second expression is the Brown–York (1993) quasi-local stress-energy tensor, which is directly defined on a two-dimensional surface. This stress-energy tensor has been studied extensively in the physics literature, especially in the presence of a cosmological constant (positive or negative), and has been shown to have many of the desired properties of the quasi-local stress-energy, as I will also review.

I also compare, in Section 9.6, the two quasi-local proposals. While quasi-local expressions have sometimes been easily dismissed (or avoided altogether) in the philosophical literature, I will argue that neither dismissal nor avoidance is justified. Also, we will see that the analysis of the Brown–York stress-energy tensor in the gauge-gravity literature gives crucial insights into the correct definition and properties of this tensor.

One obstruction to the construction of normal conservation laws for energy and momentum in curved spacetimes is the presence of a covariant derivative in the conservation law of the energy-momentum tensor.[8] One can integrate such a law to get an integral form of the conservation law that is properly covariant only if the metric admits a timelike Killing vector field, i.e. if the metric has a timelike isometry.

An important property of the quasi-local energy-momentum tensors is that they can lead to covariant integral conservation laws under weaker conditions than the energy-momentum tensor does. For, while the latter requires a timelike Killing vector field in the whole spacetime, the quasi-local expression from Section 9.4 only requires a (conformal) Killing vector field at the quasi-local surface, i.e. usually at asymptotic spatial or null infinity. I show this in Section 9.4.2.

[7] For a discussion of the literature on this point, see Section 9.2.3.

[8] Notice that this fact is not specific to general relativity: for any translation-invariant classical field theory on a curved spacetime has a covariantly conserved, two-index, energy-momentum tensor. But GR offers a specific interpretation of the source or sink terms in terms of a dynamical gravitational field.

In physics, there are well-known global expressions (called the Bondi and ADM masses) for the total energy at infinity, depending on whether they are integrated over spheres at spatial infinity (in the case of asymptotically flat and AdS cases) or null infinity (in the case of the Bondi mass). The two kinds of quasi-local quantities discussed in this chapter reproduce these expressions. But these expressions, including the quasi-local quantities, do not necessarily have well-defined local densities.[9] Chen et al. (1999, p. 2) discuss a quasi-local formula that *can* be localised: its localisation gives rise to the gravitational pseudo-tensor, which depends on the choice of coordinates. They view the interplay between the local and quasi-local quantities as follows:

> Appreciating the fundamental nonlocality of the gravitational interaction yet believing in the basically local nature of physical interactions led to the idea of *quasi-local quantities*: quantities that take on values associated with a compact orientable spatial 2-surface. (Chen and Nester 1999, p. 1280)

The plan of the chapter is as follows. In the rest of this Introduction, I summarise the discussion of energy and momentum between Hilbert, Klein, Noether, and Einstein (Section 9.1.1), illustrate Noether's theorems for Maxwell theory as my running example (Section 9.1.2), and summarise Einstein's basic proposal of pseudo-tensors as candidates for energy and momentum and its problems (Section 9.1.3). Section 9.1.3 lays out five problems that organise the subsequent discussion. Section 9.1.1 has more of a historical character; the rest of the chapter has a more philosophical and foundational character. Section 9.2 then gives Noether's two theorems and discusses two consequences of it, one of which I have not seen discussed in the literature (Section 9.2.3). Section 9.3 applies Noether's theorems to general relativity. Section 9.4 discusses the quasi-local Brown–York stress-energy tensor. Sections 9.4.1 and 9.4.2 are slightly more technical than the rest of the chapter, and can be skipped by readers interested only in the philosophical discussion. Section 9.4.3 discusses whether diffeomorphisms are redundancies of the description. Section 9.5 reviews proposals by Chang et al. (1999), Chen et al. (2017, 2018a, 2018b), and Nester (2004) that link the gravitational pseudo-tensor with a quasi-local expression that can be reproduced from the covariant Hamiltonian of general relativity, under assumptions that I will point out. Section 9.6 then compares the two proposals for quasi-local expressions and discusses the status of energy-momentum in general, and the pseudo-tensors in particular, in the philosophical literature. Sections 9.4–9.6 also discuss two themes that run through the philosophical literature on energy and momentum in general relativity: covariance and background-independence.

9.1.1 Noether's Theorem and General Relativity

Emmy Noether's (1918) two theorems, on the correspondence between symmetries of the action under infinitesimal transformations of a continuous group, and local conservation

[9] This is often further explicated by an appeal to the principle of equivalence, that a local observer cannot detect a gravitational field at a single point. This often stated, infinitesimal or strong, form of the principle of equivalence is in Pauli (1958, p. 145); see also Brown (2005, pp. 169–70). For its role in Einstein's own thinking about general relativity, and its relation with the Einstein equivalence principle, see Lehmkuhl (2022, pp. 2–7, 19–22) and Janssen (2014, pp. 174–82). The strong equivalence principle is criticised by Norton (1985, pp. 207–8, 239–2, 245) and Curiel (2019, p. 91).

laws, are better known for their applications in classical mechanics and gauge theories than in general relativity. Yet Noether's motivation to study this problem came chiefly out of discussions between David Hilbert, Felix Klein, Albert Einstein, and herself about the interpretation of *general relativity,* between 1915 and 1918: specifically, about the nature of the conservation equations in GR.[10] Noether's theorem still gives the most general and – as I will argue in Section 9.2 – best approach to the gravitational pseudo-tensor of general relativity and its associated superpotential, which was at the heart of the debate.

Since at least 1912, Einstein had been searching for an expression of the laws of gravitation, invariant under general coordinate transformations.[11] Already in his (1913, p. 318) *Entwurf* paper with Marcel Grossmann, and in his (1914, p. 99) 'Formal Foundation' paper, which amongst other things gave a detailed development of tensor calculus and worked out the geodesic equation, he discussed at length (especially in his 1914 paper) a covariant balance equation for the energy-momentum and stress tensor density, subject to external forces. When the external forces were set to zero, he got an analogue of the local conservation law of energy and momentum (1914, pp. 100, 120):

$$\partial_\mu(\mathfrak{T}^\mu{}_\nu + t^\mu{}_\nu) = 0. \tag{9.1}$$

Here, $\mathfrak{T}$ is the energy-momentum tensor density of matter,[12] and t is the 'energy tensor of the gravitational field' (later called the gravitational *pseudo-tensor*), in analogy with the Christoffel symbols, which are the components of the 'field strength of the gravitational field'. He gave an explicit expression for $t^\mu{}_\nu$ in terms of the Christoffel symbols (p. 120) and noted that it is not a tensor but is invariant only under linear (i.e. affine) transformations. At this point, Einstein wrote his tentative field equations explicitly in terms of the pseudo-tensor. Thus, from very early on, the gravitational energy-momentum pseudo-tensor was, in Einstein's mind, an important ingredient of the theory that he was seeking.

In the spring of 1915, Hilbert and Klein invited Noether to Göttingen to work with them on the implications of the general theory of relativity (Corry 1999, p. 177; Kosmann-Schwarzbach 2011, p. 45).[13] Although Noether was probably not an expert on variational methods (however, Hilbert *was* an expert on that),[14] she was well acquainted with the related methods for using formal differential operators to generate algebraic and differential invariants (Rowe 2019a, p. 6). Also, shortly after her arrival in 1915, Einstein gave a series of lectures on general relativity – he stayed at Hilbert's house during his visit. Thus, by mid-1915, Hilbert himself was working intensely to understand Einstein's papers. Early on, Hilbert saw a connection between two topics: (i) the theory of invariants and geometry and (ii) the problem of extending the special theory of relativity to general coordinate systems. This was of course a seed for Noether's theorems (and Hilbert himself would soon go on

[10] For some of the scientific context of Noether's theorems, including her own biography, see Kosmann-Schwarzbach, Chapter 1, this volume. A clear exposition of Noether's theorems in their original form is in Rowe (1999, pp. 223–6). Norton (1993) is a useful reference about the history of some issues in general relativity that are complementary to the ones discussed in this section, in particular about general covariance.

[11] See e.g. Sauer (2005, pp. 806–8) and Kosmann-Schwarzbach (2011, p. 37).

[12] Compare this equation with Eq. (9.21).

[13] For the prehistory of this invitation, see Sauer (1999, pp. 539–40). The pre-history of Hilbert's acquaintance with Einstein is on pp. 527–39.

[14] See Sauer (1999, pp. 535, 558, 569).

to give a special case of Noether's second theorem). Likewise, Klein would regard general relativity from the point of view of his Erlangen programme, which he had laid out with Sophus Lie, of geometry as the search for invariants of a group.[15]

In two of his papers from November 1915, Einstein gave a very general expression (1915c, p. 784) for the gravitational pseudo-tensor in terms of the Lagrangian density[16] and found the correct gravitational field equations (1915a, p. 845).

As is well known, on 20 November 1915, i.e. five days before Einstein submitted the paper that contained the final form of his field equations (1915a),[17] Hilbert submitted his 'Die Grundlagen der Physik', which is an attempt to axiomatise general relativity (and make a connection with Gustav Mie's unification programme).[18] Hilbert reproduced Einstein's field equations from his action and, among other results, he derived the four contracted Bianchi identities from the invariance of the action under arbitrary transformations of coordinates. Thus, in this paper Hilbert derived a special case of Noether's second theorem. At this point, however, the nature of these relations was unclear. Even less clear was the meaning of the 'energy equation', based on an 'energy vector', that he derived.[19]

Noether also began to work on this problem. In 1916, she wrote an unpublished manuscript in which she studied Hilbert's energy vector and used Hilbert's work to analyse Einstein's conservation law in general relativity. Although the manuscript is not extant, its content is partly known from two sources: namely, the exchange of letters between Klein and Hilbert published in the *Nachrichten von der Gesellschaft der Wissenschaften zu Göttingen, Mathematisch-Physikalische Klasse,* in 1917, and a partial transcription by R. J. Humm, who studied at Berlin and Göttingen at the time (Rowe 2019a, 2019b). In this manuscript, she got the contracted Bianchi identities, and in fact she anticipated the main points of Klein's influential lecture of 22 January 1918 for the Göttingen *Gesellschaft.*

Einstein (1916b, p. 1111) endorsed Hilbert's action in another paper, published in November 1916, and solved the conservation equation, Eq. (9.1), in terms of what is now called a 'superpotential',[20] i.e. a quantity whose divergence is identically zero.[21]

On 22 January 1918, Klein delivered his famous lecture at the *Mathematische Gesellschaft zu Göttingen* (i.e. the Göttingen Mathematical Society), which elicited a reaction from Hilbert one week later (Rowe 1999, p. 212; 2019a, p. 8). Klein's lecture would be

[15] See Rowe (1999, p. 215; 2019b, p. 72) and Kosmann-Schwarzbach (2011, pp. 26, 40, 64).

[16] See Eq. (9.47), which is only a minor generalisation of this equation through the addition of a total-derivative term. However, at this point Einstein's Lagrangian was incomplete, since its variation gave only (part of) the the Ricci tensor, rather than the Einstein tensor. A problem in this paper, to which he returned in (1915d, pp. 799–800), was the incompatibility between his restriction to coordinate systems where $\sqrt{-g} = 1$ and the requirement that the energy-momentum tensor has non-zero trace. The full, covariant equations without any restrictions on the trace of the energy-momentum tensor appeared in Einstein (1915a,b, p. 845).

[17] For the now-accepted resolution of the priority issue, see Corry et al. (1997, pp. 1271–3), Sauer (1999: pp. 539–57), and Renn and Stachel (2007, pp. 872–933).

[18] Hilbert's reference to Mie in the introduction is as follows: 'I would like to formulate – in the sense of the axiomatic method – a new system of basic equations of physics, out of two simple axioms, which are of ideal beauty, and which, I believe, contain the solution of the problems by both Einstein and Mie' (1915, p. 395; see also pp. 396, 407). For more on Mie's influence on this paper by Hilbert, see Sauer (1999, pp. 538, 552–5), Rowe (2019a, pp. 5–7), Renn and Stachel (2007, pp. 857–67, 871–4, 899-900), and Corry (1999, p. 176).

[19] Hilbert would later rewrite his two 'Grundlagen der Physik' papers (1916) and (1917), publishing them in sequence in (Hilbert 1924).

[20] See Eq. (9.22) below.

[21] Such an expression already appeared in Einstein (1914, p. 1077), but this was only for his old theory, in which $\sqrt{-g} = 1$.

published in the *Nachrichten von der Gesellschaft der Wissenschaften zu Göttingen,* in the unusual form of an exchange of letters with Hilbert (also, Klein and Hilbert lived a short distance from one another in the Wilhelm Weber Strasse!).[22] In it, he explicitly acknowledged Noether's help and her earlier contribution (Kosmann-Schwarzbach 2011, p. 66):

> You know that Miss Noether advises me continually regarding my work, and that in fact it is only thanks to her that I have understood these questions. When I was speaking recently to Miss Noether about my result concerning your energy vector, she was able to inform me that she had derived the same result on the basis of developments of your note... more than a year ago, and that she had then put all of that in a manuscript (which I was subsequently able to read). She simply did not set it out as forcefully as I recently did at the Mathematical Society.

Like Noether had done before, Klein compares Hilbert's energy equation and Einstein's conservation law, Eq. (9.1). Klein's main point was the disanalogy between Hilbert's energy equation (and Einstein's conservation law) and the conservation laws in classical mechanics. While, in classical mechanics, energy conservation follows as a consequence of the equations of motion, in general relativity, Eq. (9.1) is an *identity,* i.e. valid without requiring the equations of motion to be satisfied. He explained it thus:

> Thus the above statement ... cannot be regarded analogous to the conservation law of energy in ordinary mechanics. For when we write the latter as $d(T + U)/dt = 0$, this differential equation is not satisfied identically [as the Bianchi identity is], but only as a consequence of the differential equations of mechanics.

He concluded the note to Hilbert with a sceptical remark: 'After all this, I can hardly believe that it is appropriate to characterise the very arbitrarily built quantities $t^{\mu}{}_{\nu}$ as "energy components of the gravitational field" '. This agreed with Noether's own view, which she had explained to Hilbert more than a year before.

In his reply,[23] Hilbert completely agreed with Klein's comments about the lack of an analogy between the energy equation and conservation of energy in classical mechanics. He took this one step further, by expressing his belief that no true law of energy conservation exists in general relativity, and claiming that it is possible to *prove a theorem* to this effect – to which Klein responds that he 'would be very interested in seeing the mathematical proof to which you alluded to in your reply' (p. 482). Such a proof would be provided not by Hilbert, but by Noether's second theorem, in the following year.

Meanwhile, in 1917–18, Einstein and Klein corresponded about Klein's (1917) note, which Einstein (1918b, p. 673) praised for its clarity.[24] Einstein disagreed that his conservation law for general relativity was an identity, since the equations of motion *are* used. To this, Klein (1918a,b, p. 504) replied by saying that the divergence in Eq. (9.1) consists of two terms:[25]

[22] Klein (1917). For a reason why they chose this unusual form of publication, see Rowe (1999, p. 212).
[23] Klein (1917, p. 477).
[24] For an account of the correspondence, see Rowe (1999, pp. 209–13).
[25] This can be seen by expressing the pseudo-tensor t in terms of two contributions: the Einstein tensor and the superpotential: see Eq. (9.24) below.

1. the divergence of the Einstein field equations;
2. the divergence of the superpotential.

Klein's point was that the vanishing of (2) is a mathematical identity, while (1) contains the *divergence* of the equations of motion, rather than the equations of motion themselves – and so, that this term is also zero, even before one takes its divergence in the pseudo-tensor. (This differs from classical mechanics, where the equations of motion, rather than their divergence, appear in the conservation equation). Einstein (1918c, p. 513) disagreed: for the equations of motion were after all used, and the conservation equation used at least *part* of the field equations (i.e. their contraction with the divergence).[26]

In the winter and spring of 1918, Noether understood the reasons for the difficulties in the interpretation of the conservation laws of general relativity. What she set out to investigate was indeed Klein and Hilbert's central problem of the analogy between the energy equations of classical mechanics and general relativity. On 23 July 1918, she presented her work at the Göttingen Mathematical Society. In short, her second theorem could account for the four identities that Hilbert had pointed out, so that the conservation laws were 'improper', and thus disanalogous to classical mechanics.

Noether had a deeper reason for the difference between general relativity and other theories: her theorems allowed her to 'deduce ... the proof of an assertion of Hilbert concerning the relationship between the lack of a proper law of energy and "general relativity"' (Kosmann-Schwarzbach 2011, p. 19). Namely, given an action invariant under the group of translations, the energy relations are improper if and only if the action is invariant under an infinite group which contains the group of translations as a subgroup.[27]

Thus, Noether interpreted her theorems along Hilbertian lines: 'As Hilbert expresses his assertion, the lack of a proper energy law constitutes a characteristic of the general theory of relativity'.

The paper by Klein (1918b) clarifies the relationship between Einstein's and Hilbert's theories, and in particular concerning their conservation laws. In this paper, he derives what Brading and Brown (2003) call the 'boundary theorem', from which it follows that the current is proportional to the field equations (and to the superpotential), so that its divergence is identically zero – as I mentioned above. In doing so, he made essential use of Noether's work (Kosmann-Schwarzbach 2011, p. 68):[28]

> I would be remiss if I did not thank Miss Noether once again for her active participation in my new work. She has by herself completely set out in proper form the mathematical ideas that I use in connection with the physics problems related to [the action], which will be presented in a note to appear shortly in these *Nachrichten*.

Thus by 1918, Noether's theorem was, in the eyes of Hilbert, Klein, and Noether, the proof that there is no proper conservation law in general relativity: which Klein interpreted –

[26] The above summary of the argument in Klein (1918a, p. 504) and Einstein (1918; 1918a, p. 513) follows Brading (2005, p. 126), which indeed seems to me a correct interpretation both of Klein's letter and of Einstein (1918b, p. 513) reply.

[27] Noether's use of the word 'subgroup' here appears to be stronger than the modern use. Her statement assumes not just the group of translations being a subgroup of an infinite group, but also its being *embedded* in a specific way: see Section 9.2.3. I thank Bartlomiej Czech for pushing me on this point.

[28] Klein (1918b, p. 189).

with Hilbert's and Noether's approval, one should add – as the energy equations' lacking physical content.

I will return to Einstein's proposal for a conservation law in general relativity, and its problems, in more detail in Section 9.1.3. In the next section, I first illustrate Noether's two theorems in a well-known example.

9.1.2 Illustrating Noether's Theorems in the Maxwell Theory

In this paper, I will use the Maxwell theory of electromagnetism as a running example to illustrate various aspects of Noether's theorems, since this theory already exemplifies some of the issues that appear in general relativity.[29]

Thus, let me recall how the theorems work for elementary electromagnetism without sources: take the following action, defined on Minkowski spacetime:

$$S[A] = \int \mathrm{d}^4x\, \mathcal{L}(A;x) = -\frac{1}{4}\int \mathrm{d}^4x\, F_{\mu\nu}F^{\mu\nu}$$
$$F_{\mu\nu} := \partial_\mu A_\nu - \partial_\nu A_\mu. \tag{9.2}$$

The field equations, derived by setting the variation of the action to zero, are of course:

$$\partial_\nu F^{\mu\nu} = 0. \tag{9.3}$$

I will give Noether's theorems in general form in Section 9.2. Here I give the specific cases that apply to the Maxwell theory.

Noether's first theorem asserts that the conservation of the energy-momentum tensor density $T_{\mu\nu}$ follows from the *global* translational symmetry of the action in Minkowski spacetime (and vice versa), i.e. under a *constant* translation $\delta x^\mu = \epsilon^\mu$.[30] Defining the variation

$$\delta A_\mu := A'_\mu(x) - A_\mu(x) = -\epsilon^\nu\, \partial_\nu A_\mu, \tag{9.4}$$

Noether's first theorem then gives:

$$T^\mu{}_\nu := \delta^\mu_\nu\, \mathcal{L} - \frac{\partial\mathcal{L}}{\partial(\partial_\mu A_\lambda)}\partial_\nu A_\lambda = F^{\mu\lambda}\,\partial_\nu A_\lambda - \frac{1}{4}\,\delta^\mu_\nu\, F^{\lambda\sigma}F_{\lambda\sigma}. \tag{9.5}$$

One readily finds the following conservation law:

$$\partial_\mu T^\mu{}_\nu = \left(\partial_\mu\, F^{\mu\lambda}\right)\partial_\nu\, A_\lambda \overset{\text{Eq. (9.3)}}{=} 0. \tag{9.6}$$

The term between brackets vanishes only if the equations of motion, from Eq. (9.3), are satisfied: and thus the energy momentum is conserved if and only if (a linear combination of) the equations of motion are satisfied, i.e. it is a weak conservation law.

[29] For an introduction to Noether's theorem for a physics audience, see e.g. Bañados and Reyes (2016). For a mathematical treatment, see Olver (1986, pp. 272–83, 334–6). Bessel-Hagen (1921, pp. 44–51) was the first to apply Noether's first theorem to electrodynamics, using the generalised theorem involving quasi-symmetries. He credits Klein (p. 33) with suggesting the problem: 'Klein expressed a desire that Miss Emmy Noether's theorems on invariant variational problems, proved approximately two years ago, should be applied to the Maxwell equations'. I thank Harvey R. Brown for recommending the work by Bessel-Hagen.

[30] Under such a translation, we have a new field $A'_\mu(x')$, where $x' = x + \epsilon$, and $A'_\mu(x')$ is related to $A_\mu(x)$ by the usual tensor transformation rule for the components of one-forms, linearised in ϵ. In the variation, we evaluate A'_μ at x.

Notice that the first term in the energy-momentum tensor, Eq. (9.5), is not gauge invariant (it is also not symmetric in its two indices upon lowering one of them).[31] This lack of gauge invariance is unsurprising, given that we used a variation of the Lagrangian, Eq. (9.4), that is not gauge invariant. One can get a gauge-invariant energy-momentum tensor by constructing a gauge-invariant variation of the gauge potential, A: namely, by allowing this variation to contain a gauge transformation, i.e. $\delta A_\mu = -\epsilon^\nu\,\partial_\nu A_\mu + \partial_\mu\lambda$, where λ is the gauge parameter. There is a unique gauge-invariant variation of this form, viz. $\delta A_\mu = F_{\mu\nu}\,\epsilon^\nu$, where $\lambda = \epsilon^\nu A_\nu$. In effect, this adds an 'improvement term' to the energy-momentum tensor, i.e. a term that is a total derivative in an antisymmetric tensor U, without modifying the equations of motion, such that the energy-momentum tensor is still conserved:[32]

$$T'^\mu{}_\nu := T^\mu{}_\nu + \partial_\lambda\, U^{[\mu\lambda]}{}_\nu. \tag{9.7}$$

The above variation gives $U^{[\mu\lambda]}{}_\nu = -F^{\mu\lambda}\,A_\nu$, which brings the energy-momentum tensor to the standard form, and is automatically symmetric in its indices:

$$T'^\mu{}_\nu = F^{\mu\lambda}\,F_{\nu\lambda} - \frac{1}{4}\,\delta^\mu_\nu\,F^{\lambda\sigma}\,F_{\lambda\sigma}. \tag{9.8}$$

This is of course the energy-momentum tensor that also appears on the right-hand side of the Einstein field equations. (Thus, general relativity uniquely detects the energy-momentum tensor of its sources: it fixes the improvement term.)

Noether's second theorem is also exemplified by the Maxwell theory,[33] which has a *local* gauge symmetry:

$$A_\mu \mapsto A_\mu + \partial_\mu\lambda\,, \tag{9.9}$$

for an arbitrary *smooth function* $\lambda : \mathbb{R}^4 \to \mathbb{R}$. Indeed, since $F_{\mu\nu}$ is itself invariant under this symmetry, the Maxwell action is trivially invariant.

It is instructive to study the gauge invariance when we couple Maxwell's action to a gauge-invariant source J_μ:

$$S[A,J] = \int \mathrm{d}^4x\,\left(-\frac{1}{4}\,F_{\mu\nu}\,F^{\mu\nu} + J^\mu A_\mu\right), \tag{9.10}$$

and require that the action is invariant up to a boundary term that does not modify the equations of motion, so that the transformed action gives the same equations of motion as the action Eq. (9.10).

The variation of this action under the gauge symmetry, Eq. (9.9), gives:

$$\delta_\lambda S[A,J] = \int \mathrm{d}^4x\; J^\mu\partial_\mu\lambda = \int \mathrm{d}^4x\;\partial_\mu(J^\mu\lambda) - \int \mathrm{d}^4x\;(\partial_\mu J^\mu)\,\lambda, \tag{9.11}$$

31 Goldberg (1958, pp. 315, 319) discusses the desirability of symmetry for the discussion of angular momentum. See also Barbashov and Nesterenko (1983, pp. 559–60).

32 Improvement terms go back to Belinfante (1939, pp. 888). See also Anderson (1967, p. 273).

33 For a discussion of Noether's theorems for Maxwell theory coupled to a complex scalar field, see Barbashov and Nesterenko (1983, pp. 550–2) and Sus (2017, pp. 273–4).

where in the last step I used partial integration. The first integral on the right contains a total derivative, which integrates to a boundary term. Although physics texts often require that λ has compact support so that it vanishes at infinity and this boundary term is zero,[34] this is not necessary. For, as we will see in Section 9.2, Noether's theorem can be generalised to admit a variation of the action with a non-zero boundary term, because such a term does not modify the equations of motion. Thus, this term does not need to be zero, and we simply keep it. The second term *must* be zero if the action is to be invariant up to a boundary term, which requires that the source is conserved:

$$\partial_\mu J^\mu = 0. \tag{9.12}$$

Writing this out in components gives the equation for the *local conservation of charge*:

$$\frac{\partial \rho}{\partial t} + \nabla \cdot \mathbf{J} = 0. \tag{9.13}$$

Integrating over the volume V, bounded by the two-surface $S = \partial V$ with outward pointing unit vector $\hat{\mathbf{n}}$, gives the time derivative of the total charge Q_V enclosed by the volume:

$$\begin{aligned} \frac{\mathrm{d}Q_V}{\mathrm{d}t} &= \frac{\mathrm{d}}{\mathrm{d}t} \int_V \mathrm{d}^3x \, \rho\,(t, \mathbf{x}) \\ &\overset{\text{Eq. (9.13)}}{=} -\int_V \mathrm{d}^3x \, \nabla \cdot \mathbf{J} = -\int_S \mathrm{d}^2x \, \hat{\mathbf{n}} \cdot \mathbf{J} = -I_S, \end{aligned} \tag{9.14}$$

where in the second line I used Gauss' theorem. I_S is the current flowing *out* of the boundary S, so that this equation expresses the conservation of charge in the volume: if the charge in V decreases, it is carried away through the boundary by a current I_S, and vice versa.

Thus, although gauge invariance is trivially satisfied by the purely electromagnetic part of the action, it implies a physical constraint on the gauge-invariant source J to which it is coupled; the Maxwell theory can only be consistently coupled to a conserved current (I will return to this at the end of Section 9.2.2).

I will now make a point that will recur in general relativity, and which was important in the discussions between Noether, Hilbert, Klein, and Einstein, reviewed in Section 9.1.1. It is sometimes said that local gauge symmetry is a 'mere redundancy', a consequence of our reformulation of the electomagnetic fields $\mathbf{E}$ and $\mathbf{B}$ in terms of the gauge potential A_μ with no physical significance. However, it is not true that gauge symmetry has no physical significance. If it did not, it would indeed be very surprising that it could entail the local charge conservation law Eq. (9.12).

The point is that the local gauge symmetry reduces the number of physical degrees of freedom of the gauge potential A_μ from three to two, thus leaving a field that only has transverse polarisations. Physically, the freedom to do a gauge transformation secures that the longitudinal polarisation of the photon does not contribute to any of the quantities that are usually deemed physical. This is of course crucial to get the correct phenomenology of the photon. Thus, despite the fact that a choice of gauge does not fix anything physical,

[34] See e.g. Maggiore (2005, p. 69).

gauge invariance does have physical content: it secures the correct number of degrees of freedom for the photon.

Recent discussions have also concluded that the alleged lack of physical significance of the conservation laws derived from the second theorem, as claimed by Klein (see Section 9.1.1), is misleading. For example, Brading and Brown (2003, p. 99) have emphasised that symmetry transformations, such as local gauge symmetries, do have an empirical significance, because they restrict the kinds of terms that can appear in the action and in the equations of motion.[35] Indeed, as the above already shows, imposing local gauge symmetry implies, through Noether's theorem, that Maxwell theory can only be coupled to a *conserved* current.

Before we proceed, let us recall how one normally gets an expression analogous to the charge conservation law, Eq. (9.14), for energy and momentum conservation in general relativity. Such an expression can be derived *in the presence of spacetime symmetries,* in particular in spacetimes with timelike Killing vectors K:[36]

$$\nabla_\mu K_\nu + \nabla_\nu K_\mu = 0. \tag{9.15}$$

The idea here is to derive an analogue of the charge conservation equation Eq. (9.14) for energy and momentum, starting from the covariant analogue of the energy and momentum conservation equation:

$$\nabla_\mu T^\mu{}_\nu = 0. \tag{9.16}$$

If K is a timelike Killing vector satisfying Eq. (9.15), then we can define a vector current $P^\mu := T^{\mu\nu}\, K_\nu$, such that:

$$\nabla_\mu P^\mu = \nabla_\mu(T^{\mu\nu}\, K_\nu) = (\nabla_\mu T^{\mu\nu})\, K_\nu + T^{\mu\nu}\,(\nabla_\mu K_\nu) \overset{\text{Eqs. (9.15), (9.16)}}{=} 0, \tag{9.17}$$

where in the last equality I used the fact that $T^{\mu\nu}$ is symmetric, and then used Eq. (9.15) to set this term to zero. We can now readily apply to P the divergence theorem for a d-dimensional region of space(time) Σ with boundary $\partial\Sigma$:

$$0 = \int_\Sigma \mathrm{d}^d x\, \sqrt{-g}\, \nabla_\mu P^\mu = \int_\Sigma \mathrm{d}^d x\, \partial_\mu(\sqrt{-g}\, T^{\mu\nu} K_\mu) = \int_{\partial\Sigma} \mathrm{d}^{d-1}x\, \sqrt{-h}\, n_\mu\, P^\mu, \tag{9.18}$$

[35] More precisely, they admit that local gauge symmetries can have *indirect,* but not *direct,* empirical significance, i.e. as in Galileo's ship-type scenarios. While, in the philosophical literature, it is widely recognised that *global* gauge (i.e. internal) symmetries applied to subsystems can have direct empirical significance thus understood, there is a debate about whether *local* gauge symmetries can have such significance. Early work by Kosso (2000, pp. 95, 97), Brading and Brown (2004, pp. 656–7), and Healey (2009) stated that local gauge (internal) symmetries have no direct empirical significance, but only indirect empirical significance (e.g. through Noether's theorems). Their reasons for this are diverse – e.g. for Brading and Brown, it is only 'relative system transformations' (i.e. subsystem transformations), but not local gauge symmetries, that have direct empirical significance. More recently, Greaves and Wallace (2014) and Teh (2016) have argued that local gauge symmetries, applied to subsystems, *can* have a direct empirical significance, through Galileo's ship-type scenarios. I will return to the discussion of the empirical significance of local symmetries, in the context of general relativity, in Section 9.4.3. For a characterisation of gauge symmetries as transformations that only alter the description, not the physical system itself, see e.g. Horský and Novotný (1969, p. 420). For a more general discussion of symmetries in terms of 'redundant structure', see e.g. Ismael and van Fraassen (2003, pp. 378–80), Caulton (2015, pp. 156–8), and Belot (2013, p. 2).

[36] I will discuss the general case in Section 9.3. Also, this equation can be relaxed to a conformal Killing equation, as in Eq. (9.63) below.

where g is the metric on Σ, h is the induced metric on $\partial\Sigma$, and n is the unit vector normal to $\partial\Sigma$.[37]

Here, the Killing vector is important because it allows us to construct the *vector* quantity P^μ from the tensor $T^\mu{}_\nu$. For, while Eq. (9.16) does not give rise to a partial derivative that can be integrated (because it has one free index), the vanishing covariant divergence of P in Eq. (9.17), which has no free indices, *does* allow for this, in Eq. (9.18). The result is a conservation law that is completely analogous to charge conservation in the Maxwell theory, as I will now show.

Take Σ to be a three-dimensional spacelike volume V with two-boundary S, and write out the index $\mu = (0, i)$ in terms of timelike and spatial components, with coordinates x^i that cover the whole boundary surface S, but are otherwise arbitrary:

$$\int_V \partial_\mu(\sqrt{-g}\, T^{\mu\nu} K_\mu) = \frac{1}{c}\frac{\mathrm{d}}{\mathrm{d}t}\int_V \mathrm{d}^3x\, \sqrt{-g}\, T^{0\nu} K_\nu + \int_S \mathrm{d}^2x\, \sqrt{h}\, N n_i\, T^{i\nu} K_\nu = 0. \quad (9.19)$$

Here, N is the lapse function, and h is the induced metric on S. In the last term, I used the divergence theorem Eq. (9.18) applied to V with boundary S, to give the integral over the surface S of the momentum density, $\mathcal{P}_S = \sqrt{h}\, N n_i\, T^{i\nu} K_\nu$, through the boundary S. The first term in the middle expression is the time derivative of the energy associated with the region V, with energy density: $\mathcal{E}_V = \sqrt{-g}\, T^{0\nu} K_\nu$. Thus, the expression can be rewritten as:

$$\frac{1}{c}\frac{\mathrm{d}E_V}{\mathrm{d}t} = -P_S, \quad (9.20)$$

where E_V is the total energy in V and P_S is the integrated momentum flux through S. This is analogous to the charge conservation law, Eq. (9.14). Notice that this interpretation is possible thanks to the presence of a timelike Killing vector K in the spacetime, which defines the energy. As we will see in Section 9.1.3, in the absence of Killing vectors, defining an analogous expression is problematic.

9.1.3 Einstein's Proposal, and Five Problems for Energy and Momentum in GR

Having illustrated Noether's theorems in the example of Maxwell theory, I return, in this section, to Einstein's idea of energy and momentum conservation laws in general relativity.[38] I will begin by listing five problems that are prima facie associated with this notion.[39]

The first problem is the one that motivated Noether's research into invariance and symmetries: namely, Klein, Hilbert, and Noether's view that:

(i) *Energy and momentum conservation laws in GR are improper, i.e. identities with no physical content.* This is a special property of the notion of *energy* in general relativity and prompts the question whether this is the correct interpretation of Noether's theorem.

37 After the second equality sign, I used: $\sqrt{g}\,\nabla_\mu P^\mu = \partial_\mu(\sqrt{g}\, P^\mu)$. For an overview of integration on manifolds, see Wald (1984, pp. 430–4) and Landsman (2021, pp. 133–5).

38 Janssen and Renn (2007, pp. 842, 844) note that Einstein (1916b, p. 1115) is an instantiation of Noether's second theorem, predating the theorem itself.

39 See for example the discussion in Horský and Novotný (1969, pp. 426–30).

It also bears on the conceptual relation between energy in general relativity and in classical mechanics. (I will return to the resolution of this problem at the end of Section 9.2, after I discuss Noether's theorem).

In the literature, two further problems about the physical consequences of Noether's second theorem for general relativity have been discussed. Both concern the interpretation of the energy-momentum tensor:

(ii) When general relativity is coupled to matter, the Bianchi identity implies the analogue of the conservation law, Eq. (9.16). But $T^\mu{}_\nu$ is the energy-momentum tensor of *matter, rather than of the gravitational field.* So, the question is, what is the definition of the energy and momentum *of the gravitational field itself*?[40]

(iii) The conservation law, Eq. (9.16), is *not a conservation law in the usual sense.* For the energy-momentum tensor *cannot be straightforwardly integrated over space to give a conserved vector current* (though this *can* be done if there is a Killing vector, as we saw in Eqs. (9.17) and (9.18)).[41] Namely, this equation contains two additional terms that depend on the connection, i.e. it is not a pure derivative like the expression Eq. (9.12) for charge conservation in Minkowski space.[42] This suggests the question of the correct interpretation of the putative conservation law Eq. (9.16).

Einstein's (1914, 1915c, 1915d, 1916b; Einstein and Grossman 1913) idea to solve these problems was to use the particular form of the Christoffel symbols that appear in the covariant analogue of the conservation law for matter, Eq. (9.16), and to reinterpret the extra terms as a gravitational contribution to the energy and momentum. So, one can try to construct an energy-momentum tensor for *gravitation* from the Christoffel symbols – viz. introduce a *pseudo-tensor* $t^\mu{}_\nu$ such that the conservation law contains an *ordinary derivative*. This can be done by defining $t^\mu{}_\nu$ so that Eq. (9.16) takes the form of a total derivative:

$$\partial_\mu(\sqrt{g}\, T^\mu{}_\nu + t^\mu{}_\nu) = 0. \tag{9.21}$$

As previously discussed in Section 9.1.1, Einstein interpreted $t^\mu{}_\nu$ as a 'gravitational energy-momentum tensor' density. One can associate with this tensor a *superpotential*, $\mathcal{U}^{\mu\lambda}{}_\nu$, such that:

$$\sqrt{g}\, T^\mu{}_\nu + t^\mu{}_\nu = \partial_\lambda\, \mathcal{U}^{\mu\lambda}{}_\nu\,, \quad \partial_\mu\, \partial_\lambda\, \mathcal{U}^{\mu\lambda}{}_\nu = 0. \tag{9.22}$$

Einstein (1916b, p. 244) obtained the superpotential from an action that differs from the Einstein–Hilbert action by a boundary term.

Freud's (1939) expression for the superpotential is different from Einstein's, namely:

$$\mathcal{U}^{\mu\lambda}{}_\nu = -\sqrt{g}\, g^{\beta\sigma}\, \Gamma^\alpha_{\beta\gamma}\, \delta^{\mu\lambda\gamma}_{\alpha\sigma\nu}\,, \tag{9.23}$$

[40] Even though the energy-momentum tensor will have this property for any field theory in a curved spacetime, there is a specific problem to GR here: namely, how to define the energy and momentum of the gravitational field from its dynamical principles.

[41] Note that this is not an exclusive property of general relativity, since it holds for the covariant divergence of the energy-momentum tensor in any field theory on a curved spacetime. However, in general relativity, because the gravitational field is dynamical, Eq. (9.16) can be reinterpreted in terms of the gravitational energy-momentum pseudo-tensor $t^\mu{}_\nu$, as we will discuss.

[42] In particular, in Eq. (9.18), it is possible to replace the covariant derivative in the first expression by the partial derivative in the middle expression because we have the covariant divergence of a *vector* quantity, P^μ. If we had simply integrated the covariant divergence of the tensor $T^\mu{}_\nu$, this would pick up additional terms involving the connection and it would not give a boundary term. This is also the origin of the extra term $t^\mu{}_\nu$, in Eq. (9.21).

which automatically satisfies Eq. (9.22), because the symbol $\delta^{\mu\lambda\gamma}_{\alpha\sigma\nu}$ is totally antisymmetric in its upper (and lower) indices. Another well-known expression is Landau and Lifshitz's (1971, p. 306)[43] pseudo-tensor (see also Bergmann 1958, p. 289).

However, even if Einstein's strategy appears to solve the two previous problems, (ii) and (iii), of finding an expression for the energy and momentum of the gravitational field, and of recasting Eq. (9.16) with an ordinary derivative so that we can integrate to get a conserved charge, it does so at the expense of introducing two new challenges:[44]

(iv) *Non-uniqueness*: the ability to add an arbitrary superpotential $\mathcal{U}$ to the pseudo-tensor means that the pseudo-tensor is not unique (I have so far mentioned three expressions: Einstein's, Freud's, and Landau and Lifschitz's).[45] In fact, there are an infinite number of pseudo-tensors compatible with Eq. (9.21). For defining a gravitational (i.e. matter-independent) pseudo-tensor with an arbitrary superpotential:

$$t^{\mu}{}_{\nu} := -\frac{1}{\kappa}\sqrt{g}\,G^{\mu}{}_{\nu} + \partial_{\lambda}\,\mathcal{U}^{\mu\lambda}{}_{\nu}\,, \tag{9.24}$$

where $\kappa = 8\pi G_{\mathrm{N}}/c^4$, and $G^{\mu}{}_{\nu}$ is the Einstein tensor. Provided $\mathcal{U}$ is antisymmetric in its upper indices,

$$\mathcal{U}^{\mu\lambda}{}_{\nu} = \mathcal{U}^{[\mu\lambda]}{}_{\nu}$$

(see, e.g., the Freud expression, Eq. (9.23)), the gravitational pseudo-tensor thus defined automatically satisfies the conservation law Eq. (9.21) when the Einstein field equations are satisfied:[46]

Thus the definition of the pseudo-tensor, Eq. (9.24), is ambiguous. Nester (2004, p. 264) summarises the problem thus: 'there is no physical meaning here, only formal math – unless one has a good criterion for choosing the superpotential'. In my opinion, this is a serious problem that defenders of pseudo-tensors need to solve.

(v) *Lack of covariance, and lack of appropriate transformation properties*: the pseudo-tensor is not a tensor, as can be seen from its definition in Eq. (9.21). For example, it can be made to vanish at any given point in appropriately chosen coordinates.[47] It is only invariant under affine transformations.

Thus, Klein (1917) concluded: 'After all this, I can barely believe that it is useful to designate [Einstein's] very arbitrarily built quantities $t^{\mu}{}_{\nu}$ as the components of the energy of the gravitational field.' Schrödinger (1950, p. 104) called the gravitational pseudo-tensor a 'sham tensor'. Twenty-three years later, Misner, Thorne, and Wheeler (1973, p. 467) wrote:

> Anyone who looks for a magic formula for 'local gravitational energy-momentum' is looking for the right answer to the wrong question. Unhappily, enormous time and effort were devoted in the past to trying to 'answer this question' before investigators realized the futility of the enterprise.

43 For a critique of the Landau–Lifshitz pseudo-tensor, see Goldberg (1958, pp. 316, 320) and Ohanian (2010, p. 10).

44 These concerns can be found, e.g., in Hoefer (2000: pp. 191–5). My (4) corresponds to Dürr (2019a, section 3.2) label (H2), and my (5) to his (H1). I will return to Dürr's objections in Section 9.6.2.

45 Horský and Novotný (1969, pp. 426–30) is an enlightening review of various approaches. See also Komar's (1958, pp. 935–6) expressions.

46 The expression that Einstein obtained was not antisymmetric: for a discussion, see Chen et al. (2017, p. 12).

47 This argument was put forward to Einstein by Schrödinger and Bauer: for details, see Horský and Novotný (1969, p. 430). See the responses by Pitts (2010, p. 613) and Dürr (2019b, p. 8).

So, we have a strong warning from these authors that we should avoid looking for a general local expression for the energy and momentum of the gravitational field.

As I will mention in Section 9.5.2, beyond the lack of covariance, the problem is also that the pseudo-tensor does not have well-defined transformation properties: for example, it is not a geometric object, in the sense of Anderson (1967, pp. 14–16, 424), Kucharzewski and Kuczma (1964, pp. 15–18), and Trautman (1965, pp. 84–87):[48] i.e. roughly, such that its components in a given coordinate system, together with suitable transformation properties, are well-defined.[49] See the discussion in Pitts (2010, pp. 603–4, 606) and Dürr (2019a, section 9.3.3).

In Section 9.3, I will propose a solution of the non-uniqueness problem (iv) based on an extension of Noether's theorem. Although the facts that I will use are known to the experts, I have not found this solution discussed in any detail in the literature on Noether's theorems that I studied – not as a solution of the alleged problem of non-uniqueness. In Section 9.5.1, I will discuss a recent related proposal. In Section 9.5.2, I will assess how this approach deals with problem (iii) (in particular, the problem of the need for Killing vectors or similar structures) and with problem (v), i.e. covariance.

Penrose (1982) made a new proposal to solve (ii) that does not require a local expression for the energy and momentum, thereby circumventing problems (iii)–(v), because he did not use the gravitational energy-momentum pseudo-tensor or other local expressions. He argued that one can still define *quasi-local* energy-momentum and angular momentum, i.e. energy and momentum that are associated with a closed spacelike surface. It refers only to the geometry of the two-surface and the extrinsic curvature quantities for its embedding in the spacetime (Penrose 1988, p. 1). The idea has been widely accepted, and developed in various ways, in the literature on general relativity.[50] I will return to it in Section 9.4, when I discuss the Brown–York quasi-local stress-energy tensor.

9.2 Noether's Theorems and the Energy-Momentum Tensor

In this section, I give Noether's first theorem (Section 9.2.1) and her second theorem (Section 9.2.2), and I discuss two consequences (Section 9.2.3) that, in my opinion, have not been sufficiently clarified in the literature.

9.2.1 Noether's First Theorem

Noether's first theorem is a statement about the invariance of an action of the form:[51]

$$S[\phi; x] = \int \mathrm{d}^4x \; \mathcal{L}(\phi; x), \tag{9.25}$$

[48] See also Nijenhuis (1952, pp. 23–30), Schouten (1954, p. 67–8), and Bergmann (1958, p. 287).

[49] As Trautman (1965, p. 86) mentions, spinors are not geometric objects either. In Section 9.6.1 I will explain what appears to be the real worry.

[50] For some of the properties that the quasi-local energy should (presumably) have, and its relation to other definitions of energy such as the asymptotic ADM mass (for asymptotically flat spacetime regions) and Bondi mass (for asymptotically null regions), see, for example, Jaramillo and Gourgoulhon (2011, pp. 97–102, 105–7) and Christodoulou and Yau (1988, pp. 9–10).

[51] For the connection between symmetries and conservation laws *without* the appeal of Lagrangian methods, see Brown (2020, p. 3).

where ϕ is the set of fields present in the problem (labelled by a, so one writes ϕ_a if one wishes to indicate this) under an infinitesimal transformation of the dependent and/or independent variables, up to a boundary term.[52] The physical significance of this invariance of the action up to a boundary term is that the equations of motion derived from the new action, expressed in the new variables, will take the same form as in the old ones. Following Brading and Brown (2003, pp. 91–7), on which the exposition in this section is largely based,[53] I take the invariance to mean that, when the equations of motion are satisfied, the action is form-invariant up to a boundary term that depends on an arbitrary vector quantity Λ^μ:

$$\delta S = S[\phi';x'] - S[\phi;x] = \int \mathrm{d}^4x'\,\mathcal{L}(\phi',x') - \int \mathrm{d}^4x\,\mathcal{L}(\phi,x) = \int \mathrm{d}^4x\,\partial_\mu\Lambda^\mu. \quad (9.26)$$

Such a transformation, which leaves the action invariant up to a boundary term, is called a 'quasi-symmetry'.[54]

Noether's first theorem states that, if the action Eq. (9.25) is invariant with respect to a continuous group of transformations G_n that is generated by n constant parameters ξ^k (where $k=1,\ldots,n$), in the sense just indicated, then there are n linearly independent combinations of the field equations that are equal to the total divergence of a current, as follows:

$$\sum_a \left(\frac{\partial\mathcal{L}}{\partial\phi_a} - \partial_\mu\frac{\partial\mathcal{L}}{\partial(\partial_\mu\phi_a)}\right)\eta_{ak} = \partial_\mu J^\mu_k. \quad (9.27)$$

Since the index k labels the n generators of G_n, there is one such equation for each generator. The index a labels the various fields present in the problem, and η_{ak} is a field-dependent expression, defined by the variation of the fields

$$\delta_\xi\,\phi_a = \eta_{ak}\,\xi^k. \quad (9.28)$$

The current is given by the following expression:

$$J^\mu_k = -\sum_a \frac{\partial\mathcal{L}}{\partial(\partial_\mu\,\phi_a)}\frac{\partial\delta\phi_a}{\partial\xi^k} - \mathcal{L}\frac{\partial\delta x^\mu}{\partial\xi^k} + \frac{\partial\Lambda^\mu}{\partial\xi^k}. \quad (9.29)$$

The converse is also true: if the field equations satisfy Eq. (9.27), then this implies the invariance of the action under the corresponding group G_n. The theorem remains valid in the limiting case of an infinite number of parameters, $n\to\infty$.

The above current is weakly conserved, since the left-hand side of Eq. (9.27) is a linear combination of the Euler–Lagrange field equations for ϕ.

52 This boundary term was not included in Noether's paper, and so this is a slight generalisation of her two theorems, first proven by Bessel-Hagen (1921, p. 36), who attributes it directly to Noether: 'I owe this generalization to an oral communication from Emmy Noether'. See also Kosmann-Schwarzbach (Chapter 1, this volume) and Brown (2020, pp. 6, 8). This generalisation is standardly used, e.g., in Brading and Brown (2003, p. 93) and Bañados and Reyes (2016, p. 17).

53 For other treatments of Noether's theorems, closely related to the one given here, see Trautman (1962, pp. 176–9), Horský and Novotný (1969, pp. 420–4), Barbashov and Nesterenko (1983, pp. 541–4), Kastrup (1987, pp. 116–17), and Brading and Brown (2000, pp. 4–8).

54 Echoing Bessel-Hagen (1921, p. 36), Olver (1986, p. 278) calls them 'divergence symmetries'. See also Kosmann-Schwarzbach (Chapter 1, this volume).

Example: Maxwell theory. Let us illustrate the above for the Maxwell theory, familiar from Section 9.1.2.[55] Thus, we take the group generated by ξ^k to be the group of constant spacetime translations. Since A is a vector field, the index a labelling the different fields can be taken to be the spacetime index μ of the vector field components A_μ. Since ξ^k is the generator of translations, ξ is also a vector field in Minkowski space, and the index $k = 1, \ldots, n$, labelling the generators, is also a spacetime index, $\nu = 1, \ldots, 4$. Thus the group G_4 here under consideration is the group of constant spacetime translations generated by $\delta x^\mu = \epsilon^\mu$.

To see how η_{ak} depends on the vector field A, we temporarily write it as $\eta_{\mu\nu}$ (not to be confused with the Minkowski metric!) and compare the general variation, Eq. (9.28), with the variation of the gauge field, Eq. (9.4), (which is obtained by expanding the field components A_μ in terms of the constant coefficients, $\delta x^\mu = \epsilon^\mu$). The comparison gives $\eta_{\mu\nu} = -\partial_\nu A_\mu$.

Let us check that the Noether current, Eq. (9.29), reproduces the energy-momentum tensor of electromagnetism, Eq. (9.5). Since $\delta x^\mu = \epsilon^\mu$, we have $\frac{\partial \delta x^\mu}{\partial \epsilon^\nu} = \delta^\mu_\nu$. The last term in Eq. (9.29) is zero, and the first term is $-\frac{\partial \mathcal{L}}{\partial(\partial_\mu A_\lambda)}\,\eta_{\lambda\nu} = \frac{\partial \mathcal{L}}{\partial(\partial_\mu A_\lambda)}\,\partial_\nu A_\lambda = -F^{\mu\lambda}\,\partial_\nu A_\lambda$. Thus, we get $J^\mu_\nu = -F^{\mu\lambda}\,\partial_\nu A_\lambda - \mathcal{L}\,\delta^\mu_\nu = T^\mu{}_\nu$, i.e. the energy-momentum tensor Eq. (9.5). The Noether current conservation equation, Eq. (9.27), of course reproduces Eq. (9.6).

We can repeat this for the gauge-invariant variation $\delta A_\mu = F_{\mu\nu}\,\epsilon^\nu$ leading to the energy-momentum tensor with the improvement term, i.e. Eq. (9.7). Comparing this variation with Eq. (9.28), we find that $\eta_{\mu\nu} = F_{\mu\nu}$. Evaluating the current Eq. (9.29), we indeed get $J^\mu_\nu = -T'^\mu{}_\nu$, i.e. we reproduce the standard energy-momentum tensor, Eq. (9.8).

9.2.2 Noether's Second Theorem and the Boundary Theorem

In this section, I will give Noether's second theorem and the identities that it implies. I will follow the terminology in Brading and Brown (2003, pp. 100–4), who distinguish a set of formulas that they call the 'boundary theorem', due to Klein (1918b), from Noether's second theorem. (I will occasionally refer to both sets of formulas, collectively, as 'Noether's second theorem', since both are entailed by Noether's general expression for the invariance of the action under a continuous group.)

The boundary theorem and Noether's second theorem, in generalised form, assume that the action, Eq. (9.25), is invariant (up to a boundary term) with respect to a continuous group of transformations $G_{\infty n}$ that is generated by n arbitrary *functions* $\xi^k(x)$ $(k = 1, \ldots, n)$.

Before giving the theorems, it will be useful to introduce three important pieces of notation. I use the Einstein summation convention for repeated indices a and, below, also for repeated k indices:

$$E^a := \frac{\partial \mathcal{L}}{\partial \phi_a} - \partial_\mu \frac{\partial \mathcal{L}}{\partial(\partial_\mu \phi_a)}, \tag{9.30}$$

[55] Noether's theorems for Yang–Mills theory are discussed in Barbashov and Nesterenko (1983, pp. 552–6).

$$J^\mu_k := -\frac{\partial \mathcal{L}}{\partial(\partial_\mu \phi_a)} \frac{\partial \delta \phi_a}{\partial \xi^k} - \mathcal{L} \frac{\partial \delta x^\mu}{\partial \xi^k} + \frac{\partial \Lambda^\mu}{\partial \xi^k}, \tag{9.31}$$

$$U^{\mu\nu}_k := \frac{\partial \mathcal{L}}{\partial(\partial_\nu \phi_a)} \gamma^\mu_{ak} + \frac{\partial \Lambda^\mu}{\partial(\partial_\nu \xi^k)}. \tag{9.32}$$

The quantities here defined are, respectively, the Euler–Lagrange equations of motion, the Noether current, and the superpotential (more on this later). γ^ν_{ak} is a (field-dependent) variation of the field, to be defined below. Λ^μ is a boundary term that may appear in the variation of the action, Eq. (9.26). Notice that we can also apply Noether's theorem to a single term in (i.e. to a *part* of) the action, to require the invariance of that particular term. In that case, for example, E^a is simply the functional variation of that term in the action with respect to the fields ϕ_a, and likewise for the other two quantities.

Using this notation, Noether's *first* theorem takes a particularly simple form:

$$E^a \, \eta_{ak} = \partial_\mu J^\mu_k. \tag{9.33}$$

The expression γ^ν_{ak} in Eq. (9.32), together with η_{ak}, are the (field-dependent) variations of the field with respect to ξ and its first derivative (repeated k indices are again summed over) relevant to Noether's second theorem and her boundary theorem:

$$\delta\phi_a = \eta_{ak}(\phi; x)\, \xi^k(x) + \gamma^\mu_{ak}(\phi; x)\, \partial_\mu \xi^k(x). \tag{9.34}$$

This variation is analogous to Eq. (9.28), where the second term is non-zero because in Noether's second theorem, ξ is a function of x.

The boundary theorem, with this notation, takes the following simple form:

$$\partial_\mu(E^a \gamma^\mu_{ak}) = \partial_\mu J^\mu_k, \tag{9.35}$$

$$E^a \gamma^\mu_{ak} = J^\mu_k - \partial_\nu U^{\mu\nu}_k, \tag{9.36}$$

$$U^{\mu\nu}_k = -U^{\nu\mu}_k. \tag{9.37}$$

Again, the index $k = 1, \ldots, n$ labels the functions $\xi^k(x)$ that generate $G_{\infty n}$, and so there is one such equation for each generator. Note that the first equation here follows from the second and third.[56]

Noether's second theorem reads:

$$E^a \, \eta_{ak} = \partial_\mu(E^a \gamma^\mu_{ak}). \tag{9.38}$$

Again, the full statement is that, if the action, Eq. (9.25), is invariant under the group $G_{\infty n}$ of variations as indicated, then the Euler–Lagrange equations satisfy the identities, Eq. (9.38). The converse statement is here also valid.

Notice that both the boundary theorem and Noether's second theorem, Eqs. (9.35)–(9.38), take forms that differ from Noether's first theorem, Eq. (9.33). However, it is possible

[56] My notation, using the Einstein convention for the three different types of indices, also corrects errors in Brading and Brown (2003: pp. 92, 95, 97, 101) here and in their equation corresponding to my Eq. (9.29), where summation signs over a (i.e. their i) are incorrectly extended to terms that contain no index a.

to derive Noether's first theorem, which is the special case that the continuous group of transformations G_n is a subgroup of $G_{\infty n}$, i.e. $G_n \subset G_{\infty n}$. In this case, the generators ξ^k of G_n are constant, and the corresponding second term in Eq. (9.34) vanishes. Combining the first of Eq. (9.35) with Eq. (9.38), we recover Noether's first theorem, Eq. (9.33), with the k generators thus identified.

Example: Maxwell theory. It will be instructive to compare with our previous notation and reproduce the conservation law of Maxwell's action coupled to a gauge-invariant source, Eq. (9.10). The gauge symmetry modifies the field A as: $\delta A_\mu = \partial_\mu \lambda$. Since λ is a function of x, Noether's second theorem is relevant here. Comparing with Eq. (9.34), since there is a single generator, namely λ, we see that $k = 1$ and $\xi = \lambda$ (thus, we can drop the index k). As before, the index a is a spacetime index, $a = \nu$. Further, δA_μ depends only on λ's derivative and not on λ itself, so $\eta_\mu = 0$ and $\gamma_\nu^\mu = \delta_\nu^\mu$. In particular, this means that there is no non-trivial subgroup of transformations with *constant* parameters in this case, and so no Noether first theorem for the gauge transformations. Also, recall, from Eq. (9.11) that the variation of the action was zero up to a boundary term, i.e. a term with a total derivative $\int \mathrm{d}^4x\ \partial_\mu \Lambda^\mu$. We find that Λ is given by:

$$\Lambda^\mu = J^\mu \lambda\,. \tag{9.39}$$

This confirms my claim in Section 9.1.2 that, with this generalised form of Noether's theorem, the gauge parameter λ does not need to be zero at the boundary. It is straightforward to check the boundary theorem and Noether's second theorem, Eqs. (9.35)–(9.38). For the quantities defined in Eqs. (9.30) and (9.32), we get the following expressions:

$$\begin{aligned} E^\mu &= J^\mu - \partial_\nu F^{\mu\nu}, \\ U^{\mu\nu} &= F^{\mu\nu}. \end{aligned} \tag{9.40}$$

The first equation is of course the expression for the Euler-Lagrange equation derived from the action, Eq. (9.10).

With these expressions in hand, it is straightforward to verify that the boundary theorem, Eqs. (9.35)–(9.37), is trivially satisfied, given the anti-symmetry of the field tensor. Interestingly, the left-hand side of the expression from Noether's second theorem, Eq. (9.38), is zero (because η is zero), and so Noether's second theorem takes the form $\partial_\mu E^\mu = 0$. Given the anti-symmetry of the field tensor, this is identically satisfied if and only if J is conserved, *independent of the satisfaction of the equations of motion of the Maxwell theory.* Thus imposing gauge symmetry implies, through Noether's theorem, that the Maxwell theory can only be coupled to a conserved current J.

9.2.3 Two Consequences of Noether's Second Theorem

I will first make a comment that I have not found in the literature on Noether's theorems, although it may well be known to the experts.[57] These remarks are made visible by the

[57] For example, Barbashov and Nesterenko (1993, p. 562) mention, for the specific energy-momentum (pseudo-)tensors considered by Einstein and by Lorentz, that they are related by a total derivative and so lead to the same equations of motion.

simplified notation above, and they will bear on the question of energy and momentum conservation discussed in the next section. Then I will answer Klein's critique of the local energy-momentum conservation law in GR.

The superpotential is fixed, through Noether's theorem, by the boundary terms. In studies of Noether's theorem in connection with general relativity, emphasis is often laid on the integration of Eq. (9.35) (see Brading and Brown 2003, p. 102), as follows:

$$J_k^\mu = E^a \gamma_{ak}^\mu + \partial_\nu \mathcal{U}_k^{\mu\nu}, \tag{9.41}$$

where $\mathcal{U}_k^{\mu\nu}$ is an arbitrarily chosen superpotential, analogous to the one in Eq. (9.24), whose divergence is automatically zero:

$$\partial_\mu \, \partial_\nu \, \mathcal{U}_k^{\mu\nu} = 0. \tag{9.42}$$

Indeed, this is how Einstein himself arrived at his result, Eq. (9.24), for the gravitational pseudo-tensor.

However, note that Noether's theorem gives us, in Eq. (9.36), an expression for the current that is not arbitrary but fixed:

$$J_k^\mu = E^a \gamma_{ak}^\mu + \partial_\nu U_k^{\mu\nu}, \tag{9.43}$$

where U is *given* by Eq. (9.32). Since U is antisymmetric in its upper indices, it indeed satisfies Eq. (9.42).[58]

The compatibility of Eqs. (9.41) and (9.43) requires that we must have:[59] $\mathcal{U}_k^{\mu\nu} = U_k^{\mu\nu}$. Since U was defined in Eq. (9.32), and the boundary theorem depends on its having this exact form, this means that the superpotential is *not arbitrary* but is fixed through Eq. (9.32).

How exactly is the superpotential fixed by Noether's theorem? The first term in Eq. (9.32) is fixed by the choice of the Lagrangian density and of the variation of the field. The second term (depending on Λ^μ) is fixed by the boundary term that appears in the variation of the action. This means that *Noether's theorem determines the superpotential, U, once the action and its first variation are fixed.*

In the next section, I will return to the physical interpretation of this result, and to how it solves the problem (iv) of non-uniqueness in general relativity from Section 9.1.1.

But, a few paragraphs later, they categorise the non-uniqueness of the pseudo-tensor as a 'difficulty', without either mentioning the relevance of their previous remark or noting that the pseudo-tensor is unique after fixing the boundary terms of the action. Likewise, Ohanian (2010, p. 10) has a formula for the pseudo-tensor in terms of the gravitational action, but he drops a total derivative term from it, and he does not comment on the effect of the dropped boundary term on the pseudo-tensor. Similar comments apply to Goldberg (1958, p. 318), Bergmann (1958, p. 288), Komar (1958, pp. 934–5), Trautman (1962, pp. 171, 174, 190–1), and Horský–Novotný (1969, pp. 425–7), all of whom explicitly survey and compare different expressions for the pseudo-tensor, mention the lack of uniqueness, and sometimes propose a criterion for fixing a unique expression, but fail to mention the most obvious criterion: namely, that of fixing the boundary terms. That the superpotential is fixed, through Noether's theorem, by the choice boundary terms of the action, seems implicit in Wald (1993, p. 3429), but he does not discuss or mention the gravitational pseudo-tensor or the superpotential at all. In Sections 9.3 and 9.4.2, I will further comment on the physical significance of fixing the boundary terms.

58 For diffeomorphism-invariant theories, the superpotential corresponds, in Wald's (1993: p. 3429) formulation, to the Noether charge **Q**, which in a four-dimensional spacetime is a two-form.

59 This holds up to an antisymmetric, total derivative, term of the form $\partial^{[\mu} w^{\nu]}$, i.e. the superpotential obtained by integrating Eq. (9.35) must be compatible with Eq. (9.36). I thank Henrique Gomes for pointing this out. Note that this ambiguity appears only if one integrates Eq. (9.35) regardless of the other equations. Most important, this ambiguity does not affect the superpotential U of Noether's theorem, which is completely fixed by Noether's theorem, through Eq. (9.36).

A response to the Hilbert–Klein–Noether critique. The above also casts light on the Klein–Hilbert worry, problem (i) from Section 9.1.3, that energy and momentum conservation laws in GR are identities that do not require using the Einstein field equations, and hence they lack physical content. We can see that this follows from Eq. (9.43). Since the Noether current J, rather than its derivative, is linear in the equations of motion, the satisfaction of the equations of motion does not play a role in securing current conservation, i.e. in the vanishing of the current's *derivative*. In other words, by the time one wants to prove the conservation of the current, the equations of motion have 'disappeared' from it. They play a role in the current, but not in the conservation. This is unlike Noether's first theorem, Eq. (9.27), where the *divergence* of the current, rather than the current itself, is linear in the equations of motion.

Noether herself took her theorems to confirm Hilbert's view that there is no proper energy conservation law in general relativity. She wrote (see Kosmann-Schwarzbach 2011, p. 20):

> Given [an action] S invariant under the group of translations, then the energy relations [i.e. the conservation laws corresponding to translations] are improper [i.e. the divergences vanish identically] iff S is invariant under an infinite group which contains the group of translations as a subgroup[60] ... As Hilbert expresses his assertion, the lack of a proper law of energy constitutes a characteristic of the 'general theory of relativity'.

It is of course true that the details of the conservation are different in the two cases, because they satisfy different theorems. Noether's second theorem imposes more constraints.

However, as we also saw above, the boundary theorem, combined with Noether's second theorem, does imply an identity that looks exactly like Noether's first theorem, Eq. (9.27). The two theorems differ in that, in the case of the boundary theorem, the equations of motion appear in the form of the current itself, while they do not appear in Noether's first theorem. Thus, the disappearance of the equations of motion from the divergence of the Noether current in the boundary theorem is not a fact that is additional to its disappearance from the current itself (the two are connected through Noether's second theorem, Eq. (9.38)).

Let me make this reply to Klein (in particular) more precise. Recall his points (1) and (2), from Section 9.1.1, that the divergence of the total energy-momentum tensor density, i.e. $\tau^{\mu}{}_{\nu} := \sqrt{g}\, T^{\mu}{}_{\nu} + t^{\mu}{}_{\nu}$, has no physical meaning because it is an improper conservation law, i.e. it consists of two terms:[61] (1) the divergence of the Einstein field equations, and (2) the divergence of the superpotential, which is identically zero. This follows from the boundary theorem, Eq. (9.36), which gives: $\partial_\mu J^\mu_k = \partial_\mu(E^a \gamma^\mu_{ak}) + \partial_\mu \partial_\nu U^{\mu\nu}_k$, where: (1) E^a appears here under the divergence, (2) the second term vanishes identically. By (1), the equations of motion play no direct role in the vanishing of the divergence (as they do

60 The group of translations being a *subgroup* of an infinite group is not sufficient to get improper energy relations. For example, in the Maxwell theory in Minkowski space, the group of translations, G_4, is a sugroup of the total group containing both translations and local gauge transformations, i.e. $G_4 \times G_{\infty 1}$, where $G_{\infty 1}$ is the group of local gauge transformations: and yet, the energy relations of the Maxwell theory are *proper.* Thus, for *improper* energy relations, it is also necessary (and this is obviously what Noether had in mind) that the four constant generators of translations, ϵ^μ, are obtained by setting four corresponding generators of the infinite group, i.e. the four functions $\xi^\mu(x)$, to be constant. The latter condition is not satisfied in the Maxwell theory, but it is satisfied in general relativity.

61 Here, and in the rest of the chapter, the notation $\sqrt{g}$ means $\sqrt{|\det(g)|}$.

in, e.g., classical mechanics).[62] Thus, the equations of motion play no direct role in the vanishing of the divergence (as they do in, e.g., classical mechanics).

But using Noether's second theorem, i.e. Eq. (9.38), for non-zero γ^{μ}_{ak}, we rewrite this equation as: $\partial_\mu J^\mu_k = E^a\eta_{ak} + \partial_\mu\partial_\nu U^{\mu\nu}_k$.[63] Since the second term is identically zero, we see that the Noether current J is conserved if and only if (a linear combination of) the equations of motion are satisfied, in complete analogy with Noether's first theorem.[64]

Slightly anticipating the results of the next section, we can show this explicitly in the case of general relativity. Rewrite: $\tau^\mu{}_\nu = 2E^\mu{}_\nu + \partial_\lambda U^{\mu\lambda}_\nu$, where $E^\mu{}_\nu = \frac{1}{2\kappa}\sqrt{g}(G^\mu{}_\nu - \kappa T^\mu{}_\nu)$ is proportional to the Einstein field equations. Using the contracted Bianchi identities, we find: $\partial_\mu\tau^\mu{}_\nu = E^{\alpha\beta}\partial_\nu g_{\alpha\beta} + \partial_\mu\partial_\lambda U^{\mu\lambda}_\nu$. Since the second term vanishes identically, the total energy-momentum (pseudo-)tensor density is conserved if and only if a certain linear combination of Einstein's equations are satisfied. This seems to be what Einstein (1918, p. 513) had in mind in his reply that *part* of the field equations are used,[65] and he was indeed right.[66]

9.3 Noether's Theorems in General Relativity

In this section, I will apply Noether's theorem to general relativity coupled to matter.[67] I will first focus on the case of pure gravity, and then discuss adding matter.

Take the action, Eq. (9.25), with Lagrangian density given by $\mathcal{L} = \mathcal{L}_{\text{E}} + \mathcal{L}_{\text{m}}$, where $\mathcal{L}_{\text{E}}$ is the Einstein Lagrangian, and $\mathcal{L}_{\text{m}}$ is the matter Lagrangian. The Einstein Lagrangian is given by:

$$\begin{aligned} 2\kappa\,\mathcal{L}_{\text{E}}(g_{\mu\nu}, \partial_\lambda g_{\mu\nu}) &:= \sqrt{g}\,g^{\mu\nu}(\Gamma^\lambda_{\mu\nu}\Gamma^\lambda_{\lambda\sigma} - \Gamma^\sigma_{\mu\lambda}\Gamma^\lambda_{\nu\sigma}) \\ &= \sqrt{g}\,R - \partial_\mu(\sqrt{g}\,(g^{\alpha\beta}\Gamma^\mu_{\alpha\beta} - g^{\mu\nu}\Gamma^\lambda_{\nu\lambda})), \end{aligned} \tag{9.44}$$

and so the Einstein action differs from the usual Einstein–Hilbert action by a total derivative term that does not affect the equations of motion.[68]

Let us first write out the notation for Noether's theorems applied to the Einstein term in the action. For, although we have a coupled gravity-matter system, we may of course require the diffeomorphism invariance of a single term.

62 Klein's (1918a, p. 504) argument is not quite as explicit as this. This reconstruction of the argument follows Brading (2005, p. 126): see footnote 26.

63 This is different from the case of the local gauge symmetry of the Maxwell theory, because γ^μ_{ak} there is indeed zero, while it is non-zero in general relativity.

64 The (indirect) physical significance of Noether's second theorem has long been recognised by philosophers. Recently, a number of authors have also emphasised the (direct) physical significance of improper conservation laws for local gauge symmetries: see Teh (2016, p. 99) and Sus (2017, pp. 274–5).

65 Also in classical mechanics, it is only *part,* i.e. a linear combination, of the equations of motion that appear on the right-hand side. Consider, as an elementary example, the Lagrangian $L = \frac{1}{2}m\dot{x}_i^2 + V(x)$, where $i = 1, 2, 3$. The time derivative of the Hamiltonian gives: $\mathrm{d}H/\mathrm{d}t = E_i\dot{x}_i$, where $E_i = m\ddot{x}_i + \partial_i V$, i.e. the conservation law is proportional to a *linear combination* of the equations of motion.

66 Klein's argument *can* go through if $E^a\gamma^\mu_{ak} = 0$ without use of the equations of motion, e.g. if $\gamma^\mu_{ak} = 0$. In that case, it is true that the first term vanishes before its derivative is taken. However, this is not the case in general relativity.

67 For a discussion of Noether's theorem and Kretschmann's objection to Einstein's treatment of general covariance, see Brown and Brading (2002, pp. 22–5). This paper also contains a treatment of Noether's theorem for electrodynamics with an arbitrary background metric.

68 There are various boundary terms that one can add, depending on the choice of boundary conditions, and these lead to different expressions for the pseudo-tensor. For a treatment of Noether's second theorem for the Einstein–Hilbert action, see, for example, Horský and Novotný (1969, p. 424). In this section, I follow some of the historical literature in using the Einstein action. I will return to the issue of boundary terms and boundary conditions in Sections 9.4 and 9.5.

The field ϕ_a is then the metric tensor, and the index a is a symmetric pair of spacetime indices, $a = \alpha\beta$, ranging from 1 to 4, i.e. the dimension of spacetime. Under a local translation $\delta x^\mu = \xi^\mu(x)$, the metric transforms as follows:

$$\delta g_{\alpha\beta} = -\nabla_\alpha \xi_\beta - \nabla_\beta \xi_\alpha = -\xi^\mu\, \partial_\mu\, g_{\alpha\beta} - 2g_{\nu(\alpha}\, \partial_{\beta)}\xi^\nu. \tag{9.45}$$

The index k is also a spacetime index μ. Comparing with Eq. (9.34), we see that η_{ak} is equal to $-\partial_\mu g_{\alpha\beta}$, and γ^μ_{ak} is equal to $-2\delta^\mu_{(\alpha}\, g_{\beta)\nu}$.

Noether's second theorem, Eq. (9.38), applied to the gravitational part of the action, gives the contracted *Bianchi identity* for the Einstein tensor:[69]

$$\partial_\mu(\sqrt{g}\, G^\mu{}_\nu) = \frac{1}{2}\, \sqrt{g}\, G^{\alpha\beta}\, \partial_\nu g_{\alpha\beta} \quad \Leftrightarrow \quad \nabla_\mu G^\mu{}_\nu = 0, \tag{9.46}$$

Noether's second theorem, applied to the *matter* part of the action, gives the covariant analogue of the energy and momentum conservation equation, i.e. Eq. (9.16). That equation is, as expected, obtained as an identity, without the use of the equations of motion.

Let us write down the Noether current, Eq. (9.31), $J^\mu_\nu =: -t^\mu{}_\nu$, for the gravitational part of the action:

$$t^\mu{}_\nu = -\frac{\partial \mathcal{L}_{\rm E}}{\partial(\partial_\mu g_{\alpha\beta})}\, \partial_\nu\, g_{\alpha\beta} + \mathcal{L}_{\rm E}\, \delta^\mu_\nu - \frac{\partial \Lambda^\mu}{\partial \xi^\nu}. \tag{9.47}$$

This gives a gravitational analogue of the matter energy-momentum tensor. In fact, as I shall discuss in a moment, it is not a tensor: it is the gravitational energy-momentum *pseudo-tensor.*

The boundary theorem gives the following two results. The first equation of the boundary theorem, Eq. (9.35), applied to the Einstein action, gives an expression for the divergence of the gravitational pseudo-tensor:

$$\partial_\mu t^\mu{}_\nu = -\frac{1}{2\kappa}\, \sqrt{g}\, G^{\alpha\beta}\, \partial_\nu g_{\alpha\beta} = -\frac{1}{\kappa}\, \partial_\mu(\sqrt{g}\, G^\mu{}_\nu), \tag{9.48}$$

where in the last step I used the Bianchi identity, Eq. (9.46). (The same result can be obtained by taking the divergence of Eq. (9.47) directly). Using Einstein's equations, $G_{\mu\nu} = \kappa\, T_{\mu\nu}$, we find Einstein's form of the conservation law, Eq. (9.21). The second equation of the boundary theorem, Eq. (9.36), reproduces the expression Eq. (9.24) for the pseudo-tensor in terms of the superpotential.

I will now argue that this form of the boundary theorem shows that the alleged non-uniqueness of the pseudo-tensor, i.e. problem (iv) in Section 9.1.3, arises only if one treats Eq. (9.21) independently of Noether's theorems. For if one takes Eq. (9.21) as the *definition*

69 This formulation follows Noether's (1918, p. 241) original methods, reviewed in Section 9.2. However, there is an important and useful reformulation of Noether's theorem, for diffeomorphism-invariant Lagrangians, in terms of differential forms: see Wald (1993, pp. 3428–9) and Wald and Zoupas (2000, pp. 2–3).

of the pseudo-tensor and integrates it, then the pseudo-tensor indeed admits an *arbitrary* superpotential, as in Eq. (9.24).[70]

But if one, correctly, applies Noether's theorem to the gravitational part of the action, Eq. (9.44), then both the pseudo-tensor and the superpotential are *fixed* by their Noether expressions in Eqs. (9.31) and (9.32), in terms of:

(1) the Lagrangian density,

(2) the boundary term Λ^{μ} that appears in the Noether variation of the action, Eq. (9.26).

Noether's expression Eq. (9.32) for the superpotential indeed shows that the superpotential is not arbitrary, but is fixed when (1) and (2), including the boundary terms of the action, are fixed. One may add boundary terms to the action without changing the equations of motion, and in this sense there *is* freedom to change the superpotential by changing the boundary terms. However, boundary terms often *do* change the boundary conditions, and so the boundary terms must be such that they give the desired boundary conditions. (Typically, the boundary conditions are not sufficient to uniquely determine the boundary terms; the finiteness of the on-shell action is also required to fix them. See Section 9.4.)

This means that the freedom to choose the superpotential has a physical meaning: it is the freedom to change the action, without affecting the equations of motion, by boundary terms. The latter have a physical meaning, because:

(a) they determine the boundary conditions in the variation of the metric that gives the equation of motion;

(b) they determine the on-shell value of the (Euclidean) action, which, for instance in thermodynamic approaches to gravity, especially in the context of black holes, is interpreted as a gravitational potential;[71]

(c) the boundary terms of the Hamiltonian are also determined by the boundary terms of the Lagrangian.[72] Thus, the significance of the boundary terms does not hinge on interpreting the Lagrangian physically.

Chang et al. (1999) notice that, in a Hamiltonian approach, the freedom to change the superpotential is related to the freedom to change the boundary conditions, i.e. analogous to my points (a) and (c) above. However, so far as I can see, they do not notice that it is in general *Noether's theorem* that completely fixes the superpotential from the variation of the action, in particular through Eq. (9.32).

In the absence of Killing vectors that allow us to construct the energy by integrating a scalar expression like Eq. (9.19), some authors use the pseudo-tensor to convert the covariant derivative in Eq. (9.16) into a partial derivative and integrate. Thus these authors define a total momentum vector P_{μ} as an integral over a spacelike volume V:[73]

[70] For example, Lam (2011: pp. 1017–18) *defines* the energy-momentum pseudo-tensor in terms of an expression that automatically satisfies the conservation equation Eq. (9.21), as in Eq. (9.24), and thus contains an arbitrary superpotential.

[71] See, e.g., Hawking (1979, pp. 772, 778–9). Wald (1993, p. 3431) and Hawking and Horowitz (1996, pp. 1494–5) contain reformulations of the first and second laws of black hole mechanics in terms of the Lagrangian and its variation. For some examples, see Wald and Zoupas (2000, p. 8–14) Jacobson et al. (1994, pp. 6592–3).

[72] For the relation between the Lagrangian and Hamiltonian approaches to GR, see, for example, Wald (1984: pp. 464–5)

[73] See Weinberg (1972: p. 167), Landau and Lifshitz (1971, p. 306), Chen et al. (2018a, p. 3; 2018b, p. 4). For the philosophical literature on this point, see Read (2020, p. 217) and Hoefer (2000, p. 194).

$$P_\mu := \int_V \mathrm{d}\Sigma_\nu \left(T^\nu{}_\mu + \frac{1}{\sqrt{g}}\, t^\nu{}_\mu\right). \tag{9.49}$$

By choosing special coordinates, where the normal vector to the boundary S of the volume V points along the time direction, these authors derive:

$$\frac{1}{c}\frac{\mathrm{d}P_\mu}{\mathrm{d}t} = -\int_S \mathrm{d}^2x\, N n_i\, (\sqrt{h}\, T^i{}_\mu + t^i{}_\mu), \tag{9.50}$$

where, as before, n_i is the outward-pointing unit vector normal to S, N is the lapse function, and h is the metric on S induced from g in these coordinates. This expression can be shown to reproduce the well-known ADM (for Arnowitt, Deser, and Misner) formulas for asymptotically flat spacetimes: see Arnowitt et al. (1962, p. 16), Wang (2015, p. 3), Chen et al. (2017, p. 200), Chen et al. (2018a, p. 3), and Chen et al. (2018b, p. 158).[74] This expression is regarded as an analogue of the charge conservation law, Eq. (9.14).

However, as I will discuss in Section 9.5, the above definition is unsatisfactory: for it transforms covariantly only under affine transformations, and the analogy with charge conservation depends on a very special embedding of the volume V in the spacetime and on the associated choice of coordinates.

9.4 Quasi-Local Stress-Energy Tensors in GR

It is time to discuss other proposals for the energy-momentum and angular momentum of the gravitational field. This section introduces two such. Section 9.4.1 introduces the Brown–York quasi-local stress-energy tensor, which in Section 9.4.2 is then modified and gets a holographic interpretation that, I will argue, casts light on the nature of energy and momentum in GR. In Section 9.4.3, I will discuss the transformations of the Brown–York stress-energy tensor under diffeomorphisms.

9.4.1 The Brown–York Quasi-Local Stress-Energy Tensor

At the end of Section 9.1.3, where I discussed five problems with the gravitational pseudo-tensor, I mentioned Penrose's idea of giving up local expressions for energy and momentum in GR, and to consider quasi-local notions instead. This section presents one such expression in some detail.

Brown and York's (1993) proposal is to define the quasi-local stress-energy tensor, associated with a spatially bounded region M, as the conjugate of the classical action with respect to an appropriately defined metric on the boundary of that region.

The action to consider contains six terms. There is the usual Einstein–Hilbert action, coupled to matter, and possibly including a cosmological constant term. Furthermore, since the spacetime region M is bounded, appropriate boundary terms are required in order to make the variational problem well-defined. Since ∂M consists of two spacelike hypersurfaces $\Sigma_\pm$ to the past and future of M, and the timelike three-surface N that spatially bounds

74 For a discussion of the history of the ADM mass formula, and various versions and derivations of it, see Ohanian (2010, p. 9).

M, there are three such boundary terms (where the first two are abbreviated into a single term; see Eq. (9.51)).[75] To make the variational problem of the Einstein–Hilbert action well-defined, a choice of boundary conditions for the metric on the boundary is required: Brown and York consider Dirichlet boundary conditions, i.e. keeping the metric on the boundary fixed.[76] The appropriate boundary term for Dirichlet boundary conditions is the extrinsic curvature, denoted K on the hypersurfaces $\Sigma_\pm$, and Θ on the three-surface N. Finally, a reference term is required that renders the value of the action and the stress-energy tensor well-defined (I will have more to say about this term later). Thus the action is:

$$S = \frac{1}{2\kappa}\int_M \mathrm{d}^4x\,\sqrt{-g}\,R + \underbrace{\frac{1}{\kappa}\int_{\Sigma_\pm}\mathrm{d}^3x\,\sqrt{h}\,K}_{\text{extrinsic curvature of hypersurfaces }\Sigma_\pm}\;\underbrace{-\frac{1}{\kappa}\int_N \sqrt{-\gamma}\,\Theta}_{\text{extrinsic curvature of }N} + S^{\text{ref}} + S^{\text{matter}}. \quad (9.51)$$

Here, h is the metric on the hypersurfaces $\Sigma_\pm$ embedded in M, and γ is the metric on the three-boundary N embedded in M. K and Θ are their corresponding extrinsic curvatures.

Varying this action with respect to the metric, we get:

$$\delta S = -\int_M \mathrm{d}^4x\; E^{\mu\nu}\delta g_{\mu\nu} + \frac{1}{2}\int_{\Sigma_\pm}\mathrm{d}^3x\;\sqrt{h}\;P^{ij}\,\delta h_{ij} + \frac{1}{2}\int_N \mathrm{d}^3x\;\sqrt{\gamma}\;\tau^{ij}_{\text{BY}}\,\delta\gamma_{ij}. \quad (9.52)$$

The first term contains the Einstein field equations including matter, and thus vanishes if the field equations are satisfied. The second and third terms define the momenta conjugate to the three-metrics h on $\Sigma_\pm$ and γ on N, respectively (with Dirichlet boundary conditions, as announced). τ^{ij}_{BY} is a genuine (quasi-local) energy-momentum tensor, because it contains both spacelike and timelike components. It also contains contributions from both the gravitational and the matter part of the action. The subscript 'BY'. emphasises that this is the energy-momentum tensor, as defined in the proposal by Brown and York.

Restricting for the moment to the gravitational contribution to this last term, we get the following result for the Brown–York stress-energy tensor density:

$$\tau^{ij}_{\text{BY}} = -\frac{1}{\kappa\sqrt{\gamma}}\left(\Theta\,\gamma^{ij} - \Theta^{ij}\right) + \text{(subtracted terms)}. \quad (9.53)$$

The subtracted terms subtract a potentially divergent piece, in cases where one is evaluating this tensor at infinity. These subtracted terms are obtained by subtracting to the action Eq. (9.51) the same action, evaluated on an appropriate reference spacetime. For example, if the solutions of the Einstein field equations that one considers are asymptotically flat, one usually subtracts from Eq. (9.51) the same action, but evaluated on a Minkowski solution. In this way, the 'on-shell' value of the action (i.e. the action evaluated on a particular solution) is finite, and also the Brown–York stress-energy tensor is finite. I will return to this procedure in the next section, where I will mention an alternative way to resolve this problem that does *not* appeal to the reference spacetime.

75 Brown and York's (1993, p. 1410) coordinates are such that the hypersurfaces $\Sigma_\pm$ and N are orthonormal to each other on the two-surface where they intersect. Furthermore, no additional term in the action for this two-surface is required.

76 These boundary conditions also render the gravitational part of the action well-defined in case a cosmological constant term is present.

The boundary stress-energy tensor (density) satisfies one of Einstein's equations:

$$\nabla_i \tau^{ij}_{\text{BY}} = -T^{nj}, \tag{9.54}$$

where T^{nj} is the matter energy-momentum tensor with one index projected normally, and the other tangentially, to N. This is a covariant balance equation for the quasi-local stress-energy tensor, where the contribution on the right is from the matter that passes through the boundary.

If there is no matter passing through the boundary, and if the boundary three-metric γ_{ij} possesses a Killing vector field K_i,[77] then τ^{ij} defines a conserved charge, i.e. we have the conservation law $\nabla_i(\tau^{ij}_{\text{BY}} K_j) = 0$, which is analogous to Eq. (9.17). Thus by the same manipulations as in Section 9.1.2, we derive an analogue of Eqs. (9.19) and (9.20) on the boundary.

Note that, in the Brown–York proposal, the spacetime region M can be finite and embedded in an ambient spacetime; so, the three-surface N that spatially bounds M need not be at infinity. In the next section, I will discuss the case $\Lambda < 0$, and N will be taken to be the asymptotic boundary.[78]

As any quasi-local energy-momentum must, the Brown–York expression agrees with the ADM energy at infinity (see Section 9.3) for the special case of asymptotically flat spacetimes (Brown and York 1993, p. 1408).

9.4.2 The Holographic Stress-Energy Tensor

Consider the case of pure gravity with a negative cosmological constant, $\Lambda < 0$, i.e. no matter (the Brown–York stress-energy tensor can be equally well be defined in the case $\Lambda > 0$).[79] According to AdS-CFT, the gravitational theory in the spacetime region M with boundary ∂M is dual to a quantum field theory on the boundary, ∂M.[80] As we will see, one can use gauge-gravity duality to reinterpret the Brown–York stress-energy tensor, evaluated on this asymptotic boundary, in terms of the dual CFT.

Since we are interested in the Brown–York stress-energy tensor that is claimed to represent the energy and stress of the spacetime M, it will be useful to briefly mention how it fits within gauge-gravity duality.[81]

[77] In the context of spaces with a cosmological constant, this condition can be weakened for K_i to be only a *conformal* Killing vector, i.e. one that satisfies Eq. (9.63).

[78] For studies of the Brown–York stress-energy tensor at surfaces that are not asymptotic boundaries, see, e.g., De Haro et al. (2001a, p. 3174). For gauge-gravity dualities (both AdS-CFT and dS-CFT) with finite boundaries, see Harlow and Stanford (2011, pp. 4–5) and Heemskerk and Polchinski (2011, pp. 2–6).

[79] See Balasubramanian et al. (2001, p. 3) and Anninos et al. (2011, pp. 4–5). Quite independent of quantum gravity or holographic ideas, the Brown–York energy-momentum tensor has been used to define notions of mass and momentum in spaces with a positive cosmological constant: see Balasubramanian et al. (2001, p. 4). The asymptotic diffeomorphism group at future null infinity of asymptotically de Sitter spaces, and the conservation equations for the associated charges, are considered in Anninos et al. (2011, pp. 3-6) and Strominger (2001, pp. 4–5).

[80] For a general introduction to gauge-gravity dualities, see Ammon and Erdmenger (2015). For a conceptual introduction, see De Haro et al. (2016). For a sample of the recent philosophical literature on dualities, see for example: Rickles (2011), Dieks et al. (2015), Read (2016b), Huggett and Wüthrich (2020), De Haro (2020), Read and Møller-Nielsen (2020), and Butterfield (2021).

[81] In what follows, I will restrict to the case of a spacetime with a Euclidean signature, so that M has a single boundary, N, with the topology of a three-sphere. For the case with Lorentzian signature, see Skenderis and van Rees (2008, p. 2; 2009: p. 6).

Following the recent literature,[82] a duality is, roughly, an isomorphism between two theory formulations. A theory can be construed as a triple of states, quantities, and dynamics. The statement of duality says that the sets of states, quantities, and dynamics of the two theories are isomorphic.

On the gravity side of the duality, a distinguished quantity should be the stress-energy tensor of the spacetime region M. The isomorphism should relate this to an appropriate quantity in the dual CFT. The question is: which stress-energy tensor has a dual in the CFT? If the Brown–York stress-energy tensor is to be a candidate for the energy and stress in the spacetime, then it ought to be related by the duality to a quantity in the CFT that has those properties. In fact, gauge-gravity duality gives us the following correspondence:

The Brown–York stress-energy tensor, defined by the variation of the action (evaluated on the equations of motion) with respect to a boundary three-metric on N, and with appropriate boundary terms that render the action finite, is dual to the expectation value of the stress-energy tensor of the CFT defined on the background of that metric.[83]

This correspondence follows from the identification of the gravitational action, Eq. (9.51), with a given set of matter fields, evaluated on a given solution of the Einstein field equations, with the generating functional of (connected) correlation functions in the dual CFT, usually called W_{CFT}:

$$S[\gamma, \cdots]|_{\text{EFE}} \equiv W_{\text{CFT}}[\gamma, \cdots]. \tag{9.55}$$

The left-hand side of this equation is the action, Eq. (9.51), evaluated for a metric and set of matter fields that satisfy Einstein's field equations (hence the subscript 'EFE'). (Physicists call this an 'on-shell' action.) γ is the metric on ∂M, and the dots indicate that there is also a dependence on the boundary values of the matter fields. The right-hand side of this equation is the generating functional of connected correlation functions, which is related to the CFT's partition function by: $Z_{\text{CFT}} = \exp(-W_{\text{CFT}})$. Again, this generating functional depends on the metric γ, which is now viewed as a curved background on which the CFT is defined, and other sources that the theory is defined with (these are dual to the boundary values of the matter fields on the left).

On the right-hand side, the expectation value of the stress-energy tensor of the CFT is defined by the functional derivative with respect to the metric:

$$\langle T_{ij}(x)\rangle_{\text{CFT}} = \frac{2}{\sqrt{\gamma}}\frac{\delta W_{\text{CFT}}}{\delta\gamma^{ij}(x)}. \tag{9.56}$$

But substituting Eq. (9.55) into Eq. (9.56), and comparing with Eq. (9.52), we see that we indeed have the following correspondence:

$$\tau'^{\,ij}_{\text{BY}} \equiv \langle T^{ij}\rangle_{\text{CFT}}. \tag{9.57}$$

The dash on the Brown–York stress-energy tensor indicates that this tensor is modified, by replacing the 'subtracted terms' in Eq. (9.53) by local counterterms added to the action,

[82] I here follow De Haro and Butterfield (2018, pp. 335–9), De Haro (2020, pp. 264–6), and Butterfield (2021, p. 12).

[83] I here omit technical details, such as that the boundary metric is defined only up to a conformal transformation. This point will resurface in Section 9.4.3.

according to a standard method in holographic renormalisation.[84] In what follows, I will call $\langle T^{ij}\rangle_{\text{CFT}}$ 'holographic stress-energy tensor'.

The holographic stress-energy tensor can be calculated from Eq. (9.53) for any given solution of Einstein's equations.[85]

The gravitational equation, Eq. (9.54), can be reproduced from the CFT perspective. For example, consider a two-dimensional CFT with a single (scalar) operator $\mathcal{O}$ which has a non-zero expectation value $\langle\mathcal{O}\rangle$, coupling to a source J, the invariance of the CFT partition function under infinitesimal changes of coordinates,

$$\delta\gamma_{ij} = \nabla_i\xi_j + \nabla_j\xi_i, \tag{9.58}$$

yields the following identity:

$$\nabla^j\langle T_{ij}\rangle = \langle\mathcal{O}\rangle\,\partial_i J. \tag{9.59}$$

One can check that this result is reproduced, on the gravity side with a three-dimensional spacetime, by the identity of Brown and York, Eq. (9.54), since, given the form of the normal energy-momentum tensor T^{nj} for the dual scalar field, the above indeed reproduces Eq. (9.54).[86]

Noether's theorem in connection with asymptotic symmetries is discussed in Papadimitriou and Skenderis (2005, pp. 15–19), who show, for various types of fields, that the holographic charges are the Noether charges.

Implications of gauge-gravity duality for the gravitational energy. Let us briefly discuss how these results bear on the problem of energy conservation. Gauge-gravity dualities give independent support to the claim that the modified Brown–York stress-energy tensor represents the quasi-local energy. We will consider the two main features of the holographic stress-energy tensor of the dual quantum field theory that bear on the status of the quasi-local energy in GR:

(1) The duality map, Eq. (9.57), relates the Brown–York stress-energy tensor to the holographic stress-energy tensor. This gives further support to the interpretation of the Brown–York stress-energy tensor as representing quasi-local stress and energy. For recall that a duality is an isomorphism of models of a single common core theory.[87] The interpretation of two dual models need not be the same, but if it *is* the same,[88] then the two dual quantities on either side of Eq. (9.57) receive the same interpretation. Thus the right-hand side of this equation, i.e. the holographic stress-energy tensor, can give us guidance about the interpretation of the side under discussion, i.e. the gravitational, left-hand, side. And the

[84] See Henningson and Skenderis (1998, p. 5), Balasubramanian and Kraus (1999, p. 415), De Haro et al. (2001b, p. 603). For a review of holographic renormalisation, see Skenderis (2002).

[85] In general this is tedious, but there is a systematic procedure to do this calculation. See De Haro et al. (2001b, pp. 602–7, 609–12), Skenderis (2002, pp. 5855–61). For an alternative method, see Papadimitriou and Skenderis (2004, pp. 14–18).

[86] See De Haro et al. (2001b, p. 615).

[87] This 'common core' view is discussed and/or endorsed by, e.g., Rickles (2011, p. 66), Dieks et al. (2015, p. 209), De Haro et al. (2017a, p. 116), Huggett (2017, p. 84), De Haro and Butterfield (2018, p. 321) and Le Bihan and Read (2018, p. 3).

[88] That it *is* the same depends on our developing an *internal interpretation* of the dual models: for a discussion, and conditions under which this is possible, see Dieks et al. (2015, p. 209), De Haro (2020, pp. 267–71), and Read and Møller-Nielsen (2020, pp. 284–8).

interpretation of the left-hand side is indeed as energy, momentum, and stress. This is a further argument to the effect that the Brown–York stress-energy tensor represents energy, momentum, and stress.[89]

Thus, in AdS-CFT, the modified Brown–York stress-energy tensor in spacetime with $\Lambda < 0$ is *precisely* dual to the quantity that, in the CFT, is interpreted as describing energy and momentum: namely, the expectation value of the stress-energy tensor. The match is a consequence of the general AdS-CFT correspondence: namely, the identification of the gravitational action, evaluated on solutions, with the generating functional of the CFT, Eq. (9.55), and of the conformal metric on the boundary, γ_{ij}, with the metric of the dual quantum field theory. This precise correspondence, which has been illustrated in many examples, gives additional support to the statement that the Brown–York stress-energy tensor *is* indeed to be thought of as energy and momentum, since it plays that role in the boundary theory.

(2) The boundary conservation equation, Eq. (9.54), has a similar dual interpretation, and is not analogus to the usual (disputed) conservation laws of GR.[90] This boundary conservation equation is dual to Eq. (9.59), so it can be interpreted as a balance equation for the energy and stress of the common core theory, and this again strengthens that interpretation in GR.[91]

9.4.3 Diffeomorphisms and Redundancy of the Description

In our discussion, in Section 9.5, of the two main proposals for defining energy and momentum in general relativity, diffeomorphism invariance will play a crucial role. Therefore, we should understand the transformation properties of the Brown–York stress-energy under various diffeomorphisms: namely, those that are descriptive redundancies and those that are not.[92]

The asymptotic symmetries of gravity have been a central foundational topic in general relativity since at least the work of Arnowitt et al. (1959, 1962), Bondi et al. (1962), Penrose (1963, 1964), Newman and Penrose (1966), Geroch (1972), and Ashtekar and Hansen (1978), among others. For a philosophical discussion of some aspects of asymptotic symmetries, see Belot (2018).

[89] Note that the internal interpretation of the common core theory, as currently worked out, does not support the interpretation of this energy as *gravitational* energy – even though this *can* be done on an external interpretation (see footnote 91). For understanding the nature of the common degrees of freedom depends on having a good development of this common core theory, which we do not currently have. For a more detailed discussion, see De Haro (2020, pp. 275–9).

[90] There are three important disanalogies between the boundary balance equation, Eq. (9.54), and the (disputed) conservation law, Eqs. (9.16)–(9.21), which are relevant to points (i)–(iv) from Section 9.1.3. First, Eq. (9.54) distinguishes between the quasi-local gravitational energy, τ_{BY}, and matter energy-momentum, T^{nj}. Second, the covariant derivative is calculated using the boundary, metric-compatible, connection, which is non-dynamical, and fixed by the boundary conditions. Third, the right-hand side is in general non-zero, so that this is a balance equation. I will address points (iii) and (v) from Section 9.1.3 in Section 9.6.2

[91] An internal interpretation (as in 'Eq. (9.54) is the energy balance equation') is usually *compatible* with a judiciously chosen external interpretation (as in 'Eq. (9.54) is the balance equation for the *gravitational* energy'), through the process of abstraction outlined in De Haro (2021, section 4.4). This is why, if the internal interpretation is correct, it bears on the interpretation on GR.

[92] My discussion here essentially summarises De Haro, Teh, and Butterfield (2015, 2017) and De Haro (2017b). For physics references on this topic, see Imbimbo et al. (2000, pp. 1130–2), Skenderis (2001, pp. 742–7).

A central question is which asymptotic diffeomorphisms are true symmetries of a spacetime, which asymptotic diffeomorphisms act non-trivially on the physical degrees of freedom of general relativity, and how these different kinds of diffeomorphisms are to be characterised.

In spaces with a negative cosmological constant, Brown and Henneaux (1986) discovered that the asymptotic diffeomorphism of anti-de Sitter space is the conformal group of the asymptotic boundary.[93] As we will see in this section, this asymptotic symmetry group acts non-trivially on the quasi-local energy-momentum tensor.

Thus the question of which subgroup of the diffeomorphism group is a symmetry of a spacetime is important from the point of view of quasi-local energy and momentum: for the quasi-local quantities will be either invariant or else have a well-defined transformation law under the diffeomorphisms that are true asymptotic symmetries of a spacetime, and they transform non-trivially under diffeomorphisms that are not true symmetries.

What do we mean by the 'true asymptotic symmetries' of a spacetime? What we are looking for here is a criterion that distinguishes the symmetries (diffeomorphisms) that are true redundancies of a description, from those that entail a physical change in the solution. We can define redundancies as follows (De Haro et al. 2017a, p. 69):[94]

(Redundant): If a physical theory's formulation is redundant (i.e. roughly: it uses more variables than the number of degrees of freedom of the system being described), one can often think of this in terms of an equivalence relation, 'physical equivalence', on its states; so that gauge-invariant quantities are constant on an equivalence class, and gauge-symmetries are maps leaving each class (called a 'gauge-orbit') invariant. Leibniz's criticism of Newtonian mechanics provides a putative example: he believed that shifting the entire material contents of the universe by one meter must be regarded as changing only its description, and not is physical state.

I will introduce the notion of the 'invisibility' of diffeomorphisms to make precise the idea of the asymptotic symmetries that are (Redundant). I will distinguish three relevant classes of diffeomorphisms that qualify differently under redundancy: 'invisible' diffeomorphisms are always (Redundant), 'large' diffeomorphisms are never (Redundant), and the intermediate category of 'visible' diffeomorphisms can sometimes be (Redundant) and sometimes not, depending on the interpretation that the theory is given.[95] The three classes

93 The analysis by Brown and Henneaux has been extended to the case $\Lambda > 0$, including in higher dimensions: see Strominger (2001, pp. 4–5) and Balasubramanian et al. (2001, pp. 7–8). For a more detailed analysis of diffeomorphisms in the case $\Lambda > 0$, see Anninos et al. (2011, pp. 3–4). As we will see in this section, these results can be generalised to any space with $\Lambda < 0$ that is a solution of Einstein's equations, with arbitrary metric at infinity.

94 There is an ongoing debate, following Greaves and Wallace (2014) – which itself argues against earlier work: see footnote 35 – whether local gauge symmetries express mere redundancies of the description, or whether they can be empirically significant, as in Galileo's ship scenarios: see, for example, Friederich (2015, p. 537), Murgueitio Ramírez (2020, p. 1), Murgueitio Ramírez and Teh (2020, p. 3), Gomes (2020, p. 11), and Gomes (2020, pp. 3–5, 21). The approach I adopt here is, in effect, to construct a class of diffeomorphisms (the 'visible' diffeomorphisms below) that are GR-analogues of the class of gauge symmetries that can have a direct empirical equivalence in gauge theories. The latter are called 'non-interior, boundary-preserving symmetries' by Greaves and Wallace (2014, p. 71), and their definition is further refined by Teh (2016, p. 116).

95 Belot (2018, pp. 964–7) also distinguishes local translations that are redundant, from those that are physical. He also connects this to the question of boundary conditions – although he does not appear to consider my intermediate category of 'visible diffeomorphisms'.

are distinguished by the different boundary geometric structures that they preserve.[96] They are as follows:

(1) *Invisible diffeomorphisms*: these are diffeomorphisms under which the bulk action, Eq. (9.55), evaluated on a given solution, is invariant. Also, the other quantities derived from the action, like the holographic stress-energy tensor Eq. (9.57), are invariant. These diffeomorphisms fix the boundary manifold and the conformal metric.[97]

These diffeomorphisms count as (Redundant) symmetries because all of the physical quantities, for a given solution of the equations of motion, are invariant under them. In particular, these diffeomorphisms allow us to set up an analogue of the hole argument in a space with a negative cosmological constant, which in turn leads to the conclusion that the diffeomorphisms in question are (Redundant).[98]

(2) *Visible diffeomorphisms*: these are diffeomorphisms that generate conformal transformations of the boundary metric, i.e. $\gamma_{ij} \mapsto \tilde{\gamma}_{ij} := e^{-2\xi(x)}\gamma_{ij}$. The stress-energy tensor transforms by a conformal factor under them, and in the case of even boundary dimensions the stress-energy tensor (and the action) pick up an additional, anomalous term under them:

$$\langle T_{ij}[\tilde{\gamma}](x)\rangle = e^{(d-2)\,\xi(x)}\left(\langle T_{ij}[\gamma](x)\rangle + \frac{1}{\sqrt{\gamma}}\frac{\delta\mathcal{A}[\gamma]}{\delta\gamma^{ij}(x)}\right), \tag{9.60}$$

where γ_{ij} is the metric induced on the boundary, d is the dimension of the boundary, and $\mathcal{A}[\gamma]$ is the anomalous term, which can be calculated from the action.[99] It is a local term written in terms of curvature invariants (for the explicit expressions, see Deser and Schwimmer 1993; Henningson and Skenderis 1998: section 3; De Haro et al. 2001b, section 4.2.1). The anomaly vanishes when $d = 3$, i.e. for a four-dimensional spacetime.[100] Since the anomaly term vanishes in the case of a four-dimensional spacetime, which is the case of most interest, I will not discuss it further.[101]

Visible diffeomorphisms are potentially physical symmetries, rather than redundancies of the description: namely, they act as conformal transformations of the boundary quantities. Whether they are physical depends on one's interpretation of the spacetime within one's theory.[102]

96 Class (1) is the class of 'QFT-invisible' diffeomorphisms in De Haro (2017b, pp. 1468, 1484), also called '(simple) gravity-invisible'. Class (2) is the class of 'QFT-visible' diffeomorphisms, also called 'weakly gravity-invisible' diffeomorphisms'. For the reasons why these classes are cases of (Redundant) or non-(Redundant), see De Haro, Teh, and Butterfield (2015, 2017) and De Haro (2017b).

97 It is sometimes said that the diffeomorphisms that are redundancies of the theory are those that 'go to unity' at the boundary (this being the class by which the allowed diffeomorphisms have to be quotiented in order to obtain the asymptotic symmetry group). But this is imprecise, because a diffeormorphism can go to unity at the boundary while still inducing a non-trivial transformation of the boundary metric γ_{ij}, through its dependence on the parametrisation of the 'radial direction' r, which results in a rescaling of the boundary metric, and hence a diffeomorphism of type (2) below. Thus, going to unity at the boundary is not sufficient for (Redundant). Instead, the definition of invisible diffeomorphisms makes precise the sense in which diffeomorphisms are (Redundant).

98 See De Haro et al. (2017a, p. 79).

99 To simplify the discussion, I do not distinguish between conformal transformations and Weyl rescalings of the metric. For details, see De Haro (2017b, p. 1479).

100 The anomaly can be understood as coming from the divergence of the gravitational action, Eq. (9.51), at large distances. To cancel this divergence, the action is regularised and boundary counterterms are introduced, but this renormalisation procedure breaks covariance, albeit in a controlled way, i.e. through the additional term Eq. (9.60).

101 For spacetimes of odd dimension, the anomaly is proportional to powers of the *boundary* curvature that depend on the dimension, and so it is also zero if the boundary metric is flat.

102 In De Haro et al. (2017a, p. 79), it was argued that, on an 'internal' interpretation (typically, in the context of a theory of the whole universe), two states related by visible diffeomorphisms are physically equivalent, so that the transformations count as

(3) *Large diffeomorphisms*: these are diffeomorphisms that do not preserve sufficient boundary structure, and so they map solutions of Einstein's equations to *physically inequivalent* solutions of the equations (e.g. equations with a different total mass or angular momentum). Such diffeomorphisms are therefore not (Redundant), but physical transformations that relate different states. In particular, they will change the value of the holographic stress-energy tensor.[103]

An analysis of the diffeomorphisms as detailed as the preceding one has, so far as I am aware, not been done for $\Lambda > 0$, and so one should proceed with some caution. However, much work has been done on this case as well: see Anninos et al. (2011, pp. 3–6), Strominger (2001, pp. 4–5), and Balasubramanian et al. (2001, pp. 7–8).

There are also a few general things one can say. As noted in Skenderis (2002, appendix A) and De Haro (2017b, pp. 1490–1), the geometric techniques that enabled the analysis summarised in Section 9.4.3 and just above, apply in the case $\Lambda > 0$ as well, due to the fact that the differential form of Einstein's equations in the radial variable was turned into a set of coupled algebraic relations. The results depend only on the asymptotic solutions, and can be continued from one case to the other, where the radial coordinate r goes into a timelike coordinate t (the anti-de Sitter boundary is thereby mapped to the timelike future infinity in de Sitter space).

To summarise, in case (1) we have a hole-type argument that leads us to conclude that invisible diffeomorphisms correspond to redundancies of the theory. In case (3), such an argument is not available, because the diffeomorphisms in question change the physical properties of the spacetime, and we should not characterise spacetimes related by a large diffeomorphism as physically equivalent. Case (2) is an intermediate case, where the diffeomorphisms act as conformal transformations on the boundary quantities, and their nature as (Redundant) or as non-(Redundant) depends on one's interpretation (including further use) of the solutions, say, as being cosmological models of the *whole* universe, or as being merely parts of the universe.

Let me say a bit more about this last case, (2), of visible diffeomorphisms, and how they give rise to transformations of the conformal boundary: namely, conformal transformations. Choose an open neighbourhood of the boundary, such that the boundary of M is the surface $r = 0$. A theorem of Fefferman and Graham (1985, 2012) guarantees that in such an open neighbourhood of the boundary, we can write the metric in *Poincaré normal form*:

$$\mathrm{d}s^2 = \frac{\ell^2}{r^2}\left(\mathrm{d}r^2 + \gamma_{ij}(r,x)\,\mathrm{d}x^i\,\mathrm{d}x^j\right), \tag{9.61}$$

where x^i are coordinates tangent to the boundary directions, and $\gamma_{ij}(r,x)$ is a metric along this direction, that in general depends on both r and x. We obtain the earlier-used γ_{ij} by

(Redundant) gauge symmetries. On the other hand, on an 'external interpretation' (as, e.g., in a Galileo's ship scenario, where the inside of the ship is not the whole world, but can be related to other parts of the world, e.g. the shore) these diffeomorphisms are not (Redundant), but express 'the motion of the ship relative to the quay'.

103 For example, Grumiller and Riegler (2016, eq. (3.6)) report a metric that is claimed to be physically inequivalent to the corresponding metric in normal form.

setting $r = 0$, that is, $\gamma_{ij}(x) := \gamma_{ij}(0, x)$. The diffeomorphisms we are interested in are those that 'fix the location of the boundary $r = 0$'. They take the following form:

$$\delta x^i = \xi^i(x)\,, \quad \delta r = -r\,\xi(x). \tag{9.62}$$

Requiring that these diffeomorphisms satisfy the definition of 'being visible', i.e. fix: (i) the Poincaré normal form of the metric, i.e. Eq. (9.61) and (ii) the boundary metric, implies the following condition on the diffeomorphism generated by the boundary vector field ξ:[104]

$$\nabla_i\,\xi_j(x) + \nabla_j\,\xi_i(x) = \frac{2}{d}\,\gamma_{ij}(x)\,\nabla^k\,\xi_k(x), \tag{9.63}$$

where the covariant derivatives are with respect to the boundary metric $\gamma_{ij}(x)$. This equation is the *conformal Killing equation*, which is precisely the condition for the vector field ξ to be a conformal transformation, rather than an isometry, of the boundary metric $\gamma_{ij}(x)$.[105] This gives us a very general relation between the symmetries of the metric in M, in Poincaré normal form Eq. (9.61), and the conformal group on ∂M:

The asymptotic symmetry group of the Poincaré normal form of the metric, Eq. (9.61), on the manifold M, is the conformal group on the asymptotic boundary, ∂M, with boundary metric γ.

This generalises the result by Brown and Henneaux (1986), for any solution of Einstein's equations with $\Lambda < 0$.

These diffeomorphisms give us a conformal Killing vector ξ of the boundary manifold. As I noticed after Eq. (9.54), especially in footnote 77, using the properties of the holographic stress-energy tensor, Eq. (9.56), this can be used to derive a boundary analogue of the energy conservation equation Eq. (9.20).

9.5 Gravitational Energy and Momentum, Revisited

In this section, I first review two recent discussions of pseudo-tensors that I aim to justify: one is from physics (Section 9.5.1) and the other is philosophical (Section 9.5.2). I will agree with many of these defences of the pseudo-tensor approach – but I will nevertheless, in Section 9.6, argue for the superiority, in the present state of knowledge, of the Brown–York approach.

9.5.1 Superpotentials Are Boundary Terms in the Hamiltonian of GR

In recent work, Chang, Chen, Nester, and collaborators (Chang et al. 1999; Chen et al. 2017, 2018a, 2018b) have defended a rehabilitation of pseudo-tensors that thereby casts light on the physical nature of pseudo-tensors. Their work can be seen as arguing for a resolution of problem (iv) in Section 9.1.3. Namely, they argue that pseudo-tensors:

(a) *Do provide a description of energy and momentum conservation.* They do this, in particular, by establishing a relationship between pseudo-tensors and the Hamiltonian of

104 See Brown and Henneaux (1986) and De Haro (2017b, p. 1477).

105 The conformal Killing equation reproduces the ordinary Killing equation, Eq. (9.15), in the special case $\nabla^k\xi_k = 0$.

general relativity. The superpotential, $U_\lambda^{\mu\nu}$, can be identified with a total divergence in the Hamiltonian of general relativity, and so different choices of the superpotential correspond to different boundary terms/conditions. Thus, problem (iv) is thereby resolved, because fixing the boundary terms determines the pseudo-tensor uniquely.

(b) *Give, for appropriate choices of boundary conditions, the expected values of the total energy and momentum as measured at infinity,* i.e. the ADM and Bondi formulas,[106] thus further strengthening the contention, in the resolution of problem (iv), that pseudo-tensors are physical.

(c) *Have well-defined values in each system of coordinates.* In particular, the pseudo-tensors give quasi-local quantities, which are naturally identified with the two-surfaces that bound regions of space, and give correct results at infinity, as in (b).

Here I will focus on (a), which is the most distinctive part of the approach by Chang, Chen, Nester, and collaborators. Then I will more briefly comment on (b) and (c). Let me state, in words, the main statement that goes into arguing for (a):

Pseudo-tensor–Hamiltonian correspondence: *The quasi-local energy-momentum over a hypersurface Σ with constant vector N, associated with the pseudo-tensor $t^\mu{}_\nu$ and superpotential $U_\mu^{\nu\lambda}$, corresponds to a Hamiltonian $H(\Sigma, N)$ whose boundary term is given by the expression $B(N)$ (and vice versa). The choice of the superpotential $U_\mu^{\nu\lambda}$ in the pseudo-tensor determines a boundary term in the Hamiltonian (and vice versa).*

Working this out in the forward direction, i.e. showing that the quasi-local energy-momentum gives rise to the Hamiltonian in the way stated, goes in three main steps:

(1) Follow Einstein's derivation of the pseudo-tensor $t^\mu{}_\nu$, in Section 9.1.3, and express it in terms of the superpotential $U_\nu^{\mu\lambda}$, in Eq. (9.24).

(2) 'Integrate Einstein's derivation' to get a pseudo-tensor, analogously to the steps in Eqs. (9.16)–(9.18). Here, Chen, Nester, et al. construct an energy-momentum quantity $P(N)$ by integrating the total local energy-momentum density (including its matter and gravitational contributions) over a three-dimensional spacelike hypersurface Σ. Upon the use of the Einstein field equations, the result is a boundary term, i.e. an integral over $\partial\Sigma$. Thus, in this step they associate to each pseudo-tensor, and solution of Einstein's field equations, a quasi-local quantity – and this is their preferred formulation of the problem.

The details are as follows. They define energy and momentum by integrating Eq. (9.21) with an arbitrary vector N:

$$
\begin{aligned}
P(\Sigma, N) &:= \int_\Sigma \mathrm{d}\Sigma_\nu \, (T^\nu{}_\mu + t^\nu{}_\mu) N^\mu \\
&= \frac{1}{\kappa} \int_\Sigma \mathrm{d}\Sigma_\nu \left[(-G^\nu{}_\mu + \kappa \, T^\nu{}_\mu) N^\mu + \frac{1}{2} \, \partial_\lambda (N^\mu \, U_\mu^{\nu\lambda}) \right] \\
&= \int_\Sigma \mathcal{H}_\mu \, N^\mu + \oint_{\partial\Sigma} B(N), \qquad (9.64)
\end{aligned}
$$

[106] See, for example, Jaramillo and Gourgoulhon (2011, pp. 97–102, 105–7) and Christodoulou and Yau (1988, pp. 9–10).

where I have introduced the notation $\mathcal{H}_\mu = \mathrm{d}\Sigma_\nu(\frac{1}{\kappa}\, G_\mu{}^\nu - T_\mu{}^\nu)$ and $B(N) = -\frac{1}{4\kappa}\,\mathrm{d}S_{\nu\lambda}\, N^\mu\, U_\mu^{\nu\lambda}$ for the first and second terms, respectively, in the last line. Going from the first to the second line uses Eq. (9.24), and going from the second to the third line uses Gauss' law to rewrite the total derivative, integrated over the hypersurface Σ, as an integral over the boundary $\partial\Sigma$.

In this derivation, N is taken to be an arbitrary vector that is *constant* with respect to the given system of coordinates. Such a vector always exists, although it will be different for each system of coordinates. This will be an important point of discussion in Section 9.6, in connection with point (c) above.[107] For the moment, I will set this aside.

As announced earlier, and as we see from the middle expression in Eq. (9.64), the volume term in the Hamiltonian vanishes when Einstein's field equations are satisfied, and we are left only with the boundary term $B(N)$, integrated over a two-surface. Thus we have succeeded in constructing, from the local energy and momentum, a quasi-local quantity.[108]

(3) Recognise that the expression Eq. (9.64), which is already written in suggestive notation, takes exactly the form of the Hamiltonian of general relativity, associated with a finite region Σ and constant vector N:

$$P(\Sigma, N) = -H(\Sigma, N). \tag{9.65}$$

The Hamiltonian is the generator of spacetime displacements of the hypersurface Σ along a vector field N^μ, and $\mathcal{H}_\mu$ indeed has the form in Eq. (9.64), and vanishes if the field equations are satisfied.

$B(N)$ is the boundary term of the Hamiltonian, which gives the *quasi-local energy and momentum,* i.e. it is an expression for the energy and momentum on the two-surface $\partial\Sigma$. This boundary term, as it appears in the above calculation starting from the pseudo-tensor, contains the *superpotential.* Given this correspondence between superpotentials and boundary terms, the Hamiltonian approach casts light on the physical significance of the various choices of superpotentials. In particular, the variation of the boundary term, $\delta B(N)$, must be set to zero in order to get the Einstein field equations in their Hamiltonian form. Filling in various choices of superpotential $U_\mu^{\nu\lambda}$ in this variation, one finds that each choice of superpotential leads to a specific class of boundary conditions. For example, the Freud expression, Eq. (9.23), corresponds to keeping $\sqrt{g}\, g^{\mu\nu}$ fixed at the boundary,[109] while keeping the vector field N constant. Thus, the Freud superpotential allows Dirichet boundary conditions.

107 Nester (2020, personal communication) argues that, in Eq. (9.64), 'the volume integrand is proportional to the field equations, and so vanishes on-shell. The value of the integral comes only from the two-surface boundary term. This means that we can extend our formula, modifying the vector field N within the region, without changing the value of the integral, as long as one keeps N constant on the boundary. Thus it is only necessary for the vector field to match a translational Killing field of the Minkowski coordinate reference geometry on the boundary'. Although this is of course true, the independence of the value of the vector field N in the region inside the volume V is only valid in the last two lines of Eq. (9.64), i.e. for the Hamiltonian expression. However, the step from the first to the second line of Eq. (9.64) does *not* hold if N is not constant, because the second term is no longer a total derivative. In other words, the above direct relation between the integrated total energy-momentum pseudo-tensor (i.e. energy-momentum plus gravitational pseudo-tensor) and the Hamiltonian expression holds only if N is everywhere constant in a system of coordinates adapted to the boundary. (Nester's main interest is of course the Hamiltonian formalism, and less so the pseudo-tensor. But, in so far as these papers aim to *justify* the use of pseudo-tensors through the Hamiltonian approach, my point is that this works only for *constant* vector fields N and special systems of coordinates, as just argued.)

108 This is true at least formally. See Section 9.6 for a discussion.

109 The condition is $\delta(\sqrt{g}\, g^{\mu\nu}) = 0$, which gives $\delta g^{\mu\nu} = \frac{1}{2}\, g^{\mu\nu}\, g_{\alpha\beta}\, \delta g^{\alpha\beta}$. This *allows,* but does not require, the Dirichlet boundary condition $\delta g^{\mu\nu} = 0$, and thus it is slightly more general than a Dirichlet boundary condition.

Other choices of superpotential lead to other boundary conditions, e.g. Neumann boundary conditions where the momentum conjugate to the metric is held fixed. For each choice of superpotential, there is a corresponding boundary condition, and a quasi-local energy expression.

To show that the pseudo-tensor–Hamiltonian correspondence also works in the opposite direction, one reverses the above steps: begin with the Hamiltonian, and show how its boundary terms correspond to various superpotentials in the energy-momentum pseudo-tensor.

The expression Eq. (9.64) for the Hamiltonian, and in particular the various choices of its boundary term, can be motivated further by using the Hamiltonian formalism. To do this, Chen, Nester et al. construct the Hamiltonian of general relativity in form notation. In this formalism, the various boundary terms $B(N)$, that naturally enforce Dirichlet or Neumann boundary conditions in the variation, take simple forms in terms of three geometrical quantities and their corresponding conjugate momenta: namely, the metric, and the orthonormal frame and coframe fields. These relatively simple forms of the boundary terms reproduce, when worked out in tensor components, the various expressions for the superpotential found in the literature. Thus, the form notation for the Hamiltonian explains the relatively long tensor formulas for the superpotential through simple boundary terms.

9.5.2 Recent Philosophical Discussions of Energy Conservation in GR and Pseudo-Tensors

The philosophical discussion of energy conservation in general relativity revolves around two main problems, mainly to do with problems (v) (lack of covariance, and lack of appropriate transformation properties) and (iii) (non-integrability) in Section 9.1.3. (Surprisingly, Lam (2011, p. 1017) and Read (2020, p. 215) mention the problem (iv) of non-uniqueness, which in my view is pressing, only in passing; other papers do not appear to mention it.[110]) The first is about coordinate-dependence; the second is about background-dependence. I give them new labels, (A) and (B), to cover other aspects of recent discussions:

(A) *Lack of covariance, and lack of appropriate transformation properties.* The known expressions for the local energy and momentum in general relativity are not tensorial. In particular, the pseudo-tensors, Eq. (9.24), are coordinate-dependent, and so they do not have good local (or global) definitions. (As I mentioned in Section 9.1.3, pseudo-tensors are only invariant under affine transformations.)

For example, Hoefer (2000, pp. 194–5) has stressed that the expressions for energy and momentum, even in their integrated form, Eq. (9.50), are coordinate-dependent, and so are ill-defined – as I already mentioned at the end of Section 9.3.

One immediate comment to make about (A) is that diffeomorphism invariance, rather than covariance, is most important here: and that, as we saw in Section 9.4.3, the various kinds of diffeomorphisms require a different treatment. Second, although the lack of

[110] Dürr (2019b, pp. 2, 6, 9–10) does stress the importance of this ambiguity, which he calls 'baneful' (p. 9).

covariance can be technically and physically inconvenient, the main worry is the pseudo-tensor's lack of appropriate transformation properties.

(B) *'Background-dependence'*. The conservation of energy and momentum in general relativity requires the introduction of 'background structures': such as a flat metric (usually, at infinity) or timelike Killing vectors in the spacetime (see Eq. (9.15) and the discussion following).[111] Without such additional structures, the existing expressions for the energy-momentum tensors do not express genuine conservation laws and do not give unique answers for the physical energy and momentum. While both Hoefer (2000) and Lam (2011) admit that expressions like Eq. (9.50) do yield unambiguous results when such structures are added, they ultimately reject them, because they amount to introducing undesired 'background structures'.[112]

For example, Hoefer (2000, p. 194) criticises the calculations of the energy-momentum conservation equation, Eq. (9.50), that evaluate integrals at infinity to get the total energy and mass of the spacetime, because they assume that the spacetime is asymptotically flat and that the coordinate system at infinity is Minkowskian. This does not give a genuine conservation principle for general relativity, the most important reason being that fixing the asymptotics in this way 'goes against the most important and philosophically progressive approach to spacetime physics: that of downplaying coordinate-dependent notions and effects, and stressing the intrinsic, covariant and coordinate-independent as what is important'. Thus, points (A) and (B) are, for Hoefer, really part of a *single* worry: namely, coordinate-dependence.[113]

Lam (2011, p. 1010) is more tolerant about coordinate-dependence: he concedes that 'there does not seem to be any contradiction' in relying on 'background-dependent (coordinate-dependent) entities'. However, he worries that there is no unique definition of pseudo-tensors, because the possibility to add a superpotential to them means that 'one cannot say how much energy there is in a given spatial region', i.e. problem (iv), non-uniqueness.[114]

But his main worry is about the introduction of 'background structures'. For example, although the standard integral formulation of the conservation equation with a Killing vector, Eq. (9.19), is a genuine energy-momentum conservation law for the matter fields, it requires the introduction of the Killing vector K, which Lam considers is an undesired 'non-dynamical background'. And likewise for the introduction of asymptotically flat metrics at

111 Absolute objects are defined, e.g., in Anderson (1967, pp. 83–4; 1971, p. 166). However, as we will discuss below, these background structures are not absolute objects in this sense.

112 Background-independence is obviously a large topic that I cannot enter in detail into here, nor will that be required. For, as I mentioned before, the 'background structures' discussed in this context are not the same as the ones discussed in the literature on background-independence. Thus, my comments will be restricted to the specific discussion of the role of background structures in Noether's second theorem. For more general treatments of background-independence: see, for example, Pooley (2017, pp. 24–9) and Read (2016a). For my own views about background-independence, see De Haro et al. (2017a).

113 Furthermore, Curiel (2019, p. 91) proves a theorem that there exists no local, divergence-free, covariant two-tensor other than a multiple of the Einstein tensor, and thus no such local candidates for a gravitational energy-momentum tensor.

114 Pitts (2010, pp. 605–6) proposes to restore the covariance of the gravitational energy-momentum expressions by embracing their multiplicity: namely, by constructing a single geometric object whose infinitely many components are the components of a given pseudo-tensor with respect to all possible coordinate systems, so that they transform into each other under coordinate transformations. In the text quoted, Lam is commenting on Pitts's proposal, but similar remarks could apply to the pseudo-tensor itself.

infinity in order to reproduce the ADM and Bondi formulas: these 'depend crucially on the background structures inserted in the form of asymptotic symmetries' (p. 1022).

Read (2020) has recently responded (at least partially) to the critiques (A) and (B), by arguing that *a genuine stress-energy tensor does exist in general relativity*, in both: (i) a weak sense applicable in a certain class of models of the theory (namely, those that admit appropriate Killing vectors), (ii) a strong sense, applicable in all models.

The above distinction between cases (i) and (ii) is underpinned by the existence of a timelike Killing vector in the spacetime. Namely, in response to (B), Read (2020, p. 222) argues that the Killing vector K that gives the well-defined energy conservation law Eq. (9.20), should not be seen as 'background structure'. For the existence of a Killing vector is merely a property of the metric: if the metric has an isometry, then a Killing vector exists. This simply means that the models with a Killing vector are a subset of all the dynamically possible models of the theory: no additional structures are introduced into the theory.

Read's advocacy of energy-momentum conservation appeals to functionalism (i.e., briefly, the idea that there exists in GR a quantity that fulfils the functional role of gravitational stress-energy: cf. p. 224) and to the possibility of being a realist about pseudo-tensors. The possibility of realism is motivated by Einstein's (1918b, pp. 449–51) own position that quantities can be coordinate-dependent, but still well-defined. Just as the connection coefficients are frame-dependent and are intimately tied to the gravitational field[115] (recall that the Newtonian gravitational potential arises from the Christoffel symbols), it is *prima facie* plausible to interpret the gravitational pseudo-tensor as representing gravitational energy, momentum, and stress in a given frame of reference. Thus, Read proposes that gravitational stress-energy is always defined with respect to a given frame of reference, analogously to how the relative velocity of a body is evaluated in the rest frame of another body (but see the objections in the next section).

He argues that in case (i), i.e. in those models admitting timelike Killing vectors, in which the conservation law Eq. (9.19) exists, the quantity $t^{\mu}{}_{\nu}$ does fulfil the role of a gravitational energy-momentum tensor, i.e. it fulfils a role analogous to that of gravitational energy as traditionally conceived, and bears an appropriate relation to the gravitational degrees of freedom. He proposes, based on (a) the pseudo-tensor's fulfilling this functional role, on (b) the larger explanatory apparatus that the advocates of gravitational energy have available to them, and on (c) the simple and practical character of functionalism involved: that, for this subset of models of the theory, the gravitational stress-energy does exist, and should be labelled as genuine energy, momentum, and stress. Thus, a frame-independent gravitational stress-energy tensor does, in this sense, exist in general relativity.

In case (ii), although a frame-independent stress-energy tensor is not available, one can still be a realist about the pseudo-tensor. Namely, realism about the pseudo-tensor is compatible with functionalism, since the pseudo-tensor may be defined, in each frame, such that it is conserved. In particular, he argues from the broader idea that it is acceptable to give explanations of phenomena by appealing to coordinate-dependent effects. If one accepts

[115] See Einstein (1916a, p. 802), and also the discussion in Norton (1985, p. 233).

this realist line, then the gravitational stress-enegy exists in the stronger sense, which is applicable to all the models of the theory.

While I will *in part* agree with Read's two arguments in response to (A) and (B), I will discuss, in the next section, whether these arguments conclusively establish the existence of local energy and momentum in general relativity – since indeed they do not offer a satisfactory response to (A).

9.6 A Comparison of Two Proposals

In this section, I will compare the two proposals for energy and momentum from the previous two sections. In Section 9.6.1, I discuss three worries about the gravitational pseudo-tensor that, in my opinion, should be addressed before pseudo-tensors are taken to be part of the ontology of general relativity. In Section 9.6.2, I will defend the holographic stress-energy tensor from two general worries, articulated in the philosophical literature on energy and momentum in general relativity, and already mentioned in the previous section. In Section 9.6.3, I will discuss further questions that one might have about the physical interpretation of quasi-local quantities.

9.6.1 What Else Is Required for Us to Accept Pseudo-Tensors?

In this Section I make three qualifications about Read's (2020) rebuttal of (A) and (B).[116] My second qualification is in the neighbourhood of a more technical objection by Dürr (2019a, section 3.3) (I will return to the resolution of a number of Dürr's other objections at the end of Section 9.6.2). My third qualification leads in to Section 9.6.2.

The first qualification is that, while Read (2020) frames his defence of pseudo-tensors in terms of scientific realism, this is not necessary. For also the empiricist, e.g. van Fraassen (1980), could argue along Read's functionalist lines and conclude that pseudo-tensors are *physical quantities,* which describe properties of some entities in the world – in this case, an unobservable entity, viz. gravitational energy and momentum. Agreed: this depends upon the empiricism under consideration. But the point is that the significance of pseudo-tensors goes beyond the question of whether there is a justified *realism* about them: rather, it is a matter of whether the semantics of GR must invoke such an entity.

Thus, while I agree that there is no a priori reason why coordinate-dependent quantities could not be physical, and also that, in general, *some* coordinate-dependent quantities *are* physical, I think that more work is required to admit these specific coordinate-dependent quantities – pseudo-tensors – into one's theory. This leads in to my next two points, about further conditions that are required to ascribe to pseudo-tensors a semantics that is physically acceptable.

[116] A fourth point is that pseudo-tensors are non-unique because one can add an arbitrary superpotential to them, and so it is not a priori clear how to interpret them. However, as I discussed in Section 9.3, this is clarified by Noether's second theorem, which establishes a clear relation between the superpotential and the boundary term in the first variation of the action, see Eq. (9.32).

The second qualification is that, to accept pseudo-tensors as good physical quantities, one *requires that they have well-defined transformation properties under (different kinds of) diffeomorphisms.* (Of course, pseudo-tensors transform covariantly only under *affine* transformations). Even if one is in principle sympathetic to the idea of coordinate-dependent quantities, surely some relationship between the quantities in the different coordinates, at least for special classes of coordinates, must be required before such quantities can be accepted in the theory.[117] As the analogy below should make clear, the real physical worry is that different coordinate systems would be, quite literally, incommensurable. This worry is, of course, in the neighbourhood of a closely related one: namely, that pseudo-tensors are not geometric objects[118] (see Anderson 1967, p. 424; Dürr 2019a, section 3.3). Therefore, it is not clear that the gravitational pseudo-tensor gives well-defined values for the energy in the various coordinate systems.[119] This is something that Nester's work, in my opinion, does not clarify.

Thus, Hoefer's (2000) argument that the quantities are not well-defined still seems to have *some* force, because a satisfactory definition of energy and stress would involve either giving their transformation rules or else giving a physical interpretation of the non-covariance (e.g. as in the case with boundaries; see Section 9.6.3). Lacking these, the realist story about pseudo-tensors seems less convincing.

Consider the analogy with special relativity, where the four-momentum vector p is a physical quantity that transforms covariantly under Lorentz transformations, but its components are not invariant. Choose a coordinate system where we write $(p^\mu) = (E/c, \mathbf{p})$, where E is the energy and $\mathbf{p}$ the spatial momentum as measured in those coordinates. Then we know that E^2/c^4 and $\mathbf{p}^2$ are not invariant quantities, whereas their difference, $-E^2/c^4+\mathbf{p}^2$, is invariant (and likewise for covariant quantities). The energy and the momentum transform into each other under Lorentz transformations, and so their definition depends on the coordinates chosen. Despite this coordinate-dependence, E and $\mathbf{p}$ *are* physical quantities, and we can:

(1) define energy and momentum for each choice of coordinates (these definitions being incompatible in the sense that what is defined as energy in one system is not the energy in another system), and furthermore these definitions can be made operational with respect to the corresponding frame, e.g. a preferred measuring apparatus;

(2) give well-defined transformation laws for E and $\mathbf{p}$ from one coordinate system to another: namely, the Lorentz transformation law.

[117] For example, when the number of spacetime dimensions is odd, under visible diffeomorphisms the quasi-local stress-energy tensor picks up an anomalous term that is in general non-zero, i.e. Eq. (9.60). Although this breaks covariance, it is a well-defined transformation law, and it is admissible.

[118] However, as Trautman (1965: p. 86) points out, spinors are not geometric objects either.

[119] In this context, Nester argues that the Hamiltonian is a fully covariant object, and that its significance is independent of the coordinates. But this is not entirely satisfying, for two reasons: (i) As I already mentioned in footnote 107, the direct relation Eq. (9.64) between the Hamiltonian and the pseudo-tensor only holds for everywhere *constant* vector fields N: and so, a putative covariance of the Hamiltonian does not automatically apply to the pseudo-tensor (which is expressly known to *not* be non-covariant). (ii) As I mentioned in my reply to (A) in Section 9.5.2, covariance is not what matters the most, because it can be easily achieved by introducing additional quantities. Specifically, the Hamiltonian depends on a choice of vector field N, and the Hamiltonian approach is only covariant relative to a given such choice. But since spacetimes do not come naturally equipped with vector fields, the Hamiltonian approach is dependent on such a choice, which makes it liable to either: (a) introducing background structures or (b) treating different models differently.

Property (2) seems to fail in the case of pseudo-tensors, and thus there is a disanalogy with other coordinate-dependent quantities that we are familiar with. The point is not that pseudo-tensors fail to be covariant, but rather that there is no well-defined transformation law between them.[120] Even if it were still possible to define energy and momentum in every coordinate system, as in (1), these different notions would be completely unrelated to each other – different coordinate systems would be, quite literally, incommensurable.

The point is that it is not quite clear that one can simply *drop* the requirement of well-defined transformation properties under diffeomorphisms and put nothing in its place: a coherent alternative for it seems to be required. At any rate, this is an objection that the programme needs to address.[121]

A third qualification is that, even if Read's arguments indeed allow taking pseudo-tensors as physical quantities, they do not require this. Pseudo-tensors *might* be the correct definition of energy and momentum in general relativity – or they might not, for there are alternative proposals – in particular, the quasi-local proposals. Thus the different proposals need to be compared. I do this in the following section.

9.6.2 Quasi-Local Expressions: Superpotentials vs. the Brown–York Stress-Energy Tensor

In this dection, I will defend the 'modified' version of the Brown–York quasi-local stress-energy tensor from Section 9.4 as a preferred quasi-local definition of energy, momentum, and stress in general relativity. I will partly do this by comparison with the other main proposal, from Section 9.5: namely, the pseudo-tensor, and the quasi-local Hamiltonian expression obtained from it. I will argue that the former is so far better developed, and thus better able to answer various queries. Hence my favouring, on balance, the Brown–York quasi-local stress-energy tensor. But since some of the details about the pseudo-tensor approach have not been fully worked out, the latter approach, with all its open questions, should still be considered.

By the 'modified' version of the Brown–York quasi-local stress-energy tensor', or simply the 'Brown–York stress-energy tensor', I will mean the quasi-local Brown–York stress-energy tensor from Section 9.4, modified, where this technique is available (i.e. in the case $\Lambda \neq 0$, regardless of its sign),[122] by the inclusion of counterterms: as opposed to the older reference field configuration subtraction (see (B)-(b) below). This is all, in principle, regardless of its possible holographic interpretation.

The quasi-local perspective has been almost entirely missed in recent philosophical discussions of energy conservation. Lam (2011: p. 1022) mentions the existence of quasi-local stress-energy tensors, but then dismisses them, on two grounds:

[120] As I showed in Section 9.4.3, in spaces with boundaries, different kinds of diffeomorphisms need to be distinguished, and for some diffeomorphism one may obtain a modified transformation law, involving the anomalous term in Eq. (9.60).

[121] Pitts (2010) is one such proposal. Dürr (2019a, section 3.3) notices, that, while 'nothing is *inherently* baneful about coordinate-dependent objects . . . [f]or pseudo-tensors . . . the coordinate-dependence is "vicious" (Pitts): in general, the preferred coordinate transformations do *not* pick out the characteristic invariants of the spacetime. The spacetime symmetries do not align with the pseudo-tensor's symmetry group'.

[122] The modified Brown-York stress-energy tensor, in the case $\Lambda > 0$, has been worked out in the references in footnote 79.

(i) A 'divide and conquer' argument to the effect that there are different expressions in the literature for such tensors, and that 'the links betwen [them] are not always clear ... there is for the time being no consensus on how to build a 'quasi-local' notion of energy'.

(ii) The quasi-local expressions allegedly also depend on particular background structures, 'such as the dependence on some particular embedding or on some particular boundary conditions'.

I will respond to both arguments below. Let me here say, about (i), that it is not a strong argument: for the mere *existence* of more than one expression, without further analysis or elaboration, is not a good reason to reject them collectively, but rather to further investigate which one is correct, or whether they are related. Furthermore, at least in cases with a non-zero cosmological constant, and *pace* Szabados (2009) which Lam cites, it is not really true that there is no consensus in – a large portion of – the physics community about which is the preferred expression[123] (by itself, this need of course not be a sufficient reason to join the consensus, but neither should the reasons for the consensus be ignored).

My comparison here will be along two main lines, which I will label (A) and (B), that parallel the philosophical discussion in Section 9.5.2 (and go back to points (iii) and (v) in Section 9.1.3). The first is a modification of point (A) in Section 9.5.2, namely: diffeomorphism-invariance; the second is, like (B) in Section 9.5.2, about 'background-dependence', or the requirement that the theory should not contain any 'background structures'. Then, at the end of the section, I will ask how the two approaches, the pseudo-tensor/Hamiltonian and the Brown–York approaches, are related.

(A) *The issue of diffeomorphism invariance.* I begin with the Brown–York approach. As I discussed in some detail in Section 9.4.3 for the case of spacetimes with a negative cosmological constant, the properties of the modified Brown–York stress-energy tensor under diffeomorphisms of different kinds are well-understood, and can be summarised in three points:

(i) The modified Brown–York stress-energy tensor is *invariant* under diffeomorphisms that express true redundancies (in the sense of (Redundant)), i.e. invisible diffeomorphisms.

(ii) It is *covariant* under visible diffeomorphisms, i.e. those that generate a conformal transformation on the boundary (see Eq. (9.63)).[124]

(iii) It is *not invariant* under those diffeomorphisms that express physical transformations and are not (Redundant), i.e. those that give *physically inequivalent* solutions of Einstein's field equations. This is as one would expect (see also the discussion in the next section).

Thus, the modified Brown–York stress-energy tensor has correct transformation properties under these different kinds of diffeomorphisms.

Also, up to a possible anomaly of the diffeomorphism group of the type discussed in Section 9.4.3, it is clear from the Brown–York construction that their stress-energy tensor transforms *covariantly* with respect to arbitrary reparametrisations, $\xi_i(x)$, of the boundary

[123] The literature here is too large to survey. Some examples are the textbook Ammon and Erdmenger (2015, p. 210), Anninos et al. (2011, p. 5), Balasubramanian and Kraus (1999, p. 413), and Skenderis (2001 2002).

[124] As I said before, I am ignoring the anomaly term in Eq. (9.60), which vanishes identically in four-dimensional spacetimes.

manifold (i.e. a visible diffeomorphism) for any value of Λ. This shows that the modified Brown–York stress-energy tensor is not liable to the allegation (A) from Section 9.5.2.

By contrast, in the case of pseudo-tensors and the quasi-local expressions derived from them in Section 9.5, an analysis along these lines has, so far as I know, not been carried out. There are, however, some remarks in the literature. For example, Chang et al. (1999, p. 3) write:

> the energy-momentum defined by such a pseudotensor does not really depend on the local value of the reference frame, it is actually *quasi-local* – it depends (through the superpotenial) on the values of the reference frame (and the fields) only on the boundary of a region.

The quote is perhaps somewhat optimistic. First, as I noticed in Section 9.5.1, there are assumptions that go into deriving the quasi-local expression Eq. (9.64) from the pseudo-tensor, which surely do not involve just the boundary itself, but at least an open neighbourhood of it. Furthermore, the 'dependence on the values of the reference frame' of the quasi-local energy that Chang et al. mention seems incompatible with its being covariant, since the reference frame dependence comes in through the superpotential, which is not a tensor, and also not when evaluated at the boundary. In other words, we appear to be in the situation that I discussed in Section 9.5.2, where there is a strong dependence on the frame of reference chosen at the boundary, and apparently no rule to compare one system to another.[125] At any rate, these authors do not clarify how the coordinate-dependence disappears. This contrasts with the Brown–York stress-energy tensor, which, as I discussed in (2), does have satisfactory properties under changes of frame on the boundary. Also, Chang et al. (1999) do not distinguish between the diffeomorphisms (1)–(3), whereas this is key to understanding the transformation properties of the quasi-local energy.

In Chen et al. (1999, section 2) and Nester (2004, pp. 272–5), a 'covariant Hamiltonian formalism' is worked out, in which, as I already mentioned, the various superpotentials are reproduced from relatively simple geometric expressions. Apart from the physical significance that one may ascribe to such covariance (see footnote 119), it is not clear how the quasi-local energy and momentum transform under (non-affine) changes of frame on the boundary, i.e. whether there is any way to compare quantities between different boundary frames.

Either way, although the Hamiltonian approach to pseudo-tensors seems promising, it needs more work before a full judgment about diffeomorphism invariance can be made.

(B) *'Background-dependence'*. Here, I begin with the pseudo-tensor/Hamiltonian approach. I already endorsed, in Section 9.5.2(B), Read's response to claims that pseudo-tensors require the introduction of background structures. Here I will defend the Brown–York approach, as regards background-independence. This will be three points, (a), (b), (c), of which (a) is a brief response to certain criticisms of AdS-CFT:[126]

125 One might think that the Hamiltonian formalism comes to the rescue here. However, two problems need to be addressed: first, the Hamiltonian depends on a choice of vector field N, and there is no natural way to choose this vector field in a general manifold; second, relating the pseudo-tensor to the Hamiltonian requires a *constant* vector field N, as already discussed in footnote 107.

126 For a fuller treatment and references, see De Haro et al. (2017a, pp. 113–16); cf. footnote 112.

(a) It has occasionally, and informally, been claimed that the methods in AdS-CFT are 'background-dependent', because they allegedly require the use of anti-de Sitter boundary conditions. This is part of a more general criticism, in Section 9.5.2(B), that defining asymptotic expressions for energy and momentum requires fixing the structure at infinity (typically, a flat metric). Here, there are two points to be made.

First, the AdS-CFT correspondence, and in particular the calculation of the modified Brown–York stress-energy tensor from Section 9.4.2, works for an *arbitrary*, smooth boundary metric γ: and so, the claim that AdS-CFT requires 'AdS boundary conditions' is incorrect. This means that the space is not even *asymptotically* AdS: it is asymptotically *locally* AdS, i.e. there exists, in an open neighbourhood of the boundary, always a choice of coordinates where the space looks like AdS – locally! But the space is in general *not* AdS: it is only required that the boundary metric have conformal Killing vectors, as in Eq. (9.63), if the CFT is to have a non-trivial conformal symmetry group. Thus, the boundary metric is taken to have conformal symmetries, but it does not need to have any pre-determimed form, like, e.g., being flat or being a round sphere. Examples in AdS-CFT have been worked out for boundary manifolds of many different kinds.

Second, the above objection perhaps originates in confusion about the role of boundary conditions in AdS-CFT. As I discussed in Section 9.4.1, the original Brown–York definition of the stress-energy tensor did have a dependence on a given, non-dynamical, background spacetime, through the introduction of the reference term in the action, Eq. (9.51). But as I also noted after Eq. (9.53), this is *not* how the tensor is rendered finite in AdS-CFT. Rather, one adds covariant counterterms that are *intrinsic* to the boundary. The renormalisation method, from Section 9.4.2, of adding boundary counterterms in the action, has the advantage of removing another kind of background: namely, the reference spacetime. Thus there is no reference spacetime that one needs to compare with.

(b) The role of boundary conditions in the approach that I am advocating here thus differs from the idea of a reference field configuration or background subtraction. For the boundary conditions required are merely those that are dictated by the fact that Einstein's equation is a second-order differential equation. For example, in globally hyperbolic spacetimes, one standardly specifies data on an initial Cauchy surface in order to solve Einstein's equations (think of the analogy with the hole argument!).[127] But this is not in any sense a dependence on a background: rather, the specification of the initial data is the specification of a solution among the set of all possible solutions. If the spacetime had a boundary somewhere at a finite distance, in order to specify a solution, one would, in addition, need to specify boundary conditions. This is what one does when one solves differential equations in spaces with boundaries: it has nothing to do with imposing backgrounds, but rather with choosing one's solution of interest out of the total set.

Thus, the background-independence of the modified Brown–York stress-energy tensor is 'as good as it gets' in general relativity.[128]

127 There are of course also disanalogies: see Stachel (2014, pp. 14–15, 50).

128 Janssen (2014, pp. 198–202) discusses Einstein's struggle, in his debate with de Sitter, with the need to impose boundary conditions at infinity, and in connection his (failed) attempts to incorporate Mach's principle into general relativity. See also Norton (1993, pp. 806–9). In De Haro et al. (2017a, p. 115), I argue that there are two distinct senses of 'background-independence', only one of which is implemented in general relativity – hence the 'as good as it gets'.

The situation is quite different for the pseudo-tensor and its quasi-local energy and momentum. First of all, these methods have only been worked out using reference spacetimes, both in the cases $\Lambda = 0$ and $\Lambda \neq 0$. Furthermore, to my knowledge, the asymptotics considered is only the simplest one: namely, Minkowski space and (A)dS space. Thus, only cases with a flat boundary metric γ have been considered. This may be only a matter of technical development: but it is of course yet to be seen whether the pseudo-tensor gives sensible (e.g. finite!) results in other cases.[129]

Nester (2004, p. 178) describes the problem as follows: 'The pseudotensor coordinate reference frame ambiguity now reappears in the guise of the reference field configration, which must be specified on the boundary'.

Despite their differences, both approaches emphasise the physical importance of choosing boundary terms to define energy and momentum. In the superpotential approach, boundary conditions/terms resolve the ambiguity in the specification of the superpotential. Also in the Brown–York approach are boundary terms central, since the stress-energy tensor is defined directly in terms of boundary quantities.

This dependence of the energy and momentum on boundary conditions is natural: for example, in classical mechanics, the momentum of a bullet of course depends on the initial velocity with which it is shot.

(c) Nor should the requirement that the *boundary* spacetime has conformal Killing vectors be seen as a dependence on a background, in the usual way. As Read (2020, p. 222) mentions, choosing a Killing vector field is making a choice of a model among all the possible dynamical models.

There is also a question of whether the pseudo-tensor and the Brown–York approach are directly related. It is unclear that they are, and a direct calculation of the Freud superpotential for the case $\Lambda < 0$, which includes Dirichlet boundary conditions, does not exhibit a clear or direct relationship to the Brown–York stress-energy tensor. Nester (2004, p. 178) claims that the Hamiltonian formalism establishes a relation between the two: 'Quasilocally, with certain choices for the reference fields and displacement, [our quasi-local expressions] reduce to the famous expressions of Brown and York'. However, the exact relationship remains obscure.

As already discussed in Section 9.4.2, AdS-CFT gives, through its holographic correspondence with the expectation value of the stress-energy tensor of the dual quantum field theory, Eq. (9.57), independent support to the modified Brown–York stress-energy tensor as being the gravitational energy and momentum.

The advantage of the quasi-local approach is perhaps best illustrated by mentioning how it solves several problems that Dürr (2019a, section 3.3) discusses in connection with the pseudo-tensor (and, as I said earlier, I agree with much of Dürr's critique).

129 It also seems a sign of lack of communication between at least some parts of the communities that the intrinsic renormalisation methods have not been applied, or even noticed, by those working on pseudo-tensors. For example, the review by Szabados (2009), written 12 years after gauge-gravity dualities were discovered by Maldacena, makes a single passing mention of it and takes no notice of how holographic renormalisation improves on the existing methods, by not appealing to a reference spacetime (see the discussion after Eq. (9.53)).

The first has already been discussed: it is the lack of covariance, and the pseudo-tensor's lacking appropriate transformation properties, i.e. my (v) from Section 9.1.3, which is Dürr's label (H1). The quasi-local stress-energy is both a geometric object and a (quasi-local) tensor under the relevant diffeomorphisms (see Section 9.4.3), and so this problem does not appear.

Second, asymptotic flatness (Dürr 2019a, section 3.3) is not a requirement of this approach: since, as I have stressed, the quasi-local stress-energy tensor is well-defined for both positive and negative values of Λ, and *for arbitrary boundary conditions on the metric,* i.e. not just the purely de Sitter or purely anti-de Sitter boundary conditions. In particular, this can be done for de Sitter-type solutions, which, in cosmological settings, are more realistic than the asymptotically flat ones. Furthermore, the divergences can be correctly renormalised by the method of holographic renormalisation. This is one of Dürr's main points of critique, and so this is a clear advantage of the quasi-local approach.

Third, as a general point (not specific to quasi-local energy): one of Dürr's (2019a, section 3.2) main criticisms to Read (2020) is that the latter *assumes,* rather than argues, that the ambiguity of the pseudo-tensor can be resolved. But we have seen that Read's 'assumption' is in fact vindicated by the details of Noether's second theorem. For the ambiguity is only apparent: it is fixed by a choice of boundary conditions, and of corresponding boundary terms (and this important point is common to all three approaches I have discussed).

Gauge-gravity dualities also cast further light on the interpretation of these boundary conditions. Each choice of boundary conditions corresponds to a different CFT (and also a different common core theory): namely, a CFT where a different kind of source is held fixed, with a different Lagrangian, and with a different path integral. The choices between such CFTs are different physical choices, which explains why their energies are different.

9.6.3 Interpreting Quasi-Local Quantitites

Although, in previous sections, I have addressed various objections to the quasi-local energy-momentum expressions, one can still ask about the physical meaning of such quasi-local expressions. After all, the traditional view in general relativity is that physical quantities are closely associated with geodesics and, in particular, with point coincidences, that are used to delimit these quantities. Indeed, the view goes back to Einstein himself.[130] In more substantivalist terms,[131] it is *points* that are fundamental in general relativity, and other meaningful geometric quantities ultimately reduce to quantities defined on points. But the pseudo-quantities, especially the ones that – like the Brown–York stress-energy tensor – are directly defined quasi-locally, without a local definition, would seem to give some physical significance to *spacelike two-surfaces,* possibly located at infinity: and it is not immediately clear what this means, both in terms of the theory's ontology (should we

[130] See Einstein (1916a, p. 776) and Kretschmann (1917, pp. 576–7, 579). However, elsewhere Einstein (1915b, p. 228) considers physically real events as consisting of more general *spatio-temporal coincidences.* For a discussion of Einstein's (1916a) paper and its prehistory, see Sauer (2005). The point-coincidence argument, and how Einstein sees in it an implicit objection to the hole argument, are discussed on page 816.

[131] For a review of substantivalist and anti-substantivalist arguments, see Pooley (2012, pp. 47–63).

include surfaces in the theory's ontology? And, if so, which ones?) and in empirical terms (what is the empirical content of such terms?).

I will here offer a number of arguments to the effect that surfaces and boundaries *should* be counted as part of general relativity's ontology, and that they do have an empirical meaning.

First, notice that the fact that *points* are regarded, by some substantivalists at least, as fundamental, does not exclude the reality of *finite surfaces,* which, just like the manifold itself, are made of points. Also *boundaries* can be understood in terms of neighbourhoods of the manifold,[132] and so as such what we call 'the existence of a boundary' is a topological property of the manifold, which just is part of the ontology of the theory. Thus, once we have a semantics for points and for neighbourhoods in general relativity (and the latter are required for both topological and geometric reasons), this very semantics gives a meaning to surfaces and boundaries. This point can also be argued from two philosophical directions:

(a) An ontology that accepts points as objects should surely accept sets, classes, or mereological fusions, of points. For example, this is standard in 'free lunch' supervenience metaphysics: see Lewis (1986a, pp. 211–12), Armstrong (1997, pp. 12–13), and Sider (2001, pp. xvi, 7, 59).

(b) As Butterfield (2011) argues, in his campaign against *pointillisme,* not all classical physical quantities are defined at a point (where 'defined at a point' is here taken in the sense of Lewis (1986b, pp. ix–xvi; 1994) Humean supervenience). Classical continuum mechanics gives us many examples of 'spatial extrinsicality: the way that an ascription "reaches across space"'. In particular, boundaries and quasi-local energy and momentum are as admissible in general relativity as they are in classical mechanics: 'continuum mechanics embraces spatial extrinsicality due to surface forces, acting at a point, *with respect to a surface through that point* (see Butterfield 2011, pp. 354, 356; my emphasis).

Second, there is no contradiction with any fact about general relativity if it turns out that energy and momentum can only be defined quasi-locally but not everywhere locally. There is no a priori reason why energy and momentum should exist everywhere locally on the manifold (see, e.g., Lam 2011: p. 1023). So, if some quasi-local notion of energy and momentum is well-defined, rather than being regarded with suspicion, it should be a welcome possibility.

Third, there are several physical reasons to consider boundaries, and stress-energy tensors as physical quantities defined on them. One reason is that one is interested in considering open systems, i.e. finite regions of the universe. In such cases, the non-vanishing of the right-hand side of Eq. (9.54) is interpreted as matter entering or leaving the region through the boundary. These open systems are very important for several reasons: as modelling a real system (e.g. a black hole, and the gravitational waves that it emits), or as a way to generate new solutions by gluing a given solution to other solutions through the boundary (for example, solutions with different values of the cosmological constant, with so-called domain walls in between: see, e.g., Freedman et al. (1999, pp. 376–7) and Ammon and

132 The boundary of a manifold M is the set of points p such that p has no neighbourhood homeomorphic with an open set of $\mathbb{R}^n$, where n is the dimension of the manifold.

Erdmenger (2015, pp. 302–4); for gluing of spacetimes in a cosmological context, see Carrera and Giulini (2010, pp. 170, 181)). One then needs to match the conditions of the two solutions at the boundary, and the boundary metric and its conjugate variable, viz. the stress-energy tensor, are the two canonical variables to control at the boundary.[133]

More generally, in semi-classical descriptions of quantum gravity theories, where the solutions of general relativity coupled to matter contain objects such as strings and D-branes, higher-dimensional surfaces such as the ones considered here 'stand ready' for an interpretation in terms of extended material membranes, i.e. as a natural generalisation of the notion of a geodesic. In other words, such surfaces must be considered, if general relativity is to be the semi-classical limit of these theories of quantum gravity.

9.7 Summary and Conclusion

In Section 9.1.3, I identified five putative, inter-dependent, problems for local definitions of energy and momentum in general relativity:

(i) the Hilbert–Klein–Noether worry that the conservation laws of GR are *identities*, with no physical content;

(ii) the lack of a *gravitational energy-momentum tensor*, analogous to the matter energy-momentum tensor;

(iii) the lack of integral conservation laws, because of the lack of a *timelike Killing vector field* in a general spacetime;

(iv) the *non-uniqueness* of the gravitational energy-momentum pseudo-tensor; and

(v) its lacking *covariance*, and its lacking appropriate geometric properties.

The Hilbert–Klein–Noether worry, (i), was addressed in Sections 9.2.2 and 9.2.3, where I argued that the conservation of energy and momentum in GR *does* have physical content.

The lack of a gravitational energy-momentum tensor, (ii), can be resolved in two ways: either by introducing the gravitational pseudo-tensor, or by defining quasi-local quantities. I then analysed two quasi-local proposals and specialised questions (iii)–(v) to these proposals.

About the lack of a timelike Killing vector field, (iii), in a general spacetime: while I accept that this is *prima facie* a problem for local expressions, I noted that the modified Brown–York stress-energy tensor requires only the existence of a *conformal* Killing vector field *at the quasi-local surface* (usually: the asymptotic boundary).

I also argued that the putative non-uniqueness, (iv), of the gravitational energy-momentum pseudo-tensor, i.e. the ability to add an 'arbitrary' superpotential, is fixed by Noether's theorem directly in terms of the variation of the action, as one sees from Eqs. (9.26) to (9.32). Specifically, it is fixed by a choice of boundary terms. So far as I know, this possibility has

133 If the Brown-York stress-energy tensor is evaluated not at conformal infinity, but at some finite distance, then this can have several other applications in string physics. For example, the universe is modelled in some models as the four-dimensional boundary of a five-dimensional space, placed at finite distance from the centre of the space (so that we 'live on a D-brane': see, e.g., Randall and Sundrum 1999, p. 4690). In that case, the Brown–York stress-energy tensor of the five-dimensional space acts as an extra source for the matter energy-momentum appearing in the four-dimensional Einstein field equations on the D-brane (De Haro et al. 2001a, p. 3171).

not been used in the literature before to solve the problem of non-uniqueness.[134] In more detail: one can understand the ability to add an abitrary term to the superpotential as the ability to add an arbitrary boundary term to the action without changing the equations of motion; this then agrees with results by Chang et al. (1999, p. 1898) in the Hamiltonian formalism.

An important point is that standard arguments in gauge-gravity dualities, which interpret the duality in terms of a common core shared between two isomorphic models, further support the statement that the Brown–York stress-energy tensor correctly represents the quasi-local energy and stress of general relativity, in the corresponding spacetimes.

Finally, point (v) about covariance, and the lack of appropriate transformation properties, was re-phrased in terms of two different notions: (A) diffeomorphism invariance, and (B) background-independence. This was done in Section 9.5.2 onwards.

About (A): I summarised, in Section 9.4.3, the properties of the modified Brown–York stress-energy tensor under three relevant kinds of diffeomorphisms. (1) The modified Brown–York stress-energy tensor is correctly invariant under 'invisible', i.e. (Redundant), diffeomorphisms. (2) It is correctly *non-invariant* under 'large', i.e. always non-redundant, diffeomorphisms, because they relate different models. (3) 'Visible' diffeomorphisms correctly give conformal transformations at infinity. These diffeomorphisms thus correspond to the conformal transformations of the conformal field theory. (The results were obtained for $\Lambda < 0$, but I argued that they can be extended to $\Lambda > 0$).

About (B): I agreed with Read's (2020) defence of the background-independence of expressions involving Killing vectors: although, as mentioned under (3), the Killing vectors required for the modified Brown–York stress-energy tensor to give an integral conservation law satisfy a weaker condition. Also, I noted that the Brown–York stress-energy tensor can be defined for any smooth boundary metric γ, and I clarified the role of the boundary conditions.

In sum, the modified Brown–York stress-energy tensor gives the clearest analysis and resolution, so far, of the above problems. As for admitting the gravitational pseudo-tensor itself into the ontology of general relativity: I argued that there are valid points in Read's (2020) defence of pseudo-tensors, although important questions remain open (especially about its transformation properties, and about which proposal is correct). The Chang et al. (1999) and Chen et al. (2017, 2018a) quasi-local proposal, and its relation to the Hamiltonian of GR, also seems promising, and it clarifies the differences between the different pseudo-tensors used in the literature. But it also faces additional questions (especially, about the status of the vector field N).

References

Ammon, M. and Erdmenger, J. (2015). *Gauge/Gravity Duality. Foundations and Applications*. Cambridge: Cambridge University Press.

Anderson, J. L. (1967). *Principles of Relativity Physics*. New York, London: Academic Press.

134 See footnote 57 for a brief discussion of the literature on this point.

Anderson, J. L. (1971). Covariance, invariance, and equivariance: a viewpoint. *General Relativity and Gravitation,* **2**(2), 161–72.

Anninos, D., Ng, G. S. and Strominger, A. (2011). Asymptotic symmetries and charges in De Sitter space. *Classical and Quantum Gravity*, **28**, 175019. arXiv:1009.4730 [gr-qc].

Armstrong, D. M. (1997). *A World of States of Affairs.* Cambridge: Cambridge University Press.

Arnowitt, R., Deser, S. and Misner, C. W. (1959). Dynamical structure and definition of energy in general relativity. *Physical Review*, **116**(5), 1322.

Arnowitt, R., Deser, S. and Misner, C. W. (1962). The Dynamics of General Relativity. In L. Witten, ed., *Gravitation: An Introduction to Current Research.* New York: Wiley, chapter 7. Reprinted in: R. Arnowitt, S. Deser and C. W. Misner (2008), Republication of: The dynamics of general relativity. *General Relativity and Gravitation,* **40**(9), 1997–2027. arXiv:gr-qc/0405109.

Ashtekar, A. and Hansen, R. O. (1978). A unified treatment of null and spatial infinity in general relativity. I. Universal structure, asymptotic symmetries, and conserved quantities at spatial infinity. *Journal of Mathematical Physics,* **19**, 1542–66.

Balasubramanian, V., de Boer and Minic, D. (2001). Mass, entropy, and holography in asymptotically de Sitter spaces. *Physical Review,* D, **65**, 123508, 1–15.

Balasubramanian, V. and Kraus, P. (1999). A stress tensor for anti-de Sitter gravity. *Communications in Mathematical Physics,* **208**, 413–28. arXiv: hep-th/9902121.

Bañados, M. and Reyes, I. (2016). 'A short review of Noether's theorems, gauge symmetries and boundary terms. *International Journal of Modern Physics,* D, **25**(10), 1630021.

Barbashov, B. M. and Nesterenko, V. V. (1983). Continuous symmetries in field theory. *Fortschritte der Physik,* **31**(10), 535–67.

Belinfante, F. J. (1939). On the spin angular momentum of mesons. *Physica*, **6**(9) 887–98.

Belot, G. (2013). Symmetry and equivalence. In R. Batterman, ed., *The Oxford Handbook of Philosophy of Physics.* Oxford: Oxford University Press, chapter 9.

Belot, G. (2018). Fifty million elvis fans can't be wrong. *Noûs,* **52**(4), 946–81.

Bergmann, P. G. (1958). Conservation laws in general relativity as the generators of coordinate transformations. *Physical Review,* **112**(1), 287–9.

Bessel-Hagen, E. (1921). Über die Erhaltungssätze der Elektrodynamik. *Mathematische Annalen,* **84**, pp. 258–76. English translation by M. Albinus and N. H. Ibragimov, On conservation laws of electrodynamics, *Archives of ALGA*, **3** (2006), 33–51. ALGA Publications. www.researchgate.net/publication/311456950_Archives_of_ALGA_vol_3. The page numbers in the text refer to the English translation.

Bondi, H., van der Burg, M. G. J. and Metzner, A. W. K. (1962). Gravitational waves in general relativity. VII. Waves from axi-symmetric isolated systems. *Proceedings of the Royal Society of London, Series A, Mathematical and Physical Sciences*, **269**(1336), 21–52.

Brading, K. (2005). A note on general relativity, energy conservation, and Noether's theorems. In A. J. Kox and J. Eisenstaedt, eds., *The Universe of General Relativity.* Boston and Berlin: Birkhäuser, pp. 125–35.

Brading, K. and Brown, H. R. (2000). Noether's theorems and gauge symmetries. arXiv:hep-th/0009058v1.

Brading, K. and Brown, H. R. (2003). Symmetries and Noether's theorems. In K. Brading and E. Castellani, eds., *Symmetries in Physics: Philosophical Reflections.* Cambridge: Cambridge University Press, pp. 89–109.

Brading, K. and Brown, H. R. (2004). Are gauge symmetry transformations observable? *The British Journal for the Philosophy of Science,* **55**, 645–65.

Brown, H. R. (2005). *Physical Relativity: Space-time Structure from a Dynamical Perspective.* Oxford: Oxford University Press.

Brown, H. R. (2022). Do symmetries 'explain' conservation laws? The modern converse Noether theorem vs. pragmatism. In J. Read and N. Teh, eds., *The Philosophy and Physics of Noether's Theorems*. Cambridge: Cambridge University Press (Chapter 7, this volume).

Brown, H. R. and Brading, K. A. (2002). *General covariance from the perspective of Noether's theorems*. PhilSci: 821. http://philsci-archive.pitt.edu/821.

Brown, J. D. and Henneaux, M. (1986). 'Central Charges in the Canonical Realization of Asymptotic Symmetries: An Example from Three-Dimensional Gravity', *Communications in Mathematical Physics* 104, pp. 207–226.

Brown, J. D. and York, J. W., Jr. (1993). Quasilocal energy and conserved charges derived from the gravitational action. *Physical Review D,* **47**, 1407–19.

Butterfield, J. N. (2011). Against Pointillisme: A call to arms'. In D. Dieks, W. J. Gonzalez, S. Hartmann, T. Uebel, and M. Weber, eds., *Explanation, Prediction, and Confirmation*. Dordrecht: Springer, pp. 347–65.

Butterfield, J. N. (2021). On dualities and equivalences between physical theories. In N. Huggett, B. Le Bihan and C. Wüthrich, eds., Philosophy beyond Spacetime: Implications from Quantum Gravity. Oxford: Oxford University Press.

Carrera, M. and Giulini, D. (2010). Influence of global cosmological expansion on local dynamics and kinematics. *Reviews of Modern Physics,* **82**(1), 169–208.

Caulton, A. (2015). The role of symmetry in the interpretation of physical theories. *Studies in History and Philosophy of Modern Physics,* **52**, pp. 153–162.

Chang, C.-C., Nester, J. M. and Chen, C.-M. (1999). Pseudotensors and quasilocal energy-momentum. *Physical Review Letters,* **83**(10), 1897. arXiv:gr-qc/9809040.

Chen, C.-M., Liu, J.-L. and Nester, J. M. (2018a). Gravitational energy is well-defined. *International Journal of Modern Physics D,* **27**(14), 1847017. arXiv: 1805.07692. (Page numbers refer to the arXiv version).

Chen, C.-M., Liu, J.-L. and Nester, J. M. (2018b). Quasi-local energy from a Minkowski reference. *General Relativity and Gravitation,* **50**(158), 1–14. arXiv:1811.05640 [gr-qc].

Chen, C.-M. and Nester, J. M. (1999). Quasilocal quantities for general relativity and other gravity theories. *Classical and Quantum Gravity*, **16**(4), 1279.

Chen, C.-M., Nester, J. M. and Tung, R.-S. (2017). Gravitational Energy for GR and Poincaré gauge theories: a covariant Hamiltonian approach. In W.-T. Ni, ed., *One Hundred Years of General Relativity: From Genesis and Empirical Foundations to Gravitational Waves, Cosmology and Quantum Gravity,* Vol. 1. London: World Scientific, pp. 187–261. arXiv:1507.07300 [gr-qc].

Christodoulou, D. and Yau, S.-T. (1988). Some remarks on the quasi-local mass. In J. A. Isenberg, ed., *Mathematics and General Relativity,* pp. 9–14. Providence, RI: American Mathematical Society, pp. 9–14.

Corry, L. (1999). Hilbert and physics (1900–1915). In J. Gray, ed., *The Symbolic Universe: Geometry and Physics 1890–1930*. Oxford: Oxford University Press, pp. 145–88.

Corry, L., Renn, J. and Stachel, J. (1997). Belated decision in the Hilbert–Einstein priority dispute'. *Science,* **278**(5341), 1270–3.

Curiel, E. (2019). On geometric objects, the non-existence of a gravitational stress-energy tensor, and the uniqueness of the Einstein field equation. *Studies in History and Philosophy of Modern Physics,* **66**, 90–102.

De Haro, S. (2017a). Dualities and emergent gravity: gauge/gravity duality, *Studies in History and Philosophy of Modern Physics*, **59**, 109–25. arXiv:1501.06162 [physics.hist-ph].

De Haro, S. (2017b). The invisibility of diffeomorphisms. *Foundations of Physics,* **47**, 1464–97. arXiv:1709.01949 [gr-qc].

De Haro, S. (2020). Spacetime and physical equivalence. In N. Huggett, K. Matsubara and C. Wüthrich, eds., *Beyond Spacetime. The Foundations of Quantum Gravity*. Cambridge: Cambridge University Press, pp. 257–83. arXiv:1707.06581 [hep-th].

De Haro, S. (2021). The empirical under-determination argument against scientific realism for dual theories. *Erkenntnis*. https://doi.org/10.1007/s10670-020-00342-0.

De Haro, S. and Butterfield, J. N. (2018). A schema for duality, illustrated by bosonization. In J. Kouneiher, ed., *Foundations of Mathematics and Physics One Century after Hilbert,* Kouneiher, J. (Ed.), Cham, Switzerland: Springer, pp. 305–76 arXiv:1707.06681 [physics.hist-ph].

De Haro, S. , Mayerson, D. and Butterfield, J. N. (2016). Conceptual aspects of gauge-gravity duality. *Foundations of Physics,* **46**, 1381–425.

De Haro, S., Skenderis, K. and Solodukhin, S. N. (2001a). Gravity in warped compactifications and the holographic stress-tensor. *Classical and Quantum Gravity,* **18**, pp. 3171–80. arXiv:hep-th/0011230.

De Haro, S., Skenderis, K. and Solodukhin, S. (2001b). Holographic reconstruction of spacetime and renormalization in the AdS/CFT correspondence. *Communications in Mathematical Physics*, **217**(3), pp. 595–622. arXiv:hep-th/0002230.

De Haro, S., Teh, N. and Butterfield, J. N. (2015). On the relation between dualities and gauge symmetries. *Philosophy of Science,* **83**(5), pp. 1059–69.

De Haro, S., Teh, N. and Butterfield, J. N. (2017). Comparing dualities and gauge symmetries. *Studies in History and Philosophy of Modern Physics*, **59**(2017), pp. 68–80. arXiv:1603.08334.

Dieks, D., van Dongen, J. and de Haro, S. (2015). Emergence in holographic scenarios for gravity. *Studies in History and Philosophy of Modern Physics,* **52**(B), 203–16. arXiv:1501.04278 [hep-th].

Dürr, P. M. (2019a). Against 'functional gravitational energy': a critical note on functionalism, selective realism, and geometric objects and gravitational energy. *Synthese,* **199**, 299–333. https://doi.org/10.1007/s11229-019-02503-3.

Dürr, P. M. (2019b). Fantastic beasts and where (not) to find them: energy conservation and local gravitational energy in general relativity. *Studies in History and Philosophy of Modern Physics,* **65**, 1–14.

Einstein, A. (1914). Die formale Grundlage der allgemeinen Relativitätstheorie. *Königlich Preußische Akademie der Wissenschaften* (Berlin). *Sitzungsberichte,* pp. 1030–85. In *The Collected Papers of Albert Einstein,* Vol. 6, Doc. 9. Princeton: Princeton University Press, pp. 72–130.

Einstein, A. (1915a). Die Feldgleichungen der Gravitation. *Königlich Preußische Akademie der Wissenschaften* (Berlin). *Sitzungsberichte,* pp. 844–7. In *The Collected Papers of Albert Einstein,* Vol. 6, Doc. 25. Princeton: Princeton University Press, pp. 225–9.

Einstein, A. (1915b). Letter to Paul Ehrenfest. In *The Collected Papers of Albert Einstein,* Vol. 8A, Doc. 173. Princeton: Princeton University Press, pp. 228–9.

Einstein, A. (1915c). Zur allgemeinen Relativitätstheorie. *Königlich Preußische Akademie der Wissenschaften* (Berlin). *Sitzungsberichte,* pp. 778–86. In: *The Collected Papers of Albert Einstein,* Vol. 6, Doc. 21. Princeton: Princeton University Press, pp. 214–24.

Einstein, A. (1915d). Zur allgemeinen Relativitätstheorie (Nachtrag). *Königlich Preußische Akademie der Wissenschaften* (Berlin). *Sitzungsberichte,* pp. 799–801. In *The Collected Papers of Albert Einstein,* Vol. 6, Doc. 22. Princeton: Princeton University Press, pp. 225–9.

Einstein, A. (1916a). Die Grundlage der allgemeinen Relativitätstheorie. *Annalen der Physik,* 49, pp. 769–822. In *The Collected Papers of Albert Einstein,* Vol. 6, Doc. 30. Princeton: Princeton University Press, pp. 284–339.

Einstein, A. (1916b). Hamiltonsches Prinzip und allgemeine Relativitätstheorie. *Königlich Preußische Akademie der Wissenschaften* (Berlin). *Sitzungsberichte,* pp. 1111–16. In *The Collected Papers of Albert Einstein,* Vol. 6, Doc. 41. Princeton: Princeton University Press, pp. 410–16.

Einstein, A. (1918a). Der Energiesatz in der allgemeinen Relativitätstheorie. *Königlich Preußische Akademie der Wissenschaften* (Berlin). *Sitzungsberichte.* In *The Collected Papers of Albert Einstein,* Vol. 7, Doc. 9. Princeton: Princeton University Press, pp. 63–77.

Einstein, A. (1918b). Letter to F. Klein, 13 March 1918. In *The Collected Papers of Albert Einstein,* Vol. 8B, Doc. 480. Princeton: Princeton University Press, pp. 673–5.

Einstein, A. (1918c). Letter to F. Klein, 24 March 1918. In *The Collected Papers of Albert Einstein,* Vol. 8B, Doc. 492. Princeton: Princeton University Press, pp. 697–700.

Einstein, A. and Grossman, M. (1913). Entwurf einer Verallgemeinerten Relativitïstheorie und einer Theorie der Gravitation. Teubner, Leipzig. In *The Collected Papers of Albert Einstein,* Vol. 4, Doc. 13. Princeton: Princeton University Press, pp. 302–43.

Fefferman, C. and Graham, C. R. (1985). Conformal invariants. In *Elie Cartan et les Mathématiques d'aujourd'hui*, Astérisque, **95**.

Fefferman, C. and Graham, C. R. (2012). The Ambient Metric, *Annals of Mathematics Studies*, **178**. Princeton and Oxford: Princeton University Press. http://arxiv.org/abs/0710.0919.

Freedman, D. Z., Gubser, S. S., Pilch, K. and Warner, N. P. (1999). Renormalization group flows from holography: supersymmetry and a c-theorem. *Advances in Theoretical and Mathematical Physics,* **3**(2), 363–417. arXiv:hep-th/9904017.

Freud, Ph. (1939). Über die Ausdrücke der Gesamtenergie und des Gesamtimpulses eines Materiellen Systems in der Allgemeinen Relativitätstheorie. *Annals of Mathematics,* **40**(2), 417–19.

Friederich, S. (2015). Symmetry, empirical equivalence, and identity. *The British Journal for the Philosophy of Science,* **66**, 537–59.

Geroch, R. (1972). Structure of the gravitational field at spatial infinity, *Journal of Mathematical Physics*, **13**, 956–68.

Goldberg, J. N. (1958). Conservation laws in general relativity. *Physical Review,* **111**(1), 315–20.

Gomes, H. (2020). Holism as the empirical significance of symmetries. arXiv:1910.05330 [physics.hist-ph], https://arxiv.org/abs/1910.05330.

Gomes, H. and Riello, A. (2020). Large gauge transformations, gauge invariance, and the QCD θ-term. arXiv:2007.04013 [physics.hist-ph], https://arxiv.org/abs/2007.04013.

Greaves, H. and Wallace, D. (2014). Empirical consequences of symmetries. *The British Journal for the Philosophy of Science,* **65**, pp. 59-89.

Grumiller, D. and Riegler, M. (2016). Most general AdS_3 boundary conditions. *The Journal of High-Energy Physics*, **1610**(023), 1–22. arXiv:1608.01308 [hep-th].

Harlow, D. and Stanford, D. (2011). Operator dictionaries and wave functions in AdS/CFT and dS/CFT. arXiv:1104.2621 [hep-th].

Hawking, S. W. (1979). The path-integral approach to quantum gravity. In S. W. Hawking and W. Israel, eds., *General Relativity: An Einstein Centenary Survey*. Cambridge: Cambridge University Press.

Hawking, S. W. and Horowitz, G. T. (1996). The gravitational Hamiltonian, action, entropy, and surface terms. *Classical and Quantum Gravity,* **13**, 1487–98.

Healey, R. (2009). Perfect symmetries. *The British Journal for the Philosophy of Science,* **60**, 697–720.

Heemskerk, I. and Polchinski, J. (2011). Holographic and Wilsonian renormalization groups. *Journal of High-Energy Physics,* **6**(31), 1–27. arXiv:1010.1264 [hep-th].

Henningson, M. and Skenderis, K. (1998). The holographic Weyl anomaly. *The Journal of High-Energy Physics,* **7**(23), 1–11. arXiv: hep-th/9806087.

Hilbert, D. (1916). Die Grundlagen der Physik (Erste Mitteilung). *Nachrichten von der Gesellschaft der Wissenschaften zu Göttingen, Mathematisch-physikalische Klasse,* Heft 3, pp. 395–407.

Hilbert, D. (1917). Die Grundlagen der Physik (Zweite Mitteilung). *Nachrichten von der Gesellschaft der Wissenschaften zu Göttingen, Mathematisch-physikalische Klasse,* pp. 53–76.

Hilbert, D. (1924). Die Grundlagen der Physik. *Nachrichten von der Gesellschaft der Wissenschaften zu Göttingen, Mathematisch-physikalische Klasse,* pp. 1–32.

Hoefer, C. (2000). Energy conservation in GTR. *Studies in History and Philosophy of Modern Physics,* **31**(2), 187–99.

Horský, J. and Novotný, J. (1969). Conservation laws in general relativity. *Czechoslovak Journal of Physics, B*, **19**, 419–42.

Huggett, N. (2017). Target space ≠ space. *Studies in History and Philosophy of Modern Physics,* **59**, 81–8.

Huggett, N. and Wüthrich, C. (2020). *Out of Nowhere: Duality*. http://philsci-archive.pitt.edu/17217.

Imbimbo, C., Schwimmer, A., Theisen, S., Yankielowicz, S. (2000). Diffeomorphisms and holographic anomalies, *Classical and Quantum Gravity,* **17**, pp. 1129–38. [hep-th/9910267].

Ismael, J. and van Fraassen, B. (2003). Symmetry as a guide to superfluous theoretical structure. In K. Brading and E. Castellani, eds., *Symmetries in Physics: Philosophical Reflections*. Cambridge: Cambridge University Press, pp. 371–92.

Jacobson, T., Kang, G. and Myers, R. C. (1994). On black hole entropy. *Physical Review D,* **49**, 6587–98. arXiv:gr-qc/9312023.

Janssen, M. (2014). 'No success like failure . . .': Einstein's quest for general relativity. In M. Janssen and C. Lehner, eds., *The Cambridge Companion to Einstein*. Cambridge: Cambridge University Press, pp. 167–227.

Janssen, M. and Renn, J. (2007). Untying the knot: how Einstein found his way back to field equations discarded in the Zurich notebook. In J. Renn and M. Schemel, eds., *The Genesis of General Relativity,* Vol. 2, *Einstein's Zurich Notebook: Commentary and Essays*. Dordrecht: Springer, pp. 839–925.

Jaramillo, J. L. and Gourgoulhon, E. (2011). Mass and angular momentum in general relativity. In L. Blanchet, A. Spallicci and B. Whiting, eds., *Mass and Motion in General Relativity,* Dordrecht: Springer, pp. 87–124.

Kastrup, H. (1987). The contributions of Emmy Noether, Felix Klein and Sophus Lie to the modern concept of symmetries in physical systems. In M. G. Doncel, A. Hermann, L. Michel and A. Pais, eds., *Symmetries in Physics (1600–1980): Proceedings of the 1st International Meeting on the History of Scientific Ideas, Sant Feliu de Guíxols, Catalonia, Spain.*

Klein, F. (1917). Zu Hilberts erster Note über die Grundlagen der Physik. *Nachrichten von der Gesellschaft der Wissenschaften zu Göttingen, Mathematisch-physikalische Klasse,* pp. 469–82.

Klein, F. (1918a). Letter to A. Einstein, 20 March 1918. In: *The Collected Papers of Albert Einstein,* Vol. 8, Doc. 487. Princeton: Princeton University Press, pp. 685–90.

Klein, F. (1918b). Über die Differentialgesetze für die Erhaltung von Impuls und Energie in der Einsteinschen Gravitationstheorie. *Nachrichten von der Gesellschaft der Wissenschaften zu Göttingen, Mathematisch-physikalische Klasse,* pp. 171–89.

Komar, A. (1958). Covariant conservation laws in general relativity. *Physical Review,* **113**(3), 934–36.

Kosmann-Schwarzbach, Y. (2011). *The Noether Theorems.* New York, Dordrecht, Heidelberg, London: Springer.

Kosso, P. (2000). The empirical status of symmetries in physics. *The British Journal for the Philosophy of Science,* **51**(1), 81–98.

Kretschmann, E. (1917). Über den physikalischen Sinn der Relativitätspostulate, A. Einsteins neue und seine ursprüngliche Relativitätstheorie. *Annalen der Physik,* **53**(16), 575–614.

Kucharzewski, M. and Kuczma, M. (1964). *Basic Concepts of the Theory of Geometric Objects*. Warsaw: Państwowe Wydawnictwo Naukowe.

Lam, V. (2011). Gravitational and nongravitational energy: the need for background structures. *Philosophy of Science,* **78**, 1012–23.

Landau, L. D. and Lifshitz, E. M. (1971). *The Classical Theory of Fields*, 3rd revised English ed. Oxford and New York: Pergamon Press.

Landsman, K. (2021). *Foundations of General Relativity: From Einstein to Black Holes.* Nijmegen: Radboud University Press.

Le Bihan, B. and Read, J. (2018). Duality and ontology. *Philosophy Compass*, **13**(12), e12555.

Lehmkuhl, D. (2022). The equivalence principle(s). In E. Knox and A. Wilson, eds., *Routledge Companion to the Philosophy of Physics.* New York and London: Routledge, pp. 125–44.

Lehmkuhl, D., Schiemann, G. and Scholz, E. (2017). *Towards a Theory of Spacetime Theories*. Springer.

Lewis, D. (1986a). *On the Plurality of Worlds.* Oxford: Blackwell.

Lewis, D. (1986b). *Philosophical Papers,* Vol. 2. New York, Oxford: Oxford University Press.

Lewis, D. (1994). Humean supervenience debugged. *Mind,* **103**(412), 473–90.

Maggiore, M. (2005). *A Modern Introduction to Quantum Field Theory.* Oxford: Oxford University Press.

Misner, C. W., Thorne, K. S. and Wheeler, J. A. (1973). *Gravitation.* San Francisco: W. H. Freeman and Company.

Murgueitio Ramírez, S. (2020). A puzzle concerning local symmetries and their empirical significance. Forthcoming in *The British Journal for the Philosophy of Science.* https://www.journals.uchicago.edu/doi/10.1086/714794.

Murgueitio Ramírez, S. and Teh, N. J. (2020). Abandoning Galileo's ship: The quest for non-relational empirical significance. Forthcoming in *The British Journal for the Philosophy of Science.* https://www.journals.uchicago.edu/doi/10.1086/718494.

Nester, J. M. (2004). General pseudotensors and quasilocal quantities. *Classical and Quantum Gravity,* **21**, pp. S261–S280.

Newman, E. T. and Penrose, R. (1966). Note on the Bondi–Metzner–Sachs group. *Journal of Mathematical Physics*, **7**, 863–70.

Nijenhuis, A. (1952). Theory of the geometric object. PhD thesis, University of Amsterdam.

Noether, E. (1918). Invariante Variationsprobleme. *Nachrichten von der Gesellschaft der Wissenschaften zu Göttingen, Mathematisch-physikalische Klasse,* 235–57.

Norton, J. (1985). What Was Einstein's Principle of Equivalence? *Studies in History and Philosophy of Science, Part A,* **16**(3), 203–46.

Norton, J. (1993). General covariance and the foundations of general relativity: Eight decades of dispute. *Reports on Progress in Physics,* **56**, 791–858.

Ohanian, H. C. (2010). The energy-momentum tensor in general relativity and in alternative theories of gravitation, and the gravitational vs. inertial mass. arXiv preprint arXiv:1010.5557, https://arxiv.org/abs/1010.5557.

Olver, P. J. (1986). *Applications of Lie Groups to Differential Equations.* Berlin, Heidelberg, New York: Springer-Verlag.

Papadimitriou, I. and Skenderis, K. (2004). AdS/CFT correspondence and geometry. *IRMA Lectures in Mathematics and Theoretical Physics,* **8**, 73–101. hep-th/0404176 [hep-th].

Papadimitriou, I. and Skenderis, K. (2005). Thermodynamics of asymptotically locally AdS spacetimes. arXiv:hep-th/0505190.

Pauli, W. (1958). *Theory of Relativity.* Oxford: Pergamon Press. New York: Dover Edition, 1981.

Penrose, R. (1963). Asymptotic properties of fields and space-times. *Physical Review Letters,* **10**(2), pp. 66–8.

Penrose, R. (1964). Conformal treatment of infinity. In B. DeWitt and C. DeWitt, eds., *Relativity, Groups and Topology.* New York and London: Gordon and Breach, pp. 565–84. Republished in: *General Relativity and Gravitation* (2011), **43**, 901–22.

Penrose, R. (1982). Quasi-local mass and angular momentum in general relativity. *Proceedings of the Royal Society of London, A,* **381**, 53–63.

Penrose, R. (1988). Aspects of quasi-local angular momentum. In J. A. Isenberg, ed., *Mathematics and General Relativity.* Providence, RI: American Mathematical Society, pp. 1–8.

Pitts, J. B. (2010). Gauge-invariant localization of infinitely many gravitational energies from all possible auxiliary structures. *General Relativity and Gravitation,* **42**, pp. 601–22.

Pooley, O. (2012). Substantivalist and relationist approaches to spacetime. In R. Batterman, ed., *The Oxford Handbook of Philosophy of Physics.* Oxford: Oxford University Press, chapter 15, pp. 1–48.

Pooley, O. (2017). Background independence, diffeomorphism invariance, and the meaning of coordinates. In Lehmkuhl et al. (2017), pp. 105–43.

Randall, L. and Sundrum, R. (1999). An alternative to compactification. *Physical Review Letters,* **83**(23), pp. 4690–3. arXiv:hep-th/9906064.

Read, J. (2016a). Background independence in classical and quantum gravity. Master's thesis, University of Oxford. https://ora.ox.ac.uk/objects/uuid:b22844af-aac9-4adc-a6c5-1e2815c59655

Read, J. (2016b). The Interpretation of String-Theoretic Dualities. *Foundations of Physics,* **46**, 209–35.

Read, J. (2020). Functional Gravitational Energy. *The British Journal for the Philosophy of Science,* **71**, 205–32.

Read, J. and Møller-Nielsen, T. (2020). Motivating dualities. *Synthese,* **197**, 263–91.

Renn, J. and Stachel, J. (2007). Hilbert's foundation of physics: from a theory of everything to a constituent of general relativity. In J. Renn and M. Schemel, eds., *The Genesis of General Relativity,* Vol. 4, *Gravitation in the Twilight of Classical Physics: The Promise of Mathematics*. Dordrecht: Springer, pp. 857–973.

Rickles, D. (2011). A philosopher looks at string dualities. *Studies in History and Philosophy of Modern Physics,* **42**, 54-67.

Rowe, D. E. (1999). The Göttingen response to general relativity and Emmy Noether's theorems. In J. Gray, ed., *The Symbolic Universe: Geometry and Physics 1890–1930.* Oxford: Oxford University Press, pp. 189–223.

Rowe, D. E. (2019a). Emmy Noether on energy conservation in general relativity. arXiv:1912.03269 [physics.hist-ph].

Rowe, D. E. (2019b). On Emmy Noether's role in the relativity revolution. *The Mathematical Intelligencer,* **41**, 65–72.

Sauer, T. (1999). The relativity of discovery: Hilbert's first note on the foundations of physics. *Archive for History of Exact Sciences,* **53**, 529–75.

Sauer, T. (2005). Albert Einstein, review paper on general relativity theory (1916). In I. Grattan-Guinness, ed., *Landmark Writings in Western Mathematics, 1640–1940.* Amsterdam: Elsevier, pp. 802–22.

Schouten, J. A. (1954). *Ricci-Calculus*. Berlin and Heidelberg: Springer-Verlag.

Schrödinger, E. (1950). *Space-Time Structure*. Cambridge: Cambridge University Press.

Sider, T. (2001). *Four-Dimensionalism: An Ontology of Persistence and Time.* Oxford: Oxford University Press.

Skenderis, K. (2001). Asymptotically anti-de Sitter space-times and their stress-energy tensor. *International Journal of Modern Physics. A*, **16**(5), 740–9. [hep-th/0010138].

Skenderis, K. (2002). Lecture notes on holographic renormalization. *Classical and Quantum Gravity,* **19**, 5849–76. arXiv:hep-th/0209067.

Skenderis, K. and van Rees, B. C. (2008). Real-time gauge/gravity duality. *Physical Review Letters,* **101**(081601), 1–4. arXiv:0805.0150 [hep-th].

Skenderis, K. and van Rees, B. C. (2009). Real-time gauge/gravity duality: prescription, renormalization and examples. *The Journal of High-Energy Physics,* **5**(85), 1–70. arXiv:0812.2909 [hep-th].

Stachel, J. (2014). The hole argument and some physical and philosophical implications. *Living Reviews in Relativity,* **17**(1), 1–66.

Strominger, A. (2001). The dS/CFT correspondence. *Journal of High-Energy Physics*, **0110**, 034. https://arxiv.org/abs/hep-th/0106113.

Sus, A. (2017). The physical significance of symmetries from the perspective of conservation laws. In Lehmkuhl et al. (2017), pp. 267–86.

Szabados, L. B. (2009). Quasi-local energy-momentum and angular momentum in general relativity. *Living Reviews in Relativity,* **12**, 4.

Teh, N. J. (2016). Galileo's gauge: understanding the empirical significance of gauge symmetry. *Philosophy of Science,* **83**, 93–118.

Trautman, A. (1962). Conservation laws in general relativity. In L. Witten, ed., *Gravitation: An Introduction to Current Research.* New York and London: John Wiley and Sons, pp. 169–98.

Trautman, A. (1965). Foundations and current problems in general relativity. In A. Trautman, F. A. E. Pirani and H. Bondi, eds., *Lectures on General Relativity*. Englewood Cliffs, NJ: Prentice-Hall, pp. 1–248.

van Fraassen, B. C. (1980). *The Scientific Image.* Oxford: Oxford University Press.

Wald, R. M. (1984). *General Relativity.* Chicago: University of Chicago Press.

Wald, R. M. (1993). Black hole entropy is the Noether charge. *Physical Review D,* **48**(8), pp. R3427–R3431. arXiv:gr-qc/9307038.

Wald, R. M. and Zoupas, A. (2000). General definition of 'conserved quantities' in general relativity and other theories of gravity. *Physical Review D*, **61**, 084027, pp. 1–16. arXiv:gr-qc/9911095.

Wang, M.-T. (2015). Four lectures on quasi-local mass. arXiv:1510.02931.

Weinberg, S. (1972). *Gravitation and Cosmology: Principles and Applications of the General Theory of Relativity.* New York, London, Sydney Toronto: John Wiley and Sons.

10

Geometric Objects and Perspectivalism*

JAMES READ

10.1 Introduction

> The two projects I have indicated (an infinite vocabulary for the natural series of numbers, a useless mental catalogue of all the images of his memory) are senseless, but they betray a certain stammering grandeur. They permit us to glimpse or infer the nature of Funes' vertiginous world.
>
> — Borges, *Funes the Memorious*

Famously, if one applies Noether's theorems to general relativity, one obtains infinitely many conserved quantities – one for each rigid symmetry of the Lagrangian – in light of the diffeomorphism invariance of that theory (Pitts 2009, 2010). Given this, many have repudiated the thought that such conserved quantities have physical significance, and have invoked notions such as Einstein's equivalence principle (itself a delicate business – see Lehmkuhl 2021 for a recent philosophical survey) in order to bolster this claim (see e.g. (Dürr 2019, 2021; Hoefer 2000) for defences of this view in the recent philosophical literature). In a previous paper (Read 2020), I defended, against this orthodoxy, the view that such conserved quantities – associated with the infamous gravitational stress-energy pseudo-tensors of general relativity, and more generally with 'non-geometric objects' – can have physical significance. On this front, I concur with Pitts, who writes,

> Noether's theorems do not care about the equivalence principle; they simply give results in any coordinate system. Rather than criticizing the results of Noether's theorem in terms of preconceived notions of invariance and then mysteriously invoking a principle irrelevant to Noether's theorem to reduce the puzzlement over the lack of an invariant energy complex, it is preferable to learn from the results of Noether's theorem that there is a broader notion of invariance suited to the existence of infinitely many distinct conserved energies. (Pitts 2010, p. 603)

My goal in this chapter is to bolster this view further, against recent critiques from Dürr (2021), via (a) a more detailed consideration of the metaphysics of geometric and non-geometric objects than has appeared in the literature up to this point and (b) an appeal to

* I am extremely thankful to Brian Pitts for detailed and very helpful comments on a previous draft of this paper. In addition, I am grateful to Adam Caulton, Neil Dewar, Fiona Doherty, John Dougherty, Patrick Dürr, Caspar Jacobs, Mahdi Khalili, James Ladyman, Joanna Luc, Tushar Menon, Matias Slavov, Nic Teh, the audience of the Trinity Term 2020 Oxford PoP-Grunch seminar, and the audience of the 2021 Munich workshop *Alternative Approaches to Scientific Realism*, for valuable discussions on this material.

perspectivalist and fragmentalist positions in the philosophy of science. Although a fair amount of caution is needed, particularly when dealing with the latter of these two theses (i.e., fragmentalism), (a) and (b) together open the door to a Borgesian vertiginous world of genuine physical quantities: exactly what I was after.

The prospectus is this. I begin in Section 10.2 by reminding the reader of essential aspects of the notions of natural kinds and projectable predicates. Then, in Sections 10.3 and 10.4, I introduce the distinction between geometric objects and non-geometric objects, and present a view according to which that the former can be assimilated (at least loosely) to projectable predicates (and, sometimes, to natural kinds), while the latter can be assimilated only to non-projectable predicates. I also consider the ways in which this view can be questioned, and in which even non-geometric objects might be assimilated to natural kinds. In Section 10.5, I argue that one can regard non-geometric objects, such as the gravitational stress-energy pseudo-tensors of general relativity, as having physical significance if one embraces either perspectivalism or fragmentalism in one's approach to metaphysics – although I argue that endorsement of the latter of these two positions constitutes a greater commitment than endorsement of the former. In Section 10.6, I consider the extent to which the 'dynamical approach' to spacetime theories of Brown (2005); Brown and Pooley (2001, 2006) can be allied to perspectivalism. In Section 10.7, I argue that one can embrace either perspectivalism or fragmentalism – both of which are related to (although distinct from) the programme of 'structural realism' in the philosophy of science – while rejecting other more opaque aspects of certain structural realist positions and commitments.

10.2 Natural Kinds and Projectable Predicates

There's a close connection between the notion of a natural kind and the notion of a projectable predicate. For the latter, I'll use Blackburn's (2008) definition: 'A property of predicates, measuring the degree to which past instances can be taken to be guides to future ones'. That is, projectable predicates are those which constitute legitimate bases for inductive inferences. As is extremely well-known, an (in)famous example of projectable versus non-projectable predicates is due to Goodman: 'green' is projectable; 'grue' is not (see Goodman 1955).[1] Only the former are supposed to be properties which, in some sense, 'reflect the structure of the world': electrons, quinine, and so forth.

Naturally, philosophers have sought to clarify this connection between projectable predicates and natural kinds; Quine (1969) took it to be the case that a sufficient condition for a predicate to be projectable is that it refer to natural kind properties. (This is likely not necessary, though. Consider, for example, the disjunctive predicate associated with the property 'being green or blue': This is not a natural kind property, but the predicate is still projectable.) For the purposes of this essay, I'll follow Quine in this regard.[2]

[1] One has to be careful, however, to get the definition of 'grue' correct, in order that this indeed be a non-projectable predicate – see Jackson (1975).

[2] For more on natural kinds, see e.g. Bird and Tobin (2017).

10.3 Geometric Objects

In a loose and heuristic sense, a predicate's being projectable means that one can 'transform' its application in one particular situation, to its application in another particular situation. Usually, such a transformation is a temporal translation to the future: if the predicate holds at such-and-such a time, then it holds also at such-and-such later time. But there's no reason not to generalise to richer transformations than mere temporal translations; there is thereby, I contend, at least a loose (but nevertheless illuminating) connection with the notion (familiar to philosophers of physics) of a 'geometric object'.

The formal presentation of the concept of a geometric object goes back (at least) to works by Nijenhuis (1952) and Schouten (1954) in the mid twentieth century, and was later elaborated on by, e.g., Trautman (1962, 1965). For a differentiable manifold M, consider an arbitrary point $p \in M$ and two arbitrary local coordinate systems around p. In order to specify a geometric object on M, one writes down (i) a set of components (a set of N real numbers) in each such coordinate system and (ii) a well-defined rule relating the components in the one coordinate system to the components in the other (Trautman 1965, pp. 84–85).[3] Non-geometric objects lack at least one of the above – Nijenhuis dubs geometric objects denuded of their transformation rules mere 'objects' (1952), and by 'non-geometric objects' I will mean such entities in what follows. For an excellent discussion of what constitutes a 'well-defined transformation rule', see Dürr (2021, §3.3) – but, in brief, the essential feature is this: if one transforms (the components of) a given object in a coordinate system A to (its components in) a coordinate system C, that must yield the same result as transforming (the components of) the object in A to some intermediate coordinate system B, and then to C. Examples of geometric objects abound: metric fields, n-forms, Christoffel symbols, etc. But there are also familiar examples of non-geometric objects – perhaps the most famous being gravitational stress-energy pseudo-tensors $t^{\mu}{}_{\nu}$, appearing in general relativistic local energy conservation laws (derivable from Noether's theorems – see e.g. (Brading 2005; de Haro, Chapter 9, this volume; Trautman 1962)), of the form

$$\partial_{\mu}\left(\sqrt{-g}T^{\mu}{}_{\nu} + \sqrt{-g}t^{\mu}{}_{\nu}\right) = 0. \tag{10.1}$$

(Spinors are another famous putative example of non-geometric objects – see Pitts 2012 for discussion – but I will not discuss them further in this article.) A particular pseudo-tensor (and there are many – e.g., the Einstein pseudo-tensor, Landau–Lifshitz pseudo-tensor, etc. – which differ by a choice of 'superpotential'; Trautman 1962) is a mere object in Nijenhuis's sense: it consists of a set of components in each local coordinate system, but no well-defined set of transformation rules between those components. Conservation laws such as (10.1) are often regarded as being problematic, for at least three reasons:[4]

[3] This is not necessarily a tensorial transformation law – for example, those for connection coefficients in general relativity, or for tensor densities, are not tensorial, but as nevertheless still well-defined.

[4] For some recent literature on this topic, see Curiel (2019); de Haro (Chapter 9, this volume); Dürr (2019, 2021); Hoefer (2000); Lam (2011); Pitts (2009, 2010); Read (2020). The issue is, however, goes back 100 years, to the Noether–Hilbert–Einstein–Klein correspondence. For the history of the latter, see Rowe (2019).

1. In general relativity, there are infinitely many such conservation laws: one (from Noether's first theorem) for each rigid symmetry of the Lagrangian density (see Pitts 2010, 2021).
2. Such conservation laws are closely associated to mathematical identities (see Brading 2005).
3. As mentioned, gravitational stress-energy pseudo-tensors are not geometric objects: they do not have associated transformation laws (see Dürr 2019, 2021).

I'll forego here a detailed discussion of the first and second objections.[5] On the third, the reasoning here is generally not spelled out explicitly, but it typically runs as follows. First, it is noted that pseudo-tensors are not geometric objects (again, see Dürr 2021, §3.3 for an excellent and explicit recent discussion of this point; de Haro's Chapter 9 in this volume is also recommended.) Then, it is asserted (generally without argument!) that geometric objects are the *sine qua non* of reasoning in contemporary physics. But, in light of the foregoing, we can now go further: in light of their lack of transformation laws, non-geometric objects are (in hopefully a not-too-idiosyncratic sense) non-projectable, and so should not be regarded as picking out natural kinds. The reason is that one cannot transform ('project') the components of such an object in one coordinate system to its components in another.

On Klein's Erlanger Programm to geometry, one can identify geometrical structures by way of specifying a set of transformation rules, and considering the invariants of those transformations; the objects so identified will coincide with those objects specified in a coordinate-*independent* manner on the 'Riemannian' approach to geometry currently popular amongst the 'Chicago-Irvine school' of philosophy of physics.[6,7] Thus, one can associate geometric objects as defined above with objects specified in a coordinate-independent manner (one might say that each coordinate system offers a different *perspective* on this object – see below for more on such perspectives); not so for non-geometric objects. Pitts himself uses the terminology of natural kinds here, when he writes, 'Thus the components of a geometric object form a natural kind mathematically: they constitute faces of one and the same entity by virtue of being interrelated by a coordinate transformation law' (Pitts 2010, p. 610). Suggestively, Quine also alludes to (but does not expand on) this connection in his seminal paper on natural kinds, when he writes: 'Perhaps the branches of science could be revealingly classified by looking to the relative similarity notion that is appropriate to each. Such a plan is reminiscent of Felix Klein's so-called *Erlangerprogramm* in geometry,

[5] In brief: I'm not convinced by them. On the first, Pitts one should take the mathematics seriously: infinitely many rigid symmetries of the Lagrangian density means infinitely many gravitational energies, in general relativity (in this regard, as already indicated above, I agree with Pitts 2009, 2010, 2021). On the second, it's not obvious that mathematical identities can't have physical content (see Brading 2005).

[6] For Klein's original work, see Klein (1892); for a recent philosophical discussion, see Wallace (2019). For the *locus classicus* of the Chicago-Irvine school, see Malament (2012).

[7] This statement isn't entirely uncontroversial: for example, it is far from clear that we have available a coordinate-independent conception of tensor densities, or of non-linear geometric objects more generally (for an attempt to secure this, see Pitts 2012, §2.2). That said, I will set this concern aside in what follows, for perhaps differential geometry will yield such conceptions in the future! (My thanks to Neil Dewar, John Dougherty, and Brian Pitts for very helpful discussions on this point.)

which involved characterizing the various branches of geometry by what transformations were immaterial to each' (Quine 1969, p. 55).[8] This, however, raises two important further questions:

A. Are all geometric objects to be assimilated to natural kinds, or only a subset? If the latter, how to identify this subset?
B. Are non-geometric objects not to be assimilated to natural kinds? If not, why not?

On (A), by analogy with more traditional discussions of natural kinds, one might wish to state that while all geometric objects are projectable (in the sense in which I have used the term in this section), only a subset are to be associated with natural kinds – we will see more on how to spell out this position in the following section. On (B), both negative and positive answers seem to be available. The negative answer – that non-geometric objects are not to be assimilated to natural kinds – would likely consist in the following: in light of the lack of transformation rules between the components of such objects in different coordinate systems, there is no unified, coordinate-independent entity to be associated with such objects; in this sense, non-geometric objects do not adequately limn reality.[9,10] The positive answer – that non-geometric objects can be assimilated to natural kinds – would consist in this: non-geometric objects can still limn the elements of reality; it is simply that these elements cannot be represented in a non-perspectival manner, i.e., *without* reference to a coordinate system. Pitts seems to prefer the latter of these two options when he writes,

> A pseudo-tensor $t^{\mu}{}_{\nu}$ shares with the Einstein tensor $G^{\mu\nu}$ the physically interesting property of having a single recipe for inferring its components in a coordinate system from the metric components and their partial derivatives in that coordinate system. Thus the components do form a natural kind in that physical sense. However, there is no transformation rule that allows one to infer the components with respect to one coordinate system from the components in another, so there is no mathematical unity. (Pitts 2010, pp. 610–611)

Although (excluding Pitts's work) the terminology of natural kinds has not been used in the literature on this topic up to this point, it is fair to say that almost all authors would disagree with Pitts on this front. For example, while in many respects sympathetic to Pitts's writing on pseudo-tensors and my own, de Haro writes in Chapter 9 of this volume (§6.1) that 'it is not quite clear that one can simply *drop* the requirement of well-defined transformation properties under diffeomorphisms and put nothing in its place: a coherent alternative for it seems required'. But this, of course, is not an argument against the status of non-geometric

8 This constitutes part of a broader tradition of identifying the basic elements of one's ontology with invariants – see e.g. Dirac (1958); Nozick (2001), and Weyl (1952) for further classic discussion. With this in mind, I take the connection between the notion of projectability in the sense of transformation laws, and the notion of projectability in the sense of Section 10.2, to be this: a projectable object in the 'transformation laws' sense affords an invariant description, which is a candidate natural kind. Then, *per* Quine, a predicate associated with this kind should be projectable in the 'standard inductive' sense.

9 The situation is complicated slightly by the fact that non-geometric objects sometimes have well-defined transformation rules for *restricted* classes of transformations. For example, the Einstein pseudo-tensor transforms tensorially under affine transformations (Trautman 1965).

10 In this regard, my appeal at Read (2020, p. 225) to the Erlanger Programm as a means of defending the reality of (the physical quantities represented by) pseudo-tensors was somewhat misguided (although cf. Read (2020, n. 38), in which I registered that I was construing the 'Kleinian conception' differently from how it was intended originally), for while pseudo-tensors must be characterised in coordinate-dependent terms, there are no well-defined transformation rules via which one can identify invariants.

objects as natural kinds. What one would require here in order to make progress would be an explicit list of the various criteria which one might take to be relevant to such status of non-geometric objects. Perhaps transformation rules are a part of the story here, but they need not be the whole story – one might think, for example, that various other functional desiderata contribute to an entity's being classified as a natural kind: that, indeed, was part of the moral of Read (2020).

10.4 Real Patterns and Mere Patterns

In order to make further progress in addressing the above questions, it is now worth bringing into the fold other recent work on natural kinds. Following Dennett (1991), Ladyman et al. (2007, §4.4) cash out their own 'ontic structural realist' ontology in terms of 'real patterns' – roughly speaking, patterns in (our description of) the world which are (i) projectable, and (ii) maximally information compressing.[11] In the language of the previous section: being a geometric object is sufficient for (i); a certain subset of these (e.g., the metric field of general relativity) will satisfy (ii). Ladyman et al. draw a direct connection between real patterns and natural kinds: 'We contend that everything a naturalist could legitimately want from the concept of a natural kind can be had simply by reference to real patterns' (Ladyman et al. 2007, p. 294). So, bringing the terminology of Ladyman et al. into contact with the previous section, we can say this: projectable predicates can be assimilated to geometric objects; a subset of these pick out the natural kinds/real patterns; it is the latter which constitute (for Ladyman et al.) the true ontology of the theory under consideration.

What exactly is pathological about objects such as pseudo-tensors, and non-geometric objects more generally, in the language of Ladyman et al.? These authors go on to distinguish real patterns from what they call 'mere patterns'. Of the latter, they write: 'A mere pattern is a locatable address associated with no projectable or non-redundant object' (Ladyman et al. 2007, p. 231, n. 51). In more detail, the idea here is the following. Given the structure of the world (as given by whatever scientific theory is currently under consideration),[12] one can lay down further structure – what Ladyman et al. call 'locators' (Ladyman et al. 2007, p. 120) – in order to identify salient aspects of that structure; it is via this process that the typical metaphysical notions of objects, properties, events, processes, etc., arise. However, for Ladyman et al., what is identified relative to such a locator cannot exist in anything but an ontologically 'thin' sense – the true structure of the world remains the structure as described directly by the scientific theory under consideration itself, without the aid of a locator.

Now, a canonical example of a locator is a *coordinate system* (Ladyman et al. 2007, p. 121); thus, for Ladyman et al., a structure which can be characterised only with reference to coordinate systems can (it seems) at best be a mere pattern. Recall now also that non-geometric objects such as pseudo-tensors are indeed defined only relative to coordinate

11 Again, roughly speaking, (ii) is intended to preclude disjunctive-but-projectable predicates, such as 'being green or blue'. See Ladyman et al. (2007, §4.4) for the full details of this account of real patterns.

12 Note: This is not 'our best fundamental theory of physics', given Ladyman et al.'s 'rainforest realism', according to which ontology is 'scale-relative' – see Ladyman et al. (2007, ch. 4).

systems: they do not possess associated transformation rules which relate their components in one coordinate system to their components in another, so one cannot (via the Kleinian approach to geometry) identify an invariant object therefrom. Given this, what it seems that one should say about pseudo-tensors and other non-geometric objects, in the language of Ladyman et al., is the following: they are not to be taken ontologically seriously, for they are not projectable/are not natural kinds/are mere patterns.

Returning to (A) and (B) in the previous section, we see that Ladyman et al. would maintain, via their notion of 'real patterns' (and, for what it's worth, in agreement with my own verdicts), that only a subset of the geometric objects should be identified with natural kinds. On (B), it seems that they would issue a negative answer. But again, one can ask: why should an invariant representation be the *sine qua non* of naturalness?[13,14] And even granting that non-geometric objects can at best be associated with 'mere patterns', does it necessarily follow that there is no sense in which such objects are 'physical'?[15] It is to this latter question which I now turn.

10.5 Perspectivalism and Fragmentalism

I assess in this section whether the verdict that non-geometric objects cannot have physical content is too fast, even granting that they are not to be associated with natural kinds, and are (in the parlance of Ladyman et al.) 'mere patterns'. One immediate thing to say on this is the following: there is a crucial difference between spacetime coordinate systems, and other locators which one may lay down (for example, particular gauge choices in electromagnetism). Namely, the former are associated with the *vantage points* of observers – thus, what is described using such locators has the capacity to be amenable to *direct empirical access*.[16,17] Given this, an alternative take on such non-geometric objects is that they are real, but *perspectival*. Embracing, *pace* the orthodoxy and Ladyman et al. (at least in this particular regard), the reality of (the objects represented by) peseudotensors (and also of the spacetime coordinate representations of geometric objects) is one particular sense in which I would count myself a 'perspectival realist'.[18]

13 Dürr (2021, p. 10) makes essentially the same point – that one needs clear and well-articulated criteria for naturalness – albeit from the standpoint of antirealism about gravitational-stress energy in general relativity. See also earlier in Dürr's paper: '*Only* if one is already attracted to realism about gravitational energy (undergirded by egalitarianism about GR), will one find Read's position attractive, too' (Dürr 2021, p. 2).

14 James Ladyman has suggested to me that one might be able to invoke the non-redundancy of real patterns, and redundancy of mere patterns (see again Ladyman et al. 2007, p. 231, n. 51), in order to argue that some non-geometric objects can be associated with real patterns after all (i.e. when they are non-redundant). While this would overcome the above-described tension with Pitts's verdicts, I struggle to square it with the scepticism of Ladyman et al. towards objects defined relative to locators; thus, more needs to be said if one wishes to defend the status of non-geometric objects as representing real patterns within their framework. (My thanks to James Ladyman for discussion here.)

15 Clearly, the notion of 'physical' requires elaboration; I turn to this in the following section.

16 Here, I am following quite standard practice in taking it that idealised observers can be operationalised via coordinate frames (more on the operational significance of coordinate frames below). This is not to deny that one might wish to do more to 'schematise the observer' in any given physical theory – see Curiel (forthcoming) for recent discussion – but I will set these matters aside in the remainder of this chapter.

17 There are complications here regarding the limits of indexical reference – see Cheng and Read (2021), Dasgupta (2015), and Maudlin (1993) for discussion – but I will also set these matters aside for the purposes of this chapter.

18 See e.g. Massimi (2018) for an introduction to perspectival realism more generally. As Massimi puts it, the starting point for perspectival realism is the thesis that 'one can accept and fully endorse that scientific inquiry is indeed pluralistic and that

Recently, some issues related to these have been discussed in the philosophical literature on special relativity, under the guise of 'fragmentalism': a term coined in this context by Fine (2005).[19] The issue which the fragmentialist about special relativity seeks to address is once again whether, in addition to four-dimensional, coordinate-independent structures, one should also regard frame-dependent phenomena (e.g., frame-relative facts about simultaneity) as being real. Here is how Lipman puts the view:

> The importance is that of marking a metaphysical realism about those variant matters. The relevant question is whether realism or antirealism is true about the frame-relative facts, that is, whether consideration of the special theory of relativity removes all frame-relative facts from one's metaphysical conception of reality: the Minkowskian answers yes, the fragmentalist answers no. (Lipman 2020, p. 31)

It should be of no surprise to find that I would count myself a realist about frame-relative facts. However, for all its many merits, in my view Lipman's discussion of these matters effaces an important difference which one might draw between perspectivalism on the one hand, and fragmentalism on the other, which deserves to be made explicit.

As I understand it, perspectivalism involves a commitment to the reality of frame-relative facts. Suppose that some object a has some property P in some coordinate frame F, but does not have property P in some other frame G (consider e.g. the property of 'moving with uniform velocity'). Then, using Fine's (2005, §2) 'reality operator' $\mathfrak{R}$, one might, as a perspectivalist, assent to both $\mathfrak{R}P_F a$ and $\mathfrak{R}\neg P_G a$, where subscripts on predicates indicate the coordinate frame ('perspective') from which the property ascription is made. Note that, given the perspective-relativisation of property ascription here, the totality of realist commitments for the perspectivalist is not contradictory. Here the view differs from fragmentalism in Fine's sense, according to which 'The overall collection of facts, "über-reality", includes pairs of mutually incompatible facts' (Lipman 2020, p. 23). Given this, one way to understand the difference between perspectivalism and fragmentalism is that the latter should simply drop the perspective-relativisation of properties in the foregoing, assenting, rather, to both $\mathfrak{R}Pa$ and $\mathfrak{R}\neg Pa$ – i.e. to a genuinely self-contradictory set of realist commitments.

If fragmentalism really is best understood in this way, then I'm not sure I can assent to it: it's too Hegelian for me.[20] So, while I leave open the possibility of fragmentalism

there is no unique, objective, and privileged epistemic vantage point without necessarily having to conclude that perspectives shape scientific facts or relativize truth' (Massimi 2018, p. 170).

[19] Fine's focus is principally on fragmentalism about tense and the A-series – according to this view, there are infinitely many A-series, all of which are *prima facie* incompatible, but which are reconcilable via fragmentalism. This position is closely related to Maudlin on tense, according to which there are infinitely many A-series (Maudlin 2007, ch. 4, p. 126, n. 11), but all of these can be reduced to a fundamentally B-theoretic ontology (clearly, however, the fragmentalist would not endorse this latter reduction). In Read and Qureshi-Hurst (2021, n. 70), it is argued that Maudlin's position on tense is akin to that of Pitts on gravitational energy – although there is a difference, for as we have already seen, in the case of gravitational energy, no such reduction is possible.

[20] See Moore (2012, §7.7) for a fascinating discussion of the role of contradiction in Hegel's metaphysics. As Moore writes, '*He* [Hegel] *accepts the contradiction.* The truly infinite, for Hegel, embraces a co-existence of opposed aspects, and the relevant arguments concerning the extent of the physical universe simply highlight some of these. Like all such opposed aspects, they are to be *aufgehoben* in the infinite's progress towards self-knowledge.' Later in the section, Moore continues, 'Contradiction is the motor force of change. By its very nature it propels reality to a higher stage of development in which it is *aufgehoben*.' One might wonder what the higher stage of development is in our present case – perhaps (a non-contradictory!) perspectivalist approach, or a (non-contradictory!) 'orthodox' spacetime approach. Insofar as I regard both of these latter approaches as being superior to fragmentalism, I am not sure that I disagree with a Hegelian reading of this dialectic (in both

in the above sense, and maintain that it does (as with perspectivalism) afford a means of being a realist about (say) pseudo-tensorial quantities, I commit myself only to perspectivalism in what follows.[21] Such a perspectivalist position seems, in my view, to address Dürr's concerns regarding how one can be a realist about gravitational energy while abandoning the paradigm that only geometric objects have physical content (see Dürr 2021, p. 10); it also, I hope, goes some way to addressing the call for clarity issued by de Haro (Chapter 9, §6.1, this volume).[22]

Consideration of non-geometric objects such as gravitational stress-energy pseudo-tensors might, in fact, afford a *better* context for a defence of perspectivalism/fragmentalism than the case of special relativity, considered by Fine (2005) and Lipman (2020). This is due to the fact that, in the special relativistic context, the existence of transformation rules between the components of the geometric objects under consideration (e.g., the Minkowski metric field) in each coordinate system means that it is possible to construct a coordinate-independent representation of such objects. When it comes to pseudo-tensors, however, it is *not* possible to construct such a representation. Thus, in this latter case, as already discussed above, there is no *non*-perspectivalist/fragmentalist picture of the physics which is available.[23]

Hofweber and Lange (2017) criticise Finean fragmentalism, on the grounds that it is incompatible with the fact that an invariant structure (Minkowksi spacetime) is usually taken to explain the Lorentz transformations. Thus, as Lipman puts it, the concern is that

> The fragmentalist interpretation under discussion, which takes the variant properties to be real and the grounds for the spacetime interval, cannot offer the standard explanation. The standard interpretation underwrites the standardly assumed explanatory priority of the spacetime interval, whereas the fragmentalist conflicts with it. (Lipman 2020, p. 33)

Following Lipman (2020, §4), it doesn't seem to me that these objections to fragmentalism are successful. On the one hand, why *assume* that the 'standard interpretation' is correct – especially when there exist alternative approaches, e.g. the 'dynamical approach' of Brown (2005) and Brown and Pooley (2001, 2006), according to which it is dynamical symmetries which ground invariant spacetime structure, rather than *vice versa* (more on the connections between perspectivalism/fragmentalism and the dynamical approach in the following section). Such methodologies are live, and would seem to evade Hofweber and Lange's critique. In addition, it is worth noting that the following combination of views seems to reconcile the 'standard interpretation' with fragmentalism: (i) the form of the coordinate transformations relating inertial frames of reference is grounded in/explained by invariant spacetime structure; (ii) nevertheless, frame-dependent phenomena should be regarded as being real (cf. Lipman 2020, p. 32).

the ordinary and Hegelian sense!) after all. (One might question whether the kind of contradiction at play here is the same kind of contradiction as Hegel's 'self-negating negation'; I'll set this worry aside.)

21 For complimentary arguments against fragmentalism and in favour of perspectivalism, see Slavov (2020).

22 Le Bihan and I take regarding all gravitational stress-energy pseudo-tensors to be physically meaningful as a case of what we call 'pluralism', and cite Pitts accordingly (Le Bihan and Read 2018, n. 31). It should be clear that this 'pluralism' is closely related to the perspectivalism/fragmentalism discussed here.

23 Returning to note 20: although in this case there remains the possibility for the contradiction to be *aufgehoben* in perspectivalism, the alternative possibility of an 'orthodox' spacetime approach is no longer available.

10.6 The Dynamical Approach and Perspectivalism

It is worth pausing to reflect further on the interrelations between perspectivalism (in the sense considered in this chapter up to this point) and the dynamical approach to spacetime theories, as defended in Brown (2005) and Brown and Pooley (2001, 2006). Based on the above discussion, one might think that these approaches are natural bedfellows, insofar as the dynamical approach (at least as presented in the context of special relativity – see Brown and Read (2021) for discussion of how the dynamical approach plays out differently in special versus general relativity) begins with coordinate-dependent descriptions of the relevant physics, and seeks to construct the associated spacetime structure therefrom.[24] Although such a line of thought is roughly correct, the situation here is, in fact, delicate, for several reasons.

One point stressed repeatedly by Brown in his presentation of the dynamical approach is owed to Bell (1987): that it is possible to teach special relativity, and derive paradigmatic special relativistic effects such as length contraction and time dilation, from within a *single* frame of reference (see in particular Brown 2005; Brown and Pooley 2001). Famously, Bell considers a classical model of an electron; after physically implementing an active boost, the electron is found to be Lorentz contracted (this is what Menon et al. (2020) called a 'Lorentz push'). It would be perfectly acceptable for a proponent of the dynamical approach to take the Poincaré invariant laws of special relativity to codify such physical effects *observed within a single frame*, and in turn to take spatiotemporal structure to be a codification of the symmetries of those laws. On this understanding, there is never any need for a proponent of the dynamical view to appeal to different perspectives (cf. Miller (2010) distinction between 'perspectival' and 'dynamical' interpretations of special relativistic effects, and the approving citation thereof to be found in Brown and Read 2021).

That being said, there is nothing problematic (modulo caveats to be discussed below) about signing up to the dynamical approach alongside perspectivalism. Consider again Bell's atom. From the point of view of the frame co-moving with the atom in its initial state, that atom undergoes length contraction; whereas, from the point of view of the frame co-moving with the atom in its final state, the atom decelerates, and thereby expands in length. A perspectivalist can say that the accounts offered from both of these perspectives are equally true; in addition, a fragmentalist will want to regard the accounts associated with both of these frames as being associated with (mutually inconsistent) facts. Focussing on perspectivalism in what follows given the above-discussed concerns regarding fragmentalism, there is nothing wrong with admitting these relativised facts into one's ontology, even as a proponent of the dynamical approach. Indeed, arguably proponents of the view *must* admit such facts, unless they wish to state that there is some preferred frame – something which Brown (2005) repudiates vehemently: 'I feel, from dire experience, I must

[24] Recall Brown (2005): 'The appropriate structure is Minkowski geometry *precisely because* the laws of physics of the non-gravitational interactions are Lorentz covariant' (p. 133: see also Brown and Pooley 2006, p. 10.) In Section 10.7, I mention programmes in metaphysics – namely, liberalised Humean approaches – which seem to afford the means to underwrite such claims.

emphasize from the outset that this approach does not involve postulating the existence of a hidden preferred inertial frame!' (p. vii).

There are, however, limits to what counts as an admissible perspective, on the dynamical approach. Recall that, throughout, Brown (2005) subscribes to Einstein's operational understanding of coordinates – presented in the latter's first *annus mirabilis* paper on special relativity (Einstein 1905) – according to which one requires spatial coordinates to 'match' the length of rigid measuring rods that are at rest in the system in question, and time coordinates to 'match' the tickings of clocks at rest in that system. Given this operational understanding of coordinates, it is clear that proponents of the dynamical approach who wish to subscribe to perspectivalism should not quantify unrestrictedly over frames of reference, but rather only over those frames of reference which can be suitably operationalised. Combining unrestricted perspectivalism with an operational understanding of coordinates would, arguably, bring with it consequences unacceptable to proponents of a dynamical approach. For consider some reference frame which cannot be operationalised with respect to material, physical bodies; a fallback attempt to do so might then appeal to immaterial bodies – e.g., spacetime – not congenial to the empiricist spirit of the dynamical view.

In sum, then, the dynamical approach is consistent with perspectivalism and fragmentalism (granting that the latter is consistent at all!); however, it does not sit well with an unrestricted quantification over frames of reference. This, potentially, has upshots also for debates over gravitational energy: while fans of the concept (e.g. Read 2020) could be read as perspectivalists about this quantity, insofar as they also wish to subscribe to an operational understanding of coordinates, and more generally to the dynamical approach, they should not necessarily affirm the reality of the physical quantities corresponding to the components of pseudo-tensors in *all* coordinate systems, but, rather, only the reality of the physical quantities corresponding to the components of pseudo-tensors in those coordinate systems which can be appropriately operationalised.[25]

10.7 Other Forms of Structuralism

In the sense articulated above, then, I would be happy to count myself a perspectivalist (although not a fragmentalist) about spacetime theories. In the literature, both perspectivalism and fragmentalism are often bound up with discussions of 'structural realism' – and while I find much of the structuralist *opera* (French 2014; Ladyman et al. 2007) to be congenial and compelling, I want to indicate in this section two views in the vicinity of structuralism which I do not endorse.

The first is the aspect of the above-discussed dynamical approach to spacetime theories, which claims that it is *incoherent* for spacetime symmetries to come apart from dynamical symmetries,[26] for the two are analytically related (for claims of this kind, see e.g. Acuña 2016; Myrvold 2019). I reject this claim; rather, I side with the orthodoxy in regarding (say)

[25] I am very grateful to Mahdi Khalili for discussions on the contents of this section.

[26] The distinction between spacetime and dynamical symmetries is standard; for the definition of each, see Earman (1989, ch. 3). Famously, Earman maintains that the matching of such symmetries is not given *a priori*, but is nevertheless an 'adequacy condition' on the formulation of a given theory.

the different spacetime settings for Newtonian mechanics as coherent metaphysical possibilities. While the dynamical approach's seeking to reduce spacetime structure to dynamical symmetries is compelling and intriguing, (a) it does not undercut this metaphysical plurality, and (b) it needs to be undergirded by a precisely articulated metaphysical strategy (see e.g. the liberalised Humean approach of Huggett 2006, 2009; Pooley 2013; Stevens 2014, 2015, 2020).

Why do I reject this aspect of the dynamical view? In part, my reasons have to do with my endorsement of what Møller-Nielsen calls the 'motivational' approach to symmetries (Martens and Read 2021; Møller-Nielsen 2017; Read and Møller-Nielsen 2020a, 2020b), according to which it is insufficient to simply *declare* that one's ontological commitments are picked out by (the invariant structures associated with) the symmetries of one's physical theories (as on the opposing 'interpretational' approach to symmetries); rather, one needs to do the metaphysical hard graft of providing a positive articulation of what such a structure is supposed to be, in advance of making such declarations of physical equivalence. The issue with the above version of the dynamical approach is that it declares that spacetime symmetries *just are* dynamical symmetries, but it does not necessarily accompany this with a positive metaphysical picture of what the structure of spacetime *actually is*. Furthermore, it may be that the positive metaphysical picture of the structure of spacetime which one *does* construct is such that spacetime symmetries do not align with dynamical symmetries. Consider, for example, the development of Newtonian spacetimes. As Dasgupta writes,

> we can draw the conclusion of the inference only when we have the alternative theory in hand and have shown that all else is equal. This explains why it was rational for Newton to believe in absolute velocity even though he knew that it was variant in [Newtonian gravity] and undetectable. The reason this was rational for him was that he had no good alternative theory to hand. (Dasgupta 2016, p. 854)

Here, it strikes me as wrong to say that Newton was simply being incoherent in his postulation of the persisting points of absolute space – as the above version of the dynamical approach would have it.[27,28]

The second version of structuralism which I hesitate to accept is what Dewar (2019) calls 'external sophistication' about symmetries, according to which, by treating symmetry-related solutions of a given physical theory 'as if' they are isomorphic, one can give a positive articulation of that theory's (non-perspectival) metaphysical commitments.[29] The reason, again, is that one has to give a positive articulation of what the non-perspectival reality is supposed to be, before such declarations of equivalence are issued; moreover, the possibility that one's ontological commitments might be given by pseudo-tensors and other

[27] For recall that Newtonian spacetime has *fewer* symmetries than the Galilean symmetries of the laws of Newtonian gravity – see Earman (1989, ch. 2).

[28] Of course, one might question whether Dasgupta's Newton exegesis is accurate: see Saunders (2013) for a recent re-reading of the *Principia*. This latter work does, not undercut the philosophical point which is being made here, however.

[29] In fact, external sophistication was developed in an attempt to bolster the interpretational approach to symmetries – see Martens and Read (2021). Note that external sophistication is to be distinguished as a means of articulating the ontology of symmetry-related models of a physical theory from what Dewar calls 'internal sophistication'; I have no qualms with the latter approach. For discussion of the difference, see Jacobs (2021); Martens and Read (2021).

non-geometric objects casts doubt on whether that invariant description is invariably there to be had (see Martens and Read 2021; Møller-Nielsen 2017; Read and Møller-Nielsen 2020a, 2020b).[30]

This being said – and contrary to Martens and Read (2021), Møller-Nielsen (2017), and Read and Møller-Nielsen (2020a, 2020b) – I am no longer convinced that mathematical reformulation is necessary, even for models which are not isomorphic, in order to secure what the motivationalist calls a 'metaphysically perspicuous characterisation' of the common ontology of these models. (In this paragraph, for the sake of simplicitly, I set aside the above concern that one's best explication of the common ontology of symmetry-related models might invoke the invocation of non-geometric objects.) To illustrate this, consider kinematic shifts in Newtonian gravitation theory.[31] Before moving to the setting of Galilean spacetime,[32] one can identify the invariants of shifted models of one's original theory: these will include relative particle positions and velocities. There is an only slightly stronger version of external sophistication which should, in principle, be acceptable to motivationalists: according to this view, once one has identified the invariants associated with symmetry-related models (this step is required; in this sense, the position is stronger than external sophistication as presented in Dewar (2019) and as discussed in Martens and Read (2021)), it is possible that one can construct *directly* a metaphysically perspicuous characterisation of the common ontology of these symmetry related models, *without* recourse to a mathematical reformulation of the original models. The mathematical reformulation may constitute a crutch for securing a metaphysically perspicuous characterisation of the common ontology of these symmetry-related models, but on this view it is not *necessary* for reaching that end. In the current example, the question would be: can one construct a metaphysically perspicuous characterisation of the common ontology of shifted Newtonian models *directly* from the above-mentioned invariant quantities?[33]

Thus, while I count myself a perspectivalist in the sense that I regard frame-dependent effects as being physical and real (at least for those frames which can be appropriately operationalised), and moreover while I find much contemporary work on structuralism to be compelling and insightful, I am hesitant to sign up straightforwardly to structuralism in either of the two senses considered in this section. Before one can declare models of a given physical theory to be associated with the same (non-perspectival) reality, one has to present a positive picture of what said reality is supposed to be;[34] the relevant aspects of the dynamical approach, and external sophistication as presented in Dewar (2019), do not suffice (at least without modification of the kind discussed above) to this end.

[30] One option for Dewar here might be to appeal to the definability of objects, rather than their invariance. My thanks to Fiona Doherty for this suggestion, which I hope to pursue in a future piece.

[31] See e.g. Maudlin (1993) for a clear introduction to such shifts and their historical origins in the correspondence between Leibniz and Clarke (Alexander 1977).

[32] For background on different spacetime settings for Newtonian mechanics, see Earman (1989, ch. 2).

[33] I am grateful to Caspar Jacobs for discussion on this paragraph.

[34] That is, one has to address what French calls 'Chakravartty's challenge' (French 2014, pp. 48–49). See (Chakravartty 2007, p. 26) for the original work, which I take to be in the same spirit as Møller-Nielsen's motivationalism.

10.8 Conclusions

In this chapter, I've related the notions of natural kinds/projectable predicates/real patterns to the notion of a geometric object, and have used this to account for the widespread suspicion towards non-geometric objects. That notwithstanding, I've argued that non-geometric objects can have physical content in a perspectivalist/fragmentalist sense, and might arguably also be associated with natural kinds, at least in some cases. Although I endorse only perspectivalism, and not fragmentalism, I've set both of these views apart from aspects of *prima facie* related structuralist positions which I reject. I have also considered how perspectivalism and fragmentalism interact with the dynamical approach to spacetime theories.

The broader lesson of this piece is the following. One finds in contemporary physics multifarious 'fantastic beasts' (cf. Dürr 2019) – the pseudo-tensors arising via application of Noether's theorems to general relativity are just one example. It is incumbent upon philosophers to engage with the metaphysics of all such objects, rather than merely with more terrestrial and prosaic objects such as tensors. Moreover, the study of such objects has the potential to bolster positions such as perspectivalism and fragmentalism in the philosophy of science.

References

Acuña, P. (2016). Minkowski spacetime and Lorentz invariance: the cart and the horse or two sides of a single coin? *Studies in History and Philosophy of Modern Physics*, **55**, 1–12.

Alexander, H. G. (1977). *The Leibniz-Clarke Correspondence: Together with Extracts from Newton's Principia and Opticks*, new ed., Manchester: Manchester University Press.

Bell, J. S. (1987). How to teach special relativity. In J. S. Bell, *Speakable and Unspeakable in Quantum Mechanics*. Cambridge: Cambridge University Press, pp. 67–80.

Bird, A. and Tobin, E. (2017). Natural kinds. In E. N. Zalta, ed., *The Stanford Encyclopedia of Philosophy*.

Blackburn, S. (2008). *The Oxford Dictionary of Philosophy*, 2nd ed., Oxford: Oxford University Press.

Brading, K. (2005). A note on general relativity, energy conservation, and Noether's theorems. In A. J. Kox and J. Eisenstaedt, eds., *The Universe of General Relativity*, Einstein Studies, vol. 11. Basel: Birkhäuser, pp. 125–35.

Brown, H. R. (2005). *Physical Relativity: Spacetime Structure from a Dynamical Perspective*. Oxford: Oxford University Press.

Brown, H. R. and Pooley, O. (2001). The origins of the spacetime metric: Bell's Lorentzian pedagogy and its significance in general relativity. In C. Callender and N. Huggett, eds., *Physics Meets Philosophy at the Plank Scale*. Cambridge: Cambridge University Press.

Brown, H. R. and Pooley, O. (2006). Minkowski space-time: a glorious non-entity. In D. Dieks, ed., *The Ontology of Spacetime*. Amsterdam: Elsevier.

Brown, H. R. and Read, J. (2021). The dynamical approach to spacetime theories. In E. Knox and A. Wilson, eds., *The Routledge Companion to the Philosophy of Physics*. London: Routledge, pp. 70–85.

Chakravartty, A. (2007). *A Metaphysics for Scientific Realism*. Cambridge: Cambridge University Press.

Cheng, B. and Read, J. (2021). Shifts and reference. Unpublished manuscript.

Curiel, E. (2019). On geometric objects, the non-existence of a gravitational stress-energy tensor, and the uniqueness of the Einstein field equation. *Studies in History and Philosophy of Modern Physics*, **66**, 90–102.

Curiel, E. (forthcoming). Schematizing the observer and the epistemic content of theories. *Studies in History and Philosophy of Modern Physics*. https://arxiv.org/abs/1903.02182.

Dasgupta, S. (2015). Inexpressible ignorance. *Philosophical Review*, **124**(4), 441–80.

Dasgupta, S. (2016). Symmetry as an epistemic notion (twice over). *British Journal for the Philosophy of Science* **67**(3), 837–78.

Dennett, D. (1991). Real patterns. *Journal of Philosophy* **88**(1), 27–51.

Dewar, N. (2019). Sophistication about Symmetries. *British Journal for the Philosophy of Science*, **70**, 485–521.

De Haro, S. (2022). Noether's theorems and energy in general relativity. In J. Read and N. Teh, eds., *The Philosophy and Physics of Noether's Theorems*. Cambridge: Cambridge University Press (Chapter 9, this volume).

Dirac, P. (1958). *The Principles of Quantum Mechanics*, 4th ed., Oxford: Oxford University Press.

Dürr, P. (2019). Fantastic beasts and where (not) to find them: local gravitational energy and energy conservation in general relativity. *Studies in History and Philosophy of Modern Physics* **65**, 1–14, 2019.

Dürr, P. (2021). Against 'functional gravitational energy': a critical note on functionalism, selective realism, and geometric objects and gravitational energy. *Synthese*, **199**, 299–333.

Earman, J. (1989). *World Enough and Spacetime*. Cambridge, MA: MIT Press.

Einstein, A. (1905). Zur Elektrodynamik bewegter Körper. *Annalen der Physik*, **17**, 891–921. Reprinted as Doc. 23 in J. Stachel, et al. (eds.), *The Collected Papers of Albert Einstein*, Vol. 2. Princeton, NJ: Princeton University Press, 1987–2006.

French, S. (2014). *The Structure of the World*. Oxford: Oxford University Press.

Fine, K. (2005). Tense and reality. In K. Fine, *Modality and Tense: Philosophical Papers*. Oxford: Oxford University Press.

Goodman, N. (1955). *Fact, Fiction, and Forecast*. Cambridge, MA: Harvard University Press.

Hoefer, C. (2000). Energy Conservation in GTR. *Studies in History and Philosophy of Modern Physics* **31**(2), 187–99, 2000.

Hofweber, T. and Lange, M. (2017). Fine's fragmentalist interpretation of special relativity. *Noûs*, **51**(4), 871–83.

Huggett, N. (2006). The regularity account of relational spacetime. *Mind*, **115**, 41–73.

Huggett, N. (2009). Essay review: physical relativity and understanding space-time. *Philosophy of Science*, **76**, 404–22.

Jackson, F. (1975). Grue. *Journal of Philosophy*, **72**(5), 113–31.

Jacobs, C. (2021). Gauge and explanation: are gauge-quantities explanatory? Unpublished D.Phil. thesis, University of Oxford.

Klein, F. (1892). A comparative review of recent researches in geometry. *Bulletin of the New York Mathematical Society*, **2**, 215–49.

Ladyman, J. and Ross, D. (with D. Spurrett and J. Collier), (2007). *Every Thing Must Go: Metaphysics Naturalized*. Oxford: Oxford University Press.

Lam, V. (2011). Gravitational and nongravitational energy: the need for background structures. *Philosophy of Science*, **78**(5), 1012–24.

Lehmkuhl, D. (2021). The equivalence principle(s). In E. Knox and A. Wilson, eds., *The Routledge Companion to Philosophy of Physics*. London: Routledge, pp. 125–144.

Le Bihan, B. and Read, J. (2018). Duality and ontology. *Philosophy Compass*, **13**, e12555.

Lipman, M. A. (2020). On the fragmentalist interpretation of special relativity. *Philosophical Studies*, **117**, 21–37.

Malament, D. B. (2012). *Topics in the Foundations of General Relativity and Newtonian Gravitation Theory*. Chicago, IL: University of Chicago Press.

Martens, N. C. M. and Read, J. (2021). Sophistry about symmetries? *Synthese*, **199**, 315–44.

Massimi, M. (2018). Perspectivism. In J. Saatsi, ed., *The Routledge Handbook of Scientific Realism*. London: Routledge, pp. 164–75.

Maudlin, T. (1993). Buckets of water and waves of space: why spacetime is probably a substance. *Philosophy of Science*, **60**(2), 183–203.

Maudlin, T. (2007). *The Metaphysics within Physics*. New York, NY: Oxford University Press.

Menon, T., Linnemann, N. and Read, J. (2020). Clocks and chronogeometry: rotating spacetimes and the relativistic null hypothesis. *British Journal for the Philosophy of Science*, **71**, pp. 1287–1317.

Møller-Nielsen, T. (2017). Invariance, interpretation, and motivation. *Philosophy of Science*, **84**(5), 1253–64.

Moore, A. W. (2012). *The Evolution of Modern Metaphysics: Making Sense of Things*. Cambridge: Cambridge University Press.

Myrvold, W. C. (2019). How could relativity be anything other than physical? *Studies in History and Philosophy of Modern Physics*, **67**, 137–43.

Nijenhuis, A. (1952). Theory of the geometric object. Unpublished Ph.D. thesis, University of Amsterdam.

Nozick, R. (2001). *Invariances: The Structure of the Objective World*. Cambridge, MA: Harvard University Press.

Pitts, J. B. (2009). Gauge-invariant localization of infinitely many gravitational energies from all possible auxiliary structures, or, why pseudotensors are okay. Proceedings of the DPF-2009 Conference, Detroit, MI, July 27–31, 2009.

Pitts, J. B. (2010). Gauge-invariant localization of infinitely many gravitational energies from all possible auxiliary Structures. *General Relativity and Gravitation*, **42**, 601–22.

Pitts, J. B. (2012). The nontriviality of trivial general covariance: how electrons restrict 'time' coordinates, spinors (almost) fit into tensor calculus, and $\frac{7}{16}$ of a tetrad is surplus structure. *Studies in History and Philosophy of Modern Physics*, **43**, 1–24.

Pitts, J. B. (2021). Conservation of energy: missing features in its nature and justification and why they matter. *Foundations of Science*, **26**, 559–84.

Pooley, O. (2013). Substantivalist and relationist approaches to space-time. In R. Batterman, ed., *The Oxford Handbook of Philosophy of Physics*. Oxford: Oxford University Press.

Quine, W. V. O. (1969). Natural Kinds. In W. V. O. Quine, *Ontological Relativity and Other Essays*. New York, NY: Columbia University Press.

Read, J. (2020). Functional gravitational energy. *British Journal for the Philosophy of Science*, **71**, 205–32.

Read, J. and Møller-Nielsen, T. (2020a). Motivating dualities. *Synthese*, **197**, 263–91.

Read, J. and Møller-Nielsen, T. (2020b). Redundant epistemic symmetries. *Studies in History and Philosophy of Modern Physics*, **70**, 88–97.

Read, J. and Qureshi-Hurst, E. (2021). Getting tense about relativity. *Synthese*, **198**, 8103–25.
Rowe, D. (2019). Emmy Noether on energy conservation in general relativity. https://arxiv.org/abs/1912.03269.
Saunders, S. (2013). Rethinking Newton's *Principia*. *Philosophy of Science*, **80**, 22–48.
Schouten, J. A. (1954). *Ricci-Calculus: An Introduction to Tensor Analysis and Its Geometrical Applications*. Berlin: Springer.
Slavov, M. (2020). Eternalism and perspectival realism about the 'now'. *Foundations of Physics*, **50**, 1398–1410.
Stevens, S. (2014). The Dynamical Approach to Relativity as a Form of Regularity Relationalism. Unplublished D.Phil. thesis, University of Oxford, 2014.
Stevens, S. (2015). The dynamical approach as practical geometry. *Philosophy of Science*, **82**, 1152–62.
Stevens, S. (2020). Regularity relationalism and the constructivist project. *British Journal for the Philosophy of Science*, **71**, 353–72.
Trautman, A. (1962). Conservation laws in general relativity. In L. Witten, ed., *Gravitation: Introduction to Current Research*. San Francisco, CA: Wiley.
Trautman, A. (1965). Foundations and current problems of general relativity. In S. Deser and K. W. Ford, eds., *Lectures on General Relativity*. Englewood Cliffs, NJ: Prentice Hall.
Wallace, D. (2019). Who's afraid of coordinate systems? An essay on representation of spacetime structure. *Studies in History and Philosophy of Modern Physics* **67**, 125–36.
Weyl, H. (1952). *Symmetry*. Princeton, NJ: Princeton University Press.

11

Substantive General Covariance and the Einstein–Klein Dispute*

A Noetherian Approach

LAURENT FREIDEL AND NICHOLAS J. TEH

11.1 Introduction

In this chapter, we will explore the enduring impact of Noether's legacy on some of the deepest issues in the philosophical foundations of field theory. The import of Noether's two theorems is certainly undisputed in both the physics and the philosophy literature, but an articulation of the why and wherefore of her second theorem's import is rather more difficult to come by. Indeed, most textbooks on field theory stress only Noether's first theorem, and even in those places where Noether's second theorem is mentioned, little more physical insight is offered than the remark that the Noether identities relate the Euler–Lagrange quantities and thus express the 'redundancies' that are characteristic of a gauge theory.

The short shrift given to Noether's second theorem goes hand in hand with a glib narrative about a famous epistolary dispute between Einstein and Felix Klein (letters 480 and 492 in Einstein 1998). Contra Einstein, Klein contended that general relativity does not have any physically contentful conserved quantities; the glib narrative takes Noether's second theorem to settle the dispute decisively in favour of Klein. As we will see shortly, the truth is somewhat more complex and to a large extent vindicates Einstein's strategy for responding to Klein.

A good reason to make clear on the Einstein-Klein dispute is that it holds the key to understanding just how fundamentally Noether's second theorem (albeit not as a piece of mathematics in itself, but as taken up and deployed by physicists) has transformed our understanding of Lagrangian field theory. Nonetheless, our reason for considering this controversy is not primarily historical; we wish instead to bear witness to the continuing vitality of Noether's theorems – as a living force that energises the physical imagination and leads us to make profound new discoveries. Our witness consists in using Noether's second theorem to provide a novel and incisive analysis of what *substantive* general covariance really amounts to: the vision that Einstein so desired to contemplate but could see only

* We thank James Read for helpful comments on this chapter. NT's research is supported by JTF grant 61521. LF's research at Perimeter Institute is supported in part by the Government of Canada through the Department of Innovation, Science and Industry Canada and by the Province of Ontario through the Ministry of Colleges and Universities.

in a mirror dimly. While there has of course been much discussion of substantive general covariance – and the related notion of background independence – in the philosophy and physics literature,[1] our discussion is (to the best of our knowledge) the first to fully exploit the power of Noether's second theorem in this setting.[2] Furthermore, we conjecture in Section 11.3.5 that our Noetherian approach to substantive general covariance provides a royal road to understanding the core phenomenon underlying the elusive notion of 'background independence'.

We begin our reflections in Section 11.2 in a historico-philosophical fashion: by surveying two questions with which Einstein wrestled long and hard, and which have been re-articulated in various guises in the contemporary philosophy of physics literature. First, in Section 11.2.1, we remind the reader of the received view of general covariance (which we call Basic General Covariance) in the context of Lagrangian field theory. Section 11.2.2 then discusses the question of whether and how general covariance plays a role in extending the Principle of Relativity, and Section 11.2.3 discusses the subject of the Einstein-Klein dispute, viz. whether generally covariant theories have non-trivial conserved charges. On the basis of the previous two subsections, Section 11.2.4 formulates what we will call the 'Einstein test' for whether a theory has substantive general covariance.

Section 11.3 develops our novel contribution to the literature: we argue that the machinery of Noether's second theorem provides a powerful technique for determining whether theories have substantive general covariance. Crucially (and unlike many other treatements), we make use not just of the Noether identities, but also of the entire mathematical nexus that relates the Noether identities, the Noether current, and the corner Noether charge which is fixed by a specific choice of Lagrangian. Section 11.3.1 summarises in an elementary and conceptually transparent way the reformulation of Noether's theorems in the covariant phase space framework, much of which was already anticipated in Noether's original work (1918). Section 11.3.2 then applies this machinery to run the Einstein test on several toy theories around which much of the philosophical debate has been centred. On the basis of the results of this test, we articulate in Section 11.3.3 a proposal for what substantive general covariance à la Einstein amounts to: we give a less demanding form (which requires only a non-trivial corner charge), and a more demanding form (which requires an extension of the Principle of Relativity). Section 11.3.4 revisits the case of General Relativity and the Einstein–Klein dispute from this perspective, and Section 11.3.5 comments briefly on the relationship between our notion of substantive general covariance and the idea of 'background independence': while Pooley (2010, 2017) attempts to make the case that the connection between some substantive form of general covariance is 'messier, and less illuminating' than the quantum gravity literature might suggest, our analysis suggests that such pessimism is largely a function of considering insufficiently robust versions of substantive general covariance.

[1] See Pooley (2017), Read (2016), and the references therein.

[2] Brown and Brading (2002) and Earman (2006) also discuss Noether's second theorem in relation to general covariance, but our use of the theorem goes far beyond these discussions.

11.2 What Work Should 'General Covariance' Do?

Although the discussion that follows will apply to Lagrangian field theories quite generally (*mutatis mutandis*), we remind the reader that the themes of interest to us were first explicitly problematised in the context of Einstein's General Relativity. Thus, our meditation will originate here.

The central concept that Einstein struggled so long and hard to formulate, to interpret, and to employ in modelling certain material aspects of the world is that of 'general covariance'. Today, the descendant of this notion that is commonly invoked by physicists is more plainly referred to as the 'diffeomorphism invariance' of General Relativity, but it is important to appreciate that in the early days of the subject, Einstein's conception of general covariance was very much in flux, in terms of its mathematical definition as well as of its physical interpretation.

Here, then, are two key questions that Einstein wrestled with, which are supposed to provide desiderata for a *substantive* – or physically contentful – notion of general covariance:[3]

1. Does general covariance play a role in extending the Principle of Relativity?
2. Do generally covariant theories have non-trivial conserved charges?

As regards the second question, we emphasise that it is clear from Einstein's correspondence with Klein (see letter 480 in Einstein 1998 and our discussion in Section 11.3.4) that the charges Einstein had in mind were *corner* ones, where by 'corner' we mean the boundary of a Cauchy surface – thus, a 'corner' from the perspective of the entire spacetime manifold.[4]

Whatever *substantive* general covariance amounted to for Einstein, it was supposed to earn its spurs by means of a 'yes' answer to the above two questions. We will proceed by laying out in Section 11.2.1 a 'basic' or minimal notion of general covariance that physicists would typically invoke. Sections 11.2.2 and 11.2.3 will explain Einstein's desiderata for the work that the notion of substantive general covariance needs to do. Section 11.2.4 then uses these respective desiderata to formulate a two-part 'Einstein test' for substantive general covariance.

11.2.1 Basic General Covariance

According to standard accounts of the history of General Relativity, Einstein initially intended to use general covariance as an adequacy constraint that would single out General Relativity.[5] To achieve this goal, he initially proposed that a generally covariant theory is

[3] Einstein's struggles with general covariance are of course famously various: for instance, we leave aside in this chapter the question of the relationship between general covariance and determinism, which has been discussed in the 'Hole Problem' literature.

[4] Of course, Einstein did not use the terminology of 'corners'. What he says is that the construction of the charges is analogous to the use of Gauss' Law in electrostatics, and that the 'possibility of this physically important interpretation is based on the fact that the same quantities … which enter in the conservation law … can, by virtue of the field equations … be written as a "divergence"' (Einstein 1998), where by 'quantities' he means what we call the Noether charge J below. This means that, returning to standard modern terminology, the on-shell charge integral will descend to a corner by Stokes' theorem.

[5] For the history, see the discussion in Norton (1993) and the references therein.

one that can be formulated in such a way that the theory is expressible in any coordinate system (where the different coordinate systems/frames are related by diffeomorphisms).

The problem with this notion of general covariance, as Kretschmann (1918) pointedly remarked, is that it does not achieve Einstein's goal of singling out a particular theory – just about any theory is generally covariant according to this definition, including Newtonian mechanics, as Cartan showed in his invariant, curved geometric, formulation of Newtonian gravitation (often called Newton–Cartan theory). In light of Kretschmann's response to Einstein, it is now commonly accepted (see Norton 2003 and references therein for discussion) that Einstein's early formulation of general covariance was a misstep. Instead, one should try to provide a 'physically contentful' or substantive notion of general covariance that really does pick out some proper subset of physical theories, and perhaps General Relativity in particular.

Within the philosophical literature, one finds various attempts to articulate a version of substantive general covariance that distinguishes it from the mere freedom to express a theory's fields, equations of motion, and solutions in arbitrary frames (see Pooley 2010 for a review). For the purposes of this piece, however, it will be helpful to build up to substantive general covariance by starting with the following notion that most physicists take as standard (albeit typically under the head of 'diffeomorphism invariance'):[6]

(BGC) First, distinguish between the background (non-dynamical) and the dynamical fields of the theory. We then say that a Lagrangian of a theory is *basically generally covariant* (BGC) just in case diffeomorphisms of its dynamical fields are variational symmetries[7] of the Lagrangian.[8]

By extension, we will also use BGC to refer to the analogous case in which diffeomorphisms are replaced by internal gauge symmetries; which case we mean will be obvious from the context. The broad contrast here is between theories whose actions have only rigid symmetries (and thus do not satisfy BGC) and theories with BGC, i.e. whose actions have symmetries that are functions of spacetime (such as diffeomorphism symmetry and internal gauge symmetry).

It should be clear that BGC is just what physicists typically mean when they say that a Lagrangian theory is diffeomorphism invariant (*mutatis mutandis* for internal gauge symmetry, where the term 'gauge-invariant' is used), and it should be equally clear that physicists generally take this feature to be informative about the physical content of the system being modelled.[9] For reasons that will become clear in Section 11.3 (and which are amply discussed in Freidel et al. 2020 and 2021), we note that we do *not* assume that a theory has a unique action: thus, the Einstein–Hilbert Lagrangian should be understood as just one out of a possible class of standardly stipulated Lagrangians for General Relativity.

[6] We will see in Section 11.3.2 that only *some* cases of basic general covariance are inequivalent to the freedom to express a theory in an arbitrary frame.

[7] In other words, diffeomorphisms leave the Lagrangian invariant up to exact terms; we remind the reader that, as a consequence, such diffeomorphisms also take solutions of the theory's equations of motion (EOM) to solutions.

[8] For ease of comparison, we note that our BGC is the variational version of what Pooley calls 'GC5' (2010) and what he calls 'Diffeomorphism Invariance (final version)' (2017). Pooley (2017) also identifies the variational form with what he calls 'Background Independence (version 2).

[9] As Norton (2003) says, the feature is taken to be physically contentful given an interpretation of the formalism, including which fields in the Lagrangian count as dynamical and which as background.

Having put on the table the standard notion of BGC, let us now try to understand what work Einstein would have general covariance do in order to qualify as *substantive*.

11.2.2 Extending the Relativity Principle

We now turn to our first question, which is about the relationship between general covariance and the Principle of Relativity (PR). At one point in the development of General Relativity, Einstein thought that general covariance was simply an extension of the PR in Special Relativity. Later, as part of his response to Kretschmann's (1918) critique, he retreated to the view that general covariance is necessary but insufficient for extending PR (see Pooley 2017 for a discussion of Einstein's retreat). For our purposes, it will be more fruitful to frame the question thus: does one's preferred notion of general covariance play a distinctive role in extending the Special Relativistic PR (or, more generally, analogs of PR for internal gauge symmetry)? If 'yes', then we can consider it to be substantive in virtue of playing this role.

Here again one finds the familiar story of Einstein making a bold and potentially physically insightful conjecture, but with murky conceptual foundations, and which was thus amenable to being problematised by others. Correspondingly, there has been much controversy about whether *any* notion of general covariance can play a role in extending PR. At least within the philosophical literature, one finds various authors (e.g. Belot 2000; Norton 1993) who answer 'no' because they claim that the desired extension of the PR to General Relativity must be trivial, in the sense that the resulting 'relativity' group is just the identity. In brief, these authors argue for their claim by appealing to the following controversial assumption about PR, which we will call **Geometric PR**: in such an extension, the empirically significant group of symmetries invoked in the Principle of Relativity should be identified with the stabiliser group of the spacetime metric in General Relativity (just as it is typically identified with the stabiliser group of the Minkowski metric in Special Relativity).

If one accepts Geometric PR, then since a generic metric – which is now a dynamical field – in General Relativity has no non-trivial automorphisms, it immediately follows that the PR is trivial in General Relativity. Thus, no form of general covariance (including BGC) can play a role in enabling such an extension. We stress that this conclusion is wholly driven by the assumption of Geometric PR: if the argument works, it works regardless of what one takes substantive general covariance to be.

The reason Geometric PR is controversial is that it is far from clear that this is the most physically perspicuous and relevant interpretation of the PR. Indeed, Brown and Sypel (1995) articulate a version of the PR which they take to be both the more physically perspicuous interpretation, as well as the one that more faithfully reflects Einstein's own understanding of PR. According to Brown and Sypel, the symmetry group of PR is to be understood as the symmetry group relating inertial frames, which are in turn to be fleshed out as – in idealisation – ways of encoding the equivalence of outcomes of experiments set up with the same initial conditions in an *isolated* subsystem. Thus, when in Special Relativity one goes on to define a geometric object (the Minkowski spacetime metric) whose stabiliser group is precisely the symmetry group of the PR, one is merely codifying 'aspects of the comparative behaviour of different systems of physical rods and clocks in relative motion', where the behaviour of the target subsystem (whose inertial frames are related by these symmetries) is being measured with respect to an environment frame, from which the subsystem is dynamically isolated. Of course, once one understands this point, one can abstract away from a concrete material system and even its idealisation in terms of inertial frames, and arrive at a kind of 'iconic' or 'generic' image that we call 'Minkowski spacetime', which provides an efficacious vehicle for abstract reasoning. This generic 'type' can then be made to descend once again into a representation of a concrete empirical scenario when it is filled in with the description of the relevant subsystem-environment decomposition, initial and boundary conditions including the relevant conditions for 'isolation', and the practical knowledge of the experimentalist.

Let us call the understanding of PR that we have just articulated **Embodied PR**, because of its emphasis on how the Principle of Relativity (and its symmetries) needs to be understood not just formally, but as an embodiment of concrete empirical scenarios and their idealisations (such as boundary conditions). We will adopt this embodied understanding of PR (and of symmetries more generally) because we are interested in the physical and empirical content of symmetries, which stems from an understanding of how the formal apparatus is used to represent particular subsystem-environment relationships and the related boundary conditions.

Notice that, according to Embodied PR, one needs to be able to understand how a theory models an 'isolated subsystem' *before* one can even articulate PR within that theory. Thus, advocates of Embodied PR will not accept the above argument for Geometric PR, because it is based on a definition of PR that ignores the question of how 'dynamical isolation' is to be modelled in the extended scenario.[10] Furthermore, a moment's thought will show that 'dynamical isolation' is a much more sophisticated notion in General Relativity than it is in Special Relativity: we typically model it as the asymptotic flatness of a subsystem spacetime.

To recapitulate, we have arrived at the following way of sharpening our second question: is there a notion of general covariance that has a distinctive role to play in extending PR – in the sense of Embodied PR – from Special Relativity to General Relativity? If we frame things from Einstein's perspective, then the question becomes: can we articulate a suitably generalised version of PR if we are concerned with asymptotically flat subsystems? But it is important to note that the question can be generalised so that we consider isolated finite subsystems as well, and not just asymptotic infinity.[11]

[10] Ironically, Brading and Brown (2004) have expressed sympathy for Geometric PR on the grounds that the diffeomorphism symmetry of GR 'does not have an active interpretation in terms of isolated subsystems of the universe'. The debate about whether and in what sense local symmetry (of which the diffeomorphism symmetry of GR is an instance) can have such an interpretation has by now played out quite fully (see Teh 2016 and references therein), but our interest here is in highlighting that, on an embodied view (and again, in contrast to Geometric PR), one's conception of substantive general covariance has to do real work in adjudicating whether (non-extension) is true.

[11] Furthermore (although it lies beyond the scope of this chapter), it is also physically relevant to consider non-isolated subsystems in which charges are not conserved because of symmetry-breaking.

In Section 11.3, we will see that Noether's theorems strongly suggest that the answer to the above question is 'yes'. But before that, we will need to turn to a second related facet of Einstein's struggle with general covariance, and the one that led Noether to write her 1918 paper.

11.2.3 Non-Trivial Corner Charges

The second question of this section is: is there a notion of general covariance that yields physically meaningful corner charges (such as energy)? For instance, General Relativity is a theory whose Lagrangian has the diffeomorphism symmetries of BGC: in virtue of this, does it have physically meaningful charges? (We will see in Section 11.3 that, from the perspective of the framework developed by Noether, it is clear that this question is intimately related to our first question.)

Historically speaking, this question found its genesis in Hilbert's variational treatment of General Relativity (see Rowe 2019 and 2021 for a masterful treatment of the history, which we follow here). In addition to deriving the field equations in his treatment, Hilbert's main contribution (which he called 'the most important goal of his theory') was the formulation of an invariant 'energy vector' and a proof of its conservation. In response, Einstein provided his own somewhat different variational approach to General Relativity; in this note he derived an energy current which can be written in modern notation as $J_X = C_X + dU_X$, where C_X is the constraint associated with the gauge transformation X. C_X vanishes on-shell. U_X is referred to as the charge aspect or superpotential (these formulae are generic for a gauge theory, but we will provide explicit details for electromagnetism in the next section).[12] Einstein also stressed the importance of the conservation law $dJ_X = 0$ for the on-shell current.

This series of developments set the stage for Emmy Noether's contribution to the mathematical foundations of General Relativity. First, as Felix Klein's assistant, Noether worked out the exact relationship between Hilbert's energy vector and Einstein's energy current. This analysis, in conjunction with Klein's own work on the topic, led to a open letter from Klein to Hilbert in which Klein explained that the energy current J_X could always be decomposed as the sum of two parts, the first of which vanishes on-shell, and the second of whose divergence is identically zero (where we emphasise that the latter is an off-shell statement). In consequence, Klein asserted, the statement that $dJ_X = 0$ is merely a mathematical identity and does not have any physical content.

Upon reading this letter, Einstein wrote Klein to express admiration for his insights, but also to protest that 'I regard what you remark about my formulation of the conservation laws as incorrect.' According to Einstein, the conservation of the current was not itself a mathematical identity, but was instead the consequence of a mathematical identity *and* the equations of motion. Einstein further argued that the physical interpretation of such a conservation statement was analogous to the integral form of Gauss's law in the following

12 We direct the reader to De Haro (2021), Chapter 9 in this volume, for an explication of Einstein's notation $J = \mathfrak{T} + \mathfrak{t}$ and how he thought about the objects $\mathfrak{T}$ and $\mathfrak{t}$.

sense: since the current J_X can be shown to be on-shell exact, i.e. $J_X \approx dU_X$, one can use Stokes' theorem to write the on-shell charge (the integrated current) as a *corner* (relative to the spacetime manifold) quantity, i.e. $\int_\Sigma dU_X = \int_{\partial\Sigma} U_X$. He then proceeded to sketch a particular subsystem for whose boundary conditions (essentially a primitive version of asymptotic flatness) such a computation of charge made sense.

As regards Einstein and Klein, their further exchange only led to a stalemate, with Einstein insisting that his conception of energy conservation in General Relativity was physically contentful, and Klein denying this claim on the grounds that 'mathematical identities' cannot be physically contentful statements. But against the background of this impasse concerning physical interpretation, mathematical progress was made: Noether published her 1918 'Invariante Variationsprobleme'. Herein, she proved two theorems (referred to in the physics community as Noether's first theorem and Noether's second theorem respectively) and derived a corollary (the so-called Hilbertian assertion) that bore directly on the Einstein–Klein debate. In our terminology, the corollary states that Lagrangian field theories with BGC only have (in virtue of that BGC) trivial currents, where we remind the reader that a 'trivial current' is one that can be written as the sum of an exact form and a term that vanishes on-shell (i.e. our C_X and dU_X above respectively). Thus, the on-shell current is identically conserved and is said to have an 'improper conservation law'.

It would be a mistake to read Noether here as weighing in on the question of physical interpretation. Rather, one should understand her as offering a mathematician's conceptual analysis: given that you think the concepts of BGC and trivial currents are physically relevant, you should be aware of the following mathematical relationship between them. Read this way, Noether's corollary does not answer our second question, but it does prompt the following specific version of it: 'Does a theory with BGC have – on that basis – physically meaningful conserved charges?' where we should now understand the latter as 'physically meaningful charges that stem from trivial currents', since (as Noether showed) non-trivial currents cannot be derived from the diffeomorphism (or gauge) symmetry of BGC.

11.2.4 The Einstein Test for 'Substantive'

To sum up this section, reflection on Einstein's struggles with 'extending the principle of relativity' and 'defining conserved charges' within the context of GR leads us to what we might call the 'Einstein test' for a *substantive* notion of general covariance:

1. The 'corner charge' part of the Einstein test (corresponding to Section 11.2.3): Does the notion of substantive general covariance yield non-trivial corner charges?[13]
2. The 'Extending PR' part of the Einstein test (corresponding to Section 11.2.2): Does the notion of substantive general covariance allow us to extend the principle of relativity (in the sense of Embodied PR that we discussed above)?

[13] We note that, based on the history, one might reasonably take the corner charge part of the test to include the *conservation* of non-trivial corner charges. However, for the purposes of analytical clarity, it will be convenient for us to make the minimum criterion the existence of *non-trivial* corner charges, with conservation left as a further criterion.

Given the status of BGC as the 'received' version of general covariance (when formulated variationally) and Einstein, Klein, and Noether's own recognition of the relevance of BGC to the question of conserved charges, it is natural for us to first try to run the Einstein test on BGC. In the next section, we will see how the raw mathematical materials of Noether's theorems can be taken up – in the art of the physicist – to give a particularly trenchant formulation of this test: we will find that BGC does not pass the test, but also that the machinery of Noether's second theorem suggests a version of substantive general covariance that does.

11.3 General Covariance through a Noetherian Lens

Since Kasia Rejzner (2020) in this volume expounds on Noether's two theorems from a more technical perspective, namely the BV complex, we shall here take a more down-to-earth approach in which we work in ghost degree zero. However, we stress that the fundamental observations that we make can be reconstructed in terms of the cohomology of the BV complex – a beautiful connection with another part of Noether's career since she was one of the developers of algebraic homology groups, albeit not in the context of her two theorems. To set the stage, we note that at a highly schematic level, the chief moral of Noether's two theorems for Lagrangian field theory is that the space of such schematic theories can be partitioned according to whether or not the theories possess two features, which in turn correspond to the symmetry-structure of the theories' Lagrangians. These two features are:

- Noether currents. A *non-trivial* Noether current is one that cannot be written as the sum of an exact form and a term that vanishes on-shell. A *trivial* Noether current is one that can be so written.
- Noether identities, viz. relations between the Euler-Lagrange quantities that take the form $N^i\left[\frac{\delta L}{\delta\phi^i}\right] = 0$; we emphasise that since these are mathematical identities, they hold off-shell. A *non-trivial* Noether identity is one whose co-efficients N^i do not vanish on-shell. A *trivial* Noether identity is one whose co-efficents vanish on-shell.

Furthermore, the correspondence to the symmetry-structure is given by Noether's two theorems (and their modern interpretations), as follows:[14]

- Rigid symmetries (symmetries with a finite set of parameters) are in 1–1 correspondence with non-trivial Noether currents. Furthermore, these currents are conserved when we assume the on-shell property in computing the divergence of the current.
- Gauge symmetries (symmetries with an infinite set of parameters, i.e. symmetries in the sense of BGC) are in 1–1 correspondence with non-trivial Noether identities. Furthermore, they yield trivial currents whose on-shell expression is conserved as a mathematical identity, i.e. without using the on-shell property in computing the divergence.

[14] Here our presentation of the 'iff' statement is heuristic. Making this an honest theorem requires considerable care about how to define equivalence classes of symmetries and currents, as demonstrated by Olver (2000).

Let us now descend from this schematic discussion to a more realistic physical context in which we consider boundary conditions of subsystems. In particular, we will use the covariant phase space framework to understand how Noether's theorems get applied in such a context.

11.3.1 The Covariant Phase Space Approach to Noether

First, we sketch the spacetime geometry of the subsystems that we will consider. It will suffice to consider the elementary case of a (simply connected) spacetime M with timelike boundary ∂M, and into which we insert an initial Cauchy surface Σ; the corner $S = \partial \Sigma$ lies in ∂M. Introducing such a boundary ∂M is crucial for modelling the boundary conditions of a subsystem; furthermore, one can think of such a finite boundary model as a regularised version of a subsystem with asymptotic boundary conditions.

Consider the equivalence class $[L]$ of Lagrangians on M, where the equivalence relation is given by $L \sim L + d\,\ell$, i.e. two representatives of the class differ by a boundary Lagrangian ℓ (on ∂M). Such an equivalence class corresponds to a set of bulk equations of motion. In the literature, there are two ways of interpreting the representatives of this class, corresponding to two ways of interpreting the fundamental variational formula, a version of which appeared in Noether's 1918 paper, i.e.

$$\delta L = -\mathcal{E}\delta\phi + d\theta, \tag{11.1}$$

where we remind the reader that $\mathcal{E} = E_\phi \delta\phi$ is a field space 1-form whose co-efficient E_ϕ is the Euler–Lagrange quantities, and θ is a field space 1-form called the (pre-)symplectic potential current.

According to the first approach, θ is defined by the fundamental variational formula, and the representatives correspond to an ambiguity in the definition of θ. According to the second approach (for which see De Haro 2021; Freidel et al. 2021; and references therein), the variations are well-defined only when we know the full bulk and boundary data – corresponding to a particular representative L (including a possible boundary Lagrangian piece) – and this fully specified data picks out a unique θ by means of Anderson's (1989) homotopy operators; this unique θ is then the one that appears in the fundamental variational formula.[15] Here we will follow the second approach, because the incorporation of the boundary data is crucial to resolving the questions we have raised in Section 11.2.

We remind the reader that given the unique θ corresponding to some L, we can construct the presymplectic form $\Omega := \int_\Sigma \delta\theta$. If the Lagrangian has a symmetry ξ, we can also construct a corresponding Noether current $J_\xi := I_\xi \theta$ and Noether charge $Q := \int_\Sigma J_\xi$, ξ is the symmetry vector field on field space, and I is the field space interior product.[16] The statement of Noether's first and second theorems can then be understood in terms of these quantities.

[15] We also assume that we have fixed a particular set of coordinates in field space.

[16] We note that, below, we will adopt a standard abuse of notation in which the notation for a vector field on field space ξ is also used to represent the gauge parameter that appears in the co-efficient of that vector field.

From this perspective, the basic statement of Noether's first theorem arises when ξ is a variational symmetry of L, i.e. $L_\xi L = dl_\xi$ (where L_ξ is the field space Lie derivative). We can then use Cartan's magic formula and (11.1) to obtain the conservation statement $dJ_\xi = I_\xi \mathcal{E}$, where the Noether current J_ξ is defined as $J_\xi := I_\xi \theta - l_\xi$. Furthermore, when ξ is a rigid symmetry, then J_ξ is a non-trivial Noether current, in the sense discussed above, and so the current is conserved only on-shell.

On the other hand, when the symmetry ξ is a non-constant function of spacetime and thus has an infinite number of parameters, we find that we can write $I_\xi \mathcal{E} = dC_\xi + \mathcal{N}_\xi$, where C_ξ is a constraint term that vanishes on-shell and $\mathcal{N}_\xi$ is the sum of terms linear in ξ, each of which contains a differential operator acting on the Euler–Lagrange quantities. We can then use the locality of ξ to argue that $\mathcal{N}$ must vanish as an identity – this yields the non-trivial Noether identity (often also called a Bianchi identity) corresponding to the gauge symmetry ξ. By combining this observation with $L_\xi L = dl_\xi$ and (11.1), we can conclude that $d(J_\xi - C_\xi) = 0$, and so the Noether current takes the form $J_\xi = C_\xi + dU_\xi$, where the 'superpotential' or 'charge aspect' U_ξ depends on the gauge parameter ξ in a linear but non-trivial manner. In other words, we have just shown that gauge symmetry corresponds to a trivial Noether current, in the above sense. There are two extremely important points to emphasise about the superpotential U_ξ at this stage:

- The superpotential U_ξ is uniquely determined by our choice of Lagrangian in the equivalence class $[L]$ and the symmetry generator ξ.[17]
- For gauge theories, i.e. theories with BGC, the Noether charge $Q_\xi^\Sigma := \int_\Sigma J_\xi = \int_{\partial\Sigma} U_\xi$ is always a *corner* quantity, because the Noether current is on-shell exact, and determined by the superpotential U_ξ.

With these observations in hand, we can see that it is the 'corner charge' part of the Einstein test which is more fundamental than the 'Extending PR' part of the test: if the corner charges for a gauge theory are trivial (i.e. vanishing), then there is no sense in which the PR can be extended (from a rigid precursor theory) to that gauge theory. If, on the other hand, the corner charges are non-trivial, then subject to understanding what kinds of boundary conditions model an 'isolated system', it may be possible to obtain a generalised version of PR in that gauge theory.

11.3.2 Test Cases

With the above Noetherian machinery in place, we are now in a position to run the 'corner charge' part of the Einstein test on BGC, i.e. to pose the question: does a theory with BGC thereby have non-trivial corner charges? To that end, it will be helpful to make a connection here with some of the toy models that Pooley (2010, 2017) and others use to test the validity of different definitions of general covariance, although we warn the reader that for reasons of simplicity, we have sometimes opted for a different labelling convention from that of Pooley (2017).

[17] See Freidel et al. (2021) for discussion, and see De Haro (2021) for an explicit formula for U_ξ.

Consider first the following theory of a free scalar field on Minkowski space that we will call **SR1**:[18]

$$L(\phi,\eta) = -\frac{1}{2}\eta^{ab}\partial_a\phi\partial_b\phi, \tag{11.2}$$

where the scalar field ϕ is the dynamical variable and the Minkowski metric η is a non-dynamical (or background) structure. Our variational formula for this theory is $\delta L = E_\phi\delta\phi + \nabla_a\theta^a$, where the Euler–Lagrange expression is $E_\phi = \Box\phi$ and the pre-symplectic potential current is $\theta^a = -\nabla^a\phi\delta\phi$. We note that the conservation of the stress-energy tensor T^{ab} follows from applying Noether's first theorem to the rigid spacetime symmetries of this theory.

Clearly, this theory does not satisfy BGC, because diffeomorphisms are not a variational symmetry of ϕ. In our next example, we will modify **SR1** slightly to show that one can produce an example of a theory that satisfies BGC, but which fails the 'corner charge' part of the Einstein test, because it has only a trivial superpotential, and thus the corner charge vanishes.

11.3.2.1 BGC with Trivial Superpotential

Consider now a theory that we will call **SR2**, which is given by,

$$L^{\mathbf{SR2}}(\phi, Y;\eta) = L(\phi, Y^*\eta), \tag{11.3}$$

where $Y\colon \mathbb{R}^d \to M$ is a parametrisation field that captures our freedom to choose coordinates on the spacetime M.[19] Due to the latter interpretation of Y, it is appropriate to call **SR2** a 'Kretchmannized' version of **SR1**.[20] Our dynamical variables in this theory will be ϕ and Y, whereas the Minkowski metric η still retains its status as a non-dynamical background structure.

We can again compute the variational formula (11.1) for this theory, thus obtaining $\delta L^{\mathbf{SR2}} = E_\phi\delta\phi + E_a\chi^a + \nabla_a\theta^a$, where $\chi := \delta Y \circ Y^{-1}$, $E_a = -2\nabla_b T^{ba}$, and $\theta^a = 2T^{ab}\chi_b - \partial^a\phi\delta\phi$. Notice that in addition to the previous equation of motion $E_\phi = 0$, we now have a new equation of motion $E_a = 0$, i.e. the conservation equation for the stress-energy tensor. Furthermore, by inspecting θ, we can see that in addition to the original symplectic pair ϕ and its momentum $\nabla^a\phi$, we now have a new symplectic pair Y^a and its momentum T^{ab}.

We now draw the reader's attention to an interesting point: clearly, $L^{\mathbf{SR2}}$ is covariant in the sense of BGC (this is true even though η remains a non-dynamical field and is not varied) and thus this theory must have a non-trivial Noether identity. By turning the crank of the Noether machine, we find that this identity is $E_\phi\partial_a\phi - 2\nabla_b T^{ba} = 0$ (that this is an identity can be straightforwardly checked from the definition of T_{ab}). In other words, the content of

[18] Pooley (2017) calls the non-variational formulation **SR1**.

[19] Pooley (2017) calls this **SR5**, and his example of **SR4** is similar.

[20] We note in passing that Yang–Mills theory and General Relativity can be analogously Kretchmannized, and this leads to the construction in Donnelly and Freidel (2016).

the Noether identity **SR2** is actually the content of Noether's first theorem for **SR1** – one might call this the *transmutation* of Noether's first theorem into Noether's second theorem via Kretchmannization!

It is also interesting to compute the (trivial) Noether current, i.e $J_\xi^a = I_\xi \theta - \iota_\xi L^{\mathbf{SR2}} = 0$. In other words, the current vanishes identically off-shell, and we can deduce that the superpotential U_ξ – and thus the corner charge algebra – is trivial.

To sum up, the theory **SR2** demonstrates why BGC fails the 'corner charge' part of the Einstein test for substantive general covariance (and in consequence fails the 'extending PR' part of the test): even though this theory clearly has BGC, it nonetheless fails to deliver non-trivial corner charges.

11.3.2.2 BGC with Non-Trivial Superpotential

Next, we turn to a familiar example in which BGC does yield a non-trivial corner charge (and thus passes the 'corner charge' part of the Einstein test). Consider the Maxwell gauge theory Lagrangian

$$L = -\frac{1}{4} F \wedge \star F, \tag{11.4}$$

where we have chosen a representative $L \in [L]$ with vanishing boundary Lagrangian, i.e. $\ell = 0$. The variational formula is

$$\delta L = -E\,\delta A + d\theta, \tag{11.5}$$

where δ is the exterior derivative on the space of fields, $E = d \star F$ is the Euler–Lagrange expression, and $\theta = \star F\,\delta A$ is the presymplectic potential current.

In this case, the gauge symmetry vector field on field space has the form $\hat{X} = dX(x,t)\delta/\delta A$, and $L_{\hat{X}}$ is the field space Lie derivative with respect to this vector field, although we will now abuse notation by using X to denote both the field space vector field and the local gauge parameter. Since $L_X L = 0$, we can repeat the general analysis summarised in Section 11.3.1 to obtain

$$0 = L_X L = d(I_X \theta) - I_X(E\delta A) \tag{11.6}$$

and then by using integration by parts and the definition of the current $J_X = I_X \theta$, we have

$$d(J_X - EX) = -X dE. \tag{11.7}$$

Since X is an arbitrary gauge symmetry, we can integrate both sides over a domain with boundary, apply Stokes' theorem to the LHS, and – by assuming that X and its derivatives vanish on the boundary – deduce that $dE = 0$ as an identity. This is a simple example of a non-trivial Noether identity (here the Noether operator N is just the exterior derivative d, which does not vanish on-shell). We note that $dE = dd * F = 0$ is clearly a mathematical identity (the Bianchi identity) since it follows from $d^2 = 0$.

By applying the Noether identity $dE = 0$ and the Poincare lemma, we see from (11.7) that

$$J_X = EX + dU_X \approx dU_X, \tag{11.8}$$

where '$\approx$' denotes 'on-shell equality'.[21] In other words, we see that the current J_X that is associated with the gauge symmetry is trivial, although – as we are about to see – this of course does not imply that the Noether charge is trivial!

Given that we have a current J, we can define a charge by integrating J over a spacelike Cauchy surface Σ to obtain:

$$Q^{\Sigma} = \int_{\Sigma} J \tag{11.9}$$

We now remind the reader that a unique possibility arises for *trivial* currents (but not non-trivial currents), which is a direct consequence of the *non-trivial* Noether identity: by Stokes' theorem, a trivial on-shell current $J_X \approx dU$ can be converted into a purely boundary Noether charge

$$Q_X^{\Sigma} = \int_{\Sigma} dU_X \approx \int_{\partial\Sigma} U_X = \int_{\partial\Sigma} X(\star F), \tag{11.10}$$

where in the last equality we used the explicit form of the superpotential U_X (fixed by X and our choice of L), which – unlike the case of **SR2** – is non-trivial and leads to a non-trivial corner charge algebra.

To go further and say more about the conservation of this corner charge, one needs to impose boundary conditions on the subsystem that rule out the leakage of 'symplectic flux' from the system (Freidel et al. 2021; Harlow and Wu 2020).[22] In the context of the present example, consider the case in which the subsystem has Dirichlet boundary conditions, i.e. $\delta A|_{\partial M} = 0$. Notice that this sets the symplectic potential to zero at the boundary, which prevents symplectic flux leakage. Then, the symmetries of the subsystem need to preserve these boundary conditions, and so $dX|_{\partial M} = 0$. Evidently, when $X|_{\partial M} = 0$, the symmetry is in the kernel of Ω and so does not represent a physical symmetry generated by a charge, leaving non-zero constant transformations at the boundary X as the only remaining possibility (which are precisely the non-trivial stabilisers of the boundary condition).

In the above example, we see that the Noether charges are also the canonical conserved charges associated with our boundary condition of interest, namely the Dirichlet boundary condition. The latter are also called 'Hamiltonian charges' because the charges are Hamiltonians for the corresponding symmetry transformations in phase space. We note that the Dirichlet boundary conditions represent an *isolated* subsystem, for which one expects the relevant Noether/Hamiltonian charge to be conserved, unlike an open system which is continually experiencing an exchange of charge with its environment.

However, as we will soon see, there can be more complex scenarios in which a Hamiltonian charge associated with a particular boundary condition is the Noether charge of $L + d\ell$, but not of L. In Section 11.3.4, we will discuss how one can resolve this tension

21 This is nothing other than the $U(1)$ gauge theory version of Einstein's formula $J = \mathfrak{T} + \mathfrak{t}$, of which the reader can find an analysis in De Haro (2021). It explains why energy pseudo-tensors in GR are often said to be defined only up to a superpotential.

22 In general, one also needs to make sure that the chosen boundary configuration has stabilisers, although this condition is guaranteed in our example, since all $U(1)$ gauge field configurations have stabilisers (the constant gauge transformations).

between Noether and Hamiltonian charges, which is relevant to the 'extending PR' part of the Einstein test of substantive general covariance.

11.3.3 Substantive General Covariance

We have just seen that, due to examples such as **SR2**, BGC does not on the whole pass the 'corner charge' part of the Einstein test for substantive general covariance. On the other hand, we have also seen that central examples of theories with BGC (electromagnetism, and by a simple extension: Yang–Mills theory and General Relativity) do yield non-trivial corner charges. The general moral is that while BGC is insufficient for the existence of non-trivial corner charges, it is nonetheless necessary, because one needs a trivial Noether current in order to obtain a *corner* charge. What, then, needs to be added to BGC to obtain a non-trivial corner charge?

As it turns out, this is one of the cases where the analysis of the difficulty (by means of the Noether machine) also makes the remedy obvious: what needs to be added is precisely the requirement that the theory's Lagrangian L be such that its superpotential is non-trivial (we again remind the reader that a choice of L also fixes the superpotential). We thus propose the following definition of substantive general covariance in response to the 'corner charge' part of the Einstein test:

(**Corner-SGC**) A theory has Corner-SGC when, in addition to BGC, its Lagrangian L has a non-trivial superpotential U_ξ.[23]

It is worth pausing to emphasise that the potency of Noether's second theorem is not that of a purely mathematical result. However, just as the character of a line or a stroke can begin to suggest the animation or dynamism of a figure in painting, so too can mathematics be taken up into the mobilisation of physical thought. What Noether's second theorem does here is offer physicists a heightened awareness of what the mathematical medium of field theory (understood through the lens of the variational calculus and Noether's theorems) is capable of when it is pressed into the service of physical representation: our discussion thus far is precisely an illustration of how the second theorem can be so recruited! Through that discussion, we come to see that the substantiveness of general covariance is not just about the symmetry of a Lagrangian, but also about how – through that symmetry – some kinds of Lagrangians determine a non-trivial corner charge algebra.

We now turn to the 'Extending PR' part of the Einstein test, for which (Corner-SGC) is a necessary but insufficient condition, because of the aforementioned potential tension between the Noether charge and the Hamiltonian charge relative to some boundary condition. Note that for a gauge theory (where we mean to include both General Relativity and Yang–Mills theory), whatever the symmetries of generalised PR are, they will need to be generated by the non-trivial corner charges from (Corner-SGC). However, there is an added complication present in generalising Embodied PR: we are now considering the

[23] NB: The corresponding Noether charge will be conserved if we impose boundary conditions that rule out flux leakage, and the Lagrangian L is consistent with those boundary conditions.

symmetries of dynamically isolated subsystems, which are generated by the Hamiltonian charges for the boundary conditions that model dynamical isolation – thus, we will only succeed in extending PR if the corner charges coming from the trivial Noether current of some Lagrangian are also Hamiltonian charges. In Section 11.3.2.2, we saw an example (Maxwell gauge theory with Dirichlet boundary conditions) where this coincidence clearly obtains: we can think of it as a regularised (finite) version of a case in which the asymptotically rigid $U(1)$ symmetries of some subsystem provide a generalised version of PR relative to an environment subsystem (where the environment has been idealised away into the boundary conditions). However, as we will see in Section 11.3.4, the corner charges from some particular Lagrangian can also fail to be Hamiltonian charges (relative to some choice of dynamically isolated boundary conditions). We thus propose the following, more demanding notion of substantive general covariance in response to the 'extending PR' part of the Einstein test:

(PR-SGC) Fix boundary conditions that represent 'dynamical isolation' in the context of some theory. The theory's Lagrangian L has PR-SGC just in case, in addition to satisfying Corner-SGC, the Noether charge coming from L is also a Hamiltonian charge relative to the 'dynamically isolated' boundary conditions.

We now turn to a brief consideration of General Relativity, which illustrates the subtleties of PR-SGC.

11.3.4 Revisiting the Case of General Relativity

Freidel et al. (2021) consider the general way in which Noether charges can fail to be Hamiltonian charges, and this framework is then applied to the case of General Relativity (section 3.4 of Freidel et al. 2021). Here we only summarise the relevant features for us to revisit the Einstein–Klein dispute.

First, consider the Einstein–Hilbert Lagrangian L_{EH}. We can follow through the covariant phase space recipe sketched in Section 11.3.1 and compute the superpotential corresponding to L_{EH}, which turns out to yield what is known as the Komar charge (see (2.38) of Freidel et al. 2020 for a definition). This shows that we have satisfied (Corner-SGC).

Next, let us consider whether this Lagrangian satisfies (PR-SGC). To do so, suppose that we wish to use Dirichlet boundary conditions to model an isolated system (or a finite boundary regularisation thereof): that is to say, we wish to set the induced metric $\bar{g}_{ab}$ on the timelike boundary to a fixed value, thereby also setting $\delta\bar{g}_{ab} \overset{\partial M}{=} 0$. More specifically, we typically take $\bar{g}_{ab}$ to be the Minkowski metric, thus leading to an 'asymptotically flat' solution in the limit where we take the boundary to infinity. It is well-known that given these Dirichlet boundary conditions, one can construct a Hamiltonian charge; however, this is not the Komar charge, but a different quasi-local charge known as the Brown–York charge (see (2.31) of Freidel et al. 2020 for a definition).

Thus, we have the complication that we discussed earlier: if one chooses L_{EH} as one's Lagrangian and the Dirichlet boundary condition as the 'isolated boundary condition' in the 'Extending PR' part of the Einstein test, then one ends up with a mismatch between the

Noether charge (the Komar charge) and the Hamiltonian charge (the Brown–York charge). The resolution of this complication, discussed in Freidel et al. (2021) and Harlow and Wu (2020), is to choose a different Lagrangian – which is associated with a different boundary condition – in order to satisfy (PR-SGC). In particular, if we choose the Lagrangian $L = L_{EH} + d\ell_{GHY}$, where ℓ_{GHY} is the celebrated Gibbons–Hawking–York boundary term, then the corresponding Noether charge is precisely the Brown–York charge, so we have now brought our Lagrangian into alignment with the boundary condition – and corner charge – that we are after. Furthermore, as Harlow and Wu (2020) show, if we assume that the metric tends towards the Minkowski metric in the asymptotic limit, then we recover precisely the ADM charges for an asymptotically flat spacetime, which is precisely what one expects of an isolated subsystem in General Relativity. In particular, we note that these charges generate the 'asymptotic Poincare symmetries', which are the generalised analogs of the rigid Poincare symmetries (from the standard PR) in this context.[24] We thus see that $L = L_{EH} + d\ell_{GHY}$ satisfies (PR-SGC), but L_{EH} does not.

With these clarifications in hand, we can also return to an assessment of the dispute between Einstein and Klein that we mentioned in our Introduction (we refer the reader to De Haro 2021 for a complementary analysis of this dispute, with an emphasis on holographic renormalisation). Is it really true that Einstein was terribly confused about the status of conserved charges in General Relativity, and that Noether's second theorem served to highlight his confusion by securing Klein's claim that General Relativity does not have any physically meaninful conservation laws? To answer this question, let us consider several key points from Einstein's March 1918 response to Klein (letter 480 in Einstein 1998):

- Einstein says the importance of the quantity $J_\xi = \mathfrak{T} + \mathfrak{t}$ (the on-shell exact Noether current, cf. (11.8) in Section 11.3.1) is its physical interpretation as the energy of a point mass when we go very far away from the subsystem; in other words, when we treat the subsystem as dynamically isolated from an environment.
- He also says that the possibility of this physical interpretation is underwritten by the fact that J_ξ (on-shell) is an *exact* term, presumably because he knows that via Stokes' theorem, a bulk integral of an exact term can be transformed into a boundary integral (which is why we call such charges 'quasi-local').
- Einstein emphasises that the equations of motion are necessary in order to show that J_ξ is exact, and thus J_ξ should be regarded as physically meaningful.

And then, a little over a week later, in letter 492 to Klein, Einstein (1998) goes on to place a heuristic 'isolated' boundary condition on the subsystem that allows for the construction of a conserved charge!

We can now ask ourselves how Einstein's more physically oriented (and more mathematically imprecise) strategy looks with the clarity of hindsight. It would seem that Einstein already understood much of the intuitive physical content that can be expressed by means of Noether's second theorem: certainly, he understood that the *distinctive* qualitative feature

[24] Here we are only discussing spatial infinity; once one takes into account null infinity, one is forced to consider the infinite-dimensional BMS group if one wants a rich set of solutions.

of a gauge theory (implied by Noether's second theorem) is the on-shell exactness of the Noether current, and that this feature leads to a boundary charge; and he was undoubtedly sensitive to the need to *choose* boundary conditions in order to construct such a charge. In all this, we believe that Einstein was vindicated in resisting Klein's criticisms. On the other hand, the place where Einstein's understanding was most lacking was in his grasp of (i) the relationship between a choice of L and a choice of boundary conditions, (ii) the potential mismatch between the Noether charge corresponding to some L and the Hamiltonian charge corresponding to some choice of boundary conditions, and (iii) the potential appearance of infinite-dimensional symmetries (such as the BMS group at null infinity) in the asymptotic limit. As we discussed in Section 11.3.2, this lack is closely related to his struggle to find an appropriate extension of the Relativity Principle to the context of General Relativity.

11.3.5 Background Independence

While this is not the place for us to offer a fully developed account of 'background independence', it seems appropriate to make some brief remarks about how this notion is related to our preferred conceptualisation of substantive general covariance, especially since Pooley (2017) has attempted to pour cold water on attempts by physicists (Rovelli 2004, 2007; Smolin 2006) to find a connection between *some* form of general covariance and background independence.[25]

Roughly speaking, Pooley's narrative runs as follows. There is a standard line of thought in the quantum gravity literature according to which a theory's 'general covariance' is tightly linked to its background independence, conceived of as a theory's spacetime structure being dynamical, as opposed to playing the role of a 'fixed background'. As evidence, Pooley cites Rovelli's (2004) claim that the background independence of classical GR is 'realized by the gauge-invariance of the action under active diffeomorphisms'. But, Pooley goes on to argue, this and similar claims seem to run counter to the moral of Kretschmann's point that any theory can be formulated in a generally covariant manner – and it is so much the worse for these claims! Pooley goes on to prosecute his case in various ways, but the part of his argument of interest to us here is his attempt to show that an identification of background independence with (what we call) BGC misclassifies certain theories which are intuitively background dependent as background independent.

There are two cases of relevance for us. First, Pooley says that the theory (that we call) **SR2** clearly satisfies BGC, but seems to be the same theory as **SR1**, which is a paradigm case of background dependence. Second, Pooley introduces a theory that we shall call **SR3**, defined by the Lagrangian[26]

$$L^{\mathbf{SR3}} = \sqrt{-g}\left(-\frac{1}{2}\nabla_a\phi\nabla^a\phi + \frac{1}{4}\lambda^{abcd}R_{abcd}\right). \tag{11.11}$$

[25] According to Pooley, Smolin (2006) seems to identify background independence with BGC, whereas Rovelli (2007) seems to want a slightly more full-blooded form of general covariance.

[26] This theory was introduced by Rosen and rediscovered by Sorkin (see Pooley 2017 for the history and references). Its most careful analysis so far seems to have been by Giulini (2007).

Pooley again notes that this theory satisfies BGC, but seems to be background dependent, and so the identification of BGC with background independence seems to misclassify the example. On this basis, Pooley concludes that 'one should recognise that we are now far past the point where one might hope to articulate a simple and illuminating connection between diffeomorphism invariance and background independence.'

We of course agree with Pooley that **SR2** and **SR3** are clear cases of Lagrangians that satisfy BGC. But as we have just argued, BGC is not plausibly a *substantive* notion of general covariance (certainly, it would not have been to Einstein in light of his desiderata), and the Noetherian analysis that reveals its implausibility also suggests a minimal notion of substantive general covariance, viz. what we called Corner-SGC in Section 11.3.3; thus, it would be premature to judge on the basis of BGC that there is no illuminating connection between some form of general covariance and background independence. With this background in mind, we now offer a conjecture for a simple and illuminating connection between background independence and Corner-SGC:

Conjecture: A theory is background independent only if its Lagrangian satisfies Corner-SGC with respect to spacetime diffeomorphisms.

The intuition driving this conjecture is that a non-trivial corner charge associated with diffeomorphisms (and the Lagrangian) is required for spacetime structure to have *dynamical* degrees of freedom, and thus for the theory to be background independent.

How does Pooley's misclassification argument against BGC – using **SR2** and **SR3** as test cases – fare against our conjecture? Let us take the test cases in turn. The case of **SR2** is especially simple: as we discussed in Section 11.3.2.1, this theory does not satisfy (Corner-SGC) because its Noether current is identically zero and its corner charge is trivial. Thus, according to our conjecture, this theory counts as background dependent, and we see that Pooley's argument is no threat to the conjecture.

The case of **SR3** is somewhat more interesting. When we put this theory through the Noether machine, we find that the equations of motion are $g_{ab}\nabla^a\nabla^b\phi = 0$, $R_{abcd}(g) = 0$, and $\nabla_a\nabla_d(\lambda^{abcd} + \lambda^{acbd}) = T^{bc}$. We also obtain a set of non-trivial Noether identities and a pre-symplectic potential current

$$\theta = -\partial^a\phi\delta\phi + \frac{1}{2}(\nabla_d\lambda^{abcd}\delta g_{bc} - \lambda^{abcd}\nabla_d\delta g_{bc}) \tag{11.12}$$

from which we can compute the Noether current

$$\begin{aligned} J^a_\xi = {} & -T^a_b\xi^b + \frac{1}{2}\nabla_b(\nabla_d\lambda^{a(bc)d}\xi_c) - \frac{1}{2}\nabla_d(\lambda^{a(bc)d}\nabla_b\xi_c) \\ & - \lambda^{abcd}\nabla_d\nabla_b\xi_c + \frac{1}{4}(\nabla_d\lambda^{abcd}\nabla_c\xi_b - \lambda^{abcd}\nabla_d\nabla_c\xi_b). \end{aligned} \tag{11.13}$$

From this we can compute the superpotential 2-form $U_\xi = \epsilon_{ab}\lambda^{abcd}\nabla_c\xi_d$, where ϵ_{ab} is an area 2-form obtained by contracting ∂_a and ∂_b with the volume form. In other words, we see that this theory has a non-trivial corner Noether charge, and thus our conjecture classifies it as potentially background independent.

What then are we to make of Pooley's 'intuition' that **SR3** is background dependent? Note that the basis for this intuition seems to have two parts. The first is the uncontroversial observation that **SR3**'s equations of motion $R^{abcd}(g) = 0$ can be replicated in a non-Lagrangian theory, viz. the theory specified just by the equations of motion $\nabla_a \nabla^a \phi = 0$ and $R^a_{bcd} = 0$. The second is Pooley's claim that if the latter theory is regarded as background dependent, then the former theory should also be regarded as background dependent.

It will help to introduce an analogy to put the observation in context and to expose the defects of Pooley's line of thought here. Consider an example in which we couple some matter field degrees of freedom to a $U(1)$ BF theory, which in 4 dimensions has the Lagrangian $B \wedge dA$, where the dynamical fields are a 2-form B and a gauge field A with curvature $F = dA$; the equations of motion of the pure BF theory are $F = 0$ and $dB = 0$. We will take this example to be our first theory, to be compared to a second theory in which the same matter field degrees of freedom are set against the fixed background of a flat $U(1)$ connection, which is of course described by the equation $F = 0$; in the former case, we can think of the flatness condition as being enforced by a Lagrange multiplier field, whereas it is imposed by fiat in the latter case. If we applied Pooley's reasoning to this analogous scenario, then we would say that since the latter theory is background dependent (now in the gauge field, and not in the spacetime, sense), we should also regard the former theory as background dependent. But now it should be evident that this line of thought cannot be right, because a BF theory is a topological field theory, which is supposed to be one of the paradigm cases of background independence! The mistake here is to conflate local dynamical degrees of freedom – which both theories lack – with global dynamical degrees of freedom, which only BF theory possesses, and in virtue of which it is called 'topological'. For instance, in the case of BF (or Chern–Simons) theories of gravity, what we would standardly say is that the spacetime is dynamical in the sense that it has these global degrees of freedom.

With the analogy in hand, we are now in a position to better understand the physics of **SR3**. It is precisely a BF-like theory, and as noted by Giulini (2007), the spacetime part has global – but not local – dynamical degrees of freedom which make the solution space for **SR3** inequivalent to the theory specified just by the equations of motion $\nabla_a \nabla^a \phi = 0$ and $R^a_{bcd} = 0$. The equation of motion for λ, in which this dynamical parameter now appears as a potential for the matter field's stress-energy, and the non-trivial corner charge from U_ξ, serve to describe how the global spacetime dynamics is coupled to the dynamics of the matter field.

To sum up: we again see that in the case of **SR3**, Pooley's misclassification argument poses no threat to our conjecture; on the contrary, a careful Noetherian analysis of **SR3** indicates that Pooley has misinterpreted this theory.

11.4 Conclusion

Noether's theorems, as we said in the Introduction, do not in themselves resolve any questions about the interpretation of General Relativity, or lend clarity to the *physical* (not merely mathematical) dispute between Einstein and Klein. Their true glory lies in their being a uniquely subtle medium for physical representation – a set of mathematical materials that physicists can bend to the ends of a certain kind of representational activity.

We pay tribute to the author of these theorems by closing with a broader philosophical reflection of the kind of activity into which her theorems are taken up. When describing the productive activity of house-building, Aristotle introduces the idea of *dunamis*, i.e. the potency or capacity of the building materials, which the builder actualises in pursuing his goal of building a house. This kind of activity Aristotle calls a *kinesis*, because it aims at a further product outside itself, namely the completed house. He goes on to contrast *kinesis* with what he calls *energeia*, i.e. the kind of activity which aims at nothing beyond itself and is its own end (two of his examples are 'seeing' and 'dancing').

Evidently, Noether's theorems are a *dunamis* in Aristotle's sense: they have an incredible capacity to be actualised in representation, which Einstein merely intuited, and which we have tried to flesh out more fully in Section 11.3, although we still managed only to scratch the surface. A further interesting question is whether this activity into which Noether's theorems are taken up – the activity of using them to represent the material world – is more akin to *kinesis* or *energeia*. And here we would like to make the suggestion that *energeia* is the better description, for the kind of insight achieved by absorbing physical subject matter into the framework of Noether's theorems (e.g. in the reconceptualisation of Section 11.3.2) leads to ever new insights that energise and propel the physicist's theorising further forward – the end here is internal to the activity of doing theoretical physics, and not a further product that is external to the activity. Perhaps this is close to what Noether had in mind when she said that 'My methods are really methods of working and thinking; this is why they have crept in everywhere anonymously.'

References

Anderson, I. M. (1989). The variational bicomplex. Technical report, Utah State Technical Report, https://ncatlab.org/nlab/files/AndersonVariationalBicomplex.pdf.

Belot, G. (2000). Geometry and motion. *British Journal for the Philosophy of Science*, **51**(4), 561–95.

Brading, K. and Brown, H. R. (2004). Are gauge symmetry transformations observable? *The British Journal for the Philosophy of Science*, **55**(4), 645–65.

Brown, H. and Brading, K. (2002). General covariance from the perspective of Noether's theorems. *Diálogos. Revista de Filosofía de la Universidad de Puerto Rico*, **37**(79), 59–86.

Brown, H. R. and Sypel, R. (1995). On the meaning of the relativity principle and other symmetries. *International Studies in the Philosophy of Science*, **9**(3), 235–53.

De Haro, S. (2021). Noether's theorems and energy in general relativity. *arXiv preprint arXiv:2103.17160*.

Donnelly, W. and Freidel, L. (2016). Local subsystems in gauge theory and gravity. *Journal of High Energy Physics*, **2016**, September.

Earman, J. (2006). Two challenges to the requirement of substantive general covariance. *Synthese*, **148**(2), 443–68.

Einstein, A. (1998). *Collected Papers of Albert Einstein: The Berlin Years (Trans Suppl)*. Princeton, NJ: Princeton University Press.

Freidel, L., Geiller, M. and Pranzetti, D. (2020). Edge modes of gravity. part I. corner potentials and charges. *Journal of High Energy Physics*, **2020**(11), 1–52.

Freidel, L., Oliveri, R., Pranzetti, D. and Speziale, S. (2021). Extended corner symmetry, charge bracket and Einstein's equations. *arXiv preprint arXiv:2104.12881*.

Giulini, D. (2007). Remarks on the notions of general covariance and background independence. In I.-O. Stamatescu and E. Seiler, eds. *Approaches to Fundamental Physics*. New York: Springer, pp. 105–20.

Harlow, D. and Wu, J. (2020). Covariant phase space with boundaries. *Journal of High Energy Physics*, **2020**(10), 1–52.

Kretschmann, E. (1918). Über den physikalischen sinn der relativitätspostulate, a. einsteins neue und seine ursprüngliche relativitätstheorie. *Annalen der Physik*, **358**(16), 575–614.

Noether, E. (1918). Invariante variationsprobleme, math-phys. *Klasse*, pp. 235–57.

Norton, J. D. (1993). General covariance and the foundations of general relativity: eight decades of dispute. *Reports on Progress in Physics*, **56**(7), 791.

Norton, J. D. (2003). General covariance, gauge theories and the Kretschmann objection. In K. Brading and E. Castellani, eds., *Symmetries in Physics: Philosophical Reflections*. Cambridge: Cambridge University Press, pp. 110–23.

Olver, P. J. (2000). *Applications of Lie Groups to Differential Equations*, volume 107. New York: Springer Science & Business Media.

Pooley, O. (2010). Substantive general covariance: Another decade of dispute. In M. Suárez, M. Dorato, and M. Rádei, eds., *EPSA Philosophical Issues in the Sciences*. New York: Springer, pp. 197–209.

Pooley, O. (2017). Background independence, diffeomorphism invariance and the meaning of coordinates. In D. Lehmkuhl, G. Schiemann, and E. Scholz, *Towards a Theory of Spacetime Theories*. New York: Springer, 2017, pp. 105–43.

Read, J. (2016). *Background independence in classical and quantum gravity*. Ph.D. thesis, University of Oxford.

Rejzner, K. (2020). Bv quantization in perturbative algebraic qft: Fundamental concepts and perspectives. *arXiv preprint arXiv:2004.14272*, 2020.

Rovelli, C. (2004). *Quantum Gravity*. Cambridge: Cambridge University Press.

Rovelli, C. (2007). Quantum gravity in J. Butterfield and J. Earman, eds., *Philosophy of Physics*, Vol. 2. Amsterdam: Elsevier, pp. 1287–1330.

Rowe, D. E. (2019). Emmy Noether on energy conservation in general relativity. *arXiv preprint arXiv:1912.03269*.

Rowe, D. E. (2021). *Emmy Noether–Mathematician Extraordinaire*. New York: Springer.

Smolin, L. (2006). The case for background independence. In D. Rickles, S. French and J. T. Saatsi, eds., *The Structural Foundations of Quantum Gravity*. Oxford: Clarendon Press, pp. 196–239.

Teh, N. J. (2016). Galileo's gauge: Understanding the empirical significance of gauge symmetry. *Philosophy of Science*, **83**(1), 93–118.

12

Noether Charges, Gauge-Invariance, and Non-Separability*

HENRIQUE GOMES

12.1 Introduction

The goal of this chapter is to argue for a chain of (informal) implications for gauge theories: '*local conservation laws*' $\overset{*}{\Rightarrow}$ *conserved regional charges* $\Leftrightarrow$ *Non-separability* $\Leftrightarrow$ *direct empirical significance of symmetries*. To forge the necessary links, I will need two raw materials: (i) the relation between symmetries and conservation, which encompasses the first three arrows and (ii) the relation between non-separability and conserved charges. The link between direct empirical significance and the non-separability will here only be referred to; it is studied in depth by Gomes (2021).

I first delineate the relevant topics in the two following subsections as a prospectus to the parts of the chapter corresponding to (i) and (ii). Then Section 12.1.3 summarises the plan.

12.1.1 Noether Theorems and Gauge-Invariant Charges

The theorems proved by Emmy Noether more than a century ago have had a profound impact on both physics and mathematics. The core idea is to relate the occurrence of symmetries of the Lagrangian to conservation laws.[1] In recent philosophy of physics, this empirical significance of symmetries – that they yield conserved charges – has been called 'indirect', in contrast to the *directly* empirically significant symmetries, exemplified by the subsystem-boosts of Galileo's ship scenario (Brading and Brown 2004; Friederich 2014; Gomes 2021; Greaves and Wallace 2014; Teh 2016).

However, in the case of gauge theories, the Noether relations do not guarantee gauge-invariance of the local charges. Indeed, intuition based on the Abelian case is misleading: for in this case, the local conservation law of electromagnetism yields a well-defined gauge-invariant conserved charge in an entire region of spacetime by integration (by the Gauss law).[2]

* I would like to thank an anonymous reviewer, for very insightful and constructive remarks.

[1] A good discussion of the significance of the theorems is Brading and Brown (2000), see also the essays in Brading and Castellani (2003) and references therein for the relation between Noether and other topics in philosophy of physics, Kosmann-Schwarzbach and Schwarzbach (2011) for a thorough treatment within physics, covering the history of the theorem and its applications and (Olver 1986) for a mathematically advanced, complete analysis of the theorems.

[2] We call quantities defined in regions of spacetime 'regional'. An interior quantity is a regional quantity that does not depend on boundary data. 'Regional charges' are ones obtained from integration of densities on the given regional volume and depend solely on matter degrees of freedom, with minimal coupling. Local conservation laws are valid at each spacetime point, but may not yield *regional* conservation laws.

Here I will use the covariant symplectic formalism – which does not require a foliation of spacetime into equal-time surfaces – to show that in the non-Abelian case the standard Noether charge, although conserved, is not gauge-invariant and cannot, in general, be written solely in terms of the matter degrees of freedom. That is, beyond the lack of gauge-invariance, the generic conserved charges guaranteed by Noether's theorems will involve more than just the standard matter degrees of freedom: they will include extra contributions from the gauge field (from the potential in the non-Abelian Yang–Mills case and from the Christoffel symbols in the gravitational case).

I find it opportune to draw a parallel with another non-linear theory: general relativity. There, the analogous naive Noether charge is known as the Landau–Lifschitz pseudo-tensor; this tensor is indeed conserved, but it is coordinate-dependent and depends on the values of the Christoffel symbols in the bulk of the region. In general relativity, one can obtain coordinate-independent conserved charges that are independent of the gravitational forces. But one can do so only in very specific models of the theory (see de Haro's contribution to this volume), namely, when the background metric is 'uniform'; that is, it has associated Killing vector fields in the directions of uniformity, and for each one of these Killing directions one obtains conserved charges written as local integrals of solely the matter part of the energy-momentum tensor (minimally coupled). The same occurs for non-Abelian Yang–Mills fields: for each gauge transformation leaving the gauge-field configuration invariant over the entire region – just as, for example, Poincaré transformations leave the Minkowski *metric* invariant – there corresponds a conserved charge.[3] This analysis thereby constrains the existence of regional gauge-invariant non-Abelian charges to very specific circumstances, as it does for general relativity.[4]

Once we have established the existence of these charges, we will have forged the links *local gauge symmetries* $\Rightarrow$ *local conservation laws* $\overset{*}{\Leftrightarrow}$ *conserved regional charges*, where the asterisk signals the special conditions that allow for conserved regional charges in the sense above.

The subtle point about the gauge-invariance of charges is important in establishing the remaining links between conserved regional charges, non-separability, and Direct Empirical Significance (DES).

12.1.2 Non-Separability and Direct Empirical Significance

Gauge theory has often been described as a theory that exhibits excess structure; or, in John Earman's (2004) words, 'descriptive fluff'. But when is 'fluff' merely decorative, and when does it perform a function?

[3] The notion is itself gauge-invariant. Again resorting to the perhaps more familiar gravitational case, if a metric $g_{\mu\nu}$ has a certain Killing field X^μ, under the action of a diffeomorphism, the transformed metric will have the appropriately transformed (under the adjoint action of the diffeomorphism) vector field as its own Killing field. The same applies to non-Abelian Yang–Mills gauge-fields and their stabilising directions.

[4] Again following the parallel with the (perhaps more widely known) analogous general relativity properties, gauge-invariant charges – i.e. those obtained through volume integrals of functions of minimally coupled matter degrees of freedom – are more rigidly constrained regionally than in the asymptotic regime. In this chapter I will not treat the asymptotic regime, save to say that the analysis remains completely valid, resulting in 'extra' conserved charges that do not require uniformity over the entire spacetime: They require only asymptotic uniformity (Regge and Teitelboim 1974). See Riello (2019) for the corresponding statements in the formalism employed in this chapter). These 'extra charges', however, appear only in the singular asymptotic limit, as a result of the effectively infinite separation between spatial points (Riello 2019).

I have elsewhere argued – broadly in line with standard physics folklore – that the 'descriptive fluff' associated to gauge theory is a sign of a particular type of non-locality, one that does not introduce any superluminal signalling. I have further argued that local gauge theories such as Yang–Mills are *characterised* by this non-locality, which shows up for any choice of variables.

Even in standard electromagnetism, employing gauge-invariant electric and magnetic field variables, differential constraints on the initial data prohibit an arbitrary specification of the electric field at each point. Naively, knowing the value of the electric flux on a sphere very far away can uniquely fix the total amount of charge contained in a much smaller region, *at any instant*, even if the matter fields are unsupported at that faraway sphere. More mathematically, solving the constraint differential equations at a given point requires integration, or the inversion of differential operators, which in their turn require conditions on some surrounding boundary. This is in contrast to the initial value problem for other dynamical systems such as the Klein–Gordon scalar field, which is unconstrained.

Indeed, the lesson of the Aharonov–Bohm experimental confirmation is that observables of the theory are given by non-local equivalence classes.[5] As Earman (2019) emphasised, a certain type of non-locality of electromagnetic phenomena *is* the lesson of the experiment:

> Thus, despite the fact that non-simple connectedness (of the electron configuration space) is essential, by definition, to the AB effect, it is not essential to some of the key issues to which it served to call attention. The philosophical literature seems incapable of absorbing this fact, as if it were under the thrall of the patently invalid inference that goes: 'The AB effect uses non-simple connectedness; the AB effect reveals a [certain] kind of nonlocality; ergo the nonlocality derives from non-simple connectedness.' (p. 196)

Belot (1998) packs a similar message: 'Until the discovery of the Aharonov–Bohm effect, we misunderstood what electromagnetism was telling us about our world' (p. 532).

Thus the question is not *whether* gauge theory is non-local, but *how* non-local it is. In trying to answer this, using holonomies as a gauge-invariant basis of vacuum electromagnetism, Myrvold (2010) argued that when non-simply connected manifolds are decomposed into simply connected pieces, there is information in the physical states of the whole that is not exhausted by the information contained in the physical states of the pieces.

Similarly – but more generally – here non-separability will be characterised as a failure of global supervenience on regions (Gomes 2021). This failure occurs when the physical states of the regions fail to determine the state of the whole, and there remains a physical variety of universe states compatible with the regional physical states.

Gomes (2021) showed that such a non-separability, or failure of global supervenience, is tantamount to the direct empirical significance of symmetries. That is, certain rigid

[5] That is, if we assume the local properties of the electromagnetic field are carried by the curvature, $F = \mathrm{d}A$ for d the exterior derivative and A the gauge-potential 1-form, then, assuming the equivalence class of gauge potentials is given by $A \sim A + \mathrm{d}f$, for f some scalar function, there can be an A' such that $A' \not\sim A$, and yet $F = F'$, so that they are physically distinct but not locally distinguishable. Namely, whenever there exist 1-forms λ such that $\mathrm{d}\lambda = 0$ but $\lambda \neq \mathrm{d}f$ for any f. Such 1-forms exist in 1–1 correspondence with the first cohomology space of $\mathsf{Ker}(\mathrm{d})/\mathsf{Im}(\mathrm{d})$, and is thus a topological invariant, and thus non-local. Although we use quantum matter to detect such differences, this is only accidental; the fact is that the gauge-potential theory carries different physical quantities than the field strength theory.

gauge-symmetries, when applied to subsystems, may yield physically distinct models of the theory. These subsystem symmetries are the generators of the residual physical variety that foils global supervenience on regions (more on this below).

And although Myrvold's argument suffices to show non-separability for electromagnetism,[6] where holonomies have simple composition properties and form a basis for gauge-invariant functions, it does not suffice for non-Abelian theories. To analyse Non-Abelian theories, we require new tools.

For to characterise how states of the universe supervene on the regional states, we must first be able to characterise the physical content of each region independently of the other. To answer these questions, my main proposal will be that, given any bounded region, one can split the degrees of freedom into ones that require extra data at the boundary for their definition from those that do not require further information. This distinction makes it easier to characterise 'non-separability' and also non-locality in gauge theories. In the Euclidean case, the split corresponds to a decomposition of the degrees of freedom of the gauge potential into 'pure-gauge' degrees of freedom and gauge-fixed, physical degrees of freedom. In the Lorentzian case, this split maps onto a Coulombic component for the electric field – which carries the boundary information – and the quasi-local radiative components of the electric field, which do not require boundary data but carry dynamical information. (This split was performed explicitly, also for non-Abelian Yang–Mills theories, by Gomes and Riello (2021); see Appendix 12.B.)

Once we have this decomposition at hand, we can investigate non-separability through a gluing/reconstruction theorem. The idea is that a manifold M is divided into a patchwork of regions, mutually exlusive, but jointly exhaustive.[7] Although the Coulombic information is essential for specifying the complete state of a single region, this data becomes superfluous once one has the (interior) radiative degrees of freedom of all the regions. In other words, knowledge of the pure radiative field content of each and every region can suffice to reconstruct the complete state of the field over the entire manifold; in this case knowledge of the regional Coulombic degrees of freedom adds no relevant information. In this sense, from the global perspective, these Coulombic degrees of freedom are indeed 'descriptive fluff'. From the perspective of an individual region however, we must retain these degrees of freedom for a full description.

But there is a caveat to the claims of reconstruction above: the reconstruction goes through without the aid of Coulombic data only for simply connected unbounded manifolds, and if *one of the two* conditions is met: (a) the regional field configurations have no isometries or (b) no charged matter is present anywhere.

When the fields have regional isometries *and* charged matter is present (i.e. when neither (a) nor (b) obtains), the radiative information of the two regions does *not* suffice to determine the entire global state. That is: there is gauge-invariant information in the global state which is *not* contained in the union of the regional, *strictly interior*, information.

[6] The argument was applied in a vacuum, where it can only detect non-separability for non-simply connected universes. But adding charges, one obtains non-separability even for simply connected regions.

[7] For example, in the simple case consisting of only two regions, this would correspond to $R_+, R_- \subset M$ such that $R_+ \cap R_- = \partial R$, $R_+ \cup R_- = M$ and $\partial M = \emptyset$.

The remaining global, gauge-invariant information – that which is *not* determined by the strictly interior content of the subsystems – is strictly *relational* and can be mapped bijectively to (subgroups of) the charge group of the theory (e.g. U(1), or SU(N), i.e. a global Lie group).

This occurrence of subgroups of the charge group arise therefore as a physical variety of global states corresponding to the same strictly interior (i.e. radiative) states. In Gomes and Riello (2021) and Gomes (2021), such a variety was taken to characterise a *direct* empirical significance of symmetries (DES), for the variety can be construed as the result of an action of a (subgroup of the) charge group on a subsystem. Since this action takes one global state to a physically different global state, the empirical significance of these subsystem symmetries is labelled 'direct'. But, most relevant for this chapter, this underdetermination, or holism, occurs precisely when non-trivial gauge-*invariant*, conserved fermionic charges also exist. In this way, the *indirect* significance of gauge symmetry associated with the Noether theorems becomes intimately connected to the *direct* significance of gauge symmetry. In this way, we forge the remaining links between *local conservation laws* $\overset{*}{\Rightarrow}$ *conserved regional charges* $\Leftrightarrow$ *Non-separability* $\Leftrightarrow$ *direct empirical significance of symmetries*.

In this chain, we restrict ourselves to spacetimes that are contractible and to standard gauge theories (including general relativity; cf. De Haro's contribution in this volume). The asterisk applies to the non-Abelian case, since local conservation laws imply a regionally conserved matter charge only for special models of the theory.

12.1.3 Road Map

The first link – that local gauge symmetries lead to local conservation laws – is the subject of Section 12.2. This link of course lies within the purview of Noether's theorems; I will summarise a version of these theorems and illustrate it with one Abelian and one non-Abelian theory.

Through these examples, we will see that the standard charges are only gauge-invariant for the Abelian theory; they are gauge-*variant* in the non-Abelian theory. Equivalently, the charge that is conserved will depend on both the matter fields and the gauge potentials.

I will then argue that a gauge-invariant regional notion of charge exists (cf. footnote 2), but it is non-trivial only in the presence of Killing fields (or regional isometries).

In Section 12.3, I will summarise a split of regional degrees of freedom into 'radiative' and 'Coulombic'.[8] The radiative degrees of freedom are regionally interior and non-local: that is, they are determinable solely from the regional content of the fields, without recourse to boundary data, but they are still non-locally defined within the region. The Coulombic degrees of freedom carry all the remaining information from outside a given region that is necessary to completely determine the state within the the region. Therefore,

[8] In this chapter I will be dealing with the dynamical case; that is, I will employ a Lorentzian structure, and a Cauchy surface in spacetime (Gomes and Riello 2021). Interstingly, some issues concerning the dynamics of the boundary do not appear in the Euclidean case (Gomes, Hopfmüllerm and Riello 2019).

Coulombic degrees of freedom are 'more non-local' than the radiative degrees of freedom. With knowledge of all the radiative degrees of freedom of all the regions, it is possible to reconstruct the entire physical state (including Coulombic components).

In the conclusions (Section 12.4), I put the results of the previous two sections together in my chain of implications. Local gauge symmetries lead to local conservation laws, which lead to a type of non-locality, which lead to a type of separability, which lead both to the gauge-invariant regional charges of 12.2 and to what I have elsewhere interpreted to be the 'direct empirical significance of symmetries' (see Section 12.1.2 and Gomes 2021).

In Appendix 12.A I will very briefly introduce the covariant symplectic formalism (Crnkovic and Witten 1987; Lee and Wald 1990). The formalism is extremely convenient for the treatment of conservation laws and provides a clear path between Lagrangian/covariant concepts and Hamiltonian/canonical ones.

12.2 Abelian and Non-Abelian Noether Regional Charges

In this section, I will give an abridged version of Noether's theorems that is sufficent for my purposes. To do that, I require the rudimentary notions of the covariant symplectic formalism, developed in Crnkovic and Witten (1987) and Lee and Wald (1990). The purpose of this section is to establish a useful, minimal version of Noether's theorem using that formalism. It can be skipped by those familiar with Noether's theorems.

First, for simplicity, suppose we are given the contractible d-dimensional spacetime manifold M and the Lagrangian configuration space of fields on M,[9] $\Phi := \{\varphi^I \in C^\infty(M, W)\}$, for W some vector space (to which the superscript I refers). These fields are not yet required to solve any equation of motion. In cases of interest, Φ has the structure of a (infinite-dimensional) manifold, which we can use to formulate, among other things, its tangent bundle, $\mathrm{T}\Phi$. An element of $\mathrm{T}_\varphi\Phi$ is the tangent to a one-parameter family of configurations, $\frac{d}{d\lambda}_{|\lambda=0}\varphi^I(x,\lambda)$. By standard convention, a general such vector based at φ is called $\delta\varphi$. In terms of a field-space basis, $\varphi^I(x)$, we can also decompose such a vector field:

$$\delta\varphi = \int \mathrm{d}x\, \delta\varphi^I(x)\frac{\delta}{\delta\varphi^I(x)} \equiv \int \delta\varphi^I \frac{\delta}{\delta\varphi^I},$$

which is understood as a derivation on function(als) on Φ, and henceforth we omit the spacetime dependence of the integral and of the component fields.[10]

Since we are heuristically taking Φ to be a bona-fide differentiable manifold, we can build up an exterior calculus on it as well. For that, we call the field-space exterior derivative by the double-struck $\mathbb{d}$, and we can extend the double-struck notation to refer to more general geometric objects in Φ, such as the Lie derivative and interior contraction, $\mathbb{L}_{\delta\varphi}, \mathbb{i}_{\delta\varphi}$. These should all be taken to represent field-space (e.g. variational) operations. So, for example, $\mathbb{L}_{\delta\varphi}F = \mathbb{i}_{\delta\varphi}\mathbb{d}F = \mathbb{d}F(\delta\varphi) = \delta\varphi[F]$ for a scalar field-space functional $F : \Phi \to \mathbb{R}$ (in some differentiability class, which we need not discuss).

[9] Also known as the *field*-space, as there is no 3+1 decomposition, usually understood to be involved in the use of the term 'configuration'.

[10] The components $\delta\varphi^I(x)$ are themselves distributional; for example, can be of the form $\kappa^I(x)\delta(x,y)$ for a smooth $\kappa^I(x)$.

12.2.1 Gauge Transformations

Now we move on to gauge theory as it is usually formulated. We are given a Lie group G (which we will explicitly set either to $U(1)$ or $\mathrm{SU}(N)$) and its Lie-algebra $\mathfrak{g}$. We have an extension of G to the group of gauge-transformations, which we denote by $\mathcal{G} = C^\infty(M, G)$, whose group operation is just that of G pointwise on M, i.e. for $g, g' \in \mathcal{G}$, $gg'(x) = g(x)g'(x)$, and thus we can define the group of infinitesimal gauge transformations, $\mathrm{Lie}(\mathcal{G}) = C^\infty(M, \mathfrak{g})$, as usual.

We also assume G has some action (a representation) on the vector space of local field-values of φ, i.e. V, and thus we again pointwise define the action $g \cdot \varphi(x) = g(x) \cdot \varphi(x) \in \Phi$. That is: we define pointwise the natural action of the group of gauge transformations on field space $\mathcal{G} \times \Phi \to \Phi$. Thus, field-space is foliated by gauge-orbits: namely, integral curves of field-space vectors $\delta_\xi \varphi$ for $\xi \in C^\infty(M, \mathfrak{g})$, such that $\delta_\xi \varphi := \frac{d}{d\lambda}_{|\lambda=0}(\exp(\lambda\xi) \cdot \varphi) \in \mathrm{T}_\varphi \Phi$. Again, if the Lagrangian in (12.18) is invariant in the direction of $\delta_\xi \varphi$, and the Euler–Lagrange equations hold, we have:

$$\mathbb{d}\mathcal{L}(\delta_\xi \varphi) = E_I \delta_\xi \varphi^I + \nabla_\mu \theta_I^\mu(\varphi)(\delta_\xi \varphi^I) \approx \nabla_\mu \theta_I^\mu(\varphi)(\delta_\xi \varphi^I) \approx 0. \tag{12.1}$$

Following (12.22) (see Appendix 12.A), using the coordinate-free notation of θ as a $d-1$ form, we have the regional conservation law:

$$\mathrm{d}(\theta_I(\varphi)(\delta_\xi \varphi^I)) \approx 0.$$

12.2.2 Noether and Conserved Currents

First, we fix our fields and the action of the gauge group. Let ψ be some complex-valued scalar field,[11] and the gauge potential be a Lie-algebra valued 1-form, $A \in \Lambda^1(M, \mathfrak{g})$. We will have the indices I refer to the vector space $\mathfrak{g}$ (since it is also the value-space of one of the fields) and omit reference to the value space of the field ψ.

We define an action of $\xi \in \mathrm{Lie}(\mathcal{G})$ on these variables as:

$$\delta_\xi A^I := \mathrm{D}\xi^I := \partial\xi^I + [A, \xi]^I \tag{12.2}$$

$$\delta_\xi \psi := -\xi\psi \tag{12.3}$$

where the square brackets delimit the $\mathfrak{g}$-commutator, within which we omit Lie indices.

The Yang–Mills Lagrangian, in form language, can be written as[12]

$$\mathcal{L}_{\mathrm{YM}} = -\frac{1}{2e^2}(F^I \wedge *F_I) + (i\overline{\psi}\gamma_\mu \mathrm{D}\psi) \wedge *\mathrm{d}x^\mu \tag{12.4}$$

where $F = \mathrm{d}A + \frac{1}{2}[A, A]$ and $*$ is the Hodge dual, γ^μ are Dirac's gamma-matrices, and $\mathrm{D}_\mu \psi = \partial_\mu \psi + A^I_\mu \tau_I \psi$, with τ^I a basis for $\mathfrak{g}$, and e is the Yang–Mills coupling constant.

[11] For definiteness, we could consider Dirac fermions valued in the fundamental representation V of the gauge group $G = \mathrm{SU}(N)$, $\psi^{B,b} \in C^\infty(M, \mathbb{C}^4 \otimes V)$. To simplify notation, we will omit the subscripts (B, b).

[12] Again, here $*$ is the Hodge-star; so $*\mathrm{d}x^\mu$ is an (d-1)-form, completed to a top d-form by the wedge product ($\wedge$) with the one-form $\mathrm{D}\psi$. Formulas in form language are given with the correct signs in even-dimensional spacetimes. I prioritised uncluttered formulas over complete generality.

Also, $\overline{\psi} = \psi^\dagger \gamma^0$, with the understanding that ψ and $\psi^\dagger$ have to be considered as two independent (complex) variables. The Lie algebra index I is lowered with the Kronecker delta.

Now, we compute the field-space differential of the Lagrangian,

$$\begin{aligned} \mathbb{d}\mathcal{L}_{\rm YM} = & \big(\mathbb{d}A^I \wedge (-e^{-2}\mathrm{D} * F_I - *J_I)\big) + \mathbb{d}\overline{\psi}\big(i\gamma_\mu \mathrm{D}\psi \wedge *\mathrm{d}x^\mu\big) + \big(-i\mathrm{D}\overline{\psi}\gamma_\mu \wedge *\mathrm{d}x^\mu\big)\mathbb{d}\psi \\ & + \mathrm{d}\big(-e^{-2}\mathbb{d}A^I \wedge *F_I + i\overline{\psi}\gamma_\mu \mathbb{d}\psi(*\mathrm{d}x^\mu)\big); \end{aligned} \tag{12.5}$$

from which the equations of motion are easily extracted:

$$\mathrm{EL}^I_A = -e^{-2}\mathrm{D} * F^I - *J^I, \qquad \mathrm{EL}_{\overline{\psi}} = i\gamma^\mu \mathrm{D}_\mu \psi, \qquad \mathrm{EL}_\psi = (\mathrm{EL}_{\overline{\psi}})^\dagger \gamma^0. \tag{12.6}$$

Here the matter, that is, the fermions ψ, carry a $\mathfrak{g}$-current density:

$$J^\mu = (\rho, J^i) \quad \text{with} \quad J^\mu_I = \overline{\psi}\gamma^\mu \tau_I \psi = \frac{\partial \mathcal{L}}{\partial \mathrm{D}_\mu \psi}\tau_I \psi \tag{12.7}$$

This is the current whose conservation we would like to enforce, but it is not the current whose conservation falls out of Noether's theorem.

From (12.5) (and (12.18) in the appendix), one finds that the symplectic potential is given by:

$$\theta_{\rm YM} = -e^{-2}\mathbb{d}A^I \wedge *F_I + i\overline{\psi}\gamma_\mu \mathbb{d}\psi * \mathrm{d}x^\mu. \tag{12.8}$$

The covariant Noether current density j_ξ is given by

$$\begin{aligned} j_\xi = \mathbb{i}_{\xi^\#}\theta_{\rm YM} &= -e^{-2}\mathrm{D}\xi^I \wedge *F_I - i\overline{\psi}\gamma_\mu \xi \psi * \mathrm{d}x^\mu \\ &= (e^{-2}\mathrm{D} * F + *J)_I \xi^I - \mathrm{d}(*F_I \xi^I) \\ &\approx -e^{-2}\mathrm{d}(*F_I \xi^I). \end{aligned} \tag{12.9}$$

12.2.2.1 Local and Regional, Abelian and Non-Abelian Conservation Laws

The conservation equation $\mathrm{d}j_\xi \approx 0$ of the standard Noether current trivially follows from (12.9), but it is not a gauge-covariant conservation law. That is, it obeys: $\partial_\mu j^\mu_I = 0$ but *not* $\mathrm{D}_\mu j^\mu_I = 0$.

To obtain the relation of the Noether to the matter current, J^μ_I, we insert a constant $\xi^I(x) = \xi^I_o$, such that $\partial_\mu \xi^I = 0$ in the first line of (12.9). Then the ξ-independent relation between the currents is:

$$j^\mu_I = -[A_\nu, F^{\mu\nu}]_I + J^\mu_I \tag{12.10}$$

To arrive at a regional conservation law, we must integrate this current over a region. But here we see that, upon integration, the conservation for j^μ would involve not only the values of the matter fields and current in the bulk of a region, J^μ_I, but *also* the values of the gauge potential and curvature there! This failure to decouple the conservation of fermionic matter from bosonic fields implies that we cannot easily relate the bulk integral of quantities involving matter fields with the boundary fluxes of quantities involving the forces. Thus, the relation (12.10) so far has no implications for non-locality: we must know the values of the gauge-fields everywhere, not only at a distant boundary, to infer the total amount of matter charge contained in a region.

To remedy this situation, we first obtain a current that obeys a local covariant conservation law. Of course, J fills this role on-shell, by applying the Bianchi identities to the equations of motion for A:[13]

$$\mathrm{D}_\mu J_I^\mu \approx 0 \tag{12.11}$$

In the Abelian case, there is no difference between j and J, of course, so the conservation equation is automatically covariant. But now, at least at first sight, we have lost the possibility of integrating this current to obtain a regional conservation law, and so no non-locality is apparent.

Crucially, to determine regional conservation laws, we need not the spatially constant gauge-transformations, generated by $\partial_\mu \xi^I = 0$, but those generated by the *covariantly constant* $\mathrm{D}_\mu \xi^I = 0$.

In other words, we can integrate $\mathrm{D}_\mu J_I^\mu$ against any Lie-algebra valued scalar $\xi \in C^\infty(M, \mathfrak{g})$ on the bounded region R, obtaining:

$$\int_N (\mathrm{D}_\mu J_I^\mu \xi^I) = -\oint_{\partial N} n_\mu J_I^\mu \xi^I + \int_N J_I^\mu \mathrm{D}_\mu \xi^I = 0 \tag{12.12}$$

But this is a bona-fide regional conservation law if ξ^I's are such that $\mathrm{D}_\mu \xi^I = 0$, for the integral will not generically reduce to a boundary flux.

This establishes some caveats to our usual understanding of the Noether conserved currents. Such caveats are ackowledged,[14] but their more radical consequences are usually left unsaid: (i) ultimately, the very concept of a non-Abelian regionally conserved charge makes sense only over a uniform background (as the notion of regionally conserved energy-momentum in general relativity) and (ii) associated charges are not invariant in the presence of boundaries to the Cauchy surface.

There is a useful parallel to be drawn here with general relativity. In general relativity, one can use the same procedure as above to find $j_\mu^I \sim \tau_{\mu\nu}$, the so-called Landau–Lifschitz pseudo-tensor (see e.g. Lifshitz 1987, ch. 11; Weinberg 2005, ch. 15.3). But what we want is a covariant local conservation law, such as $\nabla_\mu T^{\mu\nu} = 0$. Indeed, the energy-momentum

13 Also note that with ξ constant, the last line of (12.10) yields $j_\xi = j_\mu^I \xi_I \approx \partial^\nu F_{\mu\nu}^I \xi_I$, and thus (12.10) is just the equations of motion for A. While it is standard to just define J_μ^I as the source of the Yang–Mills equations of motion (12.6) and derive the covariant laws from consistency with the Bianchi identities, $\mathrm{D} * \mathrm{D} * F = 0$, we can recover a general result about the matter current J_μ^I which does not rest on the Bianchi identities for F. Suppose we take only the coupled matter part of equations (12.4), that is, with no self-action for A. We obtain from (12.6):

$$\mathbb{d}S_{\mathrm{YM}}^{\mathrm{matter}}(\delta_\xi A) = \int \mathrm{EL}_A^{\mathrm{matter}} \mathrm{D}\xi \overset{\&}{=} -\int (*\mathrm{D} * J)\xi = 0$$

which is valid for all ξ, and thus, inserting the spacetime indices: $\mathrm{D}_\mu J_I^\mu = 0$. In the equation above, & signals the only place in the argument where we use the specific matter part of the Yang–Mills Lagrangian. We did not need to do this: with minimal coupling, on-shell of the matter equations of motion, and for no kinetic term of the gauge fields, we would obtain $J_\mu^I := \frac{\partial \mathcal{L}^{\mathrm{matter}}}{\partial \mathrm{D}_\mu \psi} \tau_I \psi \approx \frac{\partial \mathcal{L}^{\mathrm{matter}}}{\partial A_\mu^I}$. The same type of relation arises in general relativity (one usually actually defines the matter energy momentum tensor as the functional derivative of the matter Lagrangian by the metric). In this way, it becomes clear that merely employing gauge variables, we can recover the necessary local conservation laws from variational principles.

14 Brading and Brown (2000) describe this finding as follows: 'we cannot follow the procedure used for global gauge symmetries in the case of local gauge symmetries to form gauge-independent currents. A current that is dependent on the gauge transformation parameter is not satisfactory – in particular, such gauge-dependent quantities are not observable.'

tensor $T_{\mu\nu}$ defined in similar fashion to J^I_μ in (12.7) does satisfy this covariant local conservation law. But this only straightforwardly yields regional conservation laws if Killing vector fields are available,[15] that is, if there is a ξ_μ such that $\nabla_{(\mu}\xi_{\nu)} = 0$; for then we can write a smeared equation in a bounded spacetime region $N \subset M$:

$$\int_N \nabla_\mu T^{\mu\nu}\xi_\nu = \oint_{\partial N} n_\mu T^{\mu\nu}\xi_\nu = 0 \qquad (12.13)$$

where n_j is the normal to ∂N; this yields a regional conservation law in the same way as above.

As in general relativity, in the non-Abelian Yang–Mills case generic configurations do not admit any such Killing fields, or directions of uniformity. Therefore, true regional conservation laws emerge only in the presence of matter charges *and* directions of uniformity in the background field.

In the chain of implications, we can formalise the asterisk as the following condition:

$$* : \quad \text{over field configurations } \tilde{A} \text{ such that there exist } \tilde{\xi}_I \text{ with } \tilde{\mathrm{D}}\tilde{\xi} = 0$$

(mutatis mutandis for general relativity). Thus we have secured the link *local conservation laws* $\overset{*}{\Rightarrow}$ *conserved regional charges*, within the special models signalled by the asterisk. Although the link would also apply more generally for any field theory employing gauge potentials and matter fields that transform as in (12.2) and (12.3) (on-shell of the matter equations of motion), we will not show this here.

As mentioned in Section 12.1.2, according to Gomes (2021), regional conservation laws are related to the type of failure of global supervenience on regions required for the existence of ‘symmetries with direct empirical significance’ (a thesis first articulated in Gomes and Riello 2021, sec 4.4). In this respect, the important point to secure this relation is *not* that the charges J be of one specific form or another; the important point is that the conserved regional charges be associated to the Killing directions ξ^I, which are rigid. In the next section, we will examine just how this fact is related to non-separability and therefore with the direct empirical significance of symmetries.

12.3 Non-Separability

Here we will show that for simply connected manifolds, conserved regional charges occur just in case there is a certain level of non-separability in the model. Following Myrvold, we

[15] Here I am focusing on notions of charge which arise from integration over volume. Common definitions of ‘quasi-local’ charges – such as the ones studied by Szabados, e.g. Brown–York – fall outside of this definition. Although there are many suggestions for notions of conserved ‘quasi-local’ energy-momentum, it is fair to say no one notion is as of yet unproblematic. As Szabados remarks in his review (Szabados 2004, p. 9):

> Contrary to the high expectations of the eighties, finding an appropriate quasi-local notion of energy-momentum has proven to be surprisingly difficult. Nowadays, the state of the art is typically postmodern: Although there are several promising and useful suggestions, we have not only no ultimate, generally accepted expression for the [quasi-local] energy-momentum and especially for the angular momentum, but there is no consensus in the relativity community … on the list of the criteria of reasonableness of such expressions. The various suggestions are based on different philosophies, approaches and give different results in the same situation.

will first investigate this matter in electromagnetism using holonomies, but adding charged scalar fields. This will make it clear, at least in electromagnetism, that there is a residual variety in the reconstructing the global state from the states of the regions if and only if charges are present within the regions.

But holonomies are not sufficient to settle the question in the non-Abelian case, and thus we will resort to a gauge-fixing, in Section 12.3.2. Once we have gauge-fixed the regional content of the field, we will use a gluing theorem to reconstruct a gauge-fixed state of the whole from that of the regions. It will be the case once more that, in the presence of regional conserved charges, one cannot reconstruct the state of the whole from the state of the parts, thus establishing the link between *regional conserved charges* $\Rightarrow$ *non-separability*.

Section 12.3.2 applies in the Euclidean signature regime. But in such a regime, matters of locality are orthogonal to matters of separability; there are in fact no matters of locality, since there is no causal propagation. But, as separability suffices to forge the link of our chain, we leave an investigation of the Lorentzian models to Appendix 12.B. There, we will locate the source of non-locality of gauge theory in the Gauss constraint. Thus, to be able to discuss the state of one region independently of the state of another – a necessary independence to analyse separability – we will split the variables into a Coulombic part, which solves the Gauss constraint as in the electrostatic case, and a radiative, dynamical part. The Coulombic part is the one that relates the charge content in the bulk of the region to boundary information. However, only the radiative part enters the reconstruction theorem, and the theorem suffers from the same degeneracy as in the Euclidean case.

12.3.1 Holonomies and Non-Separability

The holonomy interpretation of electromagnetism takes as its basic elements assignments of unit complex numbers to loops in spacetime. A loop is the image of a smooth embedding of the oriented circle, $\gamma : S^1 \to \Sigma$; the image is therefore a closed, oriented, non-intersecting curve. One can form a basis of gauge-invariant quantities for the holonomies (cf. Barrett 1991; Healey 2007, ch. 4.4; and references therein),[16]

$$hol(\gamma) := \exp\left(i \int_\gamma A\right). \tag{12.14}$$

12.3.1.1 The Basic Formalism

Let us look at this in more detail. By exponentiation (path-ordered in the non-Abelian case), we can assign a complex number (matrix element in the non-Abelian case) $hol(C)$ to the oriented embedding of the unit interval: $C : [0,1] \mapsto M$. This makes it easier to see how composition works: if the endpoint of C_1 coincides with the starting point of C_2, we define the composition $C_1 \circ C_2$ as, again, a map from $[0,1]$ into M, which takes $[0,1/2]$ to traverse C_1 and $[1/2,1]$ to traverse C_2. The inverse C^{-1} traces out the same curve with the opposite

[16] Of course, any discussion of matter charges and normalisation of action functionals would require e and $\hbar$ to appear. However, I am not treating matter, so these questions of choice of unit do not become paramount. As before, if needed, I set my units to $e = \hbar = 1$; as is the standard choice in quantum chromodynamics (or as in the so-called Hartree convention for atomic units).

orientation, and therefore $C \circ C^{-1} = C(0)$.[17] Following this composition law, it is easy to see from (12.14) that

$$hol(C_1 \circ C_2) = hol(C_1)hol(C_2), \tag{12.15}$$

with the right-hand side understood as complex multiplication in the Abelian case, and as composition of linear transformations, or multiplication of matrices, in the non-Abelian case.

For both Abelian and non-Abelian groups, given the above notion of composition, holonomies are conceived of as smooth homomorphisms from the space of loops into a suitable Lie group. One obtains a representation of these abstractly defined holonomies as holonomies of a connection on a principal fibre bundle with that Lie group as structure group; the collection of such holonomies carries the same amount of information as the gauge-field A. However, only for an Abelian theory can we cash this relation out in terms of gauge-invariant functionals. That is, while (12.14) is gauge-invariant, the non-Abelian counterpart (with a path-ordered exponential), is not.[18]

12.3.1.2 Non-Separability

As both Healey (2007, ch. 4.4) and Belot (1998, sec. 3); Belot 2003, sec. 12) have pointed out, even classical electromagnetism, in the holonomy interpretation, evinces a form of non-locality, which one might otherwise have thought was a hallmark of non-classical physics. But is it still the case that the state of a region supervenes on assignments of intrinsic properties to patches of the region (where the patches may be taken to be arbitrarily small)? This is essentially the question of *separability* of the theory (see Belot 1998, sec. 3; 2003, sec. 12; Healey 2007, ch. 2.4; Myrvold 2010).

We are not interested here in cases of 'topological holism', as related to the Aharonov–Bohm effect, which has been extensively studied. We are asking whether a vacuum, simply connected universe still displays non-separability. For this topic, we can directly follow Myrvold's definition (Myrvold 2010, p. 427) (which builds on Healey's notion of Weak Separability (Healey 2007, p. 46) and on Belot's notion of Synchronic Locality (Belot 1998, p. 540):

- *Patchy Separability for Simply Connected Regions.* For any simply connected spacetime region R, there are arbitrarily fine open coverings $\mathcal{N} = \{R_i\}$ of R such that the state of R supervenes on an assignment of qualitative intrinsic physical properties to elements of $\mathcal{N}$.

[17] It is rather intuitive that we don't want to consider curves that trace the same path back and forth; that is, *thin* curves. Therefore, we define a closed curve as *thin* if it is possible to shrink it down to a point while remaining within its image. Quotienting the space of curves by those that are thin, we obtain the space of *hoops*, and this is the actual space considered in the treatment of holonomies. I will not call attention to this finer point, since it follows from a rather intuitive understanding of the composition of curves.

[18] For non-Abelian theories the gauge-invariant counterparts of (12.14) are Wilson loops; see for example, Barrett (1991), $W(\gamma) := \mathrm{Tr}\,\mathcal{P}\exp\left(i\int_\gamma A\right)$, where one must take the trace of the (path-ordered) exponential of the gauge-potential. It is true that all the gauge-invariant content of the theory can be reconstructed from Wilson loops. But, importantly for our purposes, it is no longer true that there is a homomorphism from the composition of loops to the composition of Wilson loops. That is, it is no longer true that the counterpart (12.15) holds, $W(\gamma_1 \circ \gamma_2) \neq W(\gamma_1)W(\gamma_2)$. This is due solely to the presence of the trace. The general composition constraints – named after Mandelstam – come from generalisations of the Jacobi identity for Lie algebras and depend on N for SU(N)-theories; e.g. for $N = 2$, they apply to three paths and are: $W(\gamma_1)W(\gamma_2)W(\gamma_3) - \frac{1}{2}(W(\gamma_1\gamma_2)W(\gamma_3) + W(\gamma_2\gamma_3)W(\gamma_1) + W(\gamma_1\gamma_3)W(\gamma_2)) + \frac{1}{4}(W(\gamma_1\gamma_2\gamma_3) + W(\gamma_1\gamma_3\gamma_2) = 0$.

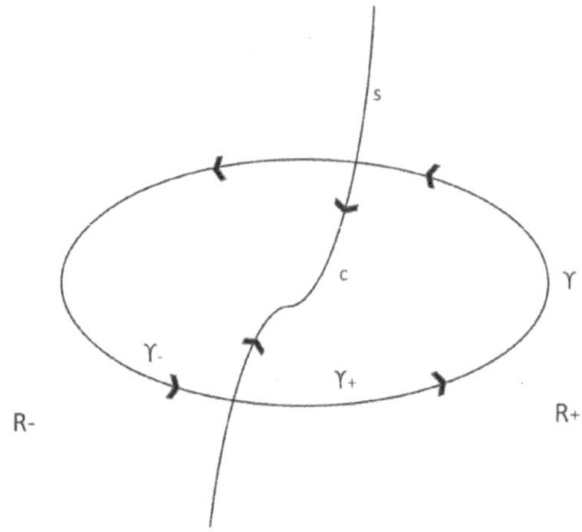

Figure 12.1 Two subregions, $R_{\pm}$, with the separating surface S. We assume no matter fields are present on S. A larger loop γ crosses both regions. But, since γ_1 and γ_2 traverse S along C in opposite directions, $\gamma = \gamma_1 \circ \gamma_2$.

In vacuum, it is easy to show that *Patchy Separability for Simply-Connected Regions* holds. In Figure 12.1, we see a loop γ not contained in either R_+ or R_-. However, we can decompose it as $\gamma = \gamma_+ \circ \gamma_-$, where each regional loop $\gamma_{\pm}$ does not enter the complementary region ($R_{\mp}$, respectively). Following (12.15), it is then true that, since holonomies form a basis of gauge-invariant quantities, the universal gauge-invariant content of the theory supervenes on the regional gauge-invariant content of the theory.

It is also easy to see how *Patchy Separability for Simply-Connected Regions* fails when charges are present within the regions but absent from the boundary S (see in particular Gomes and Riello 2021, sec. 4.3.2; footnote 70 in Gomes 2021). For, in the presence of charges, we can form *gauge-invariant* functions from a non-closed curve C' that crosses S and, for example, one positive and one negative charge, $\psi_{\pm}(x_{\pm})$, capping off C' at $x_{\pm} \in R_{\pm}$. That is, the following quantity is a gauge-invariant function:

$$Q(C', \psi_{\pm}) = \psi_-(x_-) hol(C') \psi_+(x_+)$$

for $C'(0) = x_-, C'(1) = x_+$. It is easy to check from the transformation property $\psi \mapsto g\psi$ that Q is gauge-invariant. But now, applying different rigid gauge transformations $\theta_{\pm}$ in each region will amount to choosing a different regional representation for $\psi_{\pm}$ – that is, $e^{i\theta_{\pm}}\psi_{\pm}$ – thus resulting in a relative shift $Q \to e^{i(\theta_+ - \theta_-)}Q$ (which, as stated above, is gauge-invariant and therefore cannot be undone by a gauge transformation). This is precisely what is known in the literature as "t Hooft's beam-splitter' (see Brading and Brown 2004, p. 651; 't Hooft 1980, p. 110).

If the reader is content with the Abelian case, they may feel comfortable with skipping the next subsections.

12.3.2 Gauge-Fixing Regional and Universal States

In the general non-Abelian case for bounded regions, a natural gauge-fixing suggests itself.[19] Given a field space vector $\delta A \in T_A\mathcal{A}$, where $\mathcal{A} = \Lambda^1(M, \mathfrak{g})$ is the space of (kinematical) models for the gauge-potential, we define the Landau gauge-fixing as:

$$\mathrm{D}^\mu \delta A_{\mu I} = 0 \tag{12.16}$$

In the bounded case, suppose that our patchwork of mutually exclusive, jointly exhaustive regions is given by just R_+, R_-, such that $R_+ \cap R_- = \partial R_\pm$, with s^i the normal to ∂R, and where $R_+ \cup R_- = \Sigma$. We also assume Σ to be simply-connected and without boundary, $\pi_1(\Sigma) = 0, \partial\Sigma = \emptyset$. Suppose moreover that there are is no charged matter at ∂R.

In that case, the gauge-fixing (12.16) becomes:[20]

$$\begin{cases} \mathrm{D}^\mu \delta A^\pm_{\mu I} = 0 \\ \delta A^\pm_{sI} = 0 \end{cases} \tag{12.17}$$

where, again, s is the normal to the shared boundary of the region. We call a field-space vector that satisfies these conditions, horizontal (or radiative, in the Lorentzian case), and denote them by h and $h_\pm$, in the global and regional cases, respectively.

It is easy to show that (12.16) and (12.17) are bona-fide perturbative, gauge-fixings, that are moreover covariant under a change of the perturbed configuration $A \to A + \delta_\xi A$ (and mutatis mutandi for the regional case, so that the gauge-fixing of the perturbation does not sensitively depend on the choice of representative for the perturbed configuration.

Two configurations $h_\pm$ are composable if and only if there exist gauge transformations, $\xi_\pm$, such that:

$$\delta A = (h_+ + \mathrm{D}_+\xi_+)\Theta_+ + (h_- + \mathrm{D}_-\xi_-)\Theta_-,$$

where $\Theta_\pm$ are the characteristic functions of the regions $R_\pm$, is a smooth global gauge configuration. Composability therefore requires the difference of the configurations at the boundary, $(h_+ - h_-)_{|S} = (\mathrm{D}_+\xi_+ - \mathrm{D}_-\xi_-)$, to be pure gauge. These matters are taken up in detail in Gomes (2020, 2021) and in Gomes and Riello (2021).

In general there will be many such $\xi_\pm$ if there is any. But to establish residual physical variety of the global state, we must also consider the variety of global *physical* states that will correspond to the composition, and therefore we must gauge-fix globally, with (12.16). In most cases, this global gauge-fixing will uniquely fix $\xi_\pm$; when it does not, there can be a residual physical variety, as we explain next.

[19] As we will see in Appendix 12.B, this gauge-fixing is related to the split of electric degrees of freedom into those that solve the Gauss constraint as it is solved in the static case, and those that are radiative, or dynamical. It is also naturally selected by the ultralocal inner-product in the field space of the theory, Φ.

[20] See Gomes (2020, sec. 2.2.2) for how the natural extension of the Landau gauge-fixing to the bounded case implies the Neumann boundary conditions used here.

12.3.3 Coulombic and Radiative Reconstruction Theorem

To finish this section, I will present results on 'gluing': in reassembling the regional information of the fields into the global state of the fields, employment of the pure-gauge and horizontal (alternatively, Coulombic and radiative; cf. Appendix 12.B) split of regional degrees of freedom is remarkably instructive. (To avoid having to intersperse such parentheses along the following section, the results are framed in the unifying language of 'Coulombic and radiatives' modes). Again, I will only state, and not prove any of the surprising results here (see Gomes and Riello 2021, sec. 4).

12.3.3.1 The Theorem, Loosely Stated

The reconstruction theorem states that in these circumstances, we can fully and uniquely reconstruct the state on the entire region, e.g. we can recover the full physical content of the fields,[21] solely from the *strictly interior* degrees of freedom. Those degrees of freedom that require further boundary information – the Coulombic – are superfluos for reconstruction.

As a schematic example, in the Lorentzian case (cf Appendix 12.B), we again split the electric field and the spatial gauge potential along radiative and Coulombic components. The Coulombic component of E generalises the electrostatic electric field, and is of the form $E_{\mathrm{Coul}} = \mathrm{D}\Lambda$, for some (Lie-algebra-valued) scalar potential Λ and E_{rad} being divergence-free, as in (12.16): $\mathrm{D}^i E_{iI} = 0$ (obeying the same Neumann boundary conditions as in (12.17)).

Thus the total regional field is $E_\pm = E_\pm^{\mathrm{rad}} + \mathrm{D}\Lambda_\pm$, with $\Lambda_\pm$ satisfying the Poisson equation $\mathrm{D}_\pm^2 \Lambda_\pm = \rho_\pm$, with $\rho_\pm$ related to the regional matter fields $\psi_\pm, \overline{\psi}_\pm$ by (12.7), with boundary conditions on the Poisson equation given by $s^i \mathrm{D}_i \Lambda = s^i E_i \equiv f$. It would seem the Coulombic degrees of freedom parametrised by $\Lambda_\pm$ are essential for a description of the global state. Surprisingly, we can recover the global E solely as a functional of those degrees of freedom which carry no dependence on the boundary conditions (i.e. the radiative ones).

The statement of the gluing theorem therefore is that, in the Euclidean case, if the perturbed potential $A_{\mu I}^{\pm}$ has either no Killing directions (i.e. no stabilisers), or no charged currents inside the regions $R_\pm$, then $h_\pm$ uniquely determines the global physical state h. In the Lorentzian case (for which now R and $R_\pm$ are spatial regions), the statement is that if the phase space points $(A_{iI}^{\pm}, E_i^{I\pm})$ have no stabilisers, the radiative components $E_\pm^{\mathrm{rad}}$ and the matter charges $\psi, \overline{\psi}$, suffice to determine all the remaining regional components, $E_\pm^{\mathrm{Coul}}$, and also the global ones: E^{rad} and E^{Coul}.

For the proof, which uses the remarkable properties of the Dirichlet-to-Neuman operators, see Gomes and Riello (2021, sec. 5).

21 In the non-Abelian case, we recover E up to gauge transformations, but we only recover the physical content of a perturbation of the potential, δA. We can recover the full A, and not just perturbations, in the Abelian case. This reconstruction works either in the dynamical 3+1 context in which M represents a Cauchy surface, or for the Euclidean covariant context in which M is the spacetime manifold. In the Euclidean context (described in Gomes et al. 2019), we don't talk about the phase space $\mathcal{P}$, but solely about the Lagrangian field space Φ, discussed in Section 12.A.1.

12.3.3.2 The Caveats, Loosely Stated

But, as announced, if the necessary conditions do not obtain, there are ambiguities in the reconstruction. These occur only if the underlying fields $A^{\mu I}_{\pm}$, or, in the Lorentzian setting, $(A^{iI}_{\pm}, E^{iI}_{\pm})$, admit stabilisers (or Killing vectors). Let us call such a homogeneous, or uniform potential, $\tilde{A}^{\mu I}_{\pm}$; the arising ambiguities are precisely the stabilisers of $\tilde{A}$, that we label $\chi_{\tilde{A}}$.

A gauge transformation by one such stabiliser does not affect the gauge fields (by construction, since $\mathrm{D}\xi = 0$), but does affect the matter fields, by the transformation law (12.3). Therefore, if these ambiguities exist, *and* there are non-trivial matter fields inside the regions, a corresponding ambiguity in the reconstruction slips in: the matter components *are* shifted by an element of the stabiliser. Although this rotation does not affect either regional gauge-invariant state, it has a relative effect between the regions. This is precisely the same sort of regional rigid gauge shift we saw at the end of Section 12.3.1.2, and to the 't Hooft beam-splitter. And these conditions are precisely the ones required for the existence of 'conserved regional charges'. Therefore we have established the link 'conserved regional charges'⇒'non-separability'.

It is also possible at this stage to understand why in the Abelian, but not in the non-Abelian case, we always find well-defined regional gauge-invariant charges (in the presence of charged matter), thus connecting the present discussion back to Section 12.2.2.1. Namely, for Abelian theories the existence of stabilisers χ impose no restrictions on A: for all A, the stabilisers are just the constant gauge transformations.

The same argument could be applied to general relativity: suppose there are matter fields that vanish at the intersection of two flat regions, but not in their interior. Then we could gauge-fix all the non-rigid diffeomorphisms with the metric field, but that would leave out the rigid Poincaré transformations. Then applying distinct such transformations to each region, we would have no global Poincaré transformation that would be able to efface the difference between the rigid regional transformations at the boundary. Although without the presence of matter the overall state of affairs after the tranformation would be the same as before, it would be distinct if matter was present. Those are again precisely the conditions required to obtain non-trivial regional conserved charges in general relativity.

12.3.3.3 The Link to DES

To establish the further link to DES, we note that the ambiguity corresponds to a global – or rigid, in the nomenclature of (Gomes 2021) – gauge transformation, which is necessarily a subgroup of the charge group G (see Gomes 2021, sec. 5.2.4; Gomes and Riello 2021, sec. 4) for proofs and explicit examples. In the Abelian case of electromagnetism, there is always a conserved U(1) global transformation if charged matter fields are present in the regions.[22]

[22] We note that it is uncontroversial that DES, in many of its incarnations, e.g. (Brading and Brown 2004; Friederich 2014; Greaves and Wallace 2014; Teh 2016), is conditional on the existence of universal gauge-invariant quantities that are *not* specified by the regional gauge-invariant content. It is thus easy to see that non-separability gets us almost all the way to DES; all that remains to be shown is that the variety is parametrised by the action of a rigid symmetry group, which we did obtain in the gluing theorem above. The exact entailment between the two concepts – non-separability and DES – is fleshed out in full in Gomes (2021).

To summarise, to obtain a left-over variety in the reconstruction of the whole from the radiative parts, we require non-trivial stabilisers and non-zero charged matter fields. Moreover, we can explicitly associate this physical variety with the action of a subgroup of G on a subsystem. This action can indeed be construed as endowing 'direct empirical significance' to this group of global symmetries (Gomes 2021 and Gomes and Riello 2021). Lastly, the requirements for obtaining this variety also correspond to those for obtaining non-trivially conserved regional, gauge-invariant Noether charges.

12.4 Conclusions

12.4.1 Summary

Local conservation laws are a widely celebrated consequence of symmetries. While it is true that in certain cases one can extract conserved *regional* quantities from integrating the Noether densities, such charges may be problematic in other regards. For instance, the obtained regional charges may be gauge-*variant*. Or, when we think of fermionic charges as sourcing bosonic forces, it could be that we cannot isolate a regional conservation law for the matter charges. Both of these obstacles occur for non-Abelian Yang–Mills theory and general relativity, and both can be avoided at a single stroke: with gauge-invariant conserved regional matter charges emerging from the regional volume only if the underlying configuration admits stabilisers.[23] This result provided the first link of the chain of implications proposed in this work.

The second link connected such regional charges with non-separability. This link followed without ancillary conditions and applies also to simply-connected regions (in the non-simply-connected case, the vacuum configurations are also non-separable by the same methods). We showed it first for the Abelian theory using holonomies and then in the non-Abelian case using gauge-fixings. Crucially, non-separability in the simply-connected case requires a gauge transformation that is an isomorphism of the gauge field but not of the matter field. If the matter field vanishes *anywhere*, such transformations cannot be gauge-fixed, and yet, if different such transformations are applied in each region, they will change the physical state of the union of the regions. Nonetheless, from the perspective of the region, the shifted configuration is physically indiscernible from the unshifted configuration. Thus, we obtained non-separability, and clearly, a connection with a subsystem symmetry that could be said to have empirical significance. Although Gomes (2021) fleshes out this link in greater detail, here we saw a (hopefully convincing) sketch of the link, listing its main arguments.

Although the paper was focused on Yang–Mills theories, the main arguments here can be transposed without much difficulty to general relativity. In the interest of space, I have kept this parallel mostly in the background for this paper.

[23] In Abelian theories, any gauge field configuration admits the same stabilisers: the constant gauge transformation. This property gives rise to well-defined regional electric charges independently of the background configuration in electromagnetism.

Moreover, a few words should be said about the generality of these links: in which sense are they merely kinematical, and in which do they depend on the specific form of the Lagrangian? The only step in which we explicitly used the Lagrangian was in obtaining the regionally conserved charges. But, for gauge-invariant theories that have A and ψ as variables, with the transformation properties (12.2) and (12.3), that link follows as well (on-shell of the matter equations of motion, cf. footnote 13). But I have not explored the full space of theories in which these links would hold, and therefore the implications need to be seen as 'local deductions'.

12.4.2 An Added Link to Locality

Local symmetries have another, less celebrated consequence: they yield constraints on the initial values, thus pointing to a mild form of non-locality in a gauge theory. As discussed in Appendix 12.B, one cannot freely specify the initial value of the electric field here without knowing what is the far away distribution of charges at that same instant.

The link to non-locality does not fit as neatly into the chain of deductions explored in this paper, although it is inextricably connected to them. So, lest we tie them together, we would be left with all those loose ends; thus the inclusion of the discussion in Appendix 12.B (but not on the main part of the chapter) and this extended subsection here in the conclusions.

The particular non-locality of gauge theories can be examined precisely with the aid of a decomposition of the fields into Coulombic and radiative components. The radiative component corresponds to the strictly interior field content of a region; the Coulombic component depends also on boundary information.

We can track the effect of the decomposition on the Noether charges of the theory: the emerging radiative Noether charges are identical to the gauge-invariant Noether charges discussed in the first sections. The Coulombic charges can be identified with the gauge-variant charges; they correspond to the boundary gauge group: $\mathrm{Lie}(\mathcal{G})_{\partial R}$.

In the absence of stabilisers, the radiative modes are the sole carriers of the non-redundant regional information needed to piece back together a global state. The Coulombic degrees of freedom can also be reassembled by knowledge of the regional *radiative* degrees of freedom and are, in that sense, superfluous. But they become superfluous only from the perspective of the whole; given any single region, they are functionally independent from the radiative modes (Gomes and Riello 2021, sec. 5). This analysis is in accord with the proposal of Gomes (2019), which argues for the ineluctability of redundant, pure gauge degrees of freedom for the description of regions, but not for the description of the whole.

Appendix 12.A Symplectic Formalism

12.A.1 The Covariant Symplectic Formalism Abridged

A given action

$$S[\varphi,\gamma]=\int_M \mathcal{L}\left(\varphi^I,\nabla_\mu\varphi^I,\nabla_{(\mu}\nabla_{\nu)}\varphi^I,\nabla_{(\mu_1}\cdots\nabla_{\mu_n)}\varphi^I,\gamma\right)$$

for the Lagrangian density $\mathcal{L}$ dependent on the field φ^I, its derivatives up to n-th order,[24] and some background structure γ, can be seen as a function on an appropriate n-th jet-space of Φ. Using the manifold structure of Φ we can take directional derivatives of S and indeed of $\mathcal{L}$.

Taking the functional derivative of S along some generic $\delta\varphi^I$, for which we employ the standard notation for directional derivative, $\delta\varphi[S]$, we can usefully organise the result (using the jet-bundle structure) as:

$$\mathbb{d}S(\delta\varphi) = \int \sum_{k=0}^{n} \frac{\partial \mathcal{L}}{\partial(\nabla_{(\mu_1}\cdots\nabla_{\mu_k)}\varphi^I)} \nabla_{(\mu_1}\cdots\nabla_{\mu_k)}\delta\varphi^I = \int \left(E_I\delta\varphi^I + \mathrm{d}(\theta_I(\varphi)(\delta\varphi^I))\right). \tag{12.18}$$

Here the Euler–Lagrange density is given by:

$$E_I = \sum_{k=0}^{n}(-1)^k \nabla_{\mu_1}\cdots\nabla_{\mu_k}\mathcal{L}_I^{\mu_1\cdots\mu_k}, \tag{12.19}$$

where the partial derivative functions are given by:

$$\mathcal{L}_I := \frac{\partial\mathcal{L}}{\partial\varphi^I}, \text{ for } k=0, \text{ and } \quad \mathcal{L}_I^{\mu_1\cdots\mu_k} := \frac{\partial\mathcal{L}}{\partial(\nabla_{(\mu_1}\cdots\nabla_{\mu_k)}\varphi^I)}. \tag{12.20}$$

The symplectic potential density θ is a $(d-1)$ differential form in spacetime and a 1-form in field-space (it is a local linear functional on field-space vectors $\delta\varphi$). And we can express the exterior derivative of a $d-1$ form as the divergence of the corresponding vector:[25]

$$\theta_I^\mu(\varphi)(\delta\varphi^I) := \sum_{k=1}^{n}\sum_{\ell=1}^{k}(-1)^{\ell+1}\left(\nabla_{\mu_2}\cdots\nabla_{\mu_\ell}\mathcal{L}_I^{\mu\mu_2\cdots\mu_k}\right)\left(\nabla_{\mu_{\ell+1}}\cdots\nabla_{\mu_k}\delta\varphi^I\right) \tag{12.21}$$

The idea now is simple. Suppose the Lagrangian in (12.18) is invariant in the direction of some particular $\widehat{\delta\varphi}$, and that the Euler–Lagrange equations hold (which we signal by employing the weak equality $\approx$). If we use the coordinate-free notation of θ as a $d-1$ form, we conclude that the other term of the integrand of (12.18) weakly vanishes:

$$\mathrm{d}\left(\theta_I(\varphi)(\widehat{\delta\varphi^I})\right) \approx 0, \tag{12.22}$$

where d is the spacetime exterior derivative. Thus, by integrating over a spacetime region we can obtain the usual regional conservation laws for $\theta_I(\varphi)(\widehat{\delta\varphi^I})$, which is called *the Noether current density* for $\widehat{\delta\varphi^I}$. This is the essence of Noether's theorems.

12.A.2 The Covariant Symplectic Form and the Hamiltonian

Working now in field space, we can introduce a symplectic structure as follows. Using the exterior derivative on the symplectic potential density, which is a field-space one-form θ_I^μ,

[24] We can use symmetrised derivatives because we may always rid ourselves of the anti-symmetrised ones in favor of the appearance of curvature, which here would be seen as part of the background structure. We denote spacetime indices by Greek letters, $\mu, \nu, \ldots$.

[25] This is done using the Hodge star, but we will keep notation as simple as possible and denote the vector-density form of θ with the same letter.

we obtain a symplectic-form density, $\omega^\mu_{IJ} := (\mathbb{d}\theta)^\mu_{IJ}$, which is a d-1 form in spacetime and a closed 2-form in field space (since $\mathbb{d}\mathbb{d} \equiv 0$). Choosing a closed Cauchy surface Σ (for now unbounded), we define:

$$\Omega := \mathbb{d}\Theta, \quad \omega_{IJ} = \mathbb{d}(\theta)_{IJ}; \quad \Theta(\delta\varphi) = \int_\Sigma \theta_I \delta\varphi^I, \quad \Omega(\delta\varphi_1, \delta\varphi_2) = \int_\Sigma \omega_{IJ}\delta\varphi_1^I \delta\varphi_2^J \tag{12.23}$$

where we have only kept the internal index I, which serves to remind us that we are dealing with components, also in the (suppressed) spacetime labels μ and x.

If the symplectic potential Θ is also gauge-invariant on-shell (which is always the case if $\partial\Sigma = \emptyset$, (cf. Lee and Wald 1990, eq. 3.22) we can apply the Cartan Magic formula to obtain:

$$\mathbb{L}_{\delta\varphi}\Theta \approx 0 = (\mathbb{d}\mathbb{i}_{\delta\varphi} + \mathbb{i}_{\delta\varphi}\mathbb{d})\Theta \tag{12.24}$$

and therefore we obtain $\mathbb{i}_{\delta\varphi}\Omega \approx -\mathbb{d}(\Theta(\delta\varphi))$, so that the Hamiltonian generator for the flow $\delta\varphi$ is simply given on-shell by $\Theta(\delta\varphi)$.[26] Then one usually writes the Hamiltonian generator $H[\xi]$ of the gauge transformation ξ as:

$$H[\xi] = \int_\Sigma \theta_I \delta_\xi \varphi^I. \tag{12.25}$$

Indeed, in the absence of boundaries of the Cauchy surface, we can go beyond the mere existence of a Hamiltonian generator: it is possible to show that $\mathbb{i}_{\widehat{\delta\varphi}}\Omega \approx 0$ for variations within the subset of field configurations which satisfy the equations of motion, in which case, the kernel $\widehat{\delta\varphi}$ is integrable in that subset (cf. Lee and Wald 1990, eq. 3.27).[27] More specifically, the on-shell integral manifold corresponding to the tangent vectors $\widehat{\delta\varphi}$ form 'orbits' parametrised by gauge transformations.[28] It is then possible to constitute a finite equivalence relation along elements of the gauge-orbits and thus formally reduce the pre-symplectic to the symplectic formalism.

Moreover, note that applying a Lie derivative along $\widehat{\delta\varphi}$ to a gauge-invariant density θ and using $\mathbb{i}_{\widehat{\delta\varphi}}\Omega \approx 0$, we obtain

$$\mathbb{L}_{\widehat{\delta\varphi}}\theta = \left(\mathbb{d}\mathbb{i}_{\widehat{\delta\varphi}} + \mathbb{i}_{\widehat{\delta\varphi}}\mathbb{d}\right)\theta \approx \mathbb{d}(\mathbb{i}_{\widehat{\delta\varphi}}\theta) \approx 0. \tag{12.26}$$

[26] Using this formalism it is also easy to show, assuming the Poincaré lemma applies (so at least in finite-dimensions and for simply-connected regions) that if a closed 2-form Ω is left invariant by a flow $\delta\varphi$, then it can be used to obtain the symplectic generator of that flow, $f_{\delta\varphi}$. That is, $\mathbb{L}_{\delta\varphi}\Omega = 0 = \mathbb{d}(\mathbb{i}_{\delta\varphi}\Omega) \Rightarrow \mathbb{i}_{\delta\varphi}\Omega = \mathbb{d}(f_{\delta\varphi})$.

[27] Here is a simple proof: suppose a certain class of vector fields $\widehat{\delta\varphi}$ is such that $\mathbb{i}_{\widehat{\delta\varphi}}\omega \equiv \omega_{IJ}\widehat{\delta\varphi}^J = 0$. Now, since ω is closed, using the Cartan Magic formula (which is valid on general Banach manifolds), since $\mathbb{d}\omega = 0$:

$$\mathbb{L}_{\widehat{\delta\varphi}}\omega = (\mathbb{d}\mathbb{i}_{\widehat{\delta\varphi}} + \mathbb{i}_{\widehat{\delta\varphi}}\mathbb{d})\omega = 0,$$

by which we can say that ω is $\widehat{\delta\varphi}$-invariant. Moreover, if we take the field-space commutator of $\widehat{\delta\varphi}_1, \widehat{\delta\varphi}_2$ (which we denote by a 'double-struck' square-bracket), i.e. $[\![\widehat{\delta\varphi}_1, \widehat{\delta\varphi}_2]\!] = \mathbb{L}_{\widehat{\delta\varphi}_1}\widehat{\delta\varphi}_2$, and contract with ω we obtain:

$$[\![\widehat{\delta\varphi}_1, \widehat{\delta\varphi}_2]\!]^I \omega_{IJ} = \left(\mathbb{L}_{\widehat{\delta\varphi}_1}\widehat{\delta\varphi}_2^I\right)\omega_{IJ} = \mathbb{L}_{\widehat{\delta\varphi}_1}(\widehat{\delta\varphi}_2^I\omega_{IJ}) - \widehat{\delta\varphi}_2^I \mathbb{L}_{\widehat{\delta\varphi}_1}\omega_{IJ} = 0,$$

and therefore such vector fields in the kernel of ω form an integrable distribution by the Frobenius theorem.

[28] Infinitesimally, each $\widehat{\delta\varphi}$ is of the form $\widehat{\delta\varphi} = \delta_\xi\varphi$ i.e. it is a result of an infinitesimal gauge-transformation ξ on the field space Φ.

Therefore, for bona-fide gauge transformations – those which we can treat symplectically – the associated Hamiltonian must be *on-shell* field-independent. We can therefore weakly set the Hamiltonian constraints corresponding to gauge transformations to zero: $H[\xi] \approx 0$.

Appendix 12.B Non-Locality and Coulombic and Radiative Components

12.B.1 Non-Locality

To fully comprehend the non-local content of the equations of motion of gauge theories, it is instructive to first examine an entirely local, extremely simple, example: that of a massless Klein–Gordon field – $\varphi \in C^\infty(M, \mathbb{C})$ such that $\partial^\mu \partial_\mu \varphi = 0$. Given any initial values $(\varphi, n^\mu \partial_\mu \varphi)$ of configuration and configurational 'velocity' on a Cauchy surface (or on a subset thereof), we can find a corresponding solution of the Klein–Gordon equations in the domain of dependence of the region (by the Cauchy–Kowalevsky theorem). In this case, locality is a consequence of a single equation of motion, of the same order in spatial and time derivatives.

But for gauge theories, some of the differential equations of motion may contain only spatial and no time derivatives, signalling the occurrence of initial value constraints. This is the case with the Yang–Mills theory. In other words, having chosen a Cauchy surface Σ, we are not free to choose any value we want for the doublet (E^i_I, ρ_I) on Σ.

Let us take the simple example of electromagnetism: the Gauss law $\partial^i E_i = \rho$ does not allow us to choose any value for the electric field on a given point $x \in \Sigma$ independently of the distribution of charges or of the values of E in other points: there are Coulombic constraints that must be first satisfied. These constraints must be satisfied even before any equation accounting for propagation in time is taken into account. Since the constraint is a spatial differential equation, resolving the constraint will require some amount of integration over a given region at a fixed time.

To see that this requirement must be satisfied before any dynamics takes place, we can take an even simpler example: the electrostatic regime, in which case $E_i = \partial_i \phi$, where ϕ is the Coulombic potential. We then have to satisfy a Poisson equation, $\nabla^2 \phi = \rho$, whose solution involves an integral of a Green's function over the entire Σ.

But this non-locality need not be of an extreme sort! It won't allow faster-than-light-signalling, for instance. And it doesn't necessarily involve knowledge of the charge distributions over the entire Cauchy surface; one may exchange that completely global information for a more local one: the electric flux on the boundary of a region surrounding our chosen point x. That is, given a point $x \in \Sigma$, and a closed region R around x, knowing both the charge distribution within R and boundary conditions for ϕ on ∂R, we can establish the electrostatic field at x. If one knows the 'local electric flux' through R, $\partial_n \phi = E_n := f$, where n_i is the normal to R within Σ, then one knows the value of the electric field in x.[29]

[29] While it is commonly said that if one satisfies the initial value constraint in an initial surface, the dynamical propagation equations will ensure that it continues to be satisfied throughout time, this is only assured if there is never any flux at the boundary, or if Σ is a closed surface. This is one of the consequences of the non-vanishing of the Hamiltonian generator of symmetries for bounded regions.

Thus the static electric field within a region has *regional* locality in the sense that it is determined by the distribution of charges in the region *and* boundary data. It is therefore a regional quantity, but not a strictly interior one. The boundary data represent the influence of the outside on the inside of the region. This provides a specification of degrees of freedom for each region which depends only on the interior of the region, and not on any boundary information. These are what we call 'radiative' degrees of freedom.

Any new split of degrees of freedoom must be checked for consistency in at least two ways. First, is the split kinematically consistent – i.e. is it obeyed by symplectic conjugacy relations? And second: how independent or redundant are the different types of degrees of freedom, when taken regionally and globally? That is, when does knowledge of one sort of information determine the state of the other? The first question will be taken up in Section 12.B.2.1, where we will prove consistency, and the second was taken in Section 12.3.3, where we showed that although for each region the radiative degrees of freedom is entirely functionally independent of the regional ϕ, the *collection* of radiative degrees of freedom associated to a patchwork of mutually exclusive, jointly exhaustive regions fully determines both the radiative and the Coulombic degrees of freedom of the whole. There is thus a subtle manner in which the Coulombic components are superfluous only when the entire collection of regional interior data is available, but not before. This is consistent with previous work, (Gomes 2019), arguing that gauge degrees of freedom are superfluous only from a global perspective, but not from a parochial one.

12.B.2 Coulombic and Radiative Symplectic Pairs

To establish our claims, we must further examine the source of non-locality in the dynamics of Yang–Mills theory, and that is what we will do in this section, which is based on (Gomes and Riello 2021, sec. 3). According to the above, at least in the standard variables,[30] the source lies in the Gauss laws. Thus the first order of business is to segregate the Coulombic degrees of freedom from the remaining ones in the electric field. We will see that in a concrete sense the Coulombic ones are the carriers of the boundary information of the field.

12.B.2.1 Coulombic and Radiative Decomposition

We start with the standard decomposition of the electric field:

$$E_i^I = \dot{A}_i^I - \mathrm{D}_i A_0^I. \tag{12.27}$$

Coulombic terms are usually described as those given by the gradient of the potential, i.e. $\mathrm{D}_i A_0^I$. But E_i^I can also contain Coulombic terms, i.e. terms of the form $\mathrm{D}_i \xi^I$, for some Lie-algebra-valued function $\xi \in C^\infty(\Sigma, \mathfrak{g})$.[31] Luckily for us, there is a straightfoward way to extract the strictly Coulombic component of E_i by employing a generalised Helmholtz

[30] Our conclusions are valid also using a holonomy gauge-invariant basis, e.g. with Wilson loops.

[31] Since we will assume all these functions are Lie-algebra valued, we will drop the index I to unclutter notation for what follows.

decomposition. The decomposition spits out a unique covariant gradient of a Coulombic potential and another component, covariant-divergence-free, which we will call *radiative*, and denote as E_i^{rad}.

In other words, any E_i in the region R, *without any restriction on boundary values*, can be uniquely decomposed into:

$$E_i = E_i^{\text{rad}} - \mathrm{D}_i\phi. \tag{12.28}$$

That is, both E_i^{rad} and ϕ can be uniquely written as functionals of E_i (see Gomes and Riello 2021, sec. 3.2) and, under a gauge-transformation, E_i^{rad} transforms like the electric field and ϕ tranforms like A_0.[32] For our purposes here, the precise form of these functionals is secondary; what is important to know is that the range of values ϕ can take are left unconstrained off-shell (i.e. before we consider how they react to charged matter), while E_i^{rad} obeys the following equations:

$$\begin{cases} \mathrm{D}^i E_i^{\text{rad}} = 0 & \text{in } R, \\ n^i E_i^{\text{rad}} = 0 & \text{at } \partial R. \end{cases} \tag{12.29}$$

Although the radiative components are non-local within the region, they do *not* require any further information from outside the region.[33]

In the next section, we will present the symplectic decoupling of the radiative and Coulombic degrees of freedom. Moreover, since the symplectic potential is intimately related to the Noether charges (see Section 12.2), by studying the decomposition of the symplectic potential, we will verify that their Coulombic and radiative parts correspond to the 'problematic', gauge-variant Noether charges, and the gauge-invariant, 'non-problematic' Noether charges, respectively.

12.B.2.2 Coulombic and Radiative Symplectic Potentials

The total symplectic potential, introduced in Section 12.A.1 for standard vacuum Yang–Mills for a Cauchy surface Σ without boundaries is given by (see the first term on the rhs of (12.8)):

$$\Theta = \int_\Sigma E \wedge \mathbb{d}A \tag{12.30}$$

where the phase space $\mathcal{P}$ is given in the basis (A, E). In the presence of boundaries, i.e. for $R \subset \Sigma$ such that $\partial R \neq \emptyset$, we can still translate the radiative-Coulombic decomposition of phase space variables into a decomposition of the symplectic potential into a radiative and a Coulombic component, i.e. $\Theta = \Theta^{\text{rad}} + \Theta^{\text{Coul}}$. We should note that the

32 Note that if Λ is the full Coulombic potential component of E, then $\Lambda = \phi - A_0$, and this component transforms covariantly as well (i.e. as the rest of E).

33 It is useful to distinguish 'boundary conditions' from 'boundary restrictions'. Boundary conditions such as the above are formal: they enforce no boundary restriction of the possible states. That is because, as mentioned in (12.28), any E_i can be decomposed using such a E_i^{rad}.

decomposition requires field space one forms projecting both $\delta A, \delta E$,[34] onto their radiative, i.e. non-Coulombic components, and onto the complementary Coulombic components.

The Coulombic symplectic potential is given by:

$$\Theta^{\text{Coul}} = \oint_{\partial R} f_I \phi^I. \tag{12.31}$$

Θ^{Coul} is a 1-form on phase-space, only acting as a one-form on one of the phase space directions: along disturbances in A (as in (12.30)). So in (12.31), ϕ is a field-space one-form and $\phi(\delta A)$ is the 'Coulombic extraction' of the field-space vector δA. Since a given perturbation such as $\delta A = \mathrm{D}\xi$ is pure-gauge, an extraction of the 'pure Coulombic' part of δA is an extraction of its pure-gauge component, i.e. $\phi(\delta A) = \xi$. f is the electric flux at the boundary, i.e. the normal component of the electric field there (for the complete symplectic treatment of the decomposition, see Gomes and Riello 2021, sec. 3.6). Note that the dynamical relevance of the Coulombic terms appears only at the boundary ∂R: the electric flux is conjugate to the restriction of the pure gauge (i.e. pure Coulombic) part of A to the boundary.

The radiative component of the symplectic potential is given by:[35]

$$\Theta^{\text{rad}} = \int_R \left(E^{\text{rad}}_{iI} \mathbb{d} A^{iI}_{\text{rad}} \right), \tag{12.32}$$

where $\mathbb{d} A_{\text{rad}} = \mathbb{d} A - \phi$, with ϕ as above (i.e. $\mathbb{d} A_{\text{rad}}$ only tracks variance in the physical, gauge-invariant parts of a field-space vector δA, again because $\delta A = \mathrm{D}\xi$ is symplectically orthogonal to E^{rad}_{iI}).

We find that the radiative component of the symplectic potential already has all the required properties for a formal symplectic treatment of gauge symmetries, as discussed in Section 12.2. Namely, it is gauge-invariant $\mathbb{L}_{\widehat{\delta\varphi}} \Theta^{\text{rad}} = 0$, and the corresponding Hamiltonian vanishes on-shell $\Theta^{\text{rad}}(\delta_\xi \varphi) \approx 0$.

The radiative Noether current can be seen as an 'appropriately' gauge-fixed version of the Noether current: the gauge-fixing is perturbative, covariant with respect to gauge transformations of the perturbed configurations, and does not require boundary restrictions (see footnote 33).

In sum: the pair constituting the regional radiative phase space is parametrised by a bona-fide symplectic pair. The total symplectic potential Θ decomposes into a strictly interior, radiative, gauge-*invariant* part, Θ^{rad}, and a boundary, gauge-*variant* Coulombic part, Θ^{Coul}.

The radiative degrees of freedom exhaust the interior degrees of freedom, but neither their constraints nor their observable content depend on further boundary data; i.e. gauge-invariant observables are formed on the interior of each region.[36]

[34] Where we use an affine identification of $\mathrm{T}_\varphi \mathrm{T}\Phi$ with $\mathrm{T}\Phi$ (as in e.g. $\mathrm{T}_x \mathbb{R}^n \simeq \mathbb{R}^n$).

[35] Here I limit myself to the vacuum Yang–Mills case. For the full case, see Gomes and Riello 2021, sec. 3.6 and the appendix here.

[36] Note also that the radiative degrees of freedom possess the local gauge-covariance properties of the electric field. That is, one can form gauge-invariant objects by forming local scalars (e.g. through tracing the product with another locally covariant object).

References

Barrett, J. W. (1991). Holonomy and path structures in general relativity and Yang-Mills theory. *International Journal of Theoretical Physics*, **30**(9), 1171–1215. https://doi.org/10.1007/BF00671007.

Belot, G. (1998). Understanding electromagnetism. *The British Journal for the Philosophy of Science*, **49**(4), 531–55. https://doi.org/10.1093/bjps/49.4.531.

Belot, G. (2003). Symmetry and gauge freedom. *Studies in History and Philosophy of Science Part B: Studies in History and Philosophy of Modern Physics*, **34**(2), 189–225. www.sciencedirect.com/science/article/pii/S1355219803000042.

Brading, K. and Brown, H. R. (2000). Noether's theorems and gauge symmetries. https://arxiv.org/abs/hep-th/0009058.

Brading, K. and Brown, H. R. (2004). Are Gauge Symmetry Transformations Observable? *The British Journal for the Philosophy of Science*, **55**(4), 645–65. www.jstor.org/stable/3541620.

Brading, K. and Castellani, E., eds. (2003). *Symmetries in Physics: Philosophical Reflections*. Cambridge: Cambridge University Press.

Crnkovic, C. and Witten, E. (1987). Covariant description of canonical formalism in geometrical theories. In 'three hundred years of gravitation'. In S. W. Hawking and W. Israel, eds., *Three hundred years of gravitation*. Cambridge: Cambridge University Press, pp. 676–84.

Earman, J. (2004). Curie's principle and spontaneous symmetry breaking. *International Studies in the Philosophy of Science*, **18**(2–3), 173–98. https://doi.org/10.1080/0269859042000311299.

Earman, J. (2019). The role of idealizations in the Aharonov–Bohm effect. *Synthese*, **196**(5), 1991–2019. https://doi.org/10.1007/s11229-017-1522-9.

Friederich, S. (2014). Symmetry, empirical equivalence, and identity. *The British Journal for the Philosophy of Science*, **66**(3), 537–59. https://doi.org/10.1093/bjps/axt046.

Gomes, H. (2019). Gauging the boundary in field-space. *Studies in History and Philosophy of Science Part B: Studies in History and Philosophy of Modern Physics*. www.sciencedirect.com/science/article/pii/S1355219818302144.

Gomes, H. (2020). Gauge-invariance and the direct empirical significance of symmetries. *Studies in History and Philosophy of Modern Physics*. http://philsci-archive.pitt.edu/16981.

Gomes, H. (2021). Holism as the significance of gauge symmetries. *European Journal of Philosophy of Science*, **11**, 87.

Gomes, H., Hopfmüller, F. and Riello, A. (2019). A unified geometric framework for boundary charges and dressings: Non-Abelian theory and matter. *Nuclear Physics B*, **941**, 249–315. www.sciencedirect.com/science/article/pii/S0550321319300483.

Gomes, H. and Riello, A. (2021). The quasilocal degrees of freedom in Yang–Mills theory. Forthcoming in *SciPost*. https://scipost.org/submissions/scipost_202001_00038v1/.

Greaves, H. and Wallace, D. (2014). Empirical consequences of symmetries. *British Journal for the Philosophy of Science*, **65**(1), 59–89.

Healey, R. (2007). *Gauging What's Real: The Conceptual Foundations of Gauge Theories*. Oxford: Oxford University Press.

Kosmann-Schwarzbach, Y. and Schwarzbach, B. E. (2011). *The Noether Theorems: Invariance and Conservation Laws in the Twentieth Century*. New York: Springer.

Lee, J. and Wald, R. M. (1990). Local symmetries and constraints. *Journal of Mathematical Physics*, **31**, 725–43. https://doi.org/10.1063/1.528801.

Lifshitz, L. D. L. E. (1987). *Course of Theoretical Physics, Volume 2: The Classical Theory of Fields*. Oxford: Butterworth-Heinemann.

Myrvold, W. C. (2010). Nonseparability, classical, and quantum. *The British Journal for the Philosophy of Science*, **62**(2), 417–32. https://doi.org/10.1093/bjps/axq036.

Olver, P. (1986). *Applications of Lie Groups to Differential Equations*. New York: Springer-Verlag.

Regge, T. and Teitelboim, C. (1974). Role of surface integrals in the Hamiltonian formulation of general relativity. *Annals of Physics*, **88**, 286. https://doi.org/10.1016/0003-4916(74)90404-7.

Riello, A. (2019). Soft charges from the geometry of field space. *Journal of High Energy Physics*.

Szabados, L. B. (2004). Quasi-local energy-momentum and angular momentum in GR: A review article. *Living Reviews in Relativity*, **7**(1), 4. https://doi.org/10.12942/lrr-2004-4.

Teh, N. J. (2016). Galileo's gauge: Understanding the empirical significance of gauge symmetry. *Philosophy of Science*, **83**(1), 93–118. https://doi.org/10.1086/684196.

't Hooft, G. (1980). Gauge Theories and the Forces Between Elementary Particles. *Scientific American*, **242**, 90–166.

Weinberg, S. (2005). *The Quantum Theory of Fields. Volume 2: Modern Applications*. Cambridge: Cambridge University Press.

13

Observability, Redundancy, and Modality for Dynamical Symmetry Transformations*

DAVID WALLACE

13.1 Introduction: The Dynamical Conception of Symmetry

In physical practice, symmetry is for the most part a dynamical notion: a *dynamical symmetry* is a transformation that takes solutions of the equations of motion to other solutions. Such a notion has obvious practical applications in dynamical problems, such as the solution of the equations of motion and the identification of conserved quantities, but the significance of symmetry seems to transcend this: from symmetries, it is often said, we can infer consequences about what can be observed (symmetry-variant quantities are said to be unobservable), about what parts of theories do and do not do work (symmetry-variant quantities are sometimes said to be 'redundant'; symmetry-variant structures to be 'surplus'), about representation (symmetry-related models are often taken to play the same representational role), and about modality (symmetry-related situations are at least sometimes treated as the same situation differently described).

Just what these inferences really are supposed to be, and why they are justified – if indeed they are – is a vexed question. They cannot apply without some qualifications and restrictions: in an austere sense, a quite arbitrary permutation of the space of solutions is a symmetry, and even when we restrict attention to transformations which are closer to physical practice, still problematic examples abound. To borrow an example from Belot (2013): the orbit of a planet around the Sun is well modelled by the *central-force model*, in which we idealise the Sun as at rest and exerting a centrally directed inverse-square force on the planet.[1] This theory has the familiar rotational and time-translation symmetries (boost and translational symmetry are broken by the fixed location of the Sun) but in addition, it has 'Lenz-Runge symmetries', somewhat complicated and unintuitive velocity-dependent transformations of the planet's position that do not leave its distance from the Sun invariant. Unless we make the desparate move of supposing that the distance of a planet from the Sun is neither observable nor physically significant, none of our theses seem to apply to this symmetry.

* I am grateful to Gordon Belot, Harvey R. Brown, Neil Dewar, Hilary Greaves, Eleanor Knox, Oliver Pooley, Laura Ruetsche, Simon Saunders, Nic Teh, and Jim Weatherall for the many conversations on symmetry which led to this paper. It also benefitted significantly from feedback at meetings in Louvain-la-Neuve and Hannover (2017), and in Salzburg (2019).

[1] It is even better modelled if we work in centre-of-mass coordinates, but I avoid this complication for expository convenience.

Of course one can try to rule out such counter-examples by narrowing the scope of symmetries to which the theses of unobservability, representational equivalence and the like are supposed to obtain, but it is not easy to see how this could be done, much less what principled justification there could be for doing it. To quote from two influential recent criticisms of drawing metaphysical and epistemic conclusions from the dynamical conception of symmetry (whose starting points differ sharply):

> [T]he ways of encoding the content of laws that are most appealing to mathematicians and physicists appear to lead to notions of [dynamical] symmetry that are coolly indifferent to considerations of representational equivalence (Belot 2013)

> [T]he notion [of symmetry] is often defined in purely formal, mathematical terms, so that whether a given transformation is a symmetry of a given set of laws depends just on the formal and mathematical features of those laws and their models. But why should those features of the laws have anything to do with metaphysics, with what's real? It's not obvious[.] (Dasgupta 2016)

A common response to this problem is to look for some substantive additional requirement for a transformation to count as a symmetry, along with the formal requirement that it takes solutions to solutions. There are two broad classes of strategy. The first (the 'epistemic strategy') effectively builds unobservability of symmetry transformations into the definition of a symmetry: symmetries are required to leave all observable properties of the system invariant. Ismael and van Fraassen (2003) and Dasgupta (2016) propose definitions explicitly along these lines; some advocates of the 'primitive ontology' approach to scientific metaphysics (e.g. Allori et al. 2008; Allori 2013) propose something closely related (they require that symmetries leave relevant facts about the 'primitive ontology' invariant, and if one grants their account of scientific epistemology,[2] then observations supervene only on the primitive ontology). The second (the 'representational strategy') instead builds the representational equivalence of symmetry-related models into the definition, usually by requiring that symmetries are automorphisms of the appropriate mathematical space of models (hence preserve all structure, and thus all representation-apt features, of a model). This strategy has been widely explored in the philosophy-of-spacetime literature, where symmetries are often taken to be diffeomorphisms that preserve 'absolute structure'; see, e.g., Earman (1989) and references therein. (The preservation of representational equivalence is also built directly into Healey's 2009 definition of a 'theoretical symmetry'.)

Both strategies would have the disappointing consequence of making at least some of the theses about symmetry trivial. If symmetry is *definitionally* observation-preserving, for instance, we can infer nothing *substantive* about what is observable by learning that a theory has a symmetry, since the only way to learn that would have been via deductions about observability; if symmetry is *definitionally* structure-preserving, we cannot learn about surplus structure via symmetry. But more importantly, both appear in conflict with physical practice. This is most sharply apparent for the epistemic strategy: as Dasgupta freely admits, 'on our epistemic approach, symmetry-to-reality reasoning involves not just mathematical analysis, but also considerations that reach into the philosophy of perception and mind.

[2] One should not; cf. Wallace (2020).

A far cry from the purely mathematical gloss it is often given!' Since the definition of symmetry in physics pretty clearly does not reach at all into the philosophy of perception and mind, and indeed frequently involves identifying symmetries in contexts far from the directly observable world, 'symmetry' in the epistemic-strategy sense is at any rate not what physicists mean when they talk about symmetry.

Matters are somewhat more subtle for the representational approach. It is, for example, perfectly possible to formulate Newtonian physics as a theory whose models are N-tuples of paths in $\mathbb{R}^4$, in which case the mathematical structure of a model includes many features (the x-component of the vector linking a particle to the spatial coordinate origin at time = 0, for instance) that are not invariant under the Galilean symmetries. But it is possible to reformulate Newtonian physics on a more sparsely structured space ('Galilean spacetime'; see, e.g., Weatherall 2021) for which the Galilean symmetries coincide with the automorphisms. And having made such a reformulation, we can maintain that the reformulated theory is the truer representation of the physics, and that in the original $\mathbb{R}^4$ formulation, models are representationally equivalent whenever they are symmetry-related, even though not all their mathematical structure is preserved under the symmetry transformation.

But what makes such a reformulation appropriate? The answer, in the history of physics, is pretty much always based on identifying a *dynamical* symmetry under which some of the structure of the original models is variant: that is, it is based on the assumption that structure variant under a dynamical symmetry is surplus. The Galilean-spacetime reformulation is preferable to the original formulation precisely because the automorphisms match the dynamical symmetries. Earman (1989) makes this argument explicit.

As another, more technical example, $U(1)$ gauge theory can be formulated, and historically was formulated, as a theory of a 4-vector field on spacetime. That theory has gauge symmetries as dynamical symmetries, and those symmetries do not preserve the mathematical structure of the theory: for instance, the property of having divergence zero is not gauge-invariant. The theory can be reformulated as a connection on a fibre bundle, and now the gauge symmetries are structure-preserving bundle automorphisms – but the rationale for the refomulation was precisely *because* the dynamical symmetries were not automorphisms in the original formulation.

So the representational strategy, too, does not seem to do justice to the notion of symmetry actually used in physics: insofar as it is correct, it is only because we have carried out reformulations and/or drawn conclusions about when models are representationally equivalent, based on a prior, dynamical notion of symmetry.

Only the dynamical approach, I think, can really do justice to the notion of symmetry *as we find it in physics*. And so my goal in this chapter is to defend the claim that epistemic and (indirectly) metaphysical conclusions can after all be drawn from the presence of a dynamical symmetry, at least in some circumstances, and – more importantly – to clarify how that can even be possible. To preview the core claims:

(i) Observation is itself a dynamical notion, so a dynamical symmetry can have consequences for what is observable just when that symmetry encompasses the physical processes of measurement;

(ii) Models in physics are almost always used to describe subsystems of a larger universe, so the interpretational significance of a symmetry depends not just on the dynamics of the system itself but on (explicit or tacit) assumptions about what happens to the symmetry when the system interacts with other systems.

The structure of the chapter is as follows. After outlining (Section 13.2) the main mathematical concepts I will use, in Section 13.3 I give sharp versions of four theses about symmetries in the literature: that they are unobservable, that they give rise to representationally equivalent models; that they are a guide to surplus structure; that they do not represent genuinely different possibilities. In Sections 13.4 and 13.5 I analyse the consequences of assuming that a dynamical symmetry of a system applies also to the mechanical processes used to observe that system, and I establish a version of the thesis about unobservability. In Sections 13.6 and 13.7, I use these results, and the broader framework developed in Sections 13.4 and 13.5, to assess the remaining three theses. In Sections 13.8–13.11 (which in large part can be read independently and can be skipped on a first reading) I consider four complications: the relation between symmetries and conserved quantities; time-dependent symmetries such as those found in gauge theories; quantum theory; and cosmology. Section 13.12 is the conclusion. For the most part I restrict myself to the comparatively elementary example of point-particle classical mechanics, though Sections 13.9–13.11 consider more advanced examples from quantum mechanics and field theory.

This chapter's theme – symmetries of isolated systems – is developed further in Wallace (2022, 2022), where I presuppose the results of this chapter and explore in more detail how dynamically isolated subsystems of larger systems can be modelled, how in turn isolated systems can and should be interpreted as subsystems of unspecified larger systems, and the interplay of those ideas with the interpretation of a system's symmetries (the two papers focus, respectively, on examples from particle mechanics and from field theory).

13.2 Elements of Classical Mechanics and Group Theory

For the most part, in this chapter I consider *classical* (i.e. non-quantum) systems: this is mostly for grounds of exegetical convenience, though there are a few additional subtleties in the quantum case which I discuss in Section 13.10. In this section I lay out some basic assumptions about classical mechanics which I draw upon throughout the chapter (readers already familiar with the subject might want to skim this section to become familiar with my conventions, or else skip it entirely).

Classical mechanics, in the abstract, characterises a system's dynamical history as a time-indexed collection of *configurations*. In the classic example of point-particle mechanics (with, say, N particles), a configuration is a specification of the locations of each of the point particles, and is mathematically represented by an ordered N-tuple of points in 3-dimensional Euclidean space. The collection of all such N-tuples is *configuration space*, which I denote by $\mathcal{Q}$. More abstract (say, field-theoretic, or extended-body) systems have configuration spaces with different mathematical representations: the configuration space of a rigid-body system, for instance, is given by a point representing the center-of-mass position of the system and a triple of orthogonal vectors representing its orientation.

A *history* of the system is then a smooth function $q\colon \mathbb{R} \rightarrow \mathcal{Q}$, assigning to each time t the configuration $q(t)$ of the system at that time. The dynamical equations of Newtonian mechanics distinguish dynamically possible from dynamically impossible histories; these equations are second-order in time, and so take the form[3]

$$F_J(\ddot{q}(t), \dot{q}(t), q(t)) = 0 \tag{13.1}$$

for some functions F_J. (Here as usual, $\dot{q}(t)$ and $\ddot{q}(t)$ denote, respectively, the first and second derivative of q with time.) The particular form of these equations depends on the system in question: for Newtonian point-particles moving under gravity, for instance, they encode Newton's second law and the inverse-square law. In that case, and in most elementary cases, the dynamics are *deterministic*, which means that (13.1) can be solved for $\ddot{q}$ in terms of q and $\dot{q}$:

$$\ddot{q}(t) = \mathcal{F}(q(t), \dot{q}(t)). \tag{13.2}$$

Given $q(t)$ and $\dot{q}(t)$, $\ddot{q}(t)$ is then fixed; normally[4] this allows us to integrate the equations forwards and backwards to determine a unique history. Because these equations are second-order, however, even for a deterministic system the configuration at time t is insufficient to determine future or past configurations: additional information about the rate of change of the configuration is required. The *state* of the system at time t is the ordered pair $(q(t), \dot{q}(t))$ of the system's configuration and its velocity: the space of all such pairs is the *state space* of the system, which I denote by $\mathcal{S}$.[5] In a deterministic system, the state of the system at one time uniquely determines its state at past and future times.

Given a theory in this form, a *(dynamical) configuration symmetry* is a bijection of $\mathcal{Q}$ which maps solutions to solutions: that is, if $q \rightarrow gq$ is the map, then $gq(t)$ solves the equations of motion if and only if $q(t)$ does. The bijection induces a bijection on the space of *states* (called the *tangent lift*) in a natural way:

$$\dot{q}(t_0) \rightarrow \left.\frac{\mathrm{d}}{\mathrm{d}t}(gq(t))\right|_{t=t_0} \tag{13.3}$$

(note that this depends only on $q(t_0)$ and $\dot{q}(t_0)$). For instance, spatial translation has no effect on velocity; spatial rotation rotates the velocity vectors as well as moving the configuration.

The requirement for g to be a symmetry can be characterised directly on state space too; it is a symmetry provided that $gx(t)$ solves the equations of motion on state space if and only if $x(t)$ does. The class of bijections of $\mathcal{S}$ satisfying this requirement contains some

3 For simplicity I ignore the possibility of explicit time dependence in the dynamics; it can be incorporated without substantive changes to my arguments, at the cost of a more cumbersome notation.

4 There are subtleties in doing so in certain systems (see Earman 1986 for detailed discussion), but those subtleties are normally 'unphysical' in the sense that they mark a failure of the system to correctly represent real-world physics after a certain time; in any case, I set these complications aside.

5 Technically minded readers will recognise that this simplifies some of the geometry: the state space is better described as the tangent bundle $T\mathcal{Q}$ over the configuration space $\mathcal{Q}$. These details will not matter for what follows.

transformations that are not tangent lifts of configuration symmetries (the Lenz–Runge symmetries have this feature).

The class of dynamical symmetries (whether understood as configuration symmetries or directly as state-space symmetries) forms a group in the mathematical sense: the trivial map id, taking each point to itself, is obviously a symmetry; if g, h are symmetries, then so is gh; the inverse of a symmetry is also a symmetry. It is often helpful to think of this group abstractly (i.e., just in terms of its algebraic structure) and to distinguish the group from its action via transformations. Some more terminology: given a group $\mathcal{G}$, an *action* R of the group on a space $\mathcal{S}$ is a collection[6] of bijections $R(g)\colon \mathcal{S} \to \mathcal{S}$ satisfying $R(g)R(h) = R(gh)$ for any $g, h \in \mathcal{G}$. I write e for the identity in a group (so that if R is an action, $R(e) = \mathrm{id}$). For simplicity, I normally assume that the group action is *free*: that is, if $R(g)x = x$, then $g = e$.

This notation is unavoidably somewhat abstract, but (as I noted in the introduction) for most purposes it will suffice to consider one concrete example: the case of Newtonian point particles (I will occasionally refer to other, more complicated, examples below, but readers unfamiliar with the physics of these examples may wish to skim them and focus on the Newtonian case). For definiteness, I will consider N particles moving under the inverse-square law of Coulomb electromagnetism, where the nth particle has a charge q_n and a mass m_n, where the nth particle exerts a force on the mth particle proportional to $q_n \times q_m$ divided by the square of the distance between them, and where each particle has an acceleration directly proportional to the sum of the forces upon it and inversely proportional to its own mass. There are two classes of dynamical symmetry of this theory:

Galilean spacetime symmetries: The transformation induced by translating all the particles by a common vector, or rotating them by a common amount, or by increasing their velocities by a common amount, or by translating each of their trajectories by a common amount in time, is a symmetry. (Strictly, the last two require a slight generalisation of our definition, which I omit for simplicity.)

Permutation symmetries: If any two particles have the same charge and mass, the transformation that permutes them is a symmetry.

13.3 Four Theses about Symmetry

To ensure a clear target for later analysis, here I give fairly precise formulations of four common (though contested) claims about symmetry. The first makes sharp what I mean by symmetries being unobservable.

The Unobservability Thesis: Given a family of models of a system which are related by a symmetry transformation, it is impossible to determine empirically which model in fact represents the system.

[6] Technical detail: $\mathcal{G}$ should be a Lie group, $\mathcal{S}$ should be a differentiable manifold, and the map $(g, x) \to R(g)x$ should be smooth.

The Unobservability Thesis is sometimes stated in more metaphysical terms, as the claim that quantities or properties that are 'symmetry-variant' – that is, that have values that are not invariant under a symmetry transformation of the system – are unobservable. I will argue that at least in some circumstances, and under some natural intepretations, that claim is false.

The next two theses both concern the way representation works in a theory with symmetry.

The Representational Equivalence Thesis: Given a family of models of a theory which are related by a symmetry transformation, insofar as one model successfully represents a system, so do all the others.

The Surplus Structure Thesis: Given a theory with a symmetry transformation, insofar as the symmetry falls short of being an automorphism of the mathematical structures used to define the theory's models, this points to aspects of that structure which are redundant, do no representational work, and can be removed from the theory without loss.

These two are clearly closely related: insofar as the Representational Equivalence Thesis holds, then analytically, any symmetry-variant structure does no representational work; conversely, once surplus structure is removed, symmetries preserve all mathematical structure, and plausibly mathematical models represent *only* via such structure. (Both should be distinguished from the much stronger claim that, given a symmetry, one should quotient out by that symmetry in order to understand the theory properly. That claim seems to me highly implausible in full generality in view of how ubiquitously modern mathematics deploys structures with nontrivial automorphism groups but in any case I shall have little to say about it.)

The final claim concerns the modal status of symmetries:

The Modal Equivalence Thesis: Two states of affairs related by the action of a symmetry transformation are really the same state of affairs, differently described.

Its relation to the Representational Equivalence Thesis is subtle, as we will see in Section 13.7.

13.4 Observations from Within a System

Most physical systems we study in physics are simply not rich enough to model the process of measurement – indeed, hardly any are. The most straightforward model of the Earth-Sun system, for instance, has only six degrees of freedom (three coordinates of position for the centres of mass of each of the two bodies) and it makes little sense to imagine how any physical process in that system could deserve to be called a 'measurement'. But for the moment, suppose that we *do* have a sufficiently rich system, and ask what we can infer about the measurability of a symmetry.

The standard model for measurement in physics is the measurement of one system's state by encoding it in another system's. It goes like this:

(i) The system being measured (the 'target system') has some unknown initial state, represented by a point in its state space.
(ii) The system doing the measurement (the 'measurement device') is placed in some 'ready' state, independent of what state the target system starts off in. (If it were not independent, the observer would have to have information about the result of the measurement even before making it, defeating the purpose of the measurement.)
(iii) The target system and the measurement device are then allowed to interact for a while. As a consequence, the post-interaction state of the measurement device now depends functionally on the pre-measurement state of the target system.
(iv) If that functional dependence is non-trivial (that is, if the measurement device's post-interaction state sometimes varies according to what the target system's pre-interaction state was), then the measurement has succeeded to at least some degree.

Here we are considering measurements within a single complicated system, but any such system can be decomposed into subsystems – a measurement, in this context, is a correlation established between the pre-measurement value of some 'target' quantity and the post-measurement value of a different 'readout' quantity, given that the pre-measurement value of that second quantity was in the fixed 'ready' state.

To see how this plays out in a system with a dynamical symmetry, it will be helpful to get a little more formal. Given a group $\mathcal{G}$ and an action R of $\mathcal{G}$ on a space $\mathcal{S}$, we can define the *orbits* of the action as the equivalence classes of $\mathcal{S}$ under the relation: x is related to x' if and only if $x' = R(g)x$ for some $g \in \mathcal{G}$. Suppose we arbitrarily pick a single point ϕ_O inside each orbit O: then any point $x \in O$ can be expressed as $x = R(g)\phi_O$ for some group element g, and hence, that point can be represented by the ordered pair (O, g). (Since we are assuming that the group acts freely on $\mathcal{S}$ – which is to say: $R(g)x = R(h)x$ if and only if $g = h$ – then the representation will be unique; if we had not made this assumption, multiple pairs might correspond to the same point, but each pair would still pick out a unique point.) I write $x \sim (O, g)$ to denote $x = R(g)\phi_O$.

The value of this representation is that the action of $\mathcal{G}$ has a very simple form: if $x \sim (O, g)$, so that $x = R(g)\phi_O$, then $R(h)x = R(h)R(g)\phi_O = R(hg)\phi \sim (O, hg)$. Hence the action of the group is just:

$$R(h)(O,g) = (O,hg). \tag{13.4}$$

We have effectively decomposed the space into its $\mathcal{G}$-invariant part, represented by an orbit, and its $\mathcal{G}$-variant part, represented by an element of $\mathcal{G}$ itself.

Now let's take $\mathcal{S}$ to be the state space of our system. The Unobservability Thesis concerns whether it is possible to distinguish experimentally between situations that differ only by the action of the symmetry. In this notation, this is to say: can we measure the value of g from within the system? Since our 'readout' quantity must be independent of g, it follows that it must be a function of O alone, so that the question is: is there a dynamically possible measurement process that encodes the value of g in some function of O? Intuitively, if the action of $\mathcal{G}$ is a symmetry, the answer must be 'no', but we should establish this more carefully.

Now we assume that the system has some time-independent, deterministic dynamics, with equations of motion represented by some one-parameter family $U(t)$ of bjiections of $\mathcal{S}$, satisfying $U(t+s) = U(t)U(s)$. The solutions to the dynamics are then functions $x : \mathbb{R} \to \mathcal{S}$ satisfying $U(t)x(s) = x(s+t)$. The group action is a *dynamical symmetry* if, for any $x \in \mathcal{G}$ and any path $x(t)$ in $\mathcal{S}$, $x(t)$ is a model of the theory (i.e., satisfies its equations of motion) if and only if $gx(t)$ is.[7]

Let's see how the dynamics looks in the orbit/group representation of state space. An initial state (O, g) evolves over some time t to a subsequent state (O', g'), where O' and g' are determined by O, g, and t. So there must be functions α and β such that the evolution has form

$$U(t)(O,g) = (\alpha(O,g,t), \beta(O,g,t)). \tag{13.5}$$

Suppose we evolve the system for time t and then apply a symmetry h: then the final state will be

$$hU(t)(O,g) = (\alpha(O,g,t), h\beta(O,g,t)). \tag{13.6}$$

If instead we apply the symmetry and then evolve the system, we get

$$U(t)(O,hg) = (\alpha(O,hg,t), \beta(O,hg,t)). \tag{13.7}$$

But if the action is a dynamical symmetry, these must be equal, so that we have

$$\alpha(O,hg,t) = \alpha(O,g,t) \quad \beta(O,hg,t) = h\beta(O,g,t). \tag{13.8}$$

Adjusting notation slightly, this is to say that the dynamics can be expressed as

$$U(t)(O,g) = (\alpha(O,t), g\beta(O,t)). \tag{13.9}$$

So the future evolution of O depends only on the present value of O, with no additional dependence on g. In other words: there is a self-contained dynamics for the invariant degrees of freedom of the system that is quite independent of the $\mathcal{G}$-variant features.

We can now see why no measurement internal to a system can distinguish whether a symmetry has been applied to the whole system. Such a measurement, we have argued, would have to have the form

$$(O_0, g) \to (O(g), g') \tag{13.10}$$

where O_0 is some g-independent state and $O(g)$ has some functional dependence on g. But $O(g)$ depends only on O_0 and, despite the notation, is not functionally dependent on g at all.

So the Unobservability Thesis is a straightforward consequence of the dynamics – *if* we assume that we are considering a system rich enough to model its own dynamics, and that the system is measuring itself rather than being observed from outside.

[7] This is a slightly restrictive formulation, ruling out as it does time-dependent symmetries like Galilean boosts, and time translation itself. This is for technical convenience; both can be incorporated, but at a significant cost in complexity. Specifically, we could define a symmetry group by a group $\mathcal{G}$, a time-indexed family of actions R_t of $\mathcal{G}$ on configuration space, and an action r of $\mathcal{G}$ on the real line, such that $t \to \sigma(t)$ solves the equations of motion if and only if $t \to R_t(g)\sigma(r(g)t)$ does; and we could drop the automatic presupposition of time translation symmetry by replacing the one-parameter dynamical map $U(t)$ with a two-parameter map $U(t, t_0)$. I leave it to the reader to verify that my results go through *mutatis mutandis* under this approach.

13.5 Symmetries of Systems and Subsystems

In practice, it hardly ever makes sense for the models we use in physics to imagine that the model includes its own measurement processes. In virtually all cases the measurement process, if any, is external to the system, and it is not normally modelled explicitly unless we are actually in the business of designing lab apparatus. So the consequences of the previous section for symmetries of realistic systems are indirect. Nonetheless, they are there, as we can see by considering some of the examples where dynamical symmetries clearly do *not* entail anything about observability.

Consider first the Lenz–Runge example I discussed in the Introduction, which is a symmetry of the central force problem – for concreteness, let's suppose we're using that problem to model the Earth-Sun system. A measurement of that system might, for instance, record the distance of Earth from the Sun at some given time. Without any commitment at all to the details of the physics of that measurement process, we know it must have a schematic form *something* like[8]

$$\begin{aligned}&\text{Earth at distance } r \text{ from Sun; device ready}\\ &\longrightarrow \text{Earth at distance } r \text{ from Sun; device records } r.\end{aligned}$$

Suppose that some instance of the Lenz–Runge symmetry, imagined to be applied before the measurement, changes r to $2r$. Then the measurement process will leave the device in some '$2r$' state. On the other hand, since the symmetry transformation acts only on the Earth and Sun, and is not even well-defined for whatever mixture of optics and mechanics implements the measurement process, applying the transformation *after* the measurement leaves the measurement device in the 'r' state. Which is to say, the Lenz–Runge transformation is *not* a symmetry of the combined system of measurement device plus Earth-Sun system.

A second example: (classical[9]) vacuum electrodynamics (that is, the theory of electromagnetic fields in the absence of charge) has a conformal symmetry in addition to the spacetime symmetries: it has no absolute notion of length or time, and the map $\mathbf{x}, t \rightarrow \lambda\mathbf{x}, \lambda t$, applied to spacetime points, takes solutions to solutions. The Unobservability Thesis would seem to imply that radiation frequency is undetectable, so that no observation can distinguish visible light from X-rays, and this is obviously absurd. But it is obvious *why* it is absurd: because the conformal symmetry is *only* a symmetry of radiation in the absence of matter, and ceases to be a symmetry of systems in which matter is present.

Here is the general point: we have defined symmetries as transformations on the state spaces of *isolated systems*; those systems have self-contained dynamics, and symmetries preserve the solutions of those dynamics. But while there is a temptation in the foundational literature to regard such isolated systems as proposed models of the entire Universe,[10] in real physical practice the vast majority of 'isolated system' models are used to model certain subsystems of the Universe under the idealisation that (at least on certain timescales, at

[8] Nothing about the concept of measurement actually requires the post-measurement Earth-Sun distance to equal the pre-measurement distance, but obviously any practically realistic measurement process will have this feature.

[9] Quantum electrodynamics is conformally invariant to leading order in the absence of charged particles, but that invariance is violated by quantum corrections, due to loop effects where particle-antiparticle pairs are created (and, indirectly, due to anomalous breaking of the conformal symmetry).

[10] In Wallace (2022), I call this the 'Cosmological Assumption' and critique it more thoroughly.

least to a certain degree of accuracy) they are dynamically independent of other systems. (The previous two examples illustrate this: the central force problem, and vacuum electrodynamics, are scientifically interesting because, respectively, one body's gravitational effect often dominates all others within a certain region, and because electromagnetic waves often exist for a time in regions with little matter present – not because of an abstract curiosity about distant possible worlds containing only two bodies, or containing no matter at all.)

So if an isolated system has a symmetry group $\mathcal{G}$, what happens to the symmetry when the assumption of isolation fails – for instance, when a measurement device interacts with the system? We can consider three substantively different possibilities. Firstly, $\mathcal{G}$ might be *subsystem-specific*: it has no extension to a symmetry of the combined system. This is the case for the Lenz–Runge symmetry of the central-force model: if we introduce another system (say, another planet passing close to the first, or a non-gravitational measurement device) then it doesn't really even make sense to ask how the Lenz–Runge symmetry affects this system, and even if we did come up with some arbitrary way of applying the transformation to the second planet, there is no reason at all to expect it to be a symmetry. The same applies to the conformal symmetries of vacuum electromagnetism.

Secondly, it might be that $\mathcal{G}$ is also a dynamical symmetry of the second system, with action R_2, and that the combined action $R_1 \times R_2$ makes $\mathcal{G}$ into a dynamical symmetry of the combined system: in this case, the symmetry group is *subsystem-global*. For instance, if we consider our Newtonian system of N point particles to be a subsystem of a larger system of particles, then the boost, translation, and rotation symmetries are subsystem-global: by performing them on the whole of the combined system, we still take solutions to solutions, but this is not the case if we, say, translate only the particles in the original subsystem but leave untranslated the particles in the new system with which they are interacting. (More generally, global symmetries in the usual physics sense – for instance, the global $U(1)$ symmetry of an ungauged complex scalar field – are subsystem-global.)

Thirdly, it might be that the action $R_1 \times \mathrm{id}$, where $\mathcal{G}$ acts trivially on the second system, makes $\mathcal{G}$ into a dynamical symmetry of the combined system. In this case, the symmetry group is *subsystem-local*. For instance, in the Newtonian-particle case, permutation symmetry is subsystem-local: a permutation of N identical particles is a symmetry whether or not those N particles are interacting with other particles. (For a more advanced example (cf. Wallace 2022) in gauge theory, the group of gauge transformations which vanish at the boundary of a given system is subsystem-local.)

In this third case, it will often be the case that $\mathcal{G}$ also acts on the second system as a subsystem-local symmetry. In this case the combined symmetry group of the total system will be $\mathcal{G} \times \mathcal{G}$: $\mathcal{G}$ can be applied, independently, to either system.

More complicated cases can occur, where different symmetries have different extendibility properties. Taking the second system as fixed, we can categorise individual symmetries as follows:

- g is *extendible* if there is a dynamical symmetry $\tilde{g}$ of the combined system which has the form $\tilde{g} = g \times g'$, i.e. which has a well-defined restriction to the original system that is equal to g.
- If extendible, g is *subsystem-local* if it has an extension $\tilde{g} = g \times \mathrm{id}$.

It is clear that the extendible symmetries form a group $\mathcal{G}_{ext} \subset \mathcal{G}$, and the subsystem-local symmetries form a subgroup $\mathcal{G}_{loc} \subset \mathcal{G}_{ext}$ of that group. We can then define the subsystem-global symmetry group as the quotient group $\mathcal{G}_{gl} = \mathcal{G}_{ext}/\mathcal{G}_{loc}$; elements of $\mathcal{G}_{gl}$ are equivalence classes of symmetries that differ by a subsystem-local symmetry, so that their action on the system is defined only up to a subsystem-local transformation. (In principle, this decomposition might be different when the system is coupled to a different system; in practice this does not normally occur, and for expository convenience I will usually ignore this subtlety.)

It should be clear that an inextendible symmetry has no implications at all for whether symmetry-variant quantities can be observed: by definition, as soon as an additional system is coupled to the original system, the symmetry is lost. By contrast, subsystem-local and subsystem-global symmetries have strong implications for what can and cannot be measured. For in both of these cases, by incorporating system and measurement device into a larger system, we can apply our previous results about measurements internal to a system.

Consider first the case of a subsystem-global symmetry. Using our orbit notation, we can write the combined state of target system and measurement device as $(O,g;O',g')$, with O and O' being orbits in the target system state space and the measurement-device state space respectively. The action of a symmetry transformation h is then

$$(O,g;O',g') \to (O,hg;O',hg'). \tag{13.11}$$

Importantly, since the symmetry acts simultaneously on the two systems, the symmetry-invariant information about the combined system is not exhausted by O and O' but also includes the relational quantity $g'^{-1}g$. If we now prepare the measurement device in some fixed 'ready' state (so that both O' and g' are held fixed, independent of O and g), then because this relational quantity covaries with the target system's variant quantity g, it is perfectly possible to measure g. In this sense, variant properties of a system can after all be measured. However, because the measurement proceeds *only* via the relational quantity $g'^{-1}g$, any such measurement is always reinterpretable as a measurement of that relational quantity. If the measurement device had instead been prepared in the alternative ready state (O',hg'), then the measurement process would be a measurement of $h^{-1}g$ rather than of g – and there would be no further measurement to be done that could determine whether it had been so prepared, because that pair of measurements would jointly consist of a measurement of a $\mathcal{G}$-variant quantity of the combined system from within that system, which we have seen to be impossible.

Being able to measure $\mathcal{G}$-variant quantities in even this thin sense might seem counterintuitive, but a little reflection shows that it fits our everyday and scientific practice. The speedometer of my car (which operates via interactions between my car and its environment) does indeed measure my car's velocity, in the sense that different velocities give rise to different measurements; the GPS on my phone measures my location in just the same sense. But in both cases, the measurement can equally be described as a measurement of a relative quantity (my car's velocity relative to the road; my location relative to selected coordinates on the Earth), and in neither case can the measurement determine the quantity independently of the equivalent quantity for the Earth.

(Note also that nothing in this analysis turns on whether the information about the system is encoded in the variant or the invariant degrees of freedom of the measurement device; in either case, we can establish a functional dependence on a g-variant quantity of the target system; in both cases, that functional dependence is reinterpretable as a functional dependence on an invariant relation between system and measurement device.)

What if the symmetry is subsystem-local? Then things are simpler: now the symmetry-invariant information about the combined system is exhausted by (O, O'), and (by our previous results) no measurement can determine either g or g'. So quantities variant under a subsystem-local symmetry are unobservable.

We can extend this to more complicated symmetry groups straightforwardly. Taking the symmetry group $\mathcal{G}$ as fixed, a given physical quantity (formally: a given map from $\mathcal{S}$ to some other space X, such as $\{0, 1\}$ for properties or $\mathbb{R}$ for real-valued quantities) can be classified as

- *Locally variant* if it is variant under the action of a subsystem-local element of $\mathcal{G}$;
- *Globally variant* if it is invariant under the action of a subsystem-local subgroup of $\mathcal{G}$, but variant under the action of a subsystem-global element of $\mathcal{G}$;
- *Invariant* if it is invariant under all extendible elements of $\mathcal{G}$.

We have now established a fairly precise version of the Unobservability Thesis:

- Invariant quantities are observable (or, at least, there is no symmetry-induced bar to observing them);
- Globally variant quantities are observable from a measurement device outside the system, but those observations are always reinterpretable as observations of an invariant relation between system and measurement device;
- Locally variant quantities are unobservable.

Applied to our core examples, we can see that variant spatial quantities like centre-of-mass position, orientation, and centre-of-mass velocity are observable, but only via processes reinterpretable as measurements of invariant spatial quantities; on the other hand, no measurement can distinguish two states that differ by a particle permutation. This is satisfactorily in accord with our pre-theoretic understanding in both cases.

I pause to make two observations. Firstly, our analysis does not assume that we know how to model a given measurement device. All it needs is the assumption that the measurement device, too, is bound by the symmetry. That assumption is fallible, of course – if we were to discover a set of physical processes that allowed us to probe a system but without conforming to some symmetry of the system, we could indeed measure symmetry-variant properties of that system (and not just in the thin sense discussed above). But this is an advantage of the analysis, not a weakness: it is indeed impossible to know infallibly whether a symmetry will turn out to be exact and universal, or just obeyed by some systems in some contexts, and the history of physics repeatedly displays examples of this. Consider parity, for instance: it is possible to measure the orientation of one system relative to another, but only a device governed by parity-violating interactions allows one to determine parity in a more absolute sense.[11]

[11] There remain subtleties in this case: see, e.g., Huggett (2000), Pooley (2003), and Saunders (2007) for more discussion.

Secondly, several recent authors (notably Roberts 2008; but see also Dasgupta 2016) have noted that dynamical reasons prohibit us from encoding variant features of a system in its invariant features. But this generally has not been taken to resolve the question of unobservability, unless we make a supposedly question-begging assumption – after all, variant features can be encoded in other variant features. (Roberts, for instance, then appeals to anthropic features of the sort of beings we are to complete the story.) My dynamical analysis shows that something more systematic can be said: for an isolated system there is in an important sense only one variant quantity (represented by g in my orbit-group notation), and it cannot be encoded in any way in any other independent quantity. It is true that if we have multiple systems, there are multiple independent quantities that are separately variant under the application of the symmetry to subsystems, but these quantities *can* be measured; it is just that the measurement is always reinterpretable as a measurement of an invariant relation between systems. For a collection of particles moving under Newtonian forces, for instance, the issue is not whether we can record the position of one particle in the position of another: indeed we can, and it is useful to do so. The issue is rather whether the centre-of-mass position of the whole system can be recorded in any independent degree of freedom; as we have seen, no such recording process is physically possible.

13.6 Surplus Structure and Representational Equivalence

We can now give a fairly strong argument in favor of the Surplus Structure Thesis – which, recall, is that symmetry-variant structure (e.g. an absolute rest frame) is surplus. For we have seen that no measurement process can determine whether or not a symmetry transformation has been performed on the combined system of target plus measurement device. Since the definition of symmetry-variant structure is that it is variant under such a transformation, it follows that there is no process by which it can be detected: no process, for instance, that can determine whether or not a system is moving *relative to absolute space* as opposed to moving relative to another system.

As a corollary, symmetry-variant structure can have no dynamical consequences for structure which can be detected: if surplus structure influenced the dynamical evolution of detectable quantities, it would itself be detectable indirectly, and our highly abstract analysis of measurement would pick this up.

So, consider two formulations of a theory – one containing symmetry-variant structure, one with that structure excised. (Here as usual, by 'symmetry' I mean 'subsystem-extendible dynamical symmetry'.) The second has all the resources of the first as regards describing and predicting the phenomena, which is to say that the symmetry-variant structures have no scientific work to do. There are now well-known arguments for taking the excised theory as the better representation of the physics: Ockhamist reasons based on the elimination of useless structure; semantic reasons based on the lack of any mechanism by which we could describe what the variant structure represents; pragmatic reasons based on the folly of including something demonstrably scientifically redundant in a scientific theory.

Møller-Nielsen (2017) resists these arguments on the grounds that there might in fact not exist a reformulation of the theory which eliminates the structure, so that we are stuck with the 'surplus' structure as the cost of using the theory at all. (One might think, for instance, that to move from a Newtonian to a Galilean spacetime requires mathematical creativity and is not a guaranteed possibility merely because Newtonian spacetime has symmetry-variant structure.) But I think this underestimates the powerful, general resources by which structure can be substracted from mathematical theories. For instance, in an extremely wide class of spacetime theories, methods dating back to Klein allow the automorphism group of the theory to be widened to include whatever spacetime symmetry one wishes (see Wallace 2019b and references therein; see also Dewar 2019 for a general defense of the legitimacy of this approach.). In the admittedly very abstracted setting of category theory, a class of mathematical models can be understood as *defined* by the transformations between models, and removing symmetry-variant structure is as simple as enriching the class of transformations (see, e.g., Weatherall 2016; note that this can be seen as a generalisation of the Kleinian method). In any case, the concern is fairly theoretical: I am not aware of *any* theory in extant physics (even construing 'extant' fairly broadly) which does not have a well-understood reformulation in which dynamical symmetries and automorphisms coincide, even if one eschews Kleinian and categorical tricks and insists on a more purist conception of reformulation.[12]

If the Surplus Structure Thesis is accepted on these grounds, the Representational Equivalence Thesis follows fairly directly. Suppose M and M' are symmetry-related models. Insofar as they are not isomorphic (with respect to the notion of mathematical isomorphism appropriate to the class of models from which they are drawn), this can only be because there is surplus structure. Once that structure is excised, symmetry-related models are guaranteed to be mathematically isomorphic. And since the means by which mathematical models represent the physical is entirely structural (we have no other way to make reference to mathematical objects, even assuming a background philosophy of mathematics in which that would make sense), one model is representationally adequate exactly insofar as the other is.

It is tempting to forge onto the Modal Equivalence Thesis: if symmetry-related models are representationally equivalent, how could a symmetry transformation represent anything other than a mere redescription? We will see this is far too quick, again for reasons that follow from the fact that physical theories are virtually always used to describe isolated systems in a larger universe, not the Universe as a whole.

13.7 Modal Equivalence and the Intrinsic/Extrinsic Distinction

Does a symmetry transformation, applied to a system, bring about a different possibility? The most common way to understand that question (see, e.g., Belot 2018) is as a thesis about the entire Universe: if we apply a symmetry transformation to a model of the Universe as

[12] The nearest I know to a counter-example (which lies well outside 'extant physics') occurs in supersymmetry, a theorised extension to current quantum field theory: standard '$\mathcal{N} = 1$' supersymmetry has a reformulation on 'superspace' which is analogous to the reformulation of special relativity on Minkowski spacetime, but the superspace move does not appear to be available for so-called extended supersymmetry. The issue has not been extensively explored; see Wess and Bagger (1983) and references therein for general features of supersymmetry, and Menon (2021) for some preliminary philosophical considerations.

a whole, does that bring about a new possibility? Understood that way, it seems a rather abstract question, turning as much on one's general theory of modality as anything else: conventionalists about modality, for instance (like Belot), could argue that this is partly a sociological point, to be answered by looking at the way physicists actually in fact use the concept. But in accordance with the themes of this chapter, I wish to answer it more locally, and more naturalistically, by considering symmetry transformations on isolated subsystems of a larger universe. In this case (which is, in any case, the *actual* context in which physicists use the concept), if the symmetry transformation is subsystem-global, then the answer is quite straightforwardly yes. In the notation of the previous sections, if the first system has state (O,g) and the symmetry is h, the transformed system has state (O,hg). Given any second system with state (O',g'), the result of the symmetry is to transform the invariant quantity $g^{-1}g'$ to $hg^{-1}g'$ – and this transformation is directly observable, and so clearly indicates that we are dealing with a new possibility.

How is this to be reconciled with the Representational Equivalence Thesis? As an illustration, consider our standard particle-mechanics example, and consider two systems (say, with N and M particles). Mathematically speaking, an instantaneous configuration of the first system is an ordered set of N points in Euclidean space E; an instantaneous configuration of the second is an ordered set of M points in another copy E' of Euclidean space. To compare the two systems (to check, for instance, whether they are sufficiently spatially separated that we can treat them as isolated) we need to identify E and E' – that is, specify an isomorphism between them – and there is no preferred or canonical way to do so. So given configurations q,q' of the systems separately, we have not been given enough information to describe their joint configuration: that requires, in addition, a *representational convention* as to how points in the two configuration spaces are to be compared. Such a convention is inevitably required whenever we combine subsystems into a joint system. (In practice, the convention is often given by a choice of coordinate systems, and/or of reference frames, in the two subsystems.)

Prior to stipulating any such convention, there is no sense in which (q,q') specifies a different joint configuration from $(R(g)q,q')$, since q and $R(g)q$ are representationally equivalent. Given a choice of representational convention, though, it is clear that applying the symmetry transformation to one system gives rise to a different total configuration (and that this is true independent of what the actual representational convention is). So, symmetry-related configurations can be understood as representing different possible configurations *if we hold fixed the choice of representational convention.*[13]

Introducing some metaphysical terminology, we have established, in effect, that the *intrinsic* properties of a system are invariant under symmetry transformations of that system: these properties do not depend on relations with any other systems, and so representational equivalence entails equivalence of intrinsic properties. A subsystem-global transformation brings about a change in a system's *extrinsic* properties. And this in turn gives some insight into what we are actually saying, physically, about a system when we establish that a certain transformation is a symmetry: we are identifying the fact that some properties of the system

[13] Thanks to Oliver Pooley for this observation.

are extrinsic properties. The relativity principle, for instance, can be understood physically as telling us that velocity is an extrinsic property of any system. The idea that symmetries leave intrinsic properties unchanged is far from new, and is often taken as *definitional* of a symmetry (see, e.g., Healey 2009).

Although I have carried out this analysis at the level of configurations, it applies *mutatis mutandis* at the level of states or entire histories of the system (with the caveat in the latter case that dynamically possible histories of subsystems can only be combined into dynamically possible histories of the combined system insofar as the subsystems are non-interacting).

What if the symmetry-transformation is subsystem-local? In this case, the transformation can be extended to a symmetry transformation which acts trivially on any other systems – which is to say that the symmetry transformation leaves invariant the intrinsic features not just of the system being transformed but of any larger system within which that system is embedded, up to the entire Universe, in principle. This seems to suggest that subsystem-local transformations bring about no physical change in a subsystem, and that features of the system that are variant under a subsystem-local symmetry do not represent physical features of that system.

As long as we are working at the level of histories, this seems correct: at any rate, it is difficult to see what naturalistic work is being done by a notion of modality that treats histories related by a subsystem-local symmetry as physically distinct, and it is unclear how to understand the possibility of a representational convention for the Universe as a whole which could allow us to reconcile modal inequivalence with representational equivalence. (In metaphysical terms, any new possibility would be purely haeccistic.) But matters are much subtler if we consider applying the transformation at the level of configurations (or, mutatis mutandis, states): at that level, there is a different kind of modal question we can ask for which subsystem-local transformations do real physical work. Namely, given that at time t the system has configuration q_0, and given two configurations $q, R(g)q$ related by a symmetry transformation, do these represent different configurations that the system can evolve into, or redescriptions of the same configuration?

Here too, the intrinsic/extrinsic distinction is useful. The intrinsic features of a configuration are the symmetry-invariant features; symmetry-variant features might be understandable as extrinsic features dependent on the relation with configurations of *other systems*, or as dependent on the relation with configurations of the *same* system at different times. If two particles are in circular orbits around one another, for instance, the intrinsic features of the system are time-invariant, but we can understand the system as changing *either* in the sense that its orientation relative to other systems is changing, *or* in the sense that its orientation at one time is different from its orientation at another.

We can get further insight into this by returning to the orbit/group analysis, this time applied to configurations. Given a group $\mathcal{G}$ of dynamical configuration symmetries, we can represent a configuration as a pair (o, g) where now o is an orbit in configuration space (here again, we suppose that we pick a representive point $q(o)$ in each orbit o, so that any point can be written as $gq(o)$ for some (o, g)). A point in *state* space can then be represented as $(o, g; \dot{o}, \dot{g})$, but this representation does not fully separate variant from invariant features of the state, because the quantity $g^{-1}\dot{g}$ is invariant under the group action.[14] The quantity $\xi = g^{-1}\dot{g}$ is called the *body velocity* in the physics literature, and can be understood physically as representing the rate of change of the system's variant features not as measured by an external observer, but as measured by a reference frame embedded in the system itself.

An orbit in the state space is represented by a triple $(o, \dot{o}, \xi)$, comprising the intrinsic features of the system's configuration, the rate of change of those intrinsic features, and the body velocity. In this notation (and continuing to assume deterministic dynamics), Section 13.4's dynamical analysis can be rewritten as

$$\ddot{o}(t) = \mathcal{F}_O(o(t), \dot{o}(t), \xi(t); g(t); t) \tag{13.12}$$

$$\dot{\xi}(t) = \mathcal{F}_G(o(t), \dot{o}(t), \xi(t); g(t); t) \tag{13.13}$$

$$\dot{g}(t) = g(t)\xi(t) \tag{13.14}$$

for some functions $\mathcal{F}_O, \mathcal{F}_G$. (In the physics literature this is known as *dynamical reduction*, or sometimes as *Marsden–Weinstein reduction*; see Marsden and Ratiu (1999) and references therein for details.)

Returning to the case of relations between configurations at different times, notice that once we have solved the equations of motion for the intrinsic degrees of freedom, we can go on to solve (13.14) to determine $g(t)$: that solution will have the form $g(t) = h(t)g(0)$, where $h(t)$ depends only on the intrinsic features between times 0 and t. $h(t)$ gives a dynamically determined relation between $g(0)$ and $g(t)$; given $g(0)$ and this relation, $g(t)$ is fixed for any other time. So the extrinsic features of the system at any time can be read off from the intrinsic features of the system's history and the extrinsic features at any one other time – just as the extrinsic features of a subsystem can be read off from the intrinsic features of the system and the extrinsic features of any other subsystem. Let's call extrinsic features of this kind *temporally extrinsic*, as distinct from the *subsystem-extrinsic* features that arise from considering a system's relation to other systems.

(If the symmetry is discrete rather than continuous – as in the case of permutation symmetry – the analysis goes slightly differently, because now $\dot{g}$ is not meaningful. The configuration space can still be decomposed into orbits, but the pairs $(o, \dot{o})$ fully describe orbits in state space and there is no body velocity. But it remains the case that given a solution to the intrinsic dynamics, we can determine some group element $h(t)$ such that if $g(0)$ represents the system's variant features at time 0, $g(t) = h(t)g(0)$ represents them at time t.)

Should we regard temporally extrinsic features of a system's state as physically meaningful? – or, put another way, are two states that differ only with respect to their temporally extrinsic features different representations of the same state, or representations of different possible states? The question is moot as long as we are considering subsystem-global symmetries, because in that case, any temporally extrinsic feature is also subsystem-extrinsic. But it is significant when we consider subsystem-local symmetries. For instance, given a particle permutation π, consider configuration-space trajectories $q(t)$ and $q'(t)$ where for some t_0 we have $q(t_0) = q'(t_0)$, and for some t_1, $q'(t_1) = \pi q(t_1)$. In this case, there will be

[14] Technical note: strictly $\dot{g}$ is a point in the tangent space over g, and $g^{-1}\dot{g}$ is the left translation of that point back to the tangent space over e.

no symmetry relating the two trajectories, and we can meaningfully regard $q'(t_1)$ and $q(t_1)$ as extrinsically different even though intrinsically identical.

Should we then regard πq and q as *different* possible configurations? The question seems largely conventional: it is, after all, up to us how we define configuration space (or phase space). If we want to quotient out configuration space (which amounts to working with *un*ordered N-tuples of points in Euclidean space) we can do that, albeit at the cost of making the topology a bit awkward. If we want to continue working with the unquotiented configuration space, so that points in the space represent temporally extrinsic as well as intrinsic features of configurations, we can do that too. The modal facts are in any case unambiguous at the level of entire histories, so I find it difficult to see any way of making this a matter of convention-independent fact. (We will see in Section 13.10 that the case for this apparent conventionality is sharpened in quantum theory.)

To summarise our results so far:

If $\mathcal{G}$ is a group of non-extendible dynamical symmetries, then no conclusions about observational, representational, or modal equivalence follow from the symmetry.

If $\mathcal{G}$ is a group of extendible dynamical symmetries, then

- $\mathcal{G}$-variant features of a system are unobservable from within that system.
- As a corollary, $\mathcal{G}$-variant mathematical structure in a theory is surplus and symmetry-related structures are representationally equivalent.

If in addition $\mathcal{G}$ is a group of subsystem-global dynamical symmetries, then

- $\mathcal{G}$-variant features of a system can be measured from outside that system, but any such measurement can be reinterpreted as a measurement of a $\mathcal{G}$-invariant relation between system and measurement device.
- A symmetry transformation leaves the intrinsic features of a system invariant but changes its system-extrinsic features.
- Symmetry transformations bring about new possibilities.

If instead $\mathcal{G}$ is a group of subsystem-local dynamical symmetries, then

- $\mathcal{G}$-variant features of a system are unobservable
- A symmetry transformation of an entire system (at all times) does not bring about a new possibility, and does not change any physical features of the system.
- A symmetry transformation of a system's state at a given time changes the temporally extrinsic features of the system (while leaving its intrinsic features unchanged); it appears to be conventional whether symmetry-related states represent different possibilities.

In the case where $\mathcal{G}$ contains a system-local subgroup $\mathcal{G}_{loc}$, so that we can define the system-global symmetry group as $\mathcal{G}_{gl} = \mathcal{G}/\mathcal{G}_{loc}$, and define its action on the system uniquely up to a system-local symmetry:

- $\mathcal{G}_{loc}$-variant features of a system are unobservable; features that are $\mathcal{G}_{loc}$-invariant but variant under a subsystem-global symmetry can be measured from outside that system, but any such measurement can be reinterpreted as a measurement of a $\mathcal{G}$-invariant relation between system and measurement device.

- A system-local symmetry transformation of an entire system (at all times) does not bring about a new possibility, and does not change any physical features of the system; a system-global symmetry transformation (defined only up to a system-local symmetry) changes its system-extrinsic features but not its intrinsic features.

In the remainder of the chapter I consider some applications of this framework: firstly to the relation between symmetries and conserved quantities established by Noether's theorem, then to the more technically advanced cases of time-dependent and quantum-mechanical symmetries, and finally to cosmology. The technical level of the next four sections is significantly higher than for the rest of the chapter; they can be skipped on first reading without loss of continuity.

13.8 Symmetries and Conserved Quantities

Noether's theorem, and its various generalisations, famously associate a conserved quantity with each dynamical symmetry (more precisely, with each dynamical symmetry satisfying further formal constraints, such as preserving the action, or the symplectic or Hilbert-space structure). At first sight, *this* feature of symmetries has little to do with the considerations of this chapter: Noether's theorem is an entirely formal result and does not depend on the interpretative features of symmetries I have been considering. That is, Noether's theorem will assign a conserved quantity to any symmetries with the right formal properties, whether they are extendible or inextendible, subsystem-local or subsystem-global. There is, for instance, a perfectly well-defined conserved quantity – the Lenz–Runge vector, or LR vector– associated with the Lenz–Runge symmetry of the inverse-square central-force problem (see Belot 2013 and references therein for details and further discussion).

But this is too quick. While conserved quantities like the LR vector are of genuine physical use (the LR vector provides a helpful explanation of the degeneracy of the hydrogen atom, for instance) the full power of symmetry-based conservation laws emerges only for subsystem-global symmetry groups, when a system with a symmetry can be broken down into isolated subsystems each of which separately obeys the symmetry in the idealization of perfect isolation.

Consider, for instance, the collision problem in mechanics: two systems, initially distant from one another, interact and then separate again. Long before the interaction, each system separately has rotations and spacetime translations as symmetries, and so each system separately has conserved energy and linear and angular momentum. Long after the interaction, the same is true, but during the interaction, these quantities are either undefined, or at any rate non-conserved, for each system. But the combined system comprised by the two systems *also* has translations and rotations as symmetries (because these are extendible symmetries of the individual systems) and so the *total* system has energy, linear momentum, and angular momentum, and each are conserved throughout the interaction. Furthermore, long before and long after the interaction, the conserved quantity for the combined system is just the sum of that quantity for each system separately: the energy of a system of non-interacting subsystems is just the sum of the energies of those subsystems, for instance. And so we get the conservation law

Sum of conserved quantities before systems interact
= Sum of conserved quantities after systems interact

This will hold in generality for any conserved quantities associated with a subsystem-global symmetry group (for obviously reasons, quantities like this are sometimes called *additive* conserved quantities, especially in the context of the WAY theorem, discussed in Section 13.10). It does not even rely on the decomposition into subsystems being the same before and after those subsystems interact (consider, for instance, particle-physics collisions in which particles are created or annihilated, or the collision that is conjectured[15] to have formed Earth and Moon from two planetoids in the early solar system). It is, I think, the primary context in which conservation laws are applied in physics. And it is not a consequence of Noether's theorem alone: it relies additionally on the fact that the symmetry associated to the conserved quantity is subsystem-global. It holds only trivially for subsystem-local symmetries, since then the conserved quantities can be associated separately to each subsystem and remain exactly conserved even while the subsystems are interacting. And it does not hold at all for inextendible symmetries. Given two mutually isolated subsystems both governed by an inverse-square central force law, for instance, formally there is a conserved quantity equal to the sum of the two systems' LR vectors, but it is conserved only derivatively on the separate LR vectors being conserved (any weighted sum of the two LR vectors would likewise be conserved) and there is no reason to suppose that, should the two systems interact and then separate, the sum of the post-interaction LR vectors would equal the sum of the pre-interaction LR vectors.

13.9 Time-Dependent Symmetries

Recall our original definition of a dynamical symmetry, expressed with respect to the configuration-space formulation of classical mechanics: it is specified by a group $\mathcal{G}$ and an action R of $\mathcal{G}$ on the configuration space $\mathcal{Q}$, such that $R(g)q(t)$ is a dynamically possible history if and only if $q(t)$ is. A *time-dependent* symmetry is again specified by a group and action, but now we require that for any smooth path $g(t)$ in $\mathcal{G}$ itself, $R(g(t))q(t)$ is dynamically possible if and only if $q(t)$ is. Or, put more colloquially: if the symmetry is time-dependent, then we can apply it independently at different times, subject only to the requirement that the symmetry group applied changes smoothly with time. We will assume for the moment that all elements of $\mathcal{G}$ are *small*: which is to say, each element is connected by a smooth path to the identity e. (The significance of this requirement will emerge shortly.)

If a symmetry is time-dependent, it follows that the dynamics for the theory are (formally) indeterministic. For example, suppose that $g(t) = e$ for $t < t_0$ but $g(t) \neq e$ thereafter. Then

[15] See, e.g., Canup (2012) and references therein.

if $q(t)$ is a dynamically permissible trajectory, so is $R(g(t))q(t)$ – but these two histories are identical up to time t_0 but diverge afterwards. As such, insofar as the theory's dynamics are specified by a differential equation, that equation must admit multiple solutions. (For an elementary discussion of time-dependent symmetries, with examples and references, see Wallace 2003.)

The notions of extendibility of symmetries, and of subsystem-local and subsystem-global symmetries, generalise straightforwardly to the time-dependent case. Generalising our analysis of measurement to indeterministic symmetries is a little delicate since there is no unique equation of motion, but a natural extension is to say that one system measures another if the *set* of possible post-measurement values of the measuring-device state is functionally dependent on the pre-measurement value of the target-system state. Then our previous results again generalise: subsystem-global symmetries are measurable, subsystem-local symmetries are not.

In practice, subsystem-locality and time dependence are closely related: to the best of my knowledge, there are no physically relevant examples of a time-dependent, subsystem-global symmetry group.[16] (The reason, in a nutshell, is relativity, which forces transformations of the same system at different times to be closely related to transformations of different systems at the same time.) The interesting case is time-dependent, subsystem-local symmetry.

Our previous analysis showed that state quantities variant under a time-independent symmetry could always be understood as extrinsic features of the state, representing symmetry-invariant relations either with other systems, or with states of the same system at different times (with the latter being the only available possibility if the symmetry was subsystem-local). In general this understanding is not available for quantities variant under a time-dependent symmetry: given an arbitrary $g_0 \in \mathcal{G}$, and an arbitrary time t, we can find $g(t)$ such that $g(t_0) = g_0$, and $g(t) = e$ except for values of t within some arbitrarily short distance of t_0. So just as a subsystem-local transformation of a system can be extended to a transformation that equals the identity on any other given system, so a (small) time-dependent symmetry of a *state* can be extended to a transformation that equals the identity on any other given state. For this reason, we can usefully call elements of a small time-dependent symmetry-group *time-local*.

I have already argued that when a subsystem-local transformation is applied to an entire history, there is no good reason to regard that transformation as bringing about a new possibility. Since any time-local transformation of a state can be extended to a transformation of the entire history that equals the identity except in an arbitrarily small neighbourhood of the state, it follows directly that time-local transformations of *states* also fail to give rise to new histories (and hence the formal indeterminism associated with these symmetries does not entail any physical indeterminism).

16 The closest I know to examples are the variant theories of Newtonian mechanics developed by Barbour and Bertotti (1982) and Saunders (2013), but (for reasons I develop in more detail in Wallace 2022) neither can be interpreted as representing a subsystem of a larger system.

We can also understand this through dynamical reduction. Given a system with configuration-space history $q(t) \simeq (o(t), g(t))$, a time-dependent symmetry $h(t)$ transforms it to $(o(t), h(t)g(t))$ – which is to say that the symmetry-variant part of the system's history is completely unconstrained by the dynamics. As a consequence, we can consistently reduce the dynamics to a dynamics on the space of orbits, with some differential equation

$$\ddot{o}(t) = \mathcal{F}(o(t), \dot{o}(t); t). \tag{13.15}$$

(If symmetry is the only cause of indeterminism, this equation will be deterministic.) So the symmetry-variant properties of a system have no dynamical effect either on the symmetry-invariant properties, or even on the symmetry-variant properties at later times. They are, in the fullest sense, dynamically redundant.

The requirement for symmetries to be small is doing essential work here, as the above arguments have required us to construct smooth paths from e to whatever group element we were considering. They fail for large symmetries (those which cannot be so connected to the identity); two states variant under a large element of a time-dependent symmetry group are not time-local, and quantities variant under such symmetries (but invariant under the small part of the symmetry group) can be understood as time-extrinsic just as for time-independent symmetries. Important examples arise in non-Abelian gauge theories (I discuss them further in Wallace 2022; see also Teh 2016).

All of this has a gratifyingly direct fit in the constrained Hamiltonian formalism (cf. Earman 2003; Teh 2016; and references therein). There, the phase-space dynamics is deterministic only up to *small* time-dependent symmetries, which are generated by the constraints; there, 'observables' are required to be invariant under such symmetries (but not necessarily under large elements of the symmetry group); there, phase-space points related by a small symmetry are normally interpreted as physically equivalent.

13.10 Symmetries in Quantum Theory

In the Hilbert-space formulation of quantum mechanics, a dynamical symmetry is a 1:1 transformation of Hilbert space which leaves all transition amplitudes invariant.[17] By Wigner's theorem, any such transformation must be a unitary or anti-unitary operator; writing it as $\widehat{R}$, the condition is that if $\widehat{U}(t;t_0)$ is the time evolution operator for an isolated system which has $\widehat{R}$ as a dynamical symmetry, then $\widehat{R}$ commutes with $\widehat{U}(t;t_0)$. (Any such transformation has the property that it maps solutions to solutions, but the converse is not true. In quantum theory, preserving solutions is too coarse a notion: we want to preserve the full structure of transition probabilities.)

A dynamical symmetry *group* is then a group $\mathcal{G}$ together with an action $g \rightarrow \widehat{R}(g)$ of $\mathcal{G}$ by means of (anti-)unitary operators, such that each $\widehat{R}(g)$ is a dynamical symmetry. The notions of an extendible symmetry, and of subsystem-global versus subsystem-local symmetries, carry over *mutatis mutandis*.

[17] In fact, it suffices for the transition amplitudes to be invariant up to phase. Also, and as with the classical case, I simplify slightly by leaving out velocity boosts and time translation.

The analysis of Sections 13.4 and 13.5 was extremely abstract – so abstract that it can be carried directly over to quantum theory. So given an isolated quantum system with a symmetry, that symmetry is unmeasurable from within the system; there is no in-principle difficulty in measuring it from a different system, even if the symmetry is extendible, but that measurement will always be reinterpretable as a measurement of some relational quantity. There are, however, interesting subtleties that arise from the richer structure of quantum theory.

To explore further, note that if a measurement is of some quantity invariant under a symmetry, quantum-mechanically, that means that the operator representing the measurement must be invariant under the adjoint action of the operator representing the symmetry. So given a symmetry group $\mathcal{G}$ with action $g \to \widehat{R}(g)$ as above, any internally measurable observable $\widehat{X}$ must satisfy

$$\widehat{R}(g)\widehat{X}\widehat{R}^{-1}(g) = \widehat{X} \tag{13.16}$$

for all $g \in \mathcal{G}$, and so given a quantum state ρ, the expectation value of that measurement satisfies

$$\langle\widehat{X}\rangle_\rho \equiv \mathsf{Tr}(\widehat{X}\rho) = \mathsf{Tr}(\widehat{R}(g)\widehat{X}\widehat{R}^{-1}(g)\rho) = \mathsf{Tr}(\widehat{X}\widehat{R}^{-1}(g)\rho\widehat{R}(g)) \tag{13.17}$$

which is to say that ρ can be replaced by $\widehat{R}^{-1}(g)\rho\widehat{R}(g)$ without affecting the result of any (internally) measurable quantity. If we now integrate over all such group actions, using the invariant Haar measure $\mu_{\mathcal{G}}$[18] over $\mathcal{G}$, we can define

$$\rho_{\mathcal{G}} = \int_{\mathcal{G}} \mathrm{d}\mu_{\mathcal{G}}\,\widehat{R}^{-1}(g)\rho\widehat{R}(g) \tag{13.18}$$

which, following Bartlett, Rudolph, and Spekkens (2007), we can call the $\mathcal{G}$-twirl of ρ. Then any measurement of ρ, provided it is physically possible and performed within the system, is equivalent to the same measurement of $\rho_{\mathcal{G}}$.

Now, recall that the action of a group on a Hilbert space can always be decomposed into *irreducible representations* (irreps), so that if $\widehat{R}_I$ is the Ith irrep and acts on Hilbert space $\mathcal{H}_I$, the total Hilbert space can be decomposed as

$$\mathcal{H} = \bigoplus_I \mathcal{K}_I \otimes \mathcal{H}_I \tag{13.19}$$

with $\mathcal{G}$ acting trivially on each $\mathcal{K}_I$: that is, the overall group action is

$$\widehat{R}(g) = \sum_I \mathrm{id}_I^K \otimes \widehat{R}_I(g) \tag{13.20}$$

where id_I^K denotes the identity on $\mathcal{K}$. $\rho_{\mathcal{G}}$ can be written in this notation as

$$\rho_{\mathcal{G}} = \sum_I \rho_I \otimes \mathrm{id}_I^{\mathcal{H}} \tag{13.21}$$

[18] Strictly this is defined only for compact $\mathcal{G}$, so these results need to be treated with some care if applied to non-compact groups such as translation.

where ρ_I is a density operator on $\mathcal{K}_I$ and now $\mathrm{id}_I^{\mathcal{H}}$ is the identity on $\mathcal{H}_I$. As long as we confine ourselves to internally performable measurements, then, the quantum state is empirically equivalent to a probabilistic mixture of states defined on the $\mathcal{G}$-invariant spaces $\mathcal{K}_I$.

In other language, all of this is to say (i) that as far as internal measurements are concerned, symmetries effectively define superselection rules and (ii) that within each superselection sector, the quantum state is represented for all empirical purposes by one of the invariant states ρ_I.

As long as we are considering subsystem-global symmetries, this analysis has largely practical consequences: it tells us that when analysing a measurement, we can treat the symmetry as superselected with respect to the combined system of target system plus measurement device, which interestingly constrains what measurements are possible. A very early result in this field, the WAY theorem, named for Wigner (1952) and Araki and Yanase (1960) (see Busch and Loveridge 2013 and references therein for modern presentations) established that no completely sharp measurement of a quantity was possible unless it commuted with all additive conserved quantities, and showed that the degree of accuracy that could be obtained by an unsharp measurement depended on the initial degree of asymmetry of the measurement device. More recent work (see Bartlett, Rudolph, and Spekkens 2007 and references therein) has quantified this result and developed it into a general resource theory of measurements, where the resource is the initial asymmetry. (Note the crucial role of the assumption that the conserved quantity is *additive*, which relies on the symmetry being subsystem-global – cf. Section 13.8.)

The consequences are more dramatic for a subsystem-local symmetry, for then our analysis entails that a superselection rule with respect to that symmetry effectively applies to any system even as regards measurements made outside the system. Consider permutation symmetry, for instance: no measurement, including from outside the system, can distinguish the quantum state of N identical particles from the state obtained by $\mathcal{G}$-twirling that state. So the system's state is empirically equivalent to a probabilistic mixture of states each of which transforms under some irrep of the permutation group – and which, furthermore, is invariant under the action of that irrep. At this point, normal physics practice would be to say that the actual system is in one or an other irrep, and the probability distribution over them represents our ignorance of the actual irrep. In the case of permutation, the completely symmetric and completely asymmetric irreps represent, respectively, bosonic and fermionic statistics; the higher-dimensional, mixed-symmetry irreps represent parastatistics.

In light of these results, let us return to the discussion of Section 13.5 on the redundancy of subsystem-local symmetries. There I argued that it seems to be largely conventional whether we quotient the state space of a system by a subsystem-local symmetry: it depend on whether we want the state space to represent the extrinsic relation between states at different times which differ by the action of the symmetry, or just to represent the intrinsic properties at each time. In quantum theory, quotienting amounts to replacing the Hilbert space of the system with the direct sum of the invariant $\mathcal{K}_I$, with the additional stipulation that coherent superpositions of states in different subsystems have no physical meaning: put another way, the actual Hilbert space of the quotiented system is *one* of the $\mathcal{K}_I$. But given our superselection results, this is really just the question of whether to take the state to be ρ_I or

$\rho_I \otimes \mathrm{id}_I^{\mathcal{H}}$. Even a fairly modest structuralism in our metaphysics will make this a distinction without a difference. In the specific case of permutation symmetry, this underdetermination has been extensively discussed in the literature (see French 2015 and references therein).

What is *not* conventional is the fact that two quantum states at time t which are intrinsically identical but which differ extrinsically via a symmetry transformation can both be reached from the same quantum state at some earlier (or later) time t_0, and that both possibilities contribute coherently to the overall transition amplitude. The dynamical significance of the choice of irreducible representations is that it determines any relative phase factors between the contributions. This is particularly clear if we restrict attention to Abelian symmetries (or to Abelian representations of symmetries, such as the symmetric or antisymmetric representations of the permutation group). Let x be some initial configuration, and let y and y' be configurations at some later time, related by the action of the symmetry. If we quotient out by the symmetry, y and y' are mapped to the same point, but two trajectories $x \to y$ and $x \to y'$ will not be mapped to the same trajectory, so the action for each must be included in the path integral – and in formulating the dynamics, we always have the possibility of adding some topological phase factor to their relative contributions. If we do not quotient out, then the two trajectories correspond to different transition amplitudes $\langle x|\widehat{U}|y\rangle$ and $\langle x|\widehat{U}|y'\rangle$ – but because the final state transforms under an irrep of the symmetry, it will be a coherent superposition of $|y\rangle$ and $|y'\rangle$ with equal weight, differing only by a phase factor. So the total transition amplitude between physical states will again involve a coherent sum of contributions from the two paths – only this time, the relative phase is determined by the particular representation (-1 for fermions, +1 for bosons, for instance).[19]

I end this discussion with two brief remarks in more technical areas of physics, which may be skipped by readers unfamiliar with the relevant material.

(i) The physical significance of a subsystem-local symmetry is contained (I have argued) in the existence of inequivalent trajectories between intrinsically identical initial and final states, which can be understood as an extrinsic property of those states. This takes a subtle form in the case of identical particles in two dimensions (or in systems like thin films, which can be approximated as two-dimensional), because two trajectories might be topologically inequivalent not just because they differ by a particle permutation, but because they differ in the way in which the permutation was implemented over time: in two dimensions, trajectories often cannot be deformed into one another without crossing. The concrete implication is that it is the *braid group*, not the permutation group, which determines the statistics of identical particles in two dimensions: this leads to the phenomenon of *anyons* (see Stern 2008 for a review).

(ii) In non-Abelian gauge theory (specifically, in QCD), there is a subsystem-local (but non-small) symmetry generated by topologically non-trivial, but boundary-vanishing, gauge transformations. These transformations form a group isomorphic to the integers,

[19] The path-integral analysis for non-Abelian groups – for parastatistics, for instance – is decidedly subtler, and indeed, at one time (Laidlaw and Morette DeWitt 1971) it was argued that parastatistics could not be incorporated in the path-integral formalism. See Greenberg and Mishra (2004) and references therein for further details.

so that we can write an element of the group action as $\widehat{R}(n)$, with $\widehat{R}(n+m) = \widehat{R}(n)\widehat{R}(m)$; a state $|\varphi\rangle$ in an irrep satisfies

$$\widehat{U}(n)\,|\varphi\rangle = \mathrm{e}^{-in\theta}\,|\varphi\rangle \tag{13.22}$$

for some angle θ, so that the irreps are labelled by this θ, and in the path integral, trajectories that terminate on states differing by the action of $\widehat{R}(n)$ pick up a relative phase factor $\exp in\theta$, which has empirical consequences. Experiment suggests that it is equal to zero to very high accuracy, and the so-called *strong CP problem* is the problem of identifying why this is.

It has been argued (Fort and Gambini 1991; Healey 2010) that if we worked in the quotiented configuration space right from the start, the strong CP problem would dissolve; but this seems overstated. The topologically distinct paths would still contribute to the path integral, and we would still have the option of adding a θ phase term. Of course, we could decide not to, but equally, we could decide to set $\theta = 0$ in the non-quotiented formalism. The real issue with $\theta = 0$, in either formulation, is that in quantum field theory, quantum corrections would be expected to introduce a large non-zero θ term even if it is omitted initially: obtaining $\theta = 0$ at the level of phenomenology requires a very large, and very precisely chosen, value of θ in the bare parameters of the theory. (It is a highly contested issue whether *naturalness* arguments like this should really be seen as persuasive: I make the case that they should in Wallace 2019a.)

13.11 Symmetries of the Whole Universe

I have argued in this chapter that interpretatively there is a fundamental distinction to be drawn between symmetries based on if, and if so how, they can be extended to other systems. But what if – as is extremely common in philosophical writings – we are concerned with symmetries of the entire Universe, in which case the distinction seems to evaporate?

I am highly tempted to dismiss this question as premature. Even in cosmology, we are hardly ever working with a model intended to describe the *entire Universe* (as opposed to, say, some large patch of it); even in quantum gravity, very large additional problems occur when we move from considering some isolated system (like a black hole) to considering the entire universe. Indeed, at present many high-energy physicists would claim that we have a quantum satisfactory theory of gravity, via the AdS/CFT correspondence (and its interpretation as describing a system in a box; cf. discussion and references in Wallace 2018), but there is no comparable case within that research program that we have a quantum theory of *cosmology*. Modelling the whole Universe raises conceptual and technical problems that do not occur in normal uses of physics, and we have very little empirical evidence to help us resolve them.

But suppose we throw caution to the winds and assume (following philosophy-of-physics orthodoxy) that at least some of our current theories have physically-significant interpretations in which their models are models of the whole Universe. At this point, the key result of Section 13.4 – that no measurement internal to a system can distinguish between symmetry-related states of that system – applies to any dynamical symmetry. (This assumes, of course, that we are studying a theory rich enough to represent scientists, measurements, and observations inside its own structure. Given a possible world consisting only of two point-particles interacting via a central force, for instance, there is nothing like the structure to permit us to model measurement, and so no conclusion that can be drawn about what is observable. If the reader nonetheless wants to know how to draw epistemic and/or metaphysical conclusions about that (supposedly) possible world from its dynamical symmetries, they will have to do it on their own time: I have no intuitions about the situation, and no inclination to trust any such intuitions anyway.)

Assuming our symmetries are time-independent, the discussion in Section 13.5 on subsystem-local symmetries now applies with equal force to subsystem-global symmetries: in an N-particle Newtonian theory interpreted as a theory of the whole Universe, for instance (and ignoring severe concerns about the modellability of measurement in any such theory) then configurations at different times will have a symmetry-invariant and empirically determinable relative orientation even while the overall orientation of the system remains undetectable. But by the same token, the considerations of that section suggest that possible histories differing by a symmetry transformation ought to be regarded as redescriptions of the same history; or, at any rate, if they are not to be so regarded, it must be for metaphysical, not naturalistically motivated, reasons. (This is of course the original context of the 'Leibniz shift'.)

In the quantum context, this would imply that any symmetry of the whole Universe – even a global one – ought to be regarded as defining a superselection rule. Applying this to the Hamiltonian itself (i.e., to time translation symmetry) has radical implications for dynamics in any such universe, a problem explored extensively in the quantum gravity literature (the classic discussion is Page and Wootters 1983) and related to (but distinct from) the famous 'problem of time'.

But we may be getting ahead of ourselves, for it is not at all clear that we should expect a truly cosmological theory to have any such symmetries. Modern physics provides (admittedly speculative) grounds to think that global symmetries in this sense are not exact and/or do not apply to the Universe as a whole:

(i) The apparent exact global internal symmetries of particle physics (e.g., the symmetry underlying baryon-number conservation) are so-called accidental symmetries (see, e.g., Duncan 2012, pp. 490–1) which occur as a consequence of renormalisability, but we have no reason to expect to hold exactly when non-renormalisable corrections to the dynamics are added.

(ii) Quantum gravity offers some further reasons to expect global symmetries not to be exact: the only quantities conserved in black hole evaporation are those associated with long-range forces, which in high-energy physics correspond to those associated with asymptotic local symmetries. The lack of exact global symmetries is also provable perturbatively in string theory and can be argued for via the AdS/CFT correspondence (see Banks, Johnson, and Shomer 2006 and references therein for details).

(iii) Although local gauge symmetries (and the Poincaré gauge symmetry underlying general relativity, are believed to be exact, and give rise to subsystem-global symmetries through asymptotic effects, those asymptotic effects rely on boundary conditions 'at infinity', which are normally interpreted to describe the way in which a system embeds into a larger system and which are not applicable in cosmology. The only consistent way we currently know how to avoid such boundary conditions is to assume that space is compact, in whch case there are no asymptotic symmetries.

In the end, I return to the caution with which I began this section: we just do not know, reliably, how to think about theories – and, *a fortiori*, about symmetries of those theories – when the theories are genuinely being interpreted cosmologically. Nor do we need to in order to understand the role of symmetries in non-cosmological (i.e., almost all) applications of our theories.

13.12 Conclusion

Dynamical symmetries place constraints on observability because, and to the extent that, they apply to the physical processes by which those observations are made. Although this implies fairly straightforwardly that symmetry-variant properties of a system cannot be observed by measurement processes confined to the system itself, in practice we very rarely model a measurement process explicitly in physics, and so the implications of a symmetry of a system for observations made on that system depend on whether, and how, the symmetry can be extended to other systems, including the measurement device. If the symmetry cannot be so extended, it has no particular consequences for observation; if it is subsystem-local, so that it can be extended to a symmetry which acts trivially on any other system, it is unobservable; if it is subsystem-global, so that it can be extended to a symmetry that includes the measurement device but which does not act trivially on it, then the action of the symmetry on the system alone has measurable consequences, but these can be reinterpreted as measurements of the symmetry-invariant relations between system and measurement device.

These results strongly suggest that symmetry-invariant features of a system should be interpreted as intrinsic features, that features that are variant under subsystem-local symmetries should be interpreted as descriptively redundant, and that features that are invariant under subsystem-local symmetries but variant under subsystem-global symmetries should be interpreted as extrinsic features. It follows that subsystem-local symmetry transformations should not be regarded as bringing about different states of affairs, but that subsystem-global symmetry transformations should be so regarded.

These various results extend, albeit with some subtleties, to the physically important cases of time-dependent and quantum-mechanical symmetries. They do not extend to theories which are intended to be understood as cosmological, in which the local/global distinction collapses. But cases of this sort are extremely unusual in physical practice and there are strong, if speculative, reasons to think that a genuinely cosmological physical theory should have no global symmetries. In any case, the cosmological case only strengthens the general point that normal modelling practice in physics is to treat an isolated system as modelling a small part of a larger universe, and that the epistemology and metaphysics of symmetry in physics is much clearer once this is properly allowed for.

The results of this chapter go some way towards seeing how genuinely substantive intepretative results can follow from a purely formal concept of symmetry (and do not require us to abandon that formal concept, as is the case for some non-dynamical definitions of symmetry). But a puzzle remains: my analysis turns on how a symmetry can be extended beyond a system to other systems in dynamical contact with it, and definitionally (it would seem) no formal feature of a system's symmetries can fix this question of extension to other systems. If so, then once again we seem to need to make substantial interpretative assumptions in order to get interpretative results from a theory's symmetries, and to abandon the hope of extracting such results from a formal conception of symmetry.

This problem cannot be completely solved: nothing about a theory, in isolation, can rule out the possibility that new physics, applying to new systems, will break a symmetry of the old system once the two are dynamically coupled. Nonetheless it is still possible to draw tentative conclusions about the extendibility of a theory's symmetries from the theory itself, once it is recognised that (in many cases) physical theories themselves contain rich information about their subsystems and the restriction of symmetries to those systems. I develop this observation in depth in Wallace (2022, 2022).

References

Allori, V. (2013). Primitive ontology and the structure of fundamental physical theories. In *The Wave Function: Essays on the Metaphysics of Quantum Mechanics*. New York: Oxford University Press, pp. 58–75.

Allori, V., Goldstein, S., Tumulka, R. and Zanghí, N. (2008). On the common structure of Bohmian mechanics and the Ghirardi-Rimini-Weber theory. *British Journal for the Philosophy of Science*, **59**, 353–89.

Araki, H. and Yanase, M. M. (1960). Measurement of quantum mechanical operators. *Physical Review*, **120**, 622–26.

Banks, T., Johnson, M. and Shomer, A. (2006). A note on gauge theories coupled to gravity. *Journal of High Energy Physics*, **09**, 049.

Barbour, J. and Bertotti, B. (1982). Mach's principle and the structure of dynamical theories. *Proceedings of the Royal Society of London*, **382**, 295.

Bartlett, S. D., Rudolph, T. and Spekkens, R. W. (2007). Reference frames, superselection rules, and quantum information. *Reviews of Modern Physics*, **79**, 555–609.

Belot, G. (2013). Symmetry and equivalence. In R. Batterman, ed., *The Oxford Handbook of Philosophy of Physics*. Oxford: Oxford University Press, pp. 318–39.

Belot, G. (2018). Fifty million Elvis fans can't be wrong. *Nous*, **52**(4), 946–81. https://onlinelibrary.wiley.com/doi/abs/10.1111/nous.12200.

Busch, P. and Loveridge, L. D. (2013). Quantum measurements constrained by symmetries. In C. Bai, J.-P. Gazeau, and M.-L. Ge, eds., *Symmetries and Groups in Contemporary Physics: Proceedings of the XXIX International Colloquium on Group-Theoretical Methods in Physics, Tianjin, China, 20–26 August 2012*. Singapore: World Scientific, pp. 587–92.

Canup, R. M. (2012). Forming a moon with an earth-like composition via a giant impact. *Science*, **338**, 1052–55.

Dasgupta, S. (2016). Symmetry as an epistemic notion (twice over). *British Journal for the Philosophy of Science*, **67**, 837–78.

Dewar, N. (2019). Sophistication about symmetries. *British Journal for the Philosophy of Science*, **70**, 485–521.

Duncan, T. (2012). *The Conceptual Framework of Quantum Field Theory*. Oxford: Oxford University Press.

Earman, J. (1986). *A Primer on Determinism*. Dordrecht: Reidel.

Earman, J. (1989). *World Enough and Space-Time*. Cambridge, MA: MIT Press.

Earman, J. (2003). Tracking down gauge: an ode to the constrained Hamiltonian formalism. In K. Brading and E. Castellani, eds., *Symmetry in Physics: Philosophical Reflections*. Cambridge: Cambridge University Press, pp. 140–62.

Fort, H. and Gambini, R. (1991). Lattice QED with light fermions in the P representation. *Physical Review D*, **44**, 1257–62.

French, S. (2015). Identity and individuality in quantum theory. In E. Zalta, ed., *The Stanford Encyclopedia of Philosophy* (Fall 2015 edition). https://plato.stanford.edu/archives/fall2015/entries/qt-idind/.

Greenberg, O. and Mishra, A. K. (2004). Path integrals for parastatistics. *Physical Review D*, **70**, 125013.

Healey, R. (2009). Perfect symmetries. *British Journal for the Philosophy of Science*, **60**, 697–720.

Healey, R. (2010). Gauge symmetry and the theta-vacuum. In M. Suaurez, M. Dorato, and M. Redei, eds., *EPSA Philosophical Issues in the Sciences: Launch of the European Philosophy of Science Association*, Volume 2. New York: Springer, pp. 105–16.

Huggett, N. (2000). Reflections on parity nonconservation. *Philosophy of Science*, **67**, 219–41.

Ismael, J. and van Fraassen, B. (2003). Symmetry as a guide to superfluous theoretical structure. In K. Brading and E. Castellani, eds., *Symmetries in Physics: Philosophical Reflections*. Cambridge: Cambridge University Press, pp. 371–92.

Laidlaw, M. G. G. and Morette DeWitt, C. (1971). Feynman functional integrals for systems of indistinguishable particles. *Physical Review D*, **3**, 1375–78.

Marsden, J. E. and Ratiu, T. S. (1999). *Introduction to Mechanics and Symmetry*. New York: Springer-Verlag.

Menon, T. (2021). Taking up superspace: the spacetime structure of supersymmetric field theory. In N. Huggett, B. Le Bihan, and C. Wuthrich, eds., *Philosophy beyond Spacetime: Implications from Quantum Gravity*. Oxford: Oxford University Press, pp. 103–28.

Møller-Nielsen, T. (2017). Invariance, interpretation, and motivation. *Philosophy of Science*, **84**, 1253–64.

Page, D. N. and Wootters, W. K. (1983). Evolution without evolution: Dynamics described by stationary observables. *Physical Review D*, **27**, 2885–92.

Pooley, O. (2003). Handedness, parity violation, and the reality of space. In K. Brading and E. Castellani eds., *Symmetries in Physics: Philosophical Reflections*. Cambridge: Cambridge University Press, pp. 250–81.

Roberts, J. T. (2008). A puzzle about laws, symmetries and measurability. *British Journal for the Philosophy of Science*, **59**, 143–68.

Saunders, S. (2007). Mirroring as an a priori symmetry. *Philosophy of Science*, **74**, 452–80.

Saunders, S. (2013). Rethinking Newton's *Principia*. *Philosophy of Science*, **80**, 22–48.

Stern, A. (2008). Anyons and the quantum Hall effect – a pedagogical review. *Annals of Physics*, **323**, 204–49.

Teh, N. J. (2016). Galileo's gauge: Understanding the empirical significance of gauge theory. *Philosophy of Science*, **83**, 93–118.

Wallace, D. (2003). Time-dependent symmetries: the link between gauge symmetries and indeterminism. In K. Brading and E. Castellani (Eds.), *Symmetries in Physics: Philosophical Reflections*. Cambridge: Cambridge University Press, pp. 163–73.

Wallace, D. (2018). The case for black hole thermodynamics, part II: Statistical mechanics, *Studies in the History and Philosophy of Modern Physics*, **66**, 103–117.

Wallace, D. (2019a). Naturalness and emergence. *The Monist*, **102**(4), 499–524.

Wallace, D. (2019b). Who's afraid of coordinate systems? An essay on representation of spacetime structure. *Studies in the History and Philosophy of Modern Physics*, **67**, 125–36.

Wallace, D. (2020). On the plurality of quantum theories: Quantum theory as a framework, and its implications for the quantum measurement problem. In S. French and J. Saatsi, eds., *Scientific Metaphysics and the Quantum*. Oxford: Oxford University Press, pp. 78–102.

Wallace, D. (2022). Isolated systems and their symmetries, part I: General framework and particle-mechanics examples, *Studies in the History and Philosophy of Science Part A*, **92**, 239–248.

Wallace, D. (2022). Isolated systems and their symmetries, part II: Local and global symmetries of field theories, *Studies in the History and Philosophy of Science Part A*, **92**, 249–259.

Weatherall, J. (2016). Understanding gauge. *Philosophy of Science*, **83**, 1039–49.

Weatherall, J. (2021). Classical spacetime structure. In E. Knox, and A. Wilson, eds., *The Routledge Companion to Philosophy of Physics*. London: Routledge, pp. 33–45.

Wess, J. and Bagger, J. (1983). *Supersymmetry and Supergravity*. Princeton: Princeton University Press.

Wigner, E. (1952). Die messung quantenmechanischer operatoren. *Zeitschrift fur Physik*, **133**, 101. An English translation by Paul Busch is available at https://arxiv .org/abs/1012.4372.

14

The Gauge Argument

A Noether Reason*

HENRIQUE GOMES, BRYAN W. ROBERTS, AND JEREMY BUTTERFIELD

14.1 Introduction

All interpretations of modern gauge theories adopt two core assumptions at their foundation. The first is that gauge symmetry arises when there are more variables in a theory than there are physical degrees of freedom. Hence the well-known soubriquets: gauge is 'descriptive redundancy', 'surplus structure', and 'descriptive fluff'. Correspondingly, considerable effort has been devoted to techniques for eliminating gauge redundancy in order to appropriately interpret gauge theories.[1] The second assumption is that a theory with gauge symmetry constitutes the gold standard of a modern physical theory: witness the gauge symmetries invoked in the Standard Model. This leads to a remarkable *puzzle of gauge symmetry:* why is gauge symmetry so ubiquitous? We do not aspire to give a single, ultimate answer. The purpose of this chapter is to articulate one answer to this question: namely, that gauge symmetry provides a path to building appropriate dynamical theories – and that this rationale invokes the theorems of Emmy Noether (1918).

Of course, a number of answers – alternatives to simple eliminativist interpretations of gauge – have already been articulated. For example, gauge theories provide a convenient calculational technique, as in the use of a potential A in classical electromagnetism. On the other hand, it would be odd if the 'great gauge revolution' turned out to be, *au fond*, a matter of calculational convenience; and indeed there is more to the story. More importantly, in many cases gauge symmetry cannot be eliminated without also eliminating the possibility of local Lorentz invariance and Lorentz invariant quantities.

Gauge symmetries can also encode important physical information, in spite of their being symmetries. The best-known example, which is vivid because of its experimental significance, is the effect of Aharonov and Bohm (1959), in which non-local information (of

* We would like to thank Adam Caulton, Neil Dewar, Caspar Jacobs, Ruward Mulder, James Read, and an anonymous referee for valuable comments.

[1] Cf. Earman (2002, 2003, 2004), Healey (2007), and Rosenstock and Weatherall (2016, 2018). To some extent we agree: see Gomes and Riello (2020) in response to Dougherty (2020a).

a different kind than due to quantum entanglement) can be contained in the gauge potential.[2] But we are concerned here with the two theorems of Emmy Noether (1918).[3]

Noether's first and better-known theorem (commonly called simply *Noether's theorem*) implies that global (or what we will call *rigid*) symmetries of a classical Lagrangian field theory – i.e. symmetries in which the redundancy is specified in exactly the same way at all spacetime points – correspond to charges that are conserved over time, such as energy and angular momentum. For example, the conservation of an electron's charge can be viewed as arising from the (redundant) global phases of the electron's wavefunction. But we are equally concerned with Noether's second theorem, which is about local (or what we will call *malleable*) gauge symmetries – meaning that the specified redundancy varies between spacetime points. Agreed: this theorem's physical significance is of course already well recognised, including in the philosophical literature (Brading 2002; Brading and Brown 2000, 2003). In particular, a recent line of work shows how such malleable gauge symmetries encode relationships between spatial or spacetime regions, and thus between parts and wholes in a field theory.[4]

In this paper, we will urge that these two theorems give us a further answer to the puzzle, 'Why gauge?' It is an established, indeed conventional, answer amongst practising physicists. For it is implicit in the well-known *gauge argument* or the *gauge principle* first formulated by Hermann Weyl (1929).[5] This argument begins with an assumption of local gauge symmetry, and then claims to 'derive' the form of the dynamics of quantum theory in a way that exhibits 'minimal coupling' to an electromagnetic potential. We claim that this is an instance of a much more general role for gauge, which has not been at all discussed in the philosophical literature: gauge symmetry supports theory construction, in particular by constraining the space of models to those in which charges appropriately couple to forces. Although some philosophers like Brading and Brown (2003) have pointed out the role of gauge symmetry in theory construction, it is this last coupling of charges to forces that we would like to highlight, which provides the answer to the puzzle of gauge symmetry that we will advocate here.

As experts will be quick to note: the gauge argument in its common textbook form is fraught with difficulties. However, our argument is that these difficulties can be overcome; and indeed that there is a more general gauge argument available for use in the construction of physical theories. We thus proceed in Section 14.2 to rehearse the usual gauge argument and its woes. In Section 14.3 we offer a glimmer of hope, by reconstructing the gauge

[2] Other, more recent examples include Rovelli (2014, 2020) and Gomes (2019, 2021a, 2021b), who emphasise the role of gauge in characterising the coupling of systems and regions, respectively; and Nguyen et al. (2020), who emphasise its role in defining certain local gauge fields. For a history of the early debate on the AB Effect, see Hiley (2013). Philosophers have recently focused on questions about the locality and reality of the gauge potential (cf. Belot 1998; Belot et al. 2009; Healey 1997, 1999, 2007; Lyre 2009; Mattingly 2006; Maudlin 1998; Mulder 2021; Myrvold 2011; Nounou 2003; Wallace 2014). More recently, Shech (2018) and Earman (2019) have challenged the idealisations associated with the Aharonov–Bohm effect, and Dougherty (2020b) has defended them.

[3] For details on the historical development of Noether's theorems, see Kosmann-Schwarzbach (2011). For a modern statement of the first and second theorems, cf. Olver (1993), theorems 5.58 (p. 334) and 5.66 (p. 343) respectively. See also Chapter 1, this volume.

[4] See Donnelly and Freidel (2016), Gomes (2021a, 2021b) and Gomes and Riello (2021) in response to the discussions of 'direct empirical significance' in Brading and Brown (2004) and Greaves and Wallace (2014).

[5] An English translation and extended commentary is given by O'Raifeartaigh (1997, chapter 5).

argument as a theorem of quantum theory, which, we argue, vindicates to some extent its use in the formulation of quantum electrodynamics.

The real limitation of the textbook gauge argument, as we shall see, is that it does not reflect the generality of the kind of argument that physicists typically use. Thus, in Section 14.4, we present a much more general gauge argument, which we call the *Noether gauge argument*, in the context of classical Lagrangian field theory. The key to understanding this argument is the combined use of *both* Noether's first and second theorems. In the first step, one applies Noether's first theorem to establish the conservation of charge. In the second step, one makes use of the power of Noether's second theorem, to infer specific interpretive information about how these charges couple to gauge fields. We draw out and clarify what that information is, in the presence of various kinds of symmetries that are sometimes referred to as 'gauge', in order to illustrate the precise extent to which the gauge argument can be fruitfully used to constrain physical theories.

14.2 The Gauge Argument and Its Critics

The textbook gauge argument or gauge principle uses gauge invariance to motivate a quantum theory of electromagnetism. We begin Section 14.2.1 with a brief presentation of this argument as it is usually presented. Classic textbook statements can be found in Schutz (1980, §6.14) Göckeler and Schücker (1989, §4.2), and Ryder (1996, §3.3), among many other places. Then, in Section 14.2.2, we assess it. The argument has been discussed in the form below by philosophers as well, such as Teller (1997, 2000), Brown (1999), Martin (2002), and Wallace (2009, §2). In spite of the criticisms, we will argue in Section 14.3 that a grain of truth remains in the gauge argument.

14.2.1 Beware: Dubious Arguments Ahead

We begin by describing a quantum system with the Hilbert space $L^2(\mathbb{R}^3)$ of wavefunctions, recalling that a unique pure quantum state is represented not by vector, but by a 'ray' of vectors related by a complex unit. This implies that the transformation $\psi(x) \mapsto e^{i\theta}\psi(x)$ for some $\theta \in \mathbb{R}$, referred to as a 'global phase' transformation, acts identically on rays and is in this sense an invariance of the quantum system. But now, the story goes, suppose we replace this with a 'local phase' transformation $\psi(x) \mapsto e^{i\phi(x)}\psi(x)$, in which the constant θ is replaced with a function $\phi : \mathbb{R}^3 \to \mathbb{R}$, or indeed with a smooth one-parameter family of such functions $\phi_t(x)$ for each $t \in \mathbb{R}$. This transformation is 'local' in the sense that its values vary smoothly across space and time. The corresponding Hilbert space map $W_\phi\colon \psi \mapsto e^{i\phi}\psi$ does not act identically on rays. However, one might still wish to postulate that this transformation has no 'physical effect' on the system, or is 'gauge'. Various motivations for this step are given in the textbooks, often with vague references to general covariance of the kind found in general relativity, to which we will return shortly. But to mimic the standard presentation, we will simply press forward, referring to $W_\phi : \psi \mapsto e^{i\phi}\psi$ as a *local* or *malleable gauge transformation*.

The main premise of the argument is to assume that the Schrödinger equation must be invariant under this local phase transformation. But, for the free non-relativistic Hamiltonian in the Schrödinger (position) representation, this is not the case.[6] Writing $\psi_t(x) := e^{-itH}\psi(x)$ with $H = \frac{1}{2m}P^2$, one finds that $W_\phi : \psi \mapsto e^{i\phi_t(x)}\psi$ transforms the Schrödinger equation to $i\frac{d}{dt}\left(e^{i\phi_t(x)}\psi_t(x)\right) = \frac{1}{2m}P^2 e^{i\phi_t(x)}\psi_t(x)$, which is equivalent[7] to the statement that,

$$i\frac{d}{dt}\psi_t(x) = \left(\tfrac{1}{2m}(P + \nabla\phi_t)^2 + \tfrac{d\phi_t}{dt}\right)\psi_t(x). \tag{14.1}$$

Instead of preserving the Schrödinger equation, a gauge transformation produces the additional terms $\nabla\phi_t$ and $\frac{d\phi_t}{dt}$ in the Hamiltonian.

To correct this situation, the big move of the gauge argument is to introduce a vector $A = (A_1, A_2, A_3)$ and a scalar V, which are assumed to behave under the gauge transformation as,

$$A \mapsto A + \nabla\phi_t, \qquad V \mapsto V - \tfrac{d\phi_t}{dt}. \tag{14.2}$$

This has the form of the familiar gauge freedom of the electromagnetic four-potential that leaves the electromagnetic field unchanged.

To restore invariance of the Schrödinger equation under gauge transformations, one thus apparently needs only to assume that the Hamiltonian is not free, but rather given by,

$$H = \sum_{r=1}^{3} \tfrac{1}{2m}(P_r - A_r)^2 + V, \tag{14.3}$$

which is known as the *minimally coupled* Hamiltonian. For, replacing the Hamiltonian in the Schrödinger equation with this one, we find that the transformation rules for A and V perfectly compensate for the extra terms appearing in Eq. (14.1). Thus, gauge invariance of the Schrödinger equation is obtained, provided the Hamiltonian contains interaction terms A and V that behave like the 3-vector potential A and scalar potential V for an electromagnetic field.

With an eye towards a modern gauge theory formulated as a vector bundle with a derivative operator, it is even possible to interpret the potentials A and V as associated with a change of derivative operator: writing $\partial_\mu := \left(\frac{d}{dt}, \nabla\right)$ and $A_\mu = (V, A)$, one finds that the procedure above is equivalent to replacing ∂_μ with,

$$D_\mu := \partial_\mu + iA_\mu = \left(\tfrac{d}{dt} + iV, \nabla + iA\right) = (D_t, D). \tag{14.4}$$

This is commonly referred to as a 'covariant derivative'. Then, substituting $\frac{d}{dt} \mapsto D_t$ and $\nabla \mapsto D$ into the free Schrödinger equation $i\frac{d}{dt}\psi = \frac{1}{2m}\nabla^2\psi$ and rearranging, we derive the minimally coupled Hamiltonian of Eq. (14.3). Accordingly, this choice of Hamiltonian is

6 Obvious variations of the argument exist for relativistic wave equations too (cf. Ryder 1996, §3.3).

7 The LHS is $i\frac{d}{dt}e^{i\phi_t(x)}\psi_t(x) = e^{i\phi_t(x)}\left(-\frac{d\phi}{dt} + i\frac{d}{dt}\right)\psi_t(x)$. For the RHS, use the fact that $e^{-i\phi_t(x)}Pe^{i\phi_t(x)} = P + \nabla\phi_t(x)$ (cf. footnote 11), and so $e^{-i\phi_t(x)}P^2e^{i\phi_t(x)} = (e^{-i\phi_t(x)}Pe^{i\phi_t(x)})^2 = (P + \nabla\phi_t(x))^2$. Thus the RHS is $\frac{1}{2m}P^2e^{i\phi_t(x)}\psi_t(x) = e^{i\phi_t(x)}\frac{1}{2m}(P + \nabla\phi_t(x))^2\psi_t(x)$. Multiplying both sides on the left by $e^{-i\phi_t(x)}$ and rearranging then gives the result.

sometimes advocated, for example by Lyre (2000), on the basis of a 'generalised equivalence principle', according to which electromagnetic interactions with all matter fields 'can be transformed away'.[8] In short, it appears as if minimal electromagnetic coupling has been derived out of nothing: or at least, from an assumption of gauge invariance.

14.2.2 Criticisms of the Gauge Argument

That is how the story is usually presented. We agree: it is far from water-tight. The argument begins with a system with a global symmetry, gratuitously generalises it to a local symmetry – which, to emphasise, was not required for mathematical consistency or for empirical adequacy – and then, in order to fix the ensuing non-invariance of the governing equations, proceeds to conjecture a new force of nature. To put it uncharitably: the argument fixes a problem that didn't exist by conjecturing a redundant field, and then turns this game around, claiming to come out successfully by 'retrodicting' the existence of electromagnetism. More charitably: the gauge argument suffers from at least three categories of concerns. We will set out each of these three concerns here. In Section 14.3 we will then offer a glimpse of how the first two can be answered, and in Section 14.4 present an alternative *Noether gauge argument* that answers them entirely.

The first category of concerns is the gauge argument's claim to have derived a dynamics that is specifically electromagnetic in nature. Although a formal set of operators $A_\mu = (V, A)$ have been included in the dynamics, no evidence is given that these operators take the form required for any *specific* electromagnetic potential, or that the coupling to A_μ will be proportional to a particle's charge e, or even that A_μ is non-zero. And if they could be shown to be non-zero, then, as Wallace (2009, p. 210) rightly asks: 'how do neutral particles fit into the argument?' A minimally coupled dynamics does not to apply to neutral particles, and yet since the gauge argument never mentioned or assumed anything about charge, it presumably is intended to apply to them.

This concern can be assuaged by scaling back the conclusion of the gauge argument: its aim is not to derive any particular electromagnetic interaction, but rather to *constrain* the dynamics so as to be compatible with gauge invariance. This leaves open the specific character of A_μ, and indeed even the question of whether it is zero. Although not all authors adopt this attitude towards the gauge argument, we advocate it as the preferable attitude and will develop it in more detail in the subsequent sections.

A second category of problems arises out of the freewheeling argumentative style of the gauge argument. For example, it is not a strict deductive derivation of either the electromagnetic potential or the dynamics. At best, the gauge argument appears to show that one *can* adopt a minimally coupled Hamiltonian in order to assure gauge invariance. But this does not ensure that one *must* do so: the door appears to be left open for other dynamics to be gauge invariant, but without taking the minimally coupled form that the gauge argument advocates. As Martin (2002, p. S230) writes: 'The most I think we can safely say is that

[8] This principle arises in particular on a principal fibre bundle formulation of gauge theory; for philosophical appraisals, see Lyre (2000), Healey (2007, ch. 6.3), and Weatherall (2016, §5).

the form of the dynamics characteristic of successful physical (gauge) theories is *suggested* through running the gauge argument.'

Another example of freewheeling argumentation is in the motivation for requiring the local gauge transformations $W_\phi : \psi \mapsto e^{i\phi}\psi$ to be symmetries. Sometimes a preference for this transformation over global phase transformations is dubiously motivated by a desire to avoid superluminal signalling.[9] In other cases it is motivated by the coordinate invariance of a spatial coordinate system. But as Wallace (2009, p. 210) points out, no reason is given as to why we do not similarly consider local transformations of configuration space, momentum space, or any other space to be symmetries. Nor is there any clear reason why the $U(1)$ symmetry of electromagnetism is chosen as the global symmetry motivating the move to the local symmetry, as opposed to (say) the $SU(3)$ symmetry of the strong nuclear force.

We will claim that most of these problems can be entirely solved. In the first place, the gauge argument can in fact be tightened and turned into a valid derivation, as we will show in the next section. Not only is it that one *can* adopt the minimally coupled Hamiltonian in the presence of gauge invariance, but one *must* do so, when gauge invariance is viewed in terms of a particular constraint on the 'velocity observable', in a sense we will precisely define. We will similarly argue that the postulate that local gauge transformations are symmetries in a certain sense – namely, that they are unitary operators – is not really a postulate, but a formal fact about the framework in which these transformations are presented.

Regarding the generalisation of the gauge argument to other global symmetry groups beyond electromagnetism, we wholeheartedly agree with Wallace: one should expect, and indeed we will argue in Section 14.4, that an appropriate generalisation of the gauge argument can also be applied to these more general gauge groups.

Our approach here speaks to a third category of concerns, that the gauge argument is awkwardly placed as an argument for a quantum theory of electromagnetism. The construction of a covariant derivative operator suggested by the gauge argument is most appropriately carried out not in quantum field theory, but in the classical Yang–Mills theory of principal fibre bundles. Here too, we agree with Wallace:

> In fact, it seems to me that the standard argument feels convincing only because, when using it, we forget what the wavefunction really is. It is not a complex classical field on spacetime, yet the standard argument, in effect, assumes that it is. This in turn suggests that the true home of the gauge argument is not non-relativistic quantum mechanics, but classical field theory. (Wallace 2009, p. 211)

Indeed, it is remarkable that in the presentation of the gauge argument above, the role of the 'rigid' or 'global' $U(1)$ symmetry is hardly substantial: only the local malleable symmetries play any substantial role in the argument. This is an oddity to be sure, though one that we will correct shortly.

9 For example, Ryder (1996, p. 93) writes:

> when we perform a rotation in the internal space of ϕ at one point, through an angle Λ, we must perform the same rotation at all other points at the same time. If we take this physical interpretation seriously, we see that it is impossible to fulfil, since it contradicts the letter and spirit of relativity, according to which there must be a minimum time delay equal to the time of light travel.

For a detailed critique, see Martin (2002, p. S227).

In Section 14.4, we will switch perspectives from the *verdammten Quantenspringerei* to the context of classical Lagrangian field theory and propose a framework that substantially clarifies the roles of rigid gauge symmetries, of malleable gauge symmetries, and of their relationship, which we will call the 'Noether gauge argument'.

But before we develop this argument, we would like to first show how the first two categories of concern are unwarranted: a more rigorous formulation of the textbook gauge argument is possible, which dispels any worries about a freewheeling, under-motivated argument. Its only serious shortcoming in this more rigorous form, as we will see, is its lack of generality. This will be corrected in Section 14.4, when we develop an analogous argument in classical field theory. It is not a straightforward generalisation of Section 14.3's theorem, but rather a 'cousin' of it. Namely, the theme of both results is that malleable gauge symmetries are used to constrain the dynamics.

14.3 The Gauge Argument as a Theorem

14.3.1 Gauge Transformations and Probabilities

Suppose we view the textbook gauge argument as beginning with a strongly continuous one-parameter unitary representation $t \mapsto U(t) = e^{-itH}$ on the Hilbert space $\mathcal{H} = L^2(\mathbb{R}^3)$, but we do not yet know the form of the Hamiltonian H. Here we will adopt the vector notation $x = (x_1, x_2, x_3)$, and similarly $Q = (Q_1, Q_2, Q_3)$, with the latter representing the vector position, where for each $i = 1, 2, 3$ we have a self-adjoint operator Q_i defined by $Q_i\psi(x) = x_i\psi(x)$. The aim of the gauge argument, as we see it, is then to use gauge symmetry to constrain the possible dynamics that are available.

The first step is to observe a fact in this framework – *not* an assumption, but a consequence! Namely: the local gauge transformations are symmetry operators that transform position and momentum as one would expect a gauge transformation to do.[10] That is, for any smooth function $\phi : \mathbb{R}^3 \to \mathbb{R}$, whose gradient $\nabla\phi$ we can think of as smoothly deforming space however we wish, there is a Hilbert space operator that preserves inner products, i.e. a unitary operator, given by

$$W_\phi\psi(x) := e^{i\phi(x)}\psi(x) \tag{14.5}$$

for all $x \in \mathbb{R}^3$. A similar fact applies in four dimensions: defining a one-parameter set $\phi_t(x)$ of such transformations, there is a 'malleable symmetry' or unitary W_ϕ defined by

$$W_{\phi_t}\psi(x) := e^{i\phi_t(x)}\psi(x). \tag{14.6}$$

for all $\psi \in \mathcal{H}$, and for all $x \in \mathbb{R}^3$ and $t \in \mathbb{R}$. These unitaries W_ϕ transform position $Q = x$ and momentum $P = -i\nabla$ as one would expect a gauge transformation to do:[11]

[10] By 'symmetry' we mean *unitary* operators, sometimes referred to as 'kinematic symmetries' (cf. Jauch 1968, chapter 13). Antiunitary operators are symmetries of time-reversing transformations (see Roberts 2017, 2021), but they will play no role in this discussion.

[11] The first equation follows just from ϕ being a function of position. The second equation follows from the fact that $PW^*_\phi\psi = -i\nabla\left(e^{-i\phi}\psi\right) = -i\left(e^{-i\phi}\nabla\psi - i(\nabla\phi)e^{-i\phi}\psi\right) = e^{-i\phi}(-i\nabla - \nabla\phi)\psi = W^*_\phi(P - \nabla\phi)\psi$, for all $\psi \in L^2(\mathbb{R}^3)$.

$$W_\phi Q W_\phi^* = Q \qquad\qquad W_\phi P W_\phi^* = P - \nabla\phi. \tag{14.7}$$

Some presentations (rather mysteriously) even say that these facts requirc a postulate of general relativity; as when Schutz (1980, p. 219) writes, '[t]hese turn out to be the appropriate equations for that system in *general* relativity' (cf. also Lyre 2000, p. 11).

We would like to emphasise that, for the purpose of constructing a unitary operator that implements the transformations in Eq. (14.7), *no such arguments are needed.* It is a direct mathematical consequence of the Hilbert space formalism that each malleable transformation ϕ gives rise to a unitary $W_\phi = e^{i\phi}$, and that these unitaries behave like gauge transformations with respect to position and momentum. A new physical postulate is only required when malleable transformations are used to constrain the dynamics, which we consider next.

14.3.2 A Rigorous Quantum Gauge Argument

We are now ready to view the gauge argument as providing a constraint on the possible dynamics for quantum theory, on the basis of a certain assumption of gauge invariance. The postulate of gauge invariance that we will state is in the spirit of the textbook gauge argument. We begin by clarifying exactly what we take this to mean.

As before, we begin with a unitary dynamics $t \mapsto U_t = e^{-itH}$ on the Hilbert space $\mathcal{H} = L^2(\mathbb{R}^3)$, for which we do not yet know the Hamiltonian H. The 3-velocity in this context is defined to be the rate of change of the position operator in the Heisenberg picture with respect to this unitary group. Thus, it is given by

$$\dot{Q} := \tfrac{d}{dt} U_t^* Q U_t \big|_{t=0} = i[H, Q], \tag{14.8}$$

where $Q = (Q_1, Q_2, Q_3)$ is defined as in the Schrödinger representation by $Q_r\psi(x) = x_r\psi(x)$ for each $r = 1, 2, 3$, for all wavefunctions ψ in its domain.

Our central postulate for how malleable symmetries constrain the dynamics, in pictorial terms, is that if a smooth transformation $\phi : \mathbb{R}^3 \to \mathbb{R}$ is applied, then the 3-velocity should compensate for the extra term $\nabla\phi$ that arises, as shown in Figure 14.1. One must remove the contribution of $\nabla\phi$ from the value of the velocity.[12] Thus, if $W_\phi = e^{i\phi}$ is the unitary transformation associated with ϕ, then the assumption says:

$$e^{i\phi} \dot{Q} e^{-i\phi} = \dot{Q} - \nabla\phi. \tag{14.9}$$

This is close enough to the meaning of a transformation by a smooth function that it is somewhat surprising that it provides any constraint on the dynamics, let alone enough to establish minimal coupling.

Although Eq. (14.9) is thus not *a priori*, it is natural and rich in consequences. We formalise this in the following:[13]

[12] In Eq. (14.13) of Section 14.4, we propose a similar postulate to characterise gauge invariance in the context of classical field theory; where we note that it is also possible to consider the contribution of higher-order terms, in which case this definition can be viewed as a first-order approximation.

[13] A proof is provided in the Appendix. The ideas of this proof are a turned-around cousin of Jauch (1964, 1968, §13–5), an analysis of which is in Roberts (2012, appendix A).

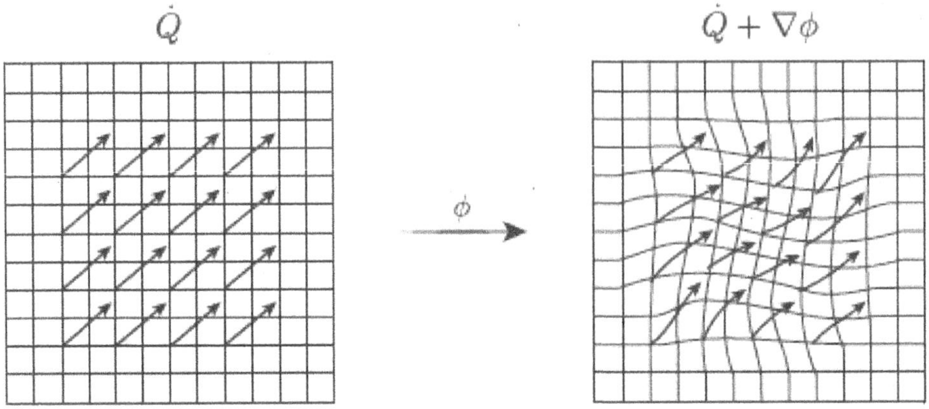

Figure 14.1 The constraint of malleable symmetries is that the 3-velocity $\dot{Q}$ must transform so as to remove the contribution of the gradient $\nabla\phi$ of a smooth function.

Theorem *In the Schrödinger (position) representation on $L^2(\mathbb{R}^3)$ with dynamics $t \mapsto U_t = e^{-itH}$, let $\dot{Q}_r := \frac{d}{dt}(U_t^* Q U_t)\big|_{t=0} = i[H, Q_r]$. Suppose that for every smooth one-parameter set of smooth functions $\phi_t : \mathbb{R}^3 \to \mathbb{R}$, at time $t = 0$,*

$$W_\phi \dot{Q} W_\phi^* = \dot{Q} - \nabla\phi, \tag{14.10}$$

where $W_\phi := e^{i\phi_t}$ and $\nabla = (\partial_x, \partial_y, \partial_z)$. Then:

(i) the Hamiltonian is constrained to be of the form,

$$H = \sum_{r=1}^{3} \tfrac{1}{2m}(P_r - A_r)^2 + V \tag{14.11}$$

for some $m > 0$ and some functions (i.e. operators) $A_r(x)$ and $V(x)$ that depend only on position, and not on momentum;

(ii) the transformed unitary group $t \mapsto \tilde{U}_t := W_\phi U_t =: e^{-itH_\phi}$ (where the last equation defines H_ϕ as the generator of $\tilde{U}_t$, by Stone's theorem) is given by,

$$H_\phi = \sum_{r=1}^{3} \tfrac{1}{2m}\left(P_r - A_r^\phi\right)^2 + V^\phi, \tag{14.12}$$

where $A_r^\phi = A_r + \nabla\phi$ and $V^\phi = V - \frac{d\phi}{dt}$; and

(iii) the Schrödinger equation is 'gauge invariant' under W_ϕ, in that if $i\frac{d}{dt}\psi(t) = H\psi(t)$, then $\psi_\phi(t) := W_\phi\psi(t)$ satisfies $i\frac{d}{dt}\psi_\phi(t) = H_\phi\psi_\phi(t)$.

In summary: we assume that the 3-velocity transforms in a given manner under smooth transformations of space ($W_\phi \dot{Q} W_\phi^* = \dot{Q} - \nabla\phi$), and prove a consequence that the dynamics *are constrained* to be given by a minimally coupled Hamiltonian: in which, moreover, the smooth transformations behave like gauge transformations of the vector and scalar potentials, in that they leave the Schrödinger equation invariant.

We take this to largely dispel the concerns about the freewheeling nature of the textbook gauge argument. Cast in this form, any Hamiltonian that is constrained by malleable transformations as in Eq. (14.10) *must* be minimally coupled; and this leads to gauge invariance as in (ii) and (iii) of the Theorem. Of course, this still leaves the specific coupling undetermined up to a choice of vector and scalar potential, including the choice of zero. But this is fully compatible with our perspective that gauge symmetry can be viewed as providing a powerful constraint on the dynamics.

There remains the question of how to view this kind of argument in the context of more general gauge groups. And there remains a question of what role the rigid gauge transformations play in constraining a theory's dynamics. Here we propose a shift to a more natural perspective for dealing with such constraints, which uses the tools of classical Lagrangian field theory. This is the subject of the remainder of this chapter.

14.4 A Noether Reason for Gauge

14.4.1 Overview

For a more general view of how gauge symmetries constrain the dynamics of a physical theory, we will now, as announced in Section 14.1, make a two-step use of the theorems of Emmy Noether (1918): the first, and then the second. We will refer to this as the *Noether gauge argument*. Agreed: this is by no means a new observation, since practising physicists use this property of gauge frequently![14] But we believe it is worth highlighting and clarifying exactly the kind of information that can be extracted in various cases, as part of our advocacy (cf. Section 14.1) that philosophical discussions of gauge should better recognise gauge's significance for theory construction.

The Noether gauge argument proceeds in two steps. First, we choose a rigid gauge symmetry associated with an arbitrary global gauge group, and propose that its action produces a variational symmetry: by Noether's first theorem, this guarantees the presence of a collection of conserved quantities. But matter fields do not exist in isolation: they couple to other 'force' fields, and possibly to long-range ones. Thus, in the second step, we introduce such a field and apply Noether's second theorem, 'loosening' the rigid symmetries to malleable ones; and we show that this provides three concrete constraints on the dynamics, viz. the vanishing of the three lines in Eq. (14.14) below. This result is limited: we make no claim that the only way to get these benefits is by malleable symmetries. However, it remains the best-known and most tractable way to achieve them.

The interpretation of these constraints can be seen on a case-by-case, or sector-by-sector, basis: we consider their implications for rigid versus malleable symmetries, as well as for A-independent versus A-dependent ones. Thus, in the following sections we will spell out the consequences of the three constraints for four different sectors of the theory. In particular, we will find through explicit computation – adopting only a minor additional assumption of non-derivative coupling – that when we couple the matter fields to force

[14] A succinct example is Avery and Schwab (2016), who write: 'Noether's second theorem, which constrains the general structure of theories with local symmetry'.

fields, gauge-invariance guarantees that the Lagrangian for these fields is massless, and so they constitute long-range interactions.[15] The generalised Gauss laws thus are guaranteed to relate the content of the matter current within a region to the flux of the other force fields at distant closed surfaces surrounding such a region.

Disclaimers: first, in the interest of clarity and pedagogy, we will not try to incorporate the full generality of Noether's theorems, which is truly extraordinary but overly complicated for our discussion. In its place we make several simplifying assumptions, both about the Lagrangian density and about the action of the gauge group, which are not, strictly speaking, necessary but which simplify our argument. Second, throughout this discussion, we will follow standard practice and distinguish two equivalence relations for classical fields on a manifold. First, we write '$=$' to denote ordinary equality between fields, irrespective of the satisfaction of the equations of motion, and refer to this as *strong* or *off-shell* equality. Second, given a fixed Lagrangian, we write '$\approx$' to denote equality between fields that holds if the Euler–Lagrange equations are satisfied for that Lagrangian, and refer to this as *weak* or *on-shell* equality.[16]

14.4.2 Field Theory and Malleable Symmetries

In the textbook gauge argument, the (rigid) gauge transformations were assumed to be the group of phase transformations $U(1)$, and the (malleable) gauge group was given by its locally varying analogues. We now lift this restriction and allow the global gauge group to be any compact Lie group G, with Lie algebra $\mathfrak{g}$. We can characterise the malleable (infinitesimal) gauge transformations induced by G (and $\mathfrak{g}$), on a local patch U, as the smooth functions from U to G (to $\mathfrak{g}$, respectively).[17] So we can write local (infinitesimal) gauge transformations as $\mathcal{G} = C^\infty(U, G)$ (or $\text{Lie}(\mathcal{G}) = C^\infty(U, \mathfrak{g})$, respectively): the group of gauge transformations has a group operation that is just that of G, pointwise on M, i.e. for $g, g' \in \mathcal{G}$, $gg'(x) = g(x)g'(x)$. The group acts as the adjoint on the algebra, $\text{Ad} : G \to GL(\mathfrak{g})$, with $(g, \xi) \mapsto \text{Ad}_g \xi := g\xi g^{-1} \in \mathfrak{g}$, and so the action of the algebra on itself is just the Lie algebra commutator, $\mathfrak{g} \times \mathfrak{g} \to \mathfrak{g}$ with $(\xi, \xi') \mapsto [\xi, \xi'] \in \mathfrak{g}$.

15 The formalism equally applies to spin-2, or gravitational, fields; but, apart from some cursory remarks, we will not discuss these.

16 This common terminology is due to Dirac (cf. Henneaux and Teitelboim 1994).

17 The precise description of gauge transformations requires a brief incursion into the mathematical theory of fibre bundles. A principal fibre bundle is a smooth manifold P that admits a smooth and free action of the group G, e.g. $G \times P \to P$; $(g, p) \mapsto g \cdot p$. Spacetime is encoded in the bundle as the collection of orbits $\mathcal{O}$ of the group: $\pi : P \to M \simeq P/G$, where π is the smooth submersion induced by the action of G. One can think of the bundle as attaching an orbit space isomorphic to G (but without a preferred identity, much like an affine space) to each point of M. Thus, $\pi^{-1}(x) =: \mathcal{O}_p = \{g \cdot p,\ g \in G\} \simeq G$ for $x \in M$, but there is no canonical isomorphism between $\pi^{-1}(x)$ and G (it requires the choice of a point p) and no decomposition $P \simeq M \times G$. To talk about gauge transformations as maps in spacetime, we must restrict their domains to subsets of spacetime. Locally, i.e. for subsets $U \subset M$, we can write $\pi^{-1}(U) \simeq U \times G$, once we have choice of trivialisation: $s : U \to P$ such that $\pi \circ s = \text{Id}_U$. Here, s can be thought of as a submanifold that intersects all the group orbits in $\pi^{-1}(U)$ only once. In terms of s we can write the trivialisation as $U \times G \to \pi^{-1}(U)$; $(x, g) \mapsto g \cdot s(x)$. Global gauge transformatios are diffeomorphisms $\tau \in \text{Diff}(P)$ such that $\tau(g \cdot p) = g \cdot \tau(p)$. Locally, by equivariance, τ maps orbits to orbits and so takes one trivialising section s to another s'; and if $\tau(s) = \tau'(s) = s'$, then $\tau(p) = \tau'(p)$ for $p \in \pi^{-1}(U)$. Therefore, locally, gauge transformations are *uniquely encoded* by the map from s to s'. It is easy to characterise these maps: since s provides a local trivialisation, we can write, for all $x \in U$, $s'(x) = g(x) \cdot s(x)$, for some $g \in C^\infty(U; G)$. Therefore, locally, gauge transformations are of the form $g \in C^\infty(U; G)$. This local construction will be mirrored for vector bundles in footnote 18.

Let φ be a map that takes each point of U into V, a vector space of dimension n which is φ's value space. We will use i to indicate components in V, e.g. φ_i. We will take φ to represent matter fields. Locally, that is, on appropriate trivialising patches $U \subset M$, φ is a V-valued smooth function on U, the class of which is $C^\infty(U; V)$.[18]

To represent the forces that are sourced by φ, we take the collection of vector-valued one forms A^a_μ, which take a vector of M at a point of U to $\mathfrak{g}$, with μ representing the spacetime components of the vector and a indicating the components in $\mathfrak{g}$.[19] These fields are associated with a dynamics by postulating a preferred real-valued action functional $S(\varphi_i, A^a_\mu)$, whose extremal values are postulated to provide the equations of motion.

We also assume G has some action (a representation) on V, the vector space of local field-values of the matter fields φ, defining this action pointwise as $g \cdot \varphi(x) = g(x) \cdot \varphi(x) \in V$. Let t_a^{ij} be the n-dimensional Hermitean matrix representation on V of $\mathfrak{g}$, i.e. $t : \mathfrak{g} \to GL(V)$, where the a are indices of the Lie algebra space, in the domain of the map, and i, j denote the matrix indices in the image of the map, acting linearly on V. Then we take the (malleable) gauge transformations, infinitesimally parametrised by $\epsilon \in \text{Lie}(\mathcal{G})$, to act on our fundamental variables as:

$$\begin{cases} \delta_\epsilon \varphi_i = \epsilon^a t_a^{ij} \varphi_j = (\epsilon t \varphi)_i \\ \delta_\epsilon A^a_\mu = \mathrm{D}_\mu \epsilon^a = \partial_\mu \epsilon^a + [\epsilon, A_\mu]^a \end{cases}. \tag{14.13}$$

where the square brackets are the Lie algebra commutators. Here the $\partial_\mu \epsilon^a$ on the right-hand side of the second line echoes the $\nabla\phi$ of Eq. (14.9), and the second term in that same side allows for a non-Abelian transformation group. These transformation rules are not as general as they could be, but neither are they arbitrary: they are the first-order terms of the Lie algebra action on the respective vector spaces – in particular, 'first-order' in the derivatives of ϵ and in powers of A and φ – and in this sense provide an appropriate approximation of any malleable gauge transformation. We focus here on this special case, Eq. 14.13, only to simplify the presentation of our argument.

Our aim now is to constrain how the matter fields φ couple to force fields. Let $\mathcal{L}(\varphi, \partial\varphi, A, \partial A)$ be the Lagrangian defining our action $S(\varphi, A)$, which, we assume for simplicity, does not depend on higher-order derivatives.[20] Variation along the directions of the gauge transformations above yields (with summation convention on all indices):

[18] Here the mathematically precise description of φ is as a section of a vector bundle $\pi : E \to M$, which one can think of as having an internal vector space isomorphic to V attached to each point of M, that is, with $\pi^{-1}(x) \simeq V$ for $x \in M$, but without a canonical decomposition $E \simeq M \times V$. Locally, i.e. for subsets $U \subset M$, we have an isomorphism $\pi^{-1}(U) \simeq U \times V$, but the isomorphism is not canonical: writing a field that is locally valued in V requires a choice of trivialisation, as in footnote 17 above (the standard example here is vector fields on M: we can only write them as maps $U \to \mathbb{R}^n$ if we fix a basis for the tangent spaces). Thus, φ is a *section* of E, i.e. $\varphi : M \to E$ such that $\pi \circ \varphi = \mathrm{Id}_M$, but locally, given a trivialisation, we can write it as function $\varphi : U \to V$. The trivialisation of the vector bundle can be 'soldered onto' the trivialisation of the principal bundle, mentioned in footnote 17, by thinking of E as an 'associated vector bundle'. The simplest way to think of this association is to see the elements $p \in P$ of the bundle as linear frames for V (at $x \in \pi(p)$). This soldering guarantees that a gauge transformation on P will also act on the matter fields in the appropriate manner, e.g. (14.13).

[19] What is a 'force' and what is 'matter' will be further distinguished by their transformation properties under a gauge transformation, in (14.13). Matter transforms linearly, whereas forces acquire derivatives of the generator as inhomogeneous terms.

[20] This can be justified by appeal to Ostrogradsky's theorem; see Swanson (2022) for a philosophical discussion.

$$\begin{aligned}&\left(\frac{\delta\mathcal{L}}{\delta\varphi_i}(t^a\varphi)_i+\frac{\delta\mathcal{L}}{\delta\partial_\mu\varphi_i}(t^a\partial_\mu\varphi)_i+\left[\frac{\delta\mathcal{L}}{\delta A_\nu},A_\nu\right]^a+\left[\frac{\delta\mathcal{L}}{\delta\partial_\nu A_\mu},\partial_\mu A_\nu\right]^a\right)\epsilon_a\\&+\left(\frac{\delta\mathcal{L}}{\delta\partial_\mu\varphi_i}(t^a\varphi)_i+\frac{\delta\mathcal{L}}{\delta A^a_\mu}+\left[\frac{\delta\mathcal{L}}{\delta\partial_\nu A_\mu},A_\nu\right]^a\right)\partial_\mu\epsilon_a\\&+\frac{\delta\mathcal{L}}{\delta\partial_\nu A^a_\mu}\partial_\mu\partial_\nu\epsilon^a=0\end{aligned}\tag{14.14}$$

Since the derivatives of ϵ are functionally independent, this equation implies that *each line must vanish separately*: the first line is a consequence of rigid symmetries, and the remaining two are of malleable ones. These are the fundamental constraints on the dynamics that we propose to analyse, and the task of the remainder of this chapter will be to unpack them.

The requirement that each of these lines vanishes provides a strong constraint on the form of the Lagrangian, and hence on the dynamics. This, we claim, provides the core of the Noether gauge argument. To extract interesting physical information from this constraint, there are four sectors to compare, arising from the use of either rigid or malleable symmetries, and either A-independent or A-dependent Lagrangians. We treat each sector in turn.

The results will be: a theory with rigid symmetries can be dynamically non-trivial and complete – i.e. it will not require further constraints – when A does not figure in the Lagrangian. With malleable symmetries and no A-dependence, the constraints demand that the dynamics be trivial, i.e. no kinetic term for the matter field can appear in the Lagrangian. When forces have their own dynamics, that is, when the Lagrangian is A-dependent, a theory with rigid symmetries may be incomplete and require further constraints to render the dynamics of A compatible with charge conservation; an example will be given. It is only in the last case, where we have malleable symmetries and A-dependence, that the equations of motion coupling forces to charges is automatically consistent with the conservation of charges (and so no further constraints are required). Thus we will see the power of malleable symmetries and A-dependence together to secure an interacting dynamics that conserves charge. And this will be our Noether gauge argument.

14.4.3 A-Independent, Rigid Symmetries

First, suppose we are as in the first step of the textbook gauge argument: there is no A in sight, and the symmetry is rigid, so that $\partial_\mu\epsilon^a=0=\partial_\mu\partial_\nu\epsilon^a$. Then the vanishing of the first line of Eq. (14.14) reduces to

$$\frac{\delta\mathcal{L}}{\delta\varphi_i}(t^a\varphi)_i+\frac{\delta\mathcal{L}}{\delta\partial_\mu\varphi_i}(t^a\partial_\mu\varphi)_i=0.\tag{14.15}$$

But by the Euler–Lagrange equations $\mathsf{E}(\varphi)_i\approx 0$, where $\mathsf{E}(\varphi)_i=\frac{\delta\mathcal{L}}{\delta\varphi_i}-\partial_\mu\frac{\delta\mathcal{L}}{\delta\partial_\mu\varphi_i}$, we have

$$\frac{\delta\mathcal{L}}{\delta\varphi_i}\approx\partial_\mu\frac{\delta\mathcal{L}}{\delta\partial_\mu\varphi_i},\tag{14.16}$$

where we again are using '$\approx$' to denote 'on-shell' equality. Applying this to Eq. (14.15), we find that

$$\partial_\mu \left(\frac{\delta \mathcal{L}}{\delta \partial_\mu \varphi_i} (t^a \varphi)_i \right) = \partial^\mu J^a_\mu(\varphi) \approx 0 \tag{14.17}$$

where we have defined the matter current as

$$J^a_\mu(\varphi) := \frac{\delta \mathcal{L}}{\delta \partial_\mu \varphi_i} (t^a \varphi)_i. \tag{14.18}$$

In summary, we have derived what is guaranteed by Noether's first theorem, that the current $J^a_\mu(\varphi)$ is conserved on-shell. Or, turning this around: symmetry requires the Lagrangian to be restricted so that $J^a_\mu(\varphi)$ defined in Eq. (14.18) is divergenceless. Having constrained the space of theories in this manner, there are no more equations to satisfy: conservation of charge is consistent with the dynamics, and no further constraints need to be imposed.

14.4.4 A-Independent, Malleable Symmetries

In the next case, suppose that we allow – in addition to Section 14.4.3's equations – the ones arising from a $\partial\epsilon \neq 0$, while still not allowing for an A in the theory. We get, in addition to Eqs. (14.17) and (14.18), from the vanishing of the second line of Eq. (14.14):

$$\frac{\delta \mathcal{L}}{\delta \partial_\mu \varphi_i} (t^a \varphi)_i = J^a_\mu(\varphi) = 0. \tag{14.19}$$

Here, the conserved currents are forced to vanish. Clearly, this condition is guaranteed for all field values if $\frac{\delta \mathcal{L}}{\delta \partial_\mu \varphi_i} = 0$, which requires a vanishing kinetic term. A careful analysis of more general cases reveals this is the only generic solution.[21]

This analysis pinpoints the obstacle appearing in the textbook gauge argument that we rehearsed in Section 14.2.1. When the matter field Lagrangian has a non-trivial kinetic term, malleable transformations cannot be variational symmetries. That is, if we impose malleable symmetries without introducing a gauge potential, we cannot consistently also allow a term in the Lagrangian including $\partial_\mu \varphi_i$. It is to allow such terms and still retain the malleable symmetries that the next two sections will introduce the gauge potential.

14.4.5 A-Dependent, Rigid Symmetries

We first proceed precisely as in the first case, introducing the A field, but still *keeping the symmetries rigid*. Using the equations of motion for A as well as those of φ, i.e. $\mathsf{E}[A] = 0$

[21] For instance, assume $\frac{\delta \mathcal{L}}{\delta \partial_\mu \varphi_i}$ depends only on $\partial_\mu \varphi_i$. Then, since $t^a_{ij} \varphi^i$ can take any value, we must have $\frac{\delta \mathcal{L}}{\delta \partial_\mu \varphi_i} = 0$. Now, suppose $\frac{\delta \mathcal{L}}{\delta \partial_\mu \varphi_i}$ depends on φ_i as well. Since φ_i has no spacetime indices to match the μ of the gradient $\partial_\mu \varphi_i$, to make a Lagrangian scalar, we would need the φ_i contribution to this term to itself be a scalar; call it $F(\varphi)$. So for example: $\frac{\delta \mathcal{L}}{\delta \partial_\mu \varphi_i} = \partial_\mu \varphi_i (\varphi_j \varphi^j)$, or more generally $\frac{\delta \mathcal{L}}{\delta \partial_\mu \varphi_i} = F'(\partial \varphi)_{i\mu} F(\varphi)$ (where we raise indices with an inner product of V); and as in the example $F(\varphi) = \varphi_j \varphi^j = 0$ iff $\varphi = 0$. But then the same argument as before suffices, since we can still allow $t^a_{ij} \varphi^i$ to take any value in V (for an appropriate, non-zero value of the scalar formed just from φ, e.g. the contraction $\varphi_j \varphi^j$). Or, in other words, for $\varphi \neq 0$, $\frac{\delta \mathcal{L}}{\delta \partial_\mu \varphi_i} (t^a \varphi)_i = 0$ iff $F^{-1}(\varphi) \frac{\delta \mathcal{L}}{\delta \partial_\mu \varphi_i} (t^a \varphi)_i = 0$ where $F^{-1}(\varphi) \frac{\delta \mathcal{L}}{\delta \partial_\mu \varphi_i}$ depends only on $\partial_\mu \varphi$; and thus we are back to the first, simple case.

as well as $\mathsf{E}[\varphi] = 0$, we get, in direct analogy to (14.17), a conserved current that is a sum of two currents,[22]

$$\partial_\mu \left(\frac{\delta \mathcal{L}}{\delta \partial_\mu \varphi_i} (t^a \varphi)_i + \left[\frac{\delta \mathcal{L}}{\delta \partial_\nu A_\mu}, A_\nu \right]^a \right) = \partial^\mu \left(J^a_\mu(\varphi) + \tilde{J}^a_\mu(A) \right) \approx 0, \tag{14.20}$$

and nothing more; there are no further conditions that the terms of the Lagrangian need to obey. (Here, the definition of $\tilde{J}^a_\mu(A))$ is given by (14.20).)

So, unlike the previous case, which admitted only a trivial kinetic term for the matter field φ, this sector will admit many possible dynamics. The problem here is of a different nature: the theories are not sufficiently constrained; the equations of motion do not automatically guarantee conservation of charges.

Let us look at an example of how things can go wrong in this intermediate sector containing forces but only rigid symmetries, for the simple, Abelian theory. In the Abelian theory, $\tilde{J}(A) \equiv 0$, since quantities trivially commute. Thus, Eq. (14.20) contains only the standard conservation of the matter charges, and the symmetries are silent about the relationship between this charge and the dynamics of the forces.

Consider a kinetic term of the form $\partial_{(\mu} A_{\nu)} \partial^{(\mu} A^{\nu)}$ where round brackets denote symmetrisation. So this differs from the standard Maxwell theory kinetic term for the gauge potential: namely, $F_{\mu\nu} F^{\mu\nu} := \partial_{[\mu} A_{\nu]} \partial^{[\mu} A^{\nu]}$, where square brackets denote *anti*-symmetrisation. But the symmetrised version is nonetheless gauge-invariant (under *rigid* transformations). Now, the Euler–Lagrange equations for this theory differ only very slightly from the Maxwell–Klein–Gordon equations. The equations of motion for A yield:

$$\partial^\mu (\partial_{(\mu} A_{\nu)}) = J_\nu \tag{14.21}$$

in contrast with the usual $\partial^\mu (\partial_{[\mu} A_{\nu]}) = J_\nu$. But clearly, unlike the usual case, the divergence of the left-hand side does *not* automatically vanish:

$$\partial^\nu \partial^\mu (\partial_{(\mu} A_{\nu)}) = \partial^\mu \partial_\mu \partial^\nu A_\nu = \Box \partial^\nu A_\nu \not\equiv 0. \tag{14.22}$$

At this point, we would have to go back to the drawing board and introduce more constraints on the theory: this theory does not couple forces to charges in a manner that guarantees charge conservation.

Thus, we glimpse our overall thesis: only by introducing malleable gauge symmetries do we restrict interactions between forces and their sources so that they are consistent with the conservation of the matter current.

Of course, in this example it is easy to see the smoking gun: the kinetic term $\partial_{(\mu} A_{\nu)} \partial^{(\mu} A^{\nu)}$ is *not* invariant under malleable transformations. According to the next section – our fourth sector – requiring this stronger form of invariance will restrict us to the space of consistent interactions. No tweaking is required.

[22] To be explicit, the A-dependent terms that appear in the first line of (14.14) are $\left[\frac{\delta \mathcal{L}}{\delta A_\nu}, A_\nu \right]^a + \left[\frac{\delta \mathcal{L}}{\delta \partial_\nu A_\mu}, \partial_\mu A_\nu \right]^a \approx \left[\partial_\mu \frac{\delta \mathcal{L}}{\delta \partial_\nu A_\mu}, A_\nu \right]^a + \left[\frac{\delta \mathcal{L}}{\delta \partial_\nu A_\mu}, \partial_\mu A_\nu \right]^a = \partial_\mu \left[\frac{\delta \mathcal{L}}{\delta \partial_\nu A_\mu}, A_\nu \right]^a$.

14.4.6 A-Dependent, Malleable Symmetries

In this fourth sector, we again obtain (14.20), from the vanishing of the first line of (14.14), since nothing changes at that level. But, from the vanishing of the second line in Eq. (14.14), we have:

$$-\frac{\delta \mathcal{L}}{\delta A^a_\mu} = \frac{\delta \mathcal{L}}{\delta \partial_\mu \varphi_i}(t^a \varphi)_i + \left[\frac{\delta \mathcal{L}}{\delta \partial_\nu A_\mu}, A_\nu\right]^a = J^a_\mu(\varphi) + \tilde{J}^a_\mu(A). \tag{14.23}$$

Once again using the Euler–Lagrange equations for A, to substitute the left-hand side, we find that

$$\mathsf{E}(A)^a_\mu = \frac{\delta \mathcal{L}}{\delta A^a_\mu} - \partial_\nu \frac{\delta \mathcal{L}}{\delta \partial_\nu A^a_\mu} \approx 0. \tag{14.24}$$

Defining $\frac{\delta \mathcal{L}}{\delta \partial_\nu A^a_\mu} =: k^a_{\mu\nu}$, we now obtain:

$$J^a_\mu(\varphi) + \tilde{J}^a_\mu(A) = -\partial^\mu k^a_{\mu\nu} + \mathsf{E}(A)^a_\mu \approx -\partial^\mu k^a_{\mu\nu} \tag{14.25}$$

This equation links both the matter and force currents to the dynamics of the force field.

We already know from the vanishing in the first line of Eq. (14.14) that the sum of the currents is divergence-free on shell (cf. Eq. (14.20)). Thus, taking the divergence on the left-hand side of (14.25), we must have $\partial^\nu \partial^\mu k^a_{\mu\nu} = 0$. Since two derivatives of a scalar field are necessarily symmetric, all we need in order to satisfy conservation is that:

$$k^a_{\mu\nu} = -k^a_{\nu\mu} \quad \text{or} \quad k^a_{\mu\nu} = k^a_{[\nu\mu]}, \tag{14.26}$$

which is just what we have from the vanishing of the *third* line of Eq. (14.14). Thus, the result of including malleable symmetries, in this simple case, restricts us to consider Lagrangians in which the derivatives of A^a_μ enter only in anti-symmetrised form: $\partial_{[\mu} A^a_{\nu]}$. This restriction excludes the previous example of Eq. (14.21).

More generally, if we try to find a Lagrangian that includes force fields without obeying the relations obtained from the malleable symmetries, the equations of motion of the force fields and those relating force fields and matter may require further constraints to be compatible with charge conservation, as we saw in the counter-example in the previous section. This is the power of local gauge symmetries: they link charge conservation – taken as empirical fact or on *a priori* grounds – with the form of the Lagrangian for the force fields.

14.4.7 Masslessness: An Invitation

There is yet more information that can be gleaned from the Noether gauge argument, which is contained in Eq. (14.25): upon integration, it yields a boundary term and a volume integral. That is, it gives a relation between a quantity at a faraway boundary – related to the flux of the force field components A^a_μ – and the matter content inside this region. In the Abelian case, for the 0th component of the equations of motion, this just gives the standard Gauss law. But more generally, being detectable at arbitrarily long distances makes the 'forces' associated to $k_{\mu\nu}$ long-range.[23]

It is common to conclude[24] on this basis that gauge invariance *forbids the presence of a mass term* for A_μ. However, like the textbook gauge argument, the general form of this argument for masslessness is often heuristic in character. In particular, it assumes that each term in the Lagrangian is independently gauge-invariant. Then it is true that, on its face, a term like $m^2 A^a_\mu A^\mu_a$ is not invariant under malleable symmetries.[25]

To show in full generality that masslessness is required would go beyond the scope of this chapter: we leave it as an exercise to the ambitious reader to explore! However, we can still improve on the standard heuristic argument without much effort in a special case that includes electromagnetism, by enforcing Eq. (14.14) off-shell, for all models (φ, A), and requiring that any mass term m be field-independent and that the only coupling between the matter field and A be just $A^a_\mu J^\mu_a$.

A term in the Lagrangian of the form $m^2 A^a_\mu A^\mu_a$ would not leave any trace in the first line of Eq. (14.14). But from the second line, again assuming no on-shell constraint between values of the matter and force field, from the second term, we obtain $m^2 A^a_\mu$, which can only cancel with something dropping out of $\left[\frac{\delta\mathcal{L}}{\delta\partial_\nu A_\mu}, A_\nu\right]^a$. Call this term $\kappa_{\mu\nu}(A, \partial A)$, which is such that $[\kappa_{\mu\nu}, A^\mu]^a \propto A^a_\nu$ for all A. Take, in this basis, A^a_ν to be constant, e.g. $\delta^a_1\delta^y_\nu$, so that $A = \tilde{A} := \sigma_1 \otimes dy$, where y is a spacetime coordinate function and σ_1 is one element of the Lie-algebra basis. This implies that the partial derivatives inside $\kappa_{\mu\nu}(\partial A, A)$ vanish, that is, $\kappa_{\mu\nu}(\partial\tilde{A}, \tilde{A}) = \kappa_{\mu\nu}(\tilde{A})$, and therefore that it is a polynomial of A with no derivatives, i.e. it is a polynomial that contains a single element of the Lie-algebra basis, σ_1, and the spacetime 1-form basis, dy. But this means that the commutator $[\kappa_{\mu\nu}, \tilde{A}^\mu]^a$ vanishes and therefore cannot be proportional to $\tilde{A}$. As a result, in order to maintain off-shell invariance under malleable transformations, a mass term for A cannot be included in the Lagrangian.

14.5 Conclusion

We have given a detailed defence of the use of gauge symmetries for theory-building, in the spirit of the textbook gauge argument.

We first showed how the textbook gauge argument can be tightened into a rigorous theorem in quantum mechanics, which begins with the assumption that malleable transformations 'appropriately' transform velocity, and concludes that the dynamics must be given by the minimally coupled Hamiltonian (Section 14.3 and the Appendix).

We then went on to defend a much more general 'Noether gauge argument' in classical field theory. In particular, gauge symmetries of various kinds were fed into the powerful theorems of Emmy Noether, in order to produce precise constraints on the possible dynamics.

[23] This is a classical treatment. Quantum mechanically, non-Abelian theories would suffer from confinement, which lies outside the scope of this discussion.

[24] For example, in the context of quantum electrodynamics, compare Weinberg (1995, p. 343).

[25] Note that the Proca action does include a mass term for the photon field, i.e. the gauge potential A, but it is only gauge-invariant with $m = 0$, in which case it just reduces to the standard Maxwell equations. For $m \neq 0$, one must have, in relation to the Maxwell equations, gauge-breaking, or 'gauge-fixing', conditions. For a discussion, see Itzykson and Zuber (1980, §3-2-3).

The result is more than a simple argument that gauge symmetry is useful for theory construction: gauge symmetry constrains how one can consistently combine charges with the fields they interact with. Noether's first theorem of course implies charge conservation; but the second theorem then implies relations between the theory's equations of motion, which amounts to a coupling constraint. In other words, converting a rigid symmetry into a malleable one enforces the compatibility between charge conservation and the dynamics of the corresponding fields: a result which has not yet been stressed by the philosophical literature.

Of course, one may still feel that gauge redundancies are like Wittgenstein's ladder: once our theories have been successfully constructed, why not throw gauge symmetries away, and move down to a description in which such redundancies have been eliminated? We would reply: on the contrary, the ladder remains invaluable in the interpretation of gauge theories. In addition to the several reasons for gauge identified in the Introduction to this chapter, we find that gauge symmetries provide an explanatory reason – a reason drawing heavily on Noether's two theorems – for the way charges couples to fields.

Appendix

Theorem *In the Schrödinger (position) representation on $L^2(\mathbb{R}^3)$ with dynamics $t \mapsto U_t = e^{-itH}$, let $\dot{Q}_r := \frac{d}{dt}(U_t^* Q U_t)\big|_{t=0} = i[H, Q_r]$. Suppose that for every smooth one-parameter set of smooth functions $\phi_t : \mathbb{R}^3 \to \mathbb{R}$, at time $t = 0$,*

$$W_\phi \dot{Q} W_\phi^* = \dot{Q} - \nabla\phi, \tag{14.10}$$

where $W_\phi := e^{i\phi_t}$ and $\nabla = (\partial_x, \partial_y, \partial_z)$. Then:

(i) the Hamiltonian is constrained to be of the form,

$$H = \sum_{r=1}^{3} \tfrac{1}{2m}(P_r - A_r)^2 + V \tag{14.11}$$

for some $m > 0$ and some functions (i.e. operators) $A_r(x)$ and $V(x)$ that depend only on position, and not on momentum;

(ii) the transformed unitary group $t \mapsto \tilde{U}_t := W_\phi U_t =: e^{-itH_\phi}$ (where the last equation defines H_ϕ as the generator of $\tilde{U}_t$, by Stone's theorem) is given by,

$$H_\phi = \sum_{r=1}^{3} \tfrac{1}{2m}\left(P_r - A_r^\phi\right)^2 + V^\phi, \tag{14.12}$$

where $A_r^\phi = A_r + \nabla\phi$ and $V^\phi = V - \frac{d\phi}{dt}$; and

(iii) the Schrödinger equation is 'gauge invariant' under W_ϕ, in that if $i\frac{d}{dt}\psi(t) = H\psi(t)$, then $\psi_\phi(t) := W_\phi\psi(t)$ satisfies $i\frac{d}{dt}\psi_\phi(t) = H_\phi\psi_\phi(t)$.

Proof We first note that if $W_\phi := e^{i\phi} := e^{i\phi(Q)}$, then

$$W_\phi Q W_\phi^* = Q, \qquad W_\phi P W_\phi^* = P - \nabla\phi. \tag{14.27}$$

The second equation follows from the observation that $PW_\phi^*\psi = -i\nabla\left(e^{-i\phi}\psi\right) = -i\left(e^{-i\phi}\nabla\psi - i(\nabla\phi)e^{-i\phi}\psi\right) = e^{-i\phi}(-i\nabla - \nabla\phi)\psi = W_\phi^*(P-\nabla\phi)\psi$, for all $\psi \in L^2(\mathbb{R}^3)$. From the second equation, and our assumption that $W_\phi \dot{Q} W_\phi^* = \dot{Q} - \nabla\phi$, it follows that $[W_\phi, m\dot{Q} - P] = 0$ for any $m \neq 0$. Since this holds in particular when $\phi(x) = ax$ for all $a \in \mathbb{R}$, it follows that $m\dot{Q} - P$ commutes with Q (Blank et al. 2008, Proposition 5.9.3). But the position operator $Q = x$ in the Schrödinger representation has a simple spectrum (Blank et al. 2008, Example 5.8.2), which is to say that its commutant $\{Q\}'$ is equal to its bicommutant $\{Q\}''$. (This is the continuous spectrum analogue for Q of all of a quantity's eigenvalues being non-degenerate.) By von Neumann's bicommutant theorem (Blank et al. 2008, theorem 5.5.6) this implies that there is a function A_r (for each $r = 1, 2, 3$) of position alone such that,

$$A_r(x) = P_r - m\dot{Q}_r. \tag{14.28}$$

Using the fact that A depends only on position and thus commutes with Q, we can now observe by direct calculation that for each $r = 1, 2, 3$:

$$\begin{aligned}
[(P-A)^2, Q] &= \underbrace{[P^2, Q]}_{-2iP} - [PA, Q] - [AP, Q] + \underbrace{[A^2, Q]}_{0} \\
&= -2iP - (P\underbrace{AQ}_{=QA} - QPA + APQ - \underbrace{QA}_{=AQ}P) \\
&= -2iP - (\underbrace{[P,Q]}_{-i}A) + A\underbrace{[P,Q]}_{-i}) \\
&= -2i(P-A),
\end{aligned} \tag{14.29}$$

and thus $i[\frac{1}{2m}(P_r - A_r)^2, Q_r] = \frac{1}{m}(P_r - A_r)$. But also, $i[H, Q_r] = \dot{Q}_r = \frac{1}{m}(P_r - A_r)$, where the second equality applies Eq. (14.28), so $[H - \frac{1}{2m}(P_r - A_r)^2, Q_r] = 0$. By the simple spectrum property, it again follows that there is a function V of position alone such that $H - \sum_r \frac{1}{2m}(P_r - A_r)^2 = V$, which establishes Eq. (14.11).

To confirm Eq. (14.12), let $\psi(t) := \tilde{U}_t\psi := W_\phi U_t =: e^{-itH_\phi}\psi$ for each $\psi \in L^2(\mathbb{R})$. Then on the one hand, we have that,

$$\tfrac{d}{dt}\psi(t) = -iH_\phi e^{-itH_\phi}\psi = -iH_\phi\psi(t). \tag{14.30}$$

On the other hand, using our definition $\tilde{U}_t\psi := W_\phi U_t\psi$, we have that,

$$\begin{aligned}
\tfrac{d}{dt}\psi(t) &= \left(W_\phi\left(\tfrac{d}{dt}U_t\right) + \left(\tfrac{d}{dt}W_\phi\right)U_t\right)\psi = -i\left(W_\phi H U_t - \tfrac{d\phi}{dt}W_\phi U_t\right)\psi \\
&= -i\left(W_\phi H W_\phi^* - \tfrac{d\phi}{dt}\right)(W_\phi U_t)\psi \\
&= -i\left(W_\phi H W_\phi^* - \tfrac{d\phi}{dt}\right)\psi(t).
\end{aligned} \tag{14.31}$$

Equations (14.30) and (14.31) together imply $H_\phi = W_\phi H W_\phi^* - \frac{d}{dt}\phi$. Moreover, since W_ϕ is a function of Q and so commutes with functions of Q, it follows that $W_\phi(P - A)W_\phi^* = P - \nabla\phi - A = P - (A + \nabla\phi)$. Therefore, using Eq. (14.11) and rearranging, we find that

$$H_\phi = W_\phi H W_\phi^* - \tfrac{d\phi}{dt} = \sum_{r=1}^{3} \tfrac{1}{2m} (P_r - (A_r + \nabla\phi))^2 + \left(V - \tfrac{d\phi}{dt}\right); \tag{14.32}$$

which proves Eq. (14.12) and claim (ii).

The final claim (iii) is now verified by confirming that if $i\frac{d}{dt}\psi = H\psi$ and $\psi_\phi := W_\phi\psi$, then we have that $i\frac{d}{dt}\psi_\phi \equiv i\frac{d}{dt}(W_\phi\psi) = W_\phi\left(-\frac{d\phi}{dt}\psi + i\frac{d}{dt}\psi\right) = W_\phi\left(H - \frac{d\phi}{dt}\right)\psi = \left(W_\phi H W_\phi^* - \frac{d\phi}{dt}\right) W_\phi\psi = H_\phi\psi_\phi$. □

References

Aharonov, Y. and Bohm, D. (1959). Significance of electromagnetic potentials in the quantum theory. *Physical Review*, **115**(3), 485.

Avery, S. G. and Schwab, B. U. (2016). Noether's second theorem and Ward identities for gauge symmetries. *Journal of High Energy Physics*, **2016**(2), 31. https://arxiv.org/abs/1510.07038.

Belot, G. (1998). Understanding electromagnetism. *The British Journal for the Philosophy of Science*, **49**(4), 531–55.

Belot, G., Earman, J., Healey, R., Maudlin, T., Nounou, A. and Struyve, W. (2009). Synopsis and discussion: philosophy of gauge theory. Unpublished manuscript. http://philsci-archive.pitt.edu/4728/.

Blank, J., Exner, P. and Havlíček, M. (2008). *Hilbert Space Operators in Quantum Physics*, 2nd ed. Dordrecht: Springer Science and Business Media B.V.

Brading, K. (2002). Which symmetry? Noether, Weyl, and conservation of electric charge. *Studies in History and Philosophy of Modern Physics*, **33**(1), 3–22.

Brading, K. and Brown, H. R. (2000). Noether's theorems and gauge symmetries. Unpublished manuscript. https://arxiv.org/abs/hep-th/0009058.

Brading, K. and Brown, H. R. (2003). Symmetries and Noether's theorems. *In* K. Brading and E. Castellani, eds., *Symmetries in Physics: Philosophical Reflections*. Cambridge: Cambridge University Press, chapter 5, pp. 89–109.

Brading, K. and Brown, H. R. (2004). Are gauge symmetry transformations observable? *The British Journal for the Philosophy of Science*, **55**(4), 645–65. http://philsci-archive.pitt.edu/1436/.

Brown, H. (1999). Aspects of objectivity in quantum mechanics. *In* J. Butterfield and C. Pagonis, eds., *From Physics to Philosophy*. Cambridge: Cambridge University Press. http://philsci-archive.pitt.edu/223/.

Donnelly, W. and Freidel, L. (2016). Local subsystems in gauge theory and gravity. *Journal of High Energy Physics*, **2016**(9), 102. https://arxiv.org/abs/1601.04744v2.

Dougherty, J. (2020a). Large gauge transformations and the strong CP problem. *Studies in History and Philosophy of Modern Physics*, **69**, 50–66.

Dougherty, J. (2020b). The non-ideal theory of the Aharonov–Bohm effect, *Synthese*, pp. 1–27. http://philsci-archive.pitt.edu/18057/.

Earman, J. (2002). Gauge matters. *Philosophy of Science*, **69**(S3), 209–20. http://philsci-archive.pitt.edu/70/.

Earman, J. (2003). Tracking down gauge: An ode to the constrained Hamiltonian formalism. *In* K. Brading and E. Castellani, eds., *Symmetries in Physics: Philosophical Reflections*. Cambridge: Cambridge University Press, chapter 8, pp. 140–62.

Earman, J. (2004). Laws, symmetry, and symmetry breaking: Invariance, conservation principles, and objectivity. *Philosophy of Science*, **71**(5), 1227–41. http://philsci-archive.pitt.edu/878/.

Earman, J. (2019). The role of idealizations in the Aharonov–Bohm effect. *Synthese*, **196**(5), 1991–2019. http://philsci-archive.pitt.edu/13359/.

Göckeler, M. and Schücker, T. (1989). *Differential Geometry, Gauge Theories, and Gravity*. Cambridge: Cambridge University Press.

Gomes, H. (2019). Gauging the boundary in field-space. *Studies in History and Philosophy of Modern Physics*, **67**, 89–110. http://philsci-archive.pitt.edu/15564/.

Gomes, H. (2021a). Gauge-invariance and the empirical significance of symmetries. Under review, http://philsci-archive.pitt.edu/16981/.

Gomes, H. (2021b). Holism as the empirical significance of symmetries. *European Journal of Philosophy of Science*, **11**, 87. https://link.springer.com/article/10.1007/s13194-021-00397-y.

Gomes, H. and Riello, A. (2020). Large gauge transformations, gauge invariance, and the QCD $theta_{YM}$-term. http://philsci-archive.pitt.edu/17446/.

Gomes, H. and Riello, A. (2021). Quasilocal degrees of freedom in Yang–Mills theory. *SciPost*. https://arxiv.org/abs/1910.04222.

Greaves, H. and Wallace, D. (2014). Empirical consequences of symmetries. *The British Journal for the Philosophy of Science*, **65**(1), 59–89. http://philsci-archive.pitt.edu/8906/.

Healey, R. (1997). Nonlocality and the Aharonov–Bohm effect. *Philosophy of Science*, **64**(1), 18–41.

Healey, R. (1999). Quantum analogies: A reply to Maudlin. *Philosophy of Science*, **66**(3): 440–7.

Healey, R. (2007). *Gauging What's Real: The Conceptual Foundations of Contemporary Gauge Theories*. New York: Oxford University Press.

Henneaux, M. and Teitelboim, C. (1994). *Quantization of Gauge Systems*. Princeton: Princeton University Press.

Hiley, B. J. (2013). The early history of the Aharonov–Bohm effect. Unpublished manuscript, https://arxiv.org/abs/1304.4736.

Itzykson, C. and Zuber, J.-B. (1980). *Quantum Field Theory*. New York: McGraw-Hill Inc.

Jauch, J. M. (1964). Gauge invariance as a consequence of Galilei-invariance for elementary particles. *Helvetica Physica Acta*, **37**, 284–92.

Jauch, J. M. (1968). *Foundations of Quantum Mechanics*. Addison-Wesley Series in Advanced Physics. Reading, MA, Menlo Park, CA, London, Don Mills, ON: Addison-Wesley Publishing Company, Inc.

Kosmann-Schwarzbach, Y. (2011). *The Noether Theorems: Invariance and Conservation Laws in the Twentieth Century*. New York: Springer Science+Business Media, LLC. Translated by Bertram E. Schwarzbach.

Lyre, H. (2000). A generalized equivalence principle. *International Journal of Modern Physics D*, **9**(06), 633–47. https://arxiv.org/abs/gr-qc/0004054.

Lyre, H. (2009). Aharonov-Bohm effect. In D. Greenberger, K. Hentschel and F. Weinert, eds., *Compendium of Quantum Physics: Concepts, Experiments, History and Philosophy*. Berlin Heidelberg: Springer-Verlag, pp. 1–3.

Martin, C. A. (2002). Gauge principles, gauge arguments and the logic of nature. *Philosophy of Science*, **69**(S3), S221–S234.

Mattingly, J. (2006). Which gauge matters? *Studies in History and Philosophy of Science Part B: Studies in History and Philosophy of Modern Physics*, **37**(2), 243–62.

Maudlin, T. (1998). Healey on the Aharonov–Bohm effect. *Philosophy of Science*, **65**(2), 361–8.

Mulder, R. A. (2021). Gauge-underdetermination and shades of locality in the Aharonov–Bohm effect. *Foundations of Physics*, **51**(2), 1–26. http://philsci-archive.pitt.edu/18767/.

Myrvold, W. C. (2011). Nonseparability, classical and quantum. *The British Journal for the Philosophy of Science*, **62**(2), 417–32. http://philsci-archive.pitt.edu/5335/.

Nguyen, J., Teh, N. J. and Wells, L. (2020). Why surplus structure is not superfluous. *The British Journal for the Philosophy of Science*, **71**(2), 665–95. http://philsci-archive.pitt.edu/14166/.

Noether, E. (1918). Invariante Variationsprobleme. *Nachrichten von der Gesellschaft der Wissenschaften zu Göttingen, Mathematisch-Physikalische Klasse* pp. 235–57. English translation by M. A. Tavel, https://arxiv.org/abs/physics/0503066.

Nounou, A. (2003). A fourth way to the Aharonov–Bohm effect. In K. Brading and E. Castellani, eds., *Symmetries in Physics: Philosophical Reflections*. Cambridge: Cambridge University Press, chapter 10, pp. 174–200.

Olver, P. J. (1993). *Applications of Lie Groups to Differential Equations*, 2nd ed. New York: Springer Verlag.

O'Raifeartaigh, L. (1997). *The Dawning of Gauge Theory*. Princeton: Princeton University Press.

Roberts, B. W. (2012). Time, symmetry and structure: A study in the foundations of quantum theory. Ph.D. thesis, University of Pittsburgh. http://d-scholarship.pitt.edu/12533/.

Roberts, B. W. (2017). Three myths about time reversal in quantum theory. *Philosophy of Science*, **84**, 1–20. Preprint, http://philsci-archive.pitt.edu/12305/.

Roberts, B. W. (2021). Time reversal. *In* E. Knox and A. Wilson, eds., *The Routledge Companion to the Philosophy of Physics*. New York: Routledge, chapter 43. http://philsci-archive.pitt.edu/15033/.

Rosenstock, S. and Weatherall, J. O. (2016). A categorical equivalence between generalized holonomy maps on a connected manifold and principal connections on bundles over that manifold. *Journal of Mathematical Physics*, **57**(10), 102902. http://philsci-archive.pitt.edu/11904/.

Rosenstock, S. and Weatherall, J. O. (2018). Erratum: 'A categorical equivalence between generalized holonomy maps on a connected manifold and principal connections on bundles over that manifold' [*J. Math. Phys.* 57, 102902 (2016)]. *Journal of Mathematical Physics*, **59**(2), 029901.

Rovelli, C. (2014). Why gauge? *Foundations of Physics*, **44**(1), 91–104. https://arxiv.org/abs/1308.5599.

Rovelli, C. (2020). Gauge is more than mathematical redundancy. *In* S. De Bianchi and C. Kiefer, eds., *One Hundred Years of Gauge Theory: Past, Present and Future Perspectives*, Cham: Springer Nature Switzerland, pp. 107–10. https://arxiv.org/abs/2009.10362v1.

Ryder, L. H. (1996). *Quantum Field Theory*. Cambridge: Cambridge University Press.

Schutz, B. F. (1980). *Geometric Methods of Mathematical Physics*. Cambridge: Cambridge University Press.

Shech, E. (2018). Idealizations, essential self-adjointness, and minimal model explanation in the Aharonov–Bohm effect. *Synthese*, **195**(11), 4839–63.

Swanson, N. (2022). On the Ostrogradski instability, or, why physics really uses second derivatives. *The British Journal for the Philosophy of Science*, **73**(1). https://doi.org/10.1093/bjps/axz042.

Teller, P. (1997). A metaphysics for contemporary field theories. *Studies in History and Philosophy of Modern Physics*, **28**(4), 507–22.

Teller, P. (2000). The gauge argument. *Philosophy of Science*, **67**, S466–S481.

Wallace, D. (2009). QFT, antimatter and symmetry. *Studies in History and Philosophy of Science Part B: Studies in History and Philosophy of Modern Physics*, **40**(3), 209–222. https://doi.org/10.1016/j.shpsb.2009.07.002

Wallace, D. (2014). Deflating the Aharonov–Bohm effect. Unpublished manuscript, https://arxiv.org/abs/1407.5073.

Weatherall, J. O. (2016). Fiber bundles, Yang–Mills theory, and general relativity. *Synthese*, **193**(8), 2389–425. http://philsci-archive.pitt.edu/11481/.

Weinberg, S. (1995). *The Quantum Theory of Fields*. Vol. 1: *Foundations*. Cambridge: Cambridge University Press.

Weyl, H. (1929). Elektron und Gravitation. I., *Zeitschrift für Physik*, **56**(5–6), 330–52.

Index

AdS-CFT correspondence, 224
Aharonov–Bohm effect, 298, 354
asymptotic diffeomorphisms, 228

background independence, 275, 291
Banach–Lie algebra, 85
boost transformations, 151
Brown–York stress-energy tensor, 199, 222, 239, 289
BRST symmetry, 100
Bryn Mawr College, 5, 46, 60
BV complex, 109
BV formalism, 100

Chevalley–Eilenberg complex, 108
Civil Service Law, Nazi Germany, 53
complex $*$-algebra, 79
complex associative algebra, 79

deformation, 101
diffeomorphism invariance, 240, 257, 276
differential invariants, 26, 134
divergence invariant, 134
dynamical approach to spacetime, 258, 324
dynamical correspondence, 69

Einstein–Klein dispute, 274
Erlangen program, 12, 41, 202, 263, 336
external sophistication, 268

factorisation algebras, 101
fragmentalism, 258

Galilean group, 135
gauge argument, 359
gauge-gravity duality, 224
general covariance, 274
geometric objects, 212, 259, 301
gravitational stress energy pseudo-tensor, 197, 259

holographic stress-energy tensor, 224
homology, 101

invariant theory, 25

Jordan algebra, 67

Klein, Felix, 6, 25, 197, 274
Koszul resolution, 108
Kretschmann, Erich, 277

Landau–Lifschitz pseudotensor, 189, 259, 297, 304
Lie's Theorem, 135

Martínez Alonso Olver (MAO) theorem, 153
Master Ward Identity, 120
mere patterns, 262
Møller map, 113
Moscow State University, 56
motivational approach to symmetries, 268

natural kinds, 258
Noether conservation law, 135, 138, 147
Noether gauge argument, 363
Noether identity, 148, 175, 282
Noether Lagrangian, 152
Noether map, 153
Noether's first theorem, 15, 25, 163, 169, 205, 212, 281, 355, 367
Noether's second theorem, 13, 25, 106, 154, 178, 199, 206, 214, 274, 355
non-geometric objects, 259
non-separability, 296

perspectivalism, 263
perturbative algebraic quantum field theory, 101
Poisson algebra, 73
Poisson bracket, 71

principle of relativity, 278
projectable predicates, 258

quasi-local definition, 222, 289
quasi-symmetry, 146, 213

real associative algebra, 76
real patterns, 262
reconstruction theorem, 299, 310
Reduction Theorem, 27, 40
regionally conserved charges, 304

self-conservation principle, 69
Somerville College Oxford, 52
structural realism, 258
superpotential, 170, 181, 198, 259, 284
symmetries: global, 171, 333, 358
  local, 100, 105, 313, 332, 358
  malleable, 355
symplectic formalism, 297, 313

time translations, 66, 76, 151

unital JB-algebras, 68
University of Göttingen, 20, 52

variational methods, 26, 41, 193

Weyl-Yang-Mills gauge principle, 161

Yang–Mills theory, 101, 185, 288, 312, 359

Printed by Integrated Books International,
United States of America